Implementing SAP® Global Trade Services

Nick Moris, Pablo LeCour, Li Yu

Implementing SAP® Global Trade Services

Editor Megan Fuerst
Copyeditor Julie McNamee
Cover Design Graham Geary
Photo Credit Shutterstock: 605340350/© Pasko Maksim
Layout Design Vera Brauner
Production Hannah Lane
Typesetting SatzPro, Germany
Printed and bound in Canada, on paper from sustainable sources

ISBN 978-1-4932-2505-7

3rd edition 2024

Library of Congress Cataloging-in-Publication Control Data
Names: Moris, Nick, author. | Lecour, Pablo, author. | Yu, Li (Expert on SAP GTS), author.
Title: Implementing SAP global trade services : edition for SAP HANA / by Nick Moris, Pablo LeCour, and Li Yu.
Description: 3rd edition. | Bonn ; Boston : Rheinwerk Publishing, 2024.
Identifiers: LCCN 2024004010 | ISBN 9781493225057 (hardcover) | ISBN 9781493225064 (ebook)
Subjects: LCSH: SAP GTS. | SAP Business information warehouse. | SAP HANA (Electronic resource) | International trade. |
Exports--Management--Computer programs. | Imports--Management--Computer programs.
Classification: LCC HF1379 .M667 2024 | DDC 658.70285/53--dc23/eng/20240312
LC record available at https://lccn.loc.gov/2024004010

Contents at a Glance

1 Introduction 19

2 SAP GTS Basics: Setup, Authorizations, and Navigation 31

3 Baseline Settings 55

4 Master Data 71

5 Classification 93

6 Compliance Management 117

7 Customs Management 189

8 Preference Management 339

9 Global Trade Reporting 443

10 Migrating from SAP GTS 11.0 to SAP GTS, Edition for SAP HANA 485

11 The Future of SAP GTS 503

Contents

Foreword 15
Acknowledgments 17

1 Introduction 19

1.1 What Is Global Trade Management? 19
1.1.1 Compliance Management 20
1.1.2 Customs Management 21
1.1.3 Preference Management 22
1.2 Global Trade Management with SAP 23
1.2.1 SAP Global Trade Services 23
1.2.2 SAP S/4HANA for International Trade 26
1.3 Summary 29

2 SAP GTS Basics: Setup, Authorizations, and Navigation 31

2.1 Connecting SAP S/4HANA to SAP GTS, Edition for SAP HANA 31
2.1.1 System Architecture 32
2.1.2 System Connection Configuration 33
2.1.3 Connecting SAP GTS to Customs Authorities 36
2.2 Connecting Using Different Interfacing Techniques 36
2.2.1 Integration with TM and EWM from a Technical Perspective 38
2.2.2 Compliance Screening 39
2.2.3 Export Declaration 40
2.2.4 Customs Warehousing Procedures 41
2.2.5 Advanced Shipping and Receiving 41
2.3 Navigating SAP GTS, Edition for SAP HANA 42
2.3.1 SAP Fiori Launchpad 43
2.3.2 Compliance Management 44
2.3.3 Customs Management 46
2.3.4 Preference Management 47
2.3.5 Products and Classification 48

2.3.6 Intrastat 49
2.3.7 Administrative Tasks 49

2.4 Using Authorizations 50

2.5 Activation of SAP Fiori Apps 52

2.6 Summary 53

3 Baseline Settings 55

3.1 Organizational Structure in SAP GTS 55
3.1.1 Foreign Trade Organization 55
3.1.2 Legal Unit 56
3.1.3 Administrative Unit 57
3.1.4 Customs ID 57
3.1.5 Customizing 57
3.1.6 Leading Practices 59

3.2 Plug-In Settings in SAP S/4HANA 59

3.3 General Settings in SAP GTS 62
3.3.1 Legal Regulation 62
3.3.2 Document Types and Mapping 65
3.3.3 Mapping Partner Roles 69
3.3.4 Number Ranges 70

3.4 Summary 70

4 Master Data 71

4.1 Business Processes 71
4.1.1 Source System 72
4.1.2 Initial Transfer from the Feeder System to SAP GTS 72

4.2 Business Partner Master Data 76
4.2.1 Customer Maintenance 77
4.2.2 Vendor Maintenance 78
4.2.3 Customs Offices Maintenance 79

4.3 Product Master Data 80
4.3.1 Tariff and Commodity Codes 81
4.3.2 Control Classes 81

4.3.3 Products 82
4.3.4 Product Classification 84

4.4 Bill of Materials Master Data 86
4.4.1 BOMs Manually Maintained in SAP GTS 86
4.4.2 BOMs Transferred from SAP S/4HANA to SAP GTS 87

4.5 License and Authorization Master Data 89

4.6 Additional Master Data 90

4.7 Customizing 91

4.8 Summary 91

5 Classification 93

5.1 Business and System Processes 94
5.1.1 Uploading Classification Codes 95
5.1.2 Classifying for Compliance Management 102
5.1.3 Classifying for Customs Management 106
5.1.4 Reclassifying Products 108
5.1.5 Tariff Number Attributes 109
5.1.6 Analysis of Customs Management Classification 110

5.2 Customizing 111
5.2.1 Setting Up Numbering Schemes 111
5.2.2 Setting Up Partner Government Agencies 114

5.3 Summary 115

6 Compliance Management 117

6.1 Customs Documents 120

6.2 Sanctioned Party List Screening 122
6.2.1 Business Process 123
6.2.2 System Process 123
6.2.3 Customizing 133
6.2.4 Master Data 141

6.3 Embargo 144
6.3.1 Business Process 145
6.3.2 System Process 145

6.3.3 Customizing 148
6.3.4 Master Data 150

6.4 Legal Control 151
6.4.1 Import Licensing 153
6.4.2 Export Licensing 157
6.4.3 Legal Control Fundamentals 161
6.4.4 Applications for Legal Control 170
6.4.5 US Reexport 172
6.4.6 Intercompany Sales Screening 178
6.4.7 Customizing 182
6.4.8 Master Data 187

6.5 Summary 188

7 Customs Management 189

7.1 Customs Processing 193
7.1.1 Manage Customs Declarations 193
7.1.2 Display Customs Declarations 196
7.1.3 Customs Declaration 197
7.1.4 Communication with Customs Authorities 199

7.2 Import and Export Processes 205
7.2.1 Import Processes 205
7.2.2 Export Processes 222
7.2.3 Customizing 227
7.2.4 Master Data 261

7.3 Special Customs Procedures 267
7.3.1 Bonded Warehouse 267
7.3.2 Inward Processing 275
7.3.3 Outward Processing 280
7.3.4 Foreign Trade Zone 282
7.3.5 Processing Trade Relief: China Localization 299
7.3.6 Customizing 308
7.3.7 Master Data 311

7.4 Excise Duty Monitoring 315
7.4.1 Business Process 315
7.4.2 System Process 317
7.4.3 Customizing 318
7.4.4 Master Data 318

7.5 US Reconciliation Process 319
7.5.1 Business and System Processes 319
7.5.2 Customizing 320

7.6 Electronic Communication 321
7.6.1 Standard Communication 321
7.6.2 Electronic Communication Transactions 323
7.6.3 Standard Broker Enablement 326

7.7 Countries Supported by SAP GTS, Edition for SAP HANA 335

7.8 Summary 337

8 Preference Management 339

8.1 Preferential Trade Agreements 339

8.2 Business Process 345
8.2.1 Obtain Preferential Status and Origin from Suppliers 347
8.2.2 Determine Preferential Status and Origin of Finished Products 348
8.2.3 Issue Certificates of Origin for Preferential Finished Products 358

8.3 System Process 359
8.3.1 Inbound, Production, and Outbound Processes 359
8.3.2 Requests and Reminders for Supplier Declarations 364
8.3.3 Maintaining Supplier Declarations 371
8.3.4 Aggregating Supplier Declarations 375
8.3.5 Preference Determination 380
8.3.6 Evaluating Preference Results 395
8.3.7 Issuing Preference Certificates 397
8.3.8 Revoking and Reissuing Supplier Declarations 401
8.3.9 Intercompany Stock Transfers 404

8.4 Customizing 408
8.4.1 Organizational Structure 409
8.4.2 Number Ranges 411
8.4.3 Free Trade Agreements, Rules of Origin, and Rule Sets 411
8.4.4 Preference Statements 413
8.4.5 Price Types Agreements 414
8.4.6 Price Categories for Value Bases 415
8.4.7 Settings for the LTSD Process 415
8.4.8 Documents of Origin 417
8.4.9 Output Management with PPF Actions 418

8.4.10 Intercompany Stock Transfer 420
8.4.11 SAP S/4HANA Plug-In 421

8.5 Management of Preference Data for Product Identifiers 422
8.5.1 Business Process 423
8.5.2 System Process 425

8.6 Master Data 430
8.6.1 Material Master Data 431
8.6.2 Business Partner Master Data 436
8.6.3 Preference Agreements and Rules of Origin 438

8.7 Summary 440

9 Global Trade Reporting 443

9.1 SAP GTS Reporting 443
9.1.1 SAP Fiori-Based Reporting 444
9.1.2 Compliance Reporting 445
9.1.3 Customs Reporting 455
9.1.4 Trade Preference Reporting 464

9.2 Embedded Analytics 469

9.3 Analytical Reporting 471
9.3.1 SAP BW/4HANA 471
9.3.2 SAP Datasphere 478
9.3.3 SAP HANA for SQL Data Warehousing 481

9.4 Summary 483

10 Migrating from SAP GTS 11.0 to SAP GTS, Edition for SAP HANA 485

10.1 Drivers for Upgrading to SAP GTS, Edition for SAP HANA 485

10.2 Upgrade Process 486
10.2.1 Prepare for Your Migration 487
10.2.2 Execute Your Migration 496

10.3 Simplification List 498
10.3.1 Obsolete Functions 498
10.3.2 Risk Management Reorganized 499

10.3.3 Product and Classification ... 499
10.3.4 Customs Management ... 500
10.3.5 Intrastat ... 501

10.4 Summary ... 501

11 The Future of SAP GTS 503

11.1 Planned Innovations ... 505
11.1.1 The Next Release of SAP GTS ... 505
11.1.2 Planned Innovations for the Next Release ... 505

11.2 The Road Ahead for SAP GTS ... 507

The Authors ... 509
Index ... 517

Foreword

The Fourth Industrial Revolution (4IR) is defined by the amalgamation of digital, biological, and physical technologies. It's also characterized by the increasing utilization of new technologies such as robotics, cloud technology, or the Internet of (Every)Things, driving the digital revolution forward. Existential threat for some, it offers formidable possibilities to those able to harness its full potential. Amid this transformative outlook, global supply chain disruptions caused by events such as the COVID-19 pandemic, wars, and other geopolitical tensions have acted as catalysts to drive this revolution forward. These disruptions compelled businesses to reevaluate their strategies and underlined the urgency of adopting digital advancements to ensure continuity and enhance resilience when braving unchartered territories.

Geopolitical shifts and technology changes are transforming international trade, creating both risks and opportunities for companies navigating this new environment. Global digital integration across devices, services, processes, and businesses continues to grow as the world becomes more politically decentralized. The simultaneous rise of economic nationalism and digital globalization is redefining international trade as we know it today. Companies must take a prudent approach to identifying and creating growth opportunities. The rules of international business are being rewritten, creating competitive advantages for companies agile enough to seize them.

Within the corporate landscape, organizations recognize that digitalization is not merely a tool for enhancing operational efficiency; it's a fundamental requirement for overhauling their business models, addressing complex business priorities, and remaining relevant. Therefore, a comprehensive and mature adoption of digital technology can create new avenues and opportunities in global trade, as well as optimize a business's ability to react promptly to trade regulations while addressing uncertainty.

One of the significant aspects of the digital revolution is the advent of intelligence-based technologies, such as artificial intelligence (AI) and blockchain technology. These new connected digital advancements have become vanguards of a new era and are driving transformative shifts in the ways companies operate and exchange information. AI assistance enables companies and authorities to process vast amounts of global trade data and extract meaningful insight. In parallel, blockchain technology decentralizes records and transactions across multiple computers, thus promising reliability, transparency, and convenience of data storing management. When implemented in the commerce context, these technologies have redefined global trade paradigms by providing a secure and streamlined system, reducing risks, and ultimately accelerating the operational process and supply chains on a global scale.

In this shifting landscape of global trade, SAP and Deloitte present this publication on SAP Global Trade Services, edition for SAP HANA. SAP GTS has pioneered trade technologies for decades. SAP GTS improves on previous systems and features a better user experience, improved usability and efficiency, embedded analytics for work, enhancements for reporting and auditing, a better enterprise search, renewed trade preference management, and a simplification of many processes.

Therefore, with great gratitude, I extend my thanks to the authors, Nick Moris, Pablo LeCour, and Li Yu of *Implementing SAP Global Trade Services* for their invaluable contribution to the body of trade knowledge and innovation through this exciting publication.

Vishal Verma
VP, Head of Solution Management—Risk, Compliance, and Tax at SAP

Acknowledgments

This publication included important contributions from many of our SAP GTS specialists around the world. These individuals have dedicated significant portions of their careers to SAP GTS implementations, gaining unique insights on leading practices to drive value for clients. Your input was invaluable to the success of this book.

First and foremost, we express heartfelt appreciation to all chapter leads and the PMO team. Thank you for the great work done on coordinating the project and managing the content of the individual chapters. We would like to thank in particular the following people:

- Mike Beni
- Anthony Clarke
- Niels De Lathouwer
- Katrien Desmyter
- Shailendra Devel
- Wouter Jansen
- Ben Lallemand
- Gilles Poncin
- Nick Robyn
- Jasper Vanmarcke
- Aranea Vos
- Iuliana Vulpe
- Marion Weynand

Additionally, we are grateful to all other team members who have contributed to the completion of the book. This would not have been possible without the contribution of the following people:

- Liora Adler
- Surbhi Agarwal
- Grégory Albert
- Mirko Ambrogetti
- Artem Andreyuk
- Razvan Arpasanu
- Charlotte Art
- Arora Bharti
- Yahya Bennaghmouch
- Elias Biesmans
- Mikail Bircan
- Anissa Blondin
- Caroline Boeckx
- Mihai-Alexandru Botezatu
- Nawfal Boujtat
- Clémence Brault
- Chloë Ceyssens
- Evan Ciampi
- Mohit Daftary
- Ellen De Backer
- Stien De Beul
- Annelies De Boever
- Marie de Ville de Goyet
- Edouard Dermonde
- Sim Dhaliwal
- Eoin Dick
- Natalie Eriksen
- Megan Gawron
- Ezaline Goffin
- Jens Goos

- Michelle Hahn
- Gregory Helssen
- Bryan Herbert
- Jaagrit Hooda
- Manish Jangra
- Noémie Janssens
- Uthaya Jeyabalan
- Bishal KC
- Gabe Kitamura
- Dimitry Knoops
- Yonghan Liu
- Yannick Maes
- Julie Mahieu
- Naomi Meta Tshibanda
- Elise Mommerency
- Shuguftha Naveed
- Tamara Orekhova
- Romina Ozoria Miranda
- Piyush Ashok Patil
- Steven Pelcovits
- Sander Phlips
- Lorana Piron
- Ruark Potgieter
- Elena-Laura Radulescu
- Sohil Shah
- Anjali Soni
- Jelle Soons
- Matheus Souza Lourenço
- Alexandre Tapia Reyes
- Afifa Tasnim
- Claudia Toma
- Shwe Uu
- Maé Van den Broeck
- Andreas Vranken
- Xenia Wegewitz
- Eric Curran Woelke
- Brad Wozniak
- Jennifer Zhuo

We would also like to extend a special thank you to Christian Stadler, SAP DE, Product Manager International Trade Management, for the contribution on the SAP GTS future outlook, and Vishal Verma, SAP DE, VP, Head of Solution Management – Risk, Compliance, and Tax at SAP, for the inspiring foreword.

Finally, our sincere thanks go to everyone at Rheinwerk Publishing, especially to Emily Nicholls and Megan Fuerst, for their support and dedication to publish this book.

Chapter 1
Introduction

This introduction provides an overview of the international trade challenges faced by global companies. We'll describe how SAP has enhanced SAP GTS to help clients respond to these challenges.

To navigate and manage global trade regulatory changes, companies require scalable global trade management systems to enable compliant operations. In this chapter, in Section 1.1, we provide an overview of the various functionalities of the SAP Global Trade Services (SAP GTS) solution focused on a brief introduction to the three main areas covered: Compliance Management, Customs Management, and Preference Management. This chapter also compares SAP S/4HANA for international trade functionality with SAP GTS, including how these components integrate into SAP S/4HANA (Section 1.2).

1.1 What Is Global Trade Management?

The global trade landscape continues to evolve with increased trade barriers through sanctions and export controls as well as new free trade agreements (FTAs). The continued use of international trade policies to drive foreign policy objectives continues to create an ever-changing regulatory landscape that requires companies to consider global trade management software solutions.

SAP GTS is a leading global trade management solution that automates, centralizes, and allows companies to manage trade compliance requirements across multiple jurisdictions. As one of the most flexible and mature global trade management solutions in the market, SAP GTS contains the functionalities needed to comply with many regulatory requirements impacting cross-border operations.

Adequately deploying SAP GTS requires in-depth knowledge of country-specific regulatory requirements, business processes, and GTS and enterprise resource planning (ERP) systems functionality. Most relevant regulations aren't written in a manner that can be easily translated to a systems control. It's therefore important to bring the right mix of skills and specialists to implement and architect SAP GTS in an efficient and effective manner.

Deloitte's[1] global SAP GTS practice is uniquely positioned to bring the collective skills and experience to effectively implement SAP GTS. Our practice has gained significant global experience implementing all SAP GTS functionalities across more than 70 countries.

In the following sections, we provide an overview of SAP GTS's main components: Compliance Management, Customs Management, and Preference Management.

1.1.1 Compliance Management

Export control regulations are restrictions to control the movement of sensitive items, typically defined as military or dual-use, which can take the form of products, software, and technology. Sanctions regimes are imposed on a country, entity, or individual to prevent either comprehensive or limited economic, financial, or trade activities. Trade regulations are complex, vary by country, and are frequently amended based on technological advances and changes to economic and political environments.

Export controls and sanctions are used as tools to drive foreign policy interests. These regulations are imposed by governments around the world and can change rapidly. Penalties for noncompliance can be severe and, in certain cases, lead to criminal proceedings.

Regulators expect companies to have adequate compliance programs to manage these regulatory requirements. A key element of a compliance program is the use of global trade management systems that can automate important controls in export and sanctions processes. SAP GTS has significant Compliance Management capabilities to help clients manage export license management (e.g., legal control), sanctioned party list (SPL) screening, and embargo checks.

The *SPL screening* functionality allows companies to screen third parties against sanctions regimes. The solution can also be configured to screen against multiple country-specific screening regulations. SAP GTS has established an ecosystem of third-party content providers from which companies can procure the latest SPL content, providing users with the flexibility to engage with the most appropriate vendors based on their business.

The *embargo* functionality allows companies to screen business partners that may be located in embargoed destinations. SAP developed a fully integrated functionality with

1 Deloitte refers to one or more of Deloitte Touche Tohmatsu Limited, a UK private company limited by guarantee ("DTTL"), its network of member firms, and their related entities. DTTL and each of its member firms are legally separate and independent entities. DTTL (also referred to as "Deloitte Global") doesn't provide services to clients. In the United States, Deloitte refers to one or more of the US member firms of DTTL, their related entities that operate using the "Deloitte" name in the United States, and their respective affiliates. Certain services may not be available to attest clients under the rules and regulations of public accounting. Please see *www.deloitte.com/about* to learn more about its global network of member firms.

procurement and sales processes to prevent shipments to/from embargoed countries. If a business partner is located in an embargoed country, SAP will block the transactions.

The *legal control* functionality allows companies to screening transactions of controlled items to comply with various export or import control regulations. The solution provides the ability to establish different legal regulations that apply to company operations in different countries, resulting in a flexible solution that allows for the management of export classification of products and export authorizations under multiple jurisdictions. The solution also allows companies, if applicable, to manage their export or import control requirements globally by applying extraterritorial trade regulations to all applicable business operations.

1.1.2 Customs Management

The United Nations Conference on Trade and Development (UNCTAD) estimates that the average customs transaction involves 20–30 different departments, 40 documents, 200 data elements (30 of which are repeated at least 30 times), and rekeying of 60–70% of all data at least once.[2]

Each transaction that involves a border-crossing movement of goods may trigger requirements for customs filing, customs duty payment, multiple reporting requirements, and records that must be maintained. The supporting documentation may consist of the commercial invoice, a packing list, a certificate of origin, an export control license, an import license, a movement certificate, and other documents. All of these documents contain data elements, such as value of the goods, description, classification codes, and others.

These data elements are critical to enable compliant customs. With the volumes of transactions most companies have, a system to manage these customs processes is necessary.

Although some companies may outsource some of these activities to third parties, the legal liability remains with the importer and exporter of record. In addition, many customs authorities around the world are encouraging companies to perform electronic customs filing. For example, in the United States, Canada, European Union (EU), and many other countries of the world, customs authorities may require electronic communication for declarations. In the case of the EU, according to Article 6 of the Union Customs Code that entered into force on May 1, 2016, all customs declarations, applications, or decisions between economic operators and customs authorities must be made using electronic data processing techniques by 2025.

2 "Roadmap for building a national trade information portal," *Transport and Trade Facilitation*, Series No 16, UNCTAD, 2022, available at *www.unctad.org/system/files/official-document/dtltlb2021d3_en.pdf*.

Customs authorities are also especially focused on whether companies employ adequate global trade management solutions to manage special customs procedures. In some cases, the use of global trade management solutions may be a prerequisite for approval to use these procedures.

The Customs Management functionality allows you to manage the following activities, among others:

- Classification of products with the appropriate Harmonized System (HS) classification codes
- Duty calculations to support the determination of the landed cost for products to be imported
- Generation of required documentation to support import and export operations
- Electronic filing of import and export data to customs brokers or directly into the systems of the customs authorities
- Management of stock accounting to support complex customs regimes such as free trade zones, bonded warehouse (BWH), and inward processing relief (IPR)

1.1.3 Preference Management

In this section, we focus on *free trade agreements (FTAs)*. FTAs are implemented by many countries to promote international trade. At the time of writing, the World Trade Organization (WTO) has been notified of more than 285 regional trade agreements (customs unions and FTAs)—a figure that increases every year. While new FTAs continue to be negotiated around the world, there have also been instances of renegotiations of FTAs, for example, the North Atlantic Free Trade Agreement (NAFTA) between the United States, Canada, and Mexico, which resulted in the United States-Mexico-Canada Agreement (USMCA).

Furthermore, unilateral preferential treatments are also possible through the Generalized System of Preferences (GSP) program, wherein many developed countries foster manufacturing and exports from less developed countries. The GSP program provides a formal system of exemption from the WTO rules, such as the most favored nation (MFN) principle, allowing countries to lower import duties on a nondiscriminatory basis for goods originating in developing countries. In these types of agreements, we notice a trend where the authorities move toward a principle of self-certification by economic operators, who make their own statements on origin (e.g., the Registered Exporter [REX] system introduced in early 2017 in the EU).

The use of FTAs can be challenging for companies as the rules to determine whether items produced meet the parameters of a specific agreement can be complex. Companies seeking to use FTAs need to be able to request declaration from vendors of the applicability of FTAs and be able to provide their customers with the required proof that their sold goods are eligible for preferential treatment in a specific trade agreement. To manage these requirements, companies need to implement global trade management solutions.

SAP GTS offers the following capabilities through the Preference Management functionality:

- Management of the request and dunning of long-term supplier declarations (LTSDs) with the vendors
- Determination of the preferential status of the finished products
- Generation of origin statements to the customers, which could consist of customer LTSDs, invoice statements, or certificates of origin

1.2 Global Trade Management with SAP

We've now covered the basics of global trade management, so let's switch gears to SAP. The following sections describe SAP's global trade management capabilities within SAP GTS and SAP S/4HANA for international trade.

1.2.1 SAP Global Trade Services

SAP GTS is the main SAP solution to manage international trade requirements across various jurisdictions. Figure 1.1 shows the evolution of the various SAP GTS versions, starting with the very limited trade capabilities in Foreign Trade in SAP R/3 to the mature global trade solution in SAP GTS, edition for SAP HANA.

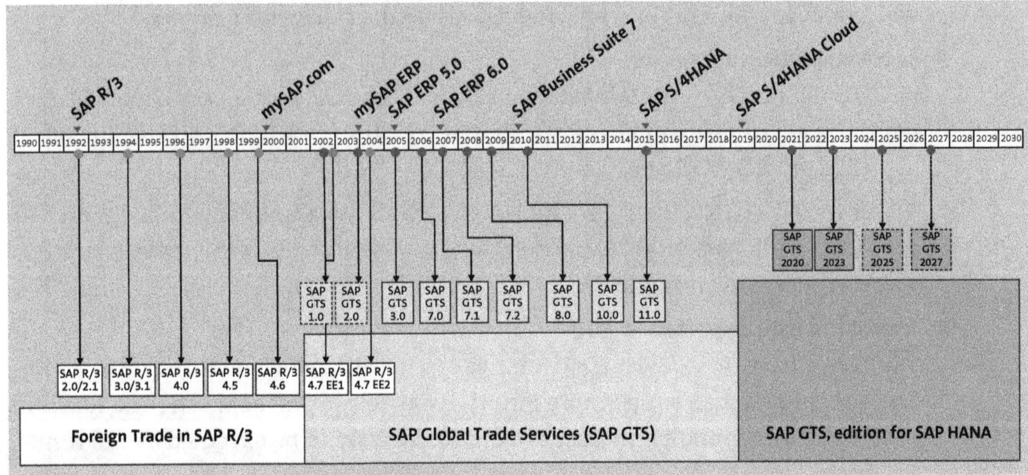

Figure 1.1 From Foreign Trade in SAP R/3 to SAP GTS, Edition for SAP HANA

Since the first release of SAP GTS 3.0, SAP provided a standard solution for Compliance Management, Customs Management, and Preference Management. Over time, SAP enhanced SAP GTS by adding functionalities and capabilities, with SAP GTS 11.0 being the latest solution before the overall transition to SAP GTS, edition for SAP HANA, which is the focus of this book.

SAP GTS, edition for SAP HANA was first released in 2020. The new edition will follow the overall release strategy for SAP HANA, with 2023 being the current release at the time of writing.

SAP GTS, edition for SAP HANA provides some new capabilities to help clients manage complex global trade requirements. Following is an overview of the new features within SAP GTS, edition for SAP HANA and the main reasons why a company should decide to migrate to the new release:

- **Future-proof roadmap**
  To remain compliant with the latest changes in the regulations, especially in the area of customs e-filing, it's mandatory to implement the regular updates provided by SAP.

  At the time of writing, SAP GTS 11.0 will be out of mainstream maintenance at the end of December 2025, meaning that the required legal updates will only be available in SAP GTS, edition for SAP HANA. An upgrade will be highly recommended to avoid disruption in your business operations or to avoid noncompliance with the trade regulations.
- **Simplified operations**
  SAP GTS, edition for SAP HANA simplifies the process steps when executing the rescreening of business partners against the updated SPLs from the authorities, which reduces the risks in errors and speeds up the rescreening process.
- **Modern platform**
  SAP GTS, edition for SAP HANA remains the solution for a company to manage their global trade processes and has a standard integration with SAP S/4HANA instances, as well as other SAP ERP environments.

  Moreover, SAP GTS, edition for SAP HANA also allows a company to integrate with non-SAP solutions and as such, centralizes and standardizes the trade operations within a global organization.
- **Improved user experience (UX)**
  SAP GTS, edition for SAP HANA comes with a new UX, which is based on SAP Fiori. SAP Fiori provides an easier, more intuitive way to run SAP applications and allows you to adapt and modify the user screens to the tasks to be executed by the respective users.
- **Redesigned apps**
  By moving to the SAP Fiori launchpad, it's possible to easily create your own apps to improve the usability of the solution.

Nevertheless, this also has provided SAP the possibility to redesign the old SAP GUI transactions and to provide more out-of-the-box apps, which are extended with every new release of SAP GTS, edition for SAP HANA. These apps provide, among other things, improved workbenches and support for more intuitive management of your trade requirements.

- **Embedded analytics capabilities**
  Together with the redesigned apps, SAP now also provides more embedded analytics capabilities. These analytics provide better insight into your trade operations and allow more flexible filtering. In Chapter 9, we've included a more complete overview of the current analytics capabilities.
- **Workbenches for status, progress, and proposal**
  SAP also upgraded the operational workbenches, for example, in the area of customs filing. In the new workbenches, a user can easily obtain the status and progress of a specific trade process, and the proposed follow-up activities are indicated, which provides a user with a more structured organization of their daily tasks.
- **Enterprise search capabilities**
  Similar to the other SAP products, SAP GTS, edition for SAP HANA comes with enterprise search. Enterprise search allows you to search through the entire SAP solution. After selecting the search icon, you can choose to do a simple or advanced search. You can search for a term without specifying any details, search in all categories, or specify a particular category.
- **Improved Preference Management**
  The Preference Management process has been thoroughly redesigned, and a number of new features have been released. The reworked solution will support the following features, among others:
  - The maintenance of LTSDs received from a vendor for which no initial request was initiated
  - The intercompany LTSD process
  - The determination of preferential status for product identifiers, such as product batches or serial numbers
- **Improved integration with SAP S/4HANA's order-to-cash process**
  SAP GTS, edition for SAP HANA provides a better visibility of the compliance status of your order-to-cash documents, and the status will appear as an indicator in your sales orders and scheduling agreements.
- **Integration with SAP S/4HANA advanced shipping and receiving (ASR)**
  Although transportation and global trade processes are very tightly interconnected, there was limited integration from SAP in the past. With SAP GTS, edition for SAP HANA, more integration points are foreseen with SAP Transportation Management

(SAP TM) and SAP Extended Warehouse Management (SAP EWM). These integrations will allow the following, for example:

  - SPL and license screening on freight orders
  - Integration of transportation data between freight orders and the customs declarations

- **Co-deployment option**
  As of SAP GTS, edition for SAP HANA 2023, SAP also supports the co-deployment of the solution with SAP S/4HANA version 2022 onwards.
  In other words, although SAP GTS, edition for SAP HANA remains a separate solution, it can be installed on the same technical infrastructure as SAP S/4HANA, which will lower the operational costs.

Although SAP GTS, edition for SAP HANA inherited almost all of the capabilities of SAP GTS 11.0, note that the following features have been decommissioned in this edition of SAP GTS, edition for SAP HANA:

- **Restitution Management**
  The restitution program has been discontinued by the EU, so the functionality is no longer supported.
- **Letter of Credit Processing**
  Letter of Credit Processing is now included in SAP S/4HANA Finance for treasury and risk management.
- **Case Management**
  The embedded Case Management is no longer supported because the workflow can be set up by more generic SAP-provided tools.

1.2.2 SAP S/4HANA for International Trade

With the introduction of SAP S/4HANA, SAP also released SAP S/4HANA for international trade, which provides the following:

- Optimized trade data models
- Enhanced synchronization of master data between SAP GTS and SAP S/4HANA for international trade
- SAP Fiori-based applications
- Business roles with dedicated SAP Fiori apps

In addition, SAP S/4HANA has enhanced and renamed SAP ERP's Foreign Trade component to SAP S/4HANA for international trade, which includes the following functions:

- Product classification for supporting export control and Intrastat
- Intrastat reporting (as was the case already for SAP R/3 Foreign Trade)

- Export license screening
- Embargo screening
- SPL watch list screening

SAP GTS remains the main global trade solution where SAP will focus on future enhancements and functionalities across customs, export controls, special customs procedures, and Preference Management. In Chapter 11, we'll discuss the SAP GTS, edition for SAP HANA future roadmap in more detail.

In the following sections, we'll describe the new functionalities in SAP S/4HANA for international trade in detail.

Classification for Commodity Codes

Every product that is imported or exported needs to be classified. The corresponding classification codes follow an internationally agreed-upon standard and establish, among other things, the applicable duties to be paid and potential required permits.

SAP S/4HANA enables you to assign commodity codes to products. The following elements are key features in SAP S/4HANA:

- A commodity code can have a validity period assigned.
- Commodity codes can be assigned to a product for a certain period, which means the classification of a product can be made effective from a specified future date.
- The maintenance of commodity codes has been improved through the development of a mass classification functionality that allows a commodity code to be assigned to multiple products.
- User-friendly SAP Fiori apps have been created to classify, reclassify, and manage products to update the material master in SAP S/4HANA.
- Additional functionality has been included to classify products by country or by group of countries via numbering schemes. The country is determined based on the country of the plant.
- Separately, in SAP GTS, edition for SAP HANA, there is more robust functionality to maintain commodity codes and to classify products. If both SAP S/4HANA and SAP GTS, edition for SAP HANA are being used, best practice is to set up the commodity codes in SAP GTS, classify the products in SAP GTS, and transfer the classification to SAP S/4HANA via the standard interface.

Intrastat

Intrastat is the official statistical reporting for the movement of goods between the member countries of the European Union. An Intrastat report provides an overview of which types of goods were traded from one EU country to another within a specific time period. It's mandatory for every natural or legal person who has a tax number in the declaration country and has a contract with a partner in another EU country to submit an Intrastat declaration when certain thresholds are met. This Intrastat declaration can be sent to the authorities responsible for Intrastat in paper or electronic format, depending on the local member state's legal requirements.

Companies needed to manage Intrastat reporting can leverage either SAP S/4HANA for international trade or SAP GTS, edition for SAP HANA.

In SAP S/4HANA, some elements regarding Intrastat are improved:

- SAP Fiori-based apps are available to manage Intrastat declarations. These apps generate Intrastat declarations in a country-specific file format.
- Default values are defined at the country level instead of at the organizational unit level, which could considerably reduce data maintenance.

Export Legal Control

SAP S/4HANA for international trade provides basic functionality to manage export licenses. The solution provides the ability to classify products for legal control (i.e., export regulations) and to create export licenses. Export licenses can be set up specifically for certain products to prevent transactions that aren't authorized because an export license isn't in place.

In SAP S/4HANA for international trade, several functionalities have been improved by leveraging SAP Fiori apps:

- Creating and maintaining licenses
- Displaying the status of trade compliance documents
- Managing blocked documents (a blocked transaction can be released by creating a new license or by assigning an existing license)

> **Note**
> More information on Compliance Management in SAP GTS can be found in Chapter 6.

Embargo

SAP S/4HANA for international trade maintains the embargo control functionality, which prevents shipments to designated embargoed countries. In SAP S/4HANA for international trade, SAP has improved some elements of this functionality:

- The embargo screening can be defined for specific partner functions.
- SAP Fiori-based apps have been developed to manage embargo-blocked documents by confirming or releasing them.
- An embargo check can be performed for a specific country within a specific date range.

Integration with SAP Watch List Screening

SAP developed a functionality to screen parties involved in a company's export, import, domestic, and payment transactions to minimize the risk to engage with sanctioned entities across applicable jurisdictions. The solution, called SAP Watch List Screening, is a software as a service (SaaS) application running on SAP Business Technology Platform (SAP BTP). This solution comes with a preconfigured comparison logic, which, unlike the SAP GTS SPL screening functionality, can't be modified.

In SAP S/4HANA, SAP Fiori apps can be used to screen and manage screening hits. An assigned user can display and resolve the screening hits by confirming or rejecting the hits. To use this functionality, the supplementary subscriptions for the user interfaces must be enabled and integrated with SAP S/4HANA Cloud.

1.3 Summary

In this chapter, we provided an overview of the global trade challenges faced by companies, as well as of the SAP solution to manage complex global trade requirements. Although SAP S/4HANA for international trade provides some limited global trade functionalities, SAP GTS, edition for SAP HANA remains the primary SAP solution to manage complex global trade scenarios and will continue to be further enhanced in future releases.

The next chapter will highlight the basics of SAP GTS, edition for SAP HANA. We'll also discuss SAP GTS, edition for SAP HANA navigation and the transition from SAP ERP to SAP S/4HANA.

> **Note**
>
> The content in this book is written based on SAP GTS, edition for SAP HANA. For ease of reading, in each chapter we will shorten SAP GTS, edition for SAP HANA to simply SAP GTS.

Chapter 2
SAP GTS Basics: Setup, Authorizations, and Navigation

This chapter discusses SAP GTS interfacing with feeder systems, configuration of user authorizations, and activation of SAP Fiori apps.

SAP GTS requires a connection with a transactional system, or *feeder system*, which will "feed" all the necessary master data and transactions. This chapter covers the most common way to connect an SAP S/4HANA system to SAP GTS using the plug-in (Section 2.1). It's important to note that SAP GTS can also be interfaced with other feeder systems, such as SAP Transportation Management (SAP TM), SAP Extended Warehouse Management (SAP EWM), SAP Customer Relationship Management (SAP CRM), and even non-SAP systems. These interfaces may require the use of web services or remote function calls (RFCs), which we'll discuss in Section 2.2.

In Section 2.3 of this chapter, we'll also describe how to navigate through different SAP Fiori apps in the SAP Fiori launchpad. The different functionalities are logically grouped in a user-friendly way based on business roles. To provide access based on business roles, we'll describe how to define and assign user authorizations, which provides the access control to the SAP GTS applications (Section 2.4). This chapter will introduce the basics of the SAP GTS authorization.

We'll also describe the different activities to follow to activate the SAP Fiori launchpad and the specific SAP Fiori apps (Section 2.5).

2.1 Connecting SAP S/4HANA to SAP GTS, Edition for SAP HANA

SAP S/4HANA is one of the most common feeder systems for SAP GTS. SAP GTS is fed with both master data (e.g., customers, vendors, materials, etc.) and transactional data (e.g., sales documents, purchase orders, deliveries, etc.). In this section, we'll examine the necessary settings to interface an SAP S/4HANA system to the SAP GTS system.

Other feeder systems or embedded add-ons, such as SAP TM and SAP EWM, can also be connected to SAP GTS (Section 2.2).

2.1.1 System Architecture

SAP GTS (technical name: GTSCORE) is typically interfaced to SAP feeder systems. Figure 2.1 shows the general exchange of data between SAP S/4HANA, some optional embedded systems such as SAP Master Data Governance (SAP MDG), and SAP GTS.

For SAP S/4HANA feeder systems, the plug-in is a part of the S4CORE foundation and is no longer required to be implemented separately to connect to SAP GTS. This applies only for SAP S/4HANA versions 1511 and onward. For earlier versions, an additional component for the plug-in (technical name: SLL_PI) remains a prerequisite.

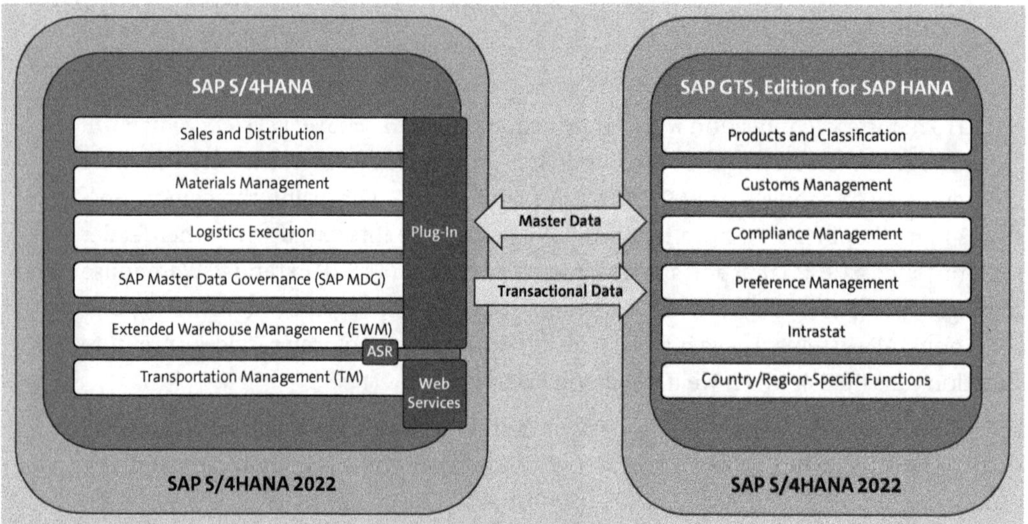

Figure 2.1 Data Exchange between SAP S/4HANA and SAP GTS

When connecting SAP GTS and an SAP S/4HANA feeder system, two different scenarios can apply:

- **Hub deployment**
  In this scenario, the feeder system and the SAP GTS system are physically separated, and SAP GTS serves as a central hub with its own dedicated instance.
- **Co-deployment**
  This scenario allows companies to install SAP GTS on the same system instance as SAP S/4HANA but in separate clients. As of SAP GTS, edition for SAP HANA 2023, co-deployment is possible on SAP S/4HANA 2022 SP02 or higher.

Note

For performance reasons, hub deployment is often favorable. The plug-in makes extensive use of RFCs. An RFC makes a call to another SAP functional module and allows you to set up a connection within different SAP modules or between an SAP system and a third-party system.

Co-deployment will lower the total cost of ownership but might have an impact on the SAP S/4HANA system availability in case of upgrades and legal changes for SAP GTS. Therefore, the innovation and patch strategy need to be aligned across different applications.

2.1.2 System Connection Configuration

The process of configuring the plug-in in SAP S/4HANA is similar to the configuration of SAP GTS. If the SAP GTS client is located in a different physical system—that is, on a separate instance—in the hub deployment scenario, then the first step is to establish the connection in both systems. We'll discuss the steps in the next sections.

Logical Systems in SAP S/4HANA and SAP GTS

Before data exchange can happen between SAP S/4HANA and SAP GTS, you must set up both SAP S/4HANA and SAP GTS as logical systems. You can do this in SAP S/4HANA via Transaction /SAPSLL/TBDLS_SPI (Define Logical System) and Transaction /SAPSLL/TOOO_SPI (Assign Logical System). In SAP GTS, you can find the **Define Logical System** and **Assign Logical System** settings in the Customizing Implementation Guide (IMG) via Transaction SPRO, as shown in Figure 2.2.

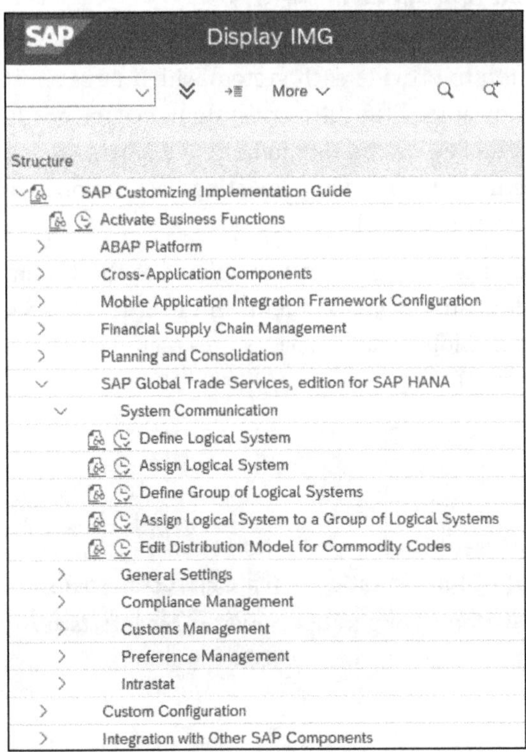

Figure 2.2 Customizing Implementation Guide (Transaction SPRO) in SAP GTS

To reduce the burden of data maintenance on the user, you may want to group logical systems in SAP GTS into a single logical system group when more than one feeder system is activated. You can define a logical system group in the Customizing IMG of SAP GTS via the **Define Group of Logical Systems** setting and assign the SAP S/4HANA feeder systems names.

Note

All SAP GTS-relevant master data (business partners, products, etc.) must be harmonized among all feeder systems in the same logical system group. If you assign several feeder systems to a single logical system group, the master record in SAP GTS is always created on the logical system group level (e.g., one product master per system group).

The *logical system* creates a unique identifier for each SAP system or non-SAP system client in the system landscape. To ensure that the logical system name is unique and the environment is simplified, SAP recommends using the naming convention *<System name>CLNT<Client key>*.

After the logical systems for SAP GTS and the feeder systems connected to SAP GTS are defined, you must then assign them to the respective clients.

Remote Function Call Destinations in SAP S/4HANA and SAP GTS

After all relevant systems have been uniquely defined in the system landscape, you must establish a bidirectional connection to provide each system with a target host and logon details. Because communication among the different systems occurs mostly via RFCs, it's typically required to set up an RFC destination in both SAP S/4HANA and SAP GTS. You can create new RFC destinations or change existing destinations via Transaction SM59 (Configuration of RFC Connections).

This **Maintain RFC Destinations for RFC Calls** setting and others related to the system connection configuration can also be done in the plug-in menu within SAP S/4HANA via Transaction /SAPSLL/MENU_LEGALR3 (Global Trade Services Area Menu) by selecting the **Basic Settings** tab shown in Figure 2.3.

Note

You can also find this configuration in the Customizing IMG:

- In SAP S/4HANA, follow this path: **SAP NetWeaver • Application Server • IDoc Interface / Application Link Enabling (ALE) • Communication • Create RFC Connections.**
- In SAP GTS, follow this path: **ABAP Platform • Application Server • IDoc Interface / Application Link Enabling (ALE) • Communication • Create RFC Connections.**

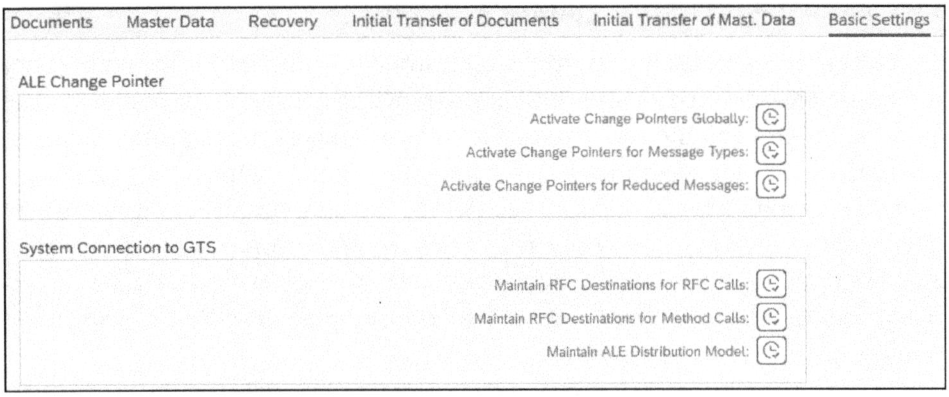

Figure 2.3 Plug-In Menu in SAP S/4HANA

In the RFC destination, you need to specify a target host and an IP address. If applicable, also specify the system number, user data, and logon data to allow for remote logon to the target system. The RFC destination contains all the technical information the requesting system needs to call the partner from within the network.

You must assign the created RFC connections to a logical system and enable communication before the application program can call on a targeted logical system. For SAP GTS, you don't need to give the RFC destination the same name as the logical target system, but you must link the previously created RFC destinations and logical systems for method calls so that you can use cross-application function calls. This is done by selecting the **Maintain RFC Destinations for Method Calls** option in the **Basic Settings** tab shown in Figure 2.3.

As a final step in system connection configuration for the feeder system, you need to define the *Application Link Enabling* (ALE) distribution model by selecting the **Maintain ALE Distribution Model** option in that same **Basic Settings** tab. This allows the data transfer through methods between the systems.

Although SAP GTS isn't based on ALE, it uses the ALE distribution model to help the SAP S/4HANA feeder system identify its SAP GTS partner system. When a user in the SAP S/4HANA feeder system carries out a transaction that involves SAP GTS functions—for example, creating a sales order that needs to be checked by SAP GTS Compliance Management—the application program (ALE) determines the appropriate receiver system by querying the distribution model.

Note

In addition to connecting using the RFC method, SAP also supports web services for the connection related to business partner transfers. The use of web services is planned to increase in the future.

2.1.3 Connecting SAP GTS to Customs Authorities

SAP GTS can be interfaced directly to certain customs authorities by using a third-party *conversion provider* such as SEEBURGER. The provider manages the communication between SAP GTS and the country-specific customs authorities. The conversion provider services include converting IDocs to XML, Electronic Data Interchange for Administration, Commerce, and Transport (EDIFACT), and other formats. You can see the interaction between SAP GTS, a conversion provider, and customs authorities in Figure 2.4. Automated Tariff and Local Customs Clearance System (ATLAS), Advanced Encryption Standard (AES), and New Computerized Transit System (NCTS) are examples of country-specific authority systems.

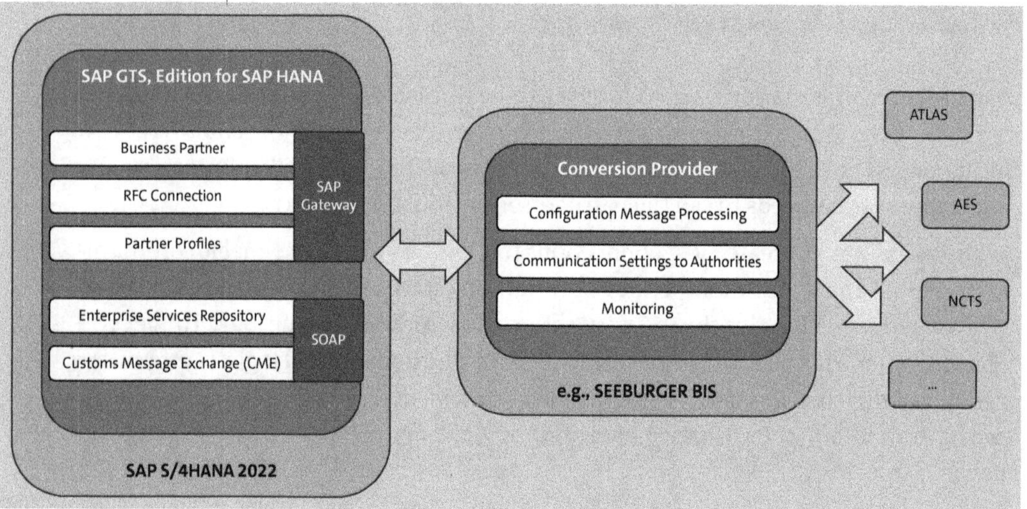

Figure 2.4 Conversion Provider

2.2 Connecting Using Different Interfacing Techniques

We'll now go into more detail on how to set up a connection with SAP GTS using different interface techniques. These techniques can be used to connect different SAP products, such as SAP CRM to SAP GTS, or to connect non-SAP feeder systems to SAP GTS.

Other common interfaces with SAP GTS are SAP TM and SAP EWM, embedded within SAP S/4HANA. These systems are integrated through web services for embedded TM and RFC connections for embedded EWM, as shown in Figure 2.5. Both techniques are quite different, but the result is similar when viewed from an SAP GTS perspective. This section will illustrate these integration techniques. Additionally, we'll provide a brief introduction to the SAP solution called *advanced shipping and receiving* (ASR), explaining its role in the integration of different SAP systems.

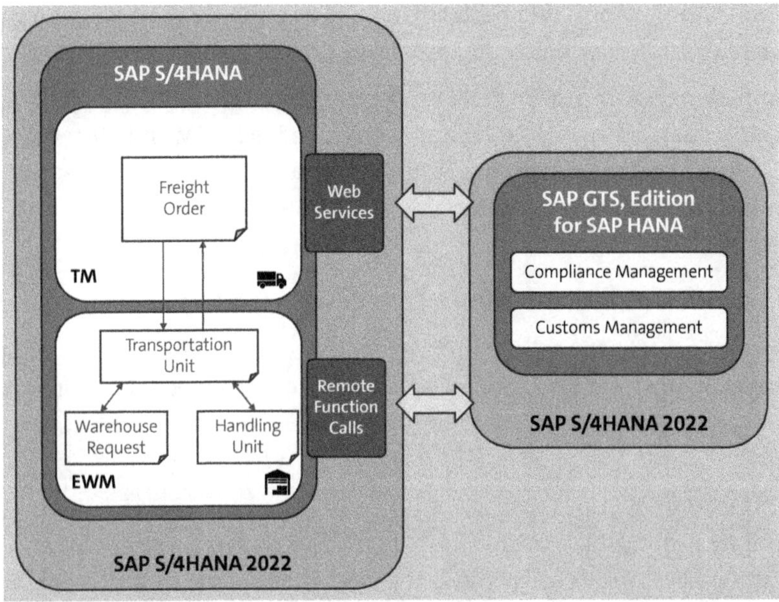

Figure 2.5 Integration between Embedded TM and EWM in SAP S/4HANA and SAP GTS

SAP TM and SAP EWM are separate modules, which can be deployed embedded in SAP S/4HANA or as standalone systems allowing decentralization and connection to multiple ERP instances (SAP S/4HANA or SAP ERP, respectively). In this section, for simplification, we'll focus on the embedded deployment options of TM and EWM with SAP S/4HANA.

Embedded TM is used to optimize transportation processes and can be used in two scenarios:

- In the shipper scenario, transactional data (sales orders and outbound deliveries) are transferred from SAP S/4HANA to the embedded TM system. This results in embedded TM transportation requirement documents (e.g., order transportation requirements and delivery transportation requirements).
- Logistics service providers (LSPs) can create forwarding orders in the embedded TM system based on the transportation requests received from their customers.

Embedded TM can also be used to assign and tender shipments to carriers. After you book the shipment via a specific carrier, the specific booking information can be housed in embedded TM. You can use this information as a base to create export declarations in SAP GTS.

Embedded EWM is focused on warehousing processes such as controlling movements to and from the warehouses or optimizing picking and replenishment processing. This module specializes in tracking goods movement (inbound and outbound) and their physical location in the warehouse. By integrating this module with SAP GTS, several new functions and automations can be introduced. For instance, companies can ensure the accurate execution of unloading procedures for all transit declarations or perform compliance checks on outbound movements.

Ultimately, the integration between SAP GTS and both embedded EWM and TM focuses on two SAP GTS functionalities, Compliance Management and Customs Management, which we'll discuss in the following sections.

2.2.1 Integration with TM and EWM from a Technical Perspective

Before discussing the new functionalities and general system benefits of integrating SAP GTS with embedded TM and EWM, we'll discuss the integration techniques in this section.

Integration via Web Services

The integration between embedded TM and SAP GTS is based on service-oriented architecture (SOA) web services. The systems can be connected either via a point-to-point connection or via middleware such as SAP Process Integration.

In the case of embedded TM, there are 12 web services (6 in SAP GTS and 6 in embedded TM) in total to support the compliance and customs functionalities. In each system, there are three types of web services for both the compliance functionality and the customs functionality:

- **Request web service**
  This web service is used to trigger the compliance screening/customs declaration from embedded TM to SAP GTS.
- **Confirmation web service**
  This web service is used to transfer back the results of the screening/export declaration.
- **Cancellation request web service**
  This web service is used to cancel the compliance/customs document in SAP GTS.

Table 2.1 lists the compliance web services you need to implement.

SAP GTS	Embedded TM
TradeComplianceCheckSUITERequest_In	TradeComplianceCheckSUITERequest_Out
TradeComplianceCheckSUITE-CancellationRequest_In	TradeComplianceCheckSUITE-CancellationRequest_Out
TradeComplianceCheckSUITE-Confirmation_Out	TradeComplianceCheckSUITE-Confirmation_In

Table 2.1 Compliance Web Services

Table 2.2 lists the customs web services you need to implement.

SAP GTS	Embedded TM
ExportDeclarationSUITERequest_In	ExportDeclarationSUITERequest_Out
ExportDeclarationSUITE-CancellationRequest_In	ExportDeclarationSUITECancellation-Request_Out
ExportDeclarationSUITE-Confirmation_Out	ExportDeclarationSUITEConfirmation_In

Table 2.2 Customs Web Services

Note that, in a typical setup, SAP GTS is still connected to the SAP S/4HANA system via the standard plug-in to accommodate the transfer of master data (material master, customer master, and vendor master) and transactional documents. Data originating from SAP S/4HANA is transferred to SAP GTS only. When TM or EWM are embedded in SAP S/4HANA, there is no transfer of data as it's the same database.

Integration via Remote Function Calls

The integration between embedded EWM and SAP GTS is through RFC connections, similar to the connection through the plug-in from your SAP S/4HANA system with SAP GTS. These RFC connections enable data transfer in both directions by using the inbound and outbound delivery processes based on the corresponding documents.

2.2.2 Compliance Screening

For compliance screening, the integration between embedded TM and SAP GTS or between embedded EWM and SAP GTS is similar to the integration with any other feeder system (e.g., SAP S/4HANA). Once a document originating from any of the other modules is saved, it's replicated in SAP GTS, and a compliance check—that is, sanctioned party list (SPL) screening and embargo screening—is performed in SAP GTS. The result of the screening (blocked/released) is transferred back to embedded TM or EWM.

Both embargo and SPL screening can be performed by mapping the SAP GTS document types against the embedded TM and EWM document types. Based on the result of the screening, the document in the original module will be blocked to prevent further processing. When you perform a recheck of a document in SAP GTS, SAP GTS transfers the compliance status back to embedded TM or EWM.

Note

SAP GTS legal control (e.g., export license checks) screening can't be performed for embedded TM transactions because the product number isn't part of the interface. For embedded EWM, SAP GTS legal control screening can be performed.

During the integration, embedded TM, embedded EWM, and SAP GTS require some configuration to activate these functionalities. For SAP GTS, the customization is similar to the integration with SAP S/4HANA. We'll explain the SAP GTS customization requirements for SAP S/4HANA in Chapter 3.

2.2.3 Export Declaration

The integration between embedded TM and SAP GTS to enable the SAP GTS export declaration functionality is similar to the billing document transfer from the SAP S/4HANA feeder system to SAP GTS. The freight order or forwarding order in embedded TM is blocked until the export declaration is completed in SAP GTS. The export declaration confirmation number (e.g., Internal Transaction Number [ITN] for US AES) received from customs authorities is transferred back to embedded TM. This transfer allows the printing of the export declaration confirmation number on the necessary documents (e.g., bill of lading).

The **Customs** tab of the freight order/freight booking, or forwarding order, contains the following:

- The top section contains global trade-relevant information (e.g., mode of transport, customs office of departure, etc.).
- The middle section contains relevant customs activities information (e.g., export).
- The bottom section, called **Item Groups**, includes information from the embedded TM freight order split into several item groups as defined by the grouping strategy.

Users can create a freight order based on multiple sales orders with a different ship-to party (i.e., consignee). A standard grouping strategy creates an item group per ship-to party to ensure that one export declaration is created per ship-to party.

Moreover, this integration makes it possible to cancel a customs declaration in SAP GTS from embedded TM. When a declaration is created, SAP GTS transfers all status information regarding the document progress to embedded TM. Once the release message from customs is received, you can commence the follow-up activities in embedded TM. This grants you the flexibility to cancel the declaration at any stage of the process or proceed with the processing part in embedded TM.

2.2.4 Customs Warehousing Procedures

The integration between SAP GTS and embedded EWM to support the SAP GTS customs warehousing functionality needs to consider the following activities:

- Discharging transit procedures in SAP GTS Customs Management
- Scrapping from the customs warehouse in SAP GTS Customs Management

First, to initiate the discharging transit procedures, an inbound delivery is created in SAP S/4HANA, which subsequently creates an inbound delivery in embedded EWM. Based on one of the documents, a transit document (T1, T2, or TIR) can be created.

Embedded EWM will then pick up this inbound delivery into its warehouse management workflow. To meet customs requirements, SAP GTS blocks the delivery from any further warehouse activities. Simultaneously, SAP GTS automatically generates a transit declaration. This declaration serves as a basis for sending the arrival Electronic Data Interchange (EDI) message to the customs authorities. Once a response from customs is received, granting unloading permission, the inbound delivery in embedded EWM is automatically released for further processing.

To achieve this, we interface SAP GTS with the *New Computerized Transit System* (NCTS), which is an electronic system for the exchange of information between a trader and the customs authorities, launched by the European Union to standardize and control the transit system.

Secondly, scrapping from the customs warehouse refers to the process of disposing of or destroying goods that are stored in a warehouse. This normally occurs when the goods are damaged, expired, or no longer usable. The integration between SAP GTS and embedded EWM helps manage this process.

When an outbound delivery for scrapping is created in embedded EWM, a customs declaration for scrapping is created automatically in SAP GTS. This scrapping outbound delivery is linked to a specific plant and storage location combination in embedded EWM, after which the scrapping reference number is transferred from SAP GTS to embedded EWM.

If SAP GTS doesn't send the scrapping number right away, the delivery process blocks the transfer request for further processing. Once the permission is received in SAP GTS, the scrapping reference number is sent to embedded EWM delivery processing, which releases the block to continue the process.

2.2.5 Advanced Shipping and Receiving

ASR is the new and simplified solution that integrates embedded TM, embedded EWM, and SAP GTS. This solution manages the communication between SAP S/4HANA components, such as embedded TM and EWM, and SAP GTS. ASR reduces the required number of business objects and data, as well as creates a more integrated solution between systems, as shown in Figure 2.6. TM and EWM are shown embedded onto a single SAP S/4HANA system (left) and TM and EWM embedded each on a dedicated SAP S/4HANA system (right).

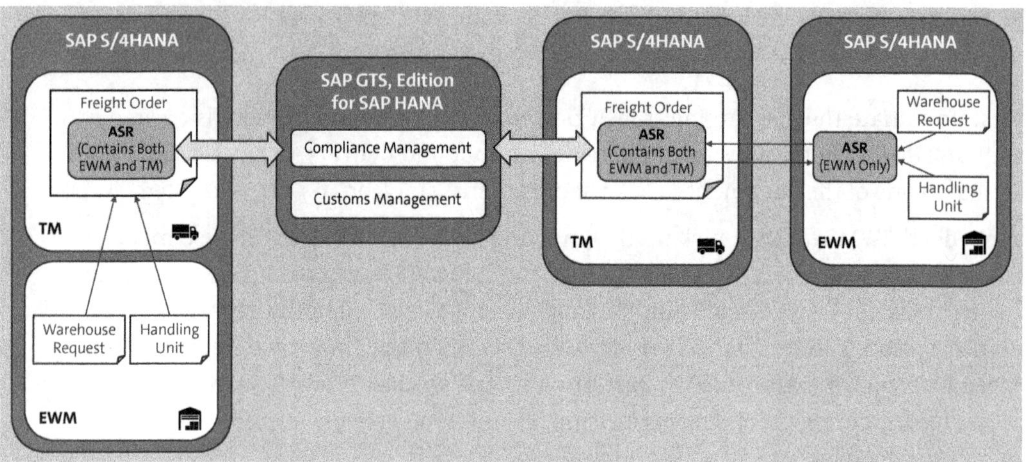

Figure 2.6 Integration between ASR and SAP GTS

Following are two of the most important aspects of this solution:

- No need to use transportation units (TUs) anymore in embedded EWM. One business object is used to manage logistics processes: the freight order of embedded TM.
- Ability to create customs invoices from embedded TM, instead of using the pro forma invoice in SAP S/4HANA. This document automatically triggers an export declaration in SAP GTS to release the goods in embedded TM or EWM.

Note

ASR doesn't fully support all essential business scenarios at the time of writing. For example, package building or delivery splits can't be managed using the current standard ASR functionality. However, SAP is actively working to improve the solution in the upcoming releases to address these limitations.

2.3 Navigating SAP GTS, Edition for SAP HANA

SAP GTS, edition for SAP HANA leverages the SAP Fiori launchpad, which is a modern and intuitive user interface (UI) with a user experience (UX) design. This new approach replaces the SAP GTS 11.0 main menu in SAP GUI. The SAP Fiori launchpad is optimized for desktops, smartphones, and tablets.

We'll explain how to navigate the standard applications in the following sections, but first, let's take a closer look at the SAP Fiori launchpad.

2.3.1 SAP Fiori Launchpad

The new UI is designed to support different user roles and provides specific dashboards tailored to what the user needs to perform their tasks. Tiles and applications, as shown in Figure 2.7, are now used to navigate SAP GTS. A *tile* is a container that represents different types of content such as key performance indicators (KPIs) or other metrics. Users can navigate SAP GTS by selecting the relevant tiles to perform specific global trade activities, such as classifying products, creating customs declarations, or monitoring compliance checks.

The SAP Fiori launchpad is designed based on spaces, pages, sections, and tiles. Spaces functions as access points for a particular business role. *Spaces* are visible at the top-level navigation, as shown in Figure 2.7 (**Compliance**, **Compliance Screening**, **Customs EMCS**, **Customs Export**, etc.). Each space comprises one or more pages. A *page* is an integral component of a space and allows the breakdown of spaces into smaller structured contexts. Pages are designed to be task-oriented such as the pages underneath the space for **Customs Import** shown in Figure 2.7 (**Declarations**, **Related Documents**, **Customs Broker**, **Monitoring**, and **Master Data**). A *section* plays an important role in organizing the content within a page. While a page bundles a larger amount of SAP Fiori tiles within a functional area, a section provides clearer subdivisions within that functional area. Also shown in Figure 2.7, the **Declarations** page contains different sections such as **Import Declarations** and **End of Transit Operations**.

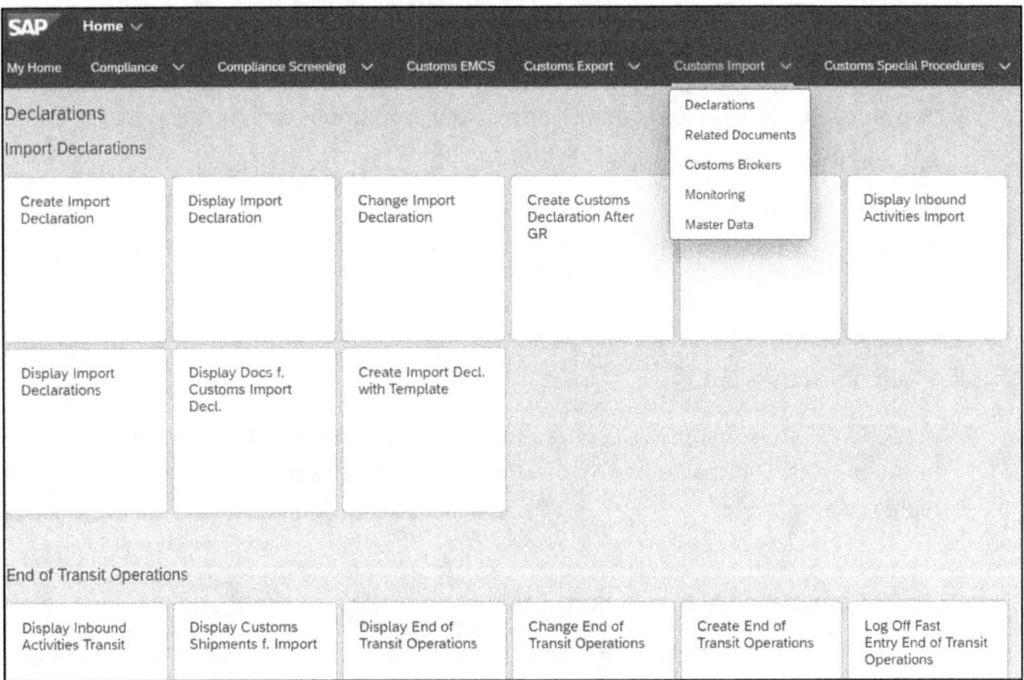

Figure 2.7 SAP Fiori Launchpad for SAP GTS, Edition for SAP HANA

The App Finder in the SAP Fiori launchpad, as shown in Figure 2.8, provides a list of all active applications for a user. The list of available applications depends on the assigned authorizations, explained later in Section 2.4.

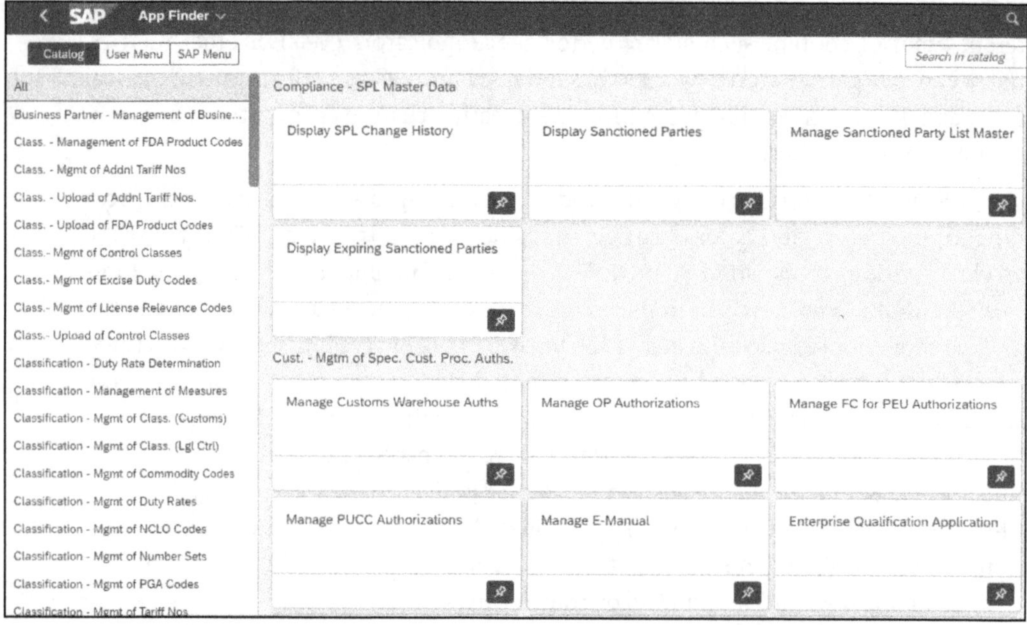

Figure 2.8 App Finder

The main functionalities in SAP GTS are grouped under different spaces based on business roles (Section 2.4) and can be summarized as follows:

- Compliance Management
- Customs Management
- Preference Management
- Products and classification
- Intrastat
- Administrative tasks

In the SAP Fiori launchpad, a user can find the spaces and SAP Fiori tiles related to the business roles. The spaces and applications visible in a user's SAP Fiori launchpad are highly customizable.

2.3.2 Compliance Management

In the SAP Fiori launchpad, a user can find all the SAP Fiori tiles related to the Compliance Management functionality of SAP GTS via the **Compliance** spaces, as shown in Figure 2.9.

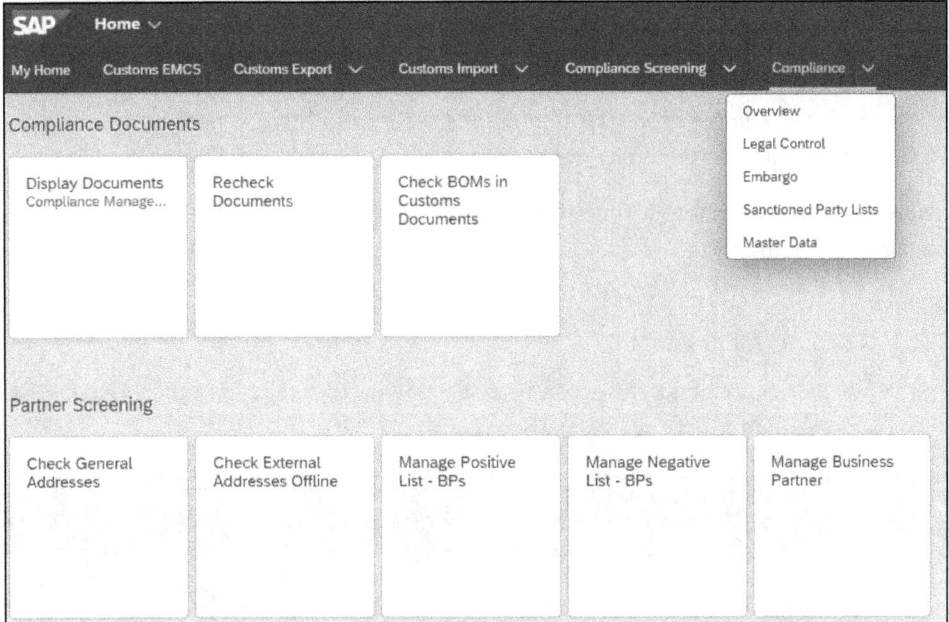

Figure 2.9 Compliance Management: Related Space, Pages, and Tiles

The activities related to Compliance Management are captured in the **Compliance** and **Compliance Screening** spaces and are subdivided into four pages, each focused on a specific compliance area or task:

- **Embargo**
  The **Embargo** page contains SAP Fiori apps to maintain embargo situations (i.e., maintain the embargo status of a specific country), to display business partners in embargoed countries, to analyze the release reason for the documents that were released after embargo screening, and to display released documents.
- **Sanctioned Party Lists**
  The **Sanctioned Party Lists** page allows a user to utilize SAP Fiori apps to display audit trails related to screened and released business partners and to analyze release reasons of blocked business partners as well as the tiles to manage the sanction list master data (e.g., **Display Sanctioned Parties**, **Manage Sanctioned Party List Master**, etc.).
- **Legal Control**
  In the **Legal Control** page, several sections subdivide its functionalities (e.g., **License Management**, **Simulation of Compliance Checks**, **Legal Control Classification**, etc.). The SAP Fiori tiles for **Legal Control** allow users to manage licenses, display licenses, maintain and display determination strategies, monitor all the documents, manage the blocked documents, and recheck documents when needed.

- **Master Data**
  Master data relevant for the compliance functionality of SAP GTS can be managed via the **Master Data** page. This contains two SAP Fiori tiles (i.e., **Manage Business Partner** and **Manage Products**) that allow users to modify business partner master data and product master data, respectively.

These examples are a few of the most relevant compliance functionalities available in the SAP Fiori launchpad.

2.3.3 Customs Management

In the SAP Fiori launchpad, a user can find all SAP Fiori tiles related to the SAP GTS Customs Management functionality, as shown in Figure 2.10.

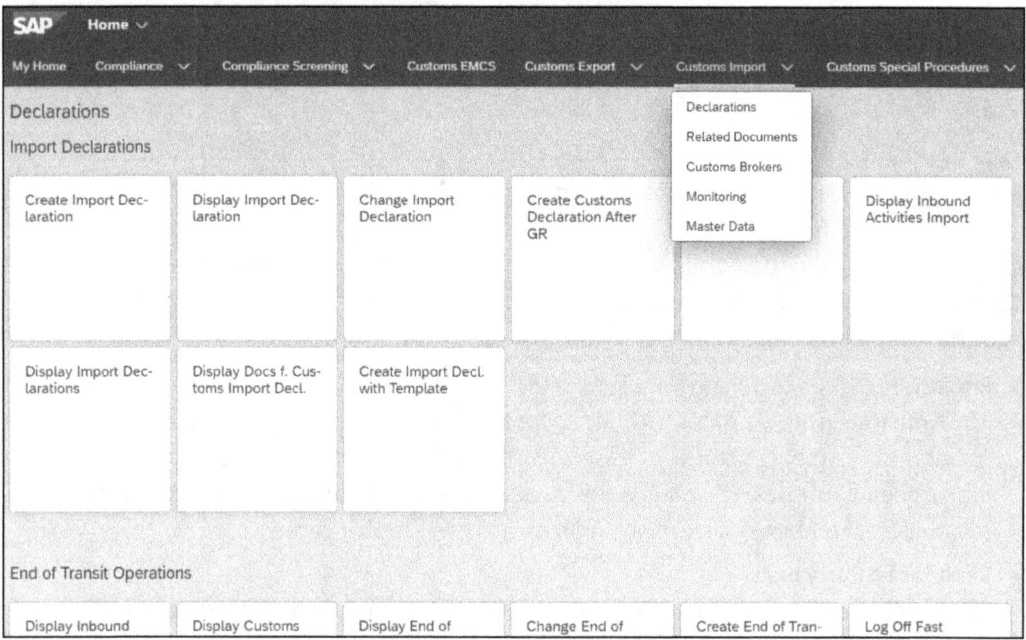

Figure 2.10 Customs Management: Related Space, Pages, and Tiles

The activities related to Customs Management are subdivided in different spaces, each focused on a specific customs area or task (i.e., **Customs Import**, **Customs Export**, **Customs EMCS**, and **Customs Special Procedures**). Within each Customs Management space, several pages bundle the SAP Fiori tiles required to perform Customs Management tasks. Here are a few examples:

- **Declarations** is an example of a page in the **Customs Import** space. This page contains SAP Fiori tiles to create, change, and display import declarations. A user can also navigate to the import declaration worklist (e.g., **Create Customs Declaration After GR**, **Create Customs Declaration Before GR**, etc.).

- Users responsible for filing import, export, transit, or EMCS declarations can use the applications in pages such as **Import Declarations**, **Export Declarations**, **Transit Declarations**, and **EMCS**.
- Special customs procedures can be managed via tiles included in pages such as **Inward Processing**, **Outward Processing**, and **Customs Warehouse**.

These examples are a few of the available customs functionalities in the SAP Fiori launchpad.

2.3.4 Preference Management

In the SAP Fiori launchpad, a user can find all SAP Fiori tiles related to the Preference Management functionality of SAP GTS, as shown in Figure 2.11.

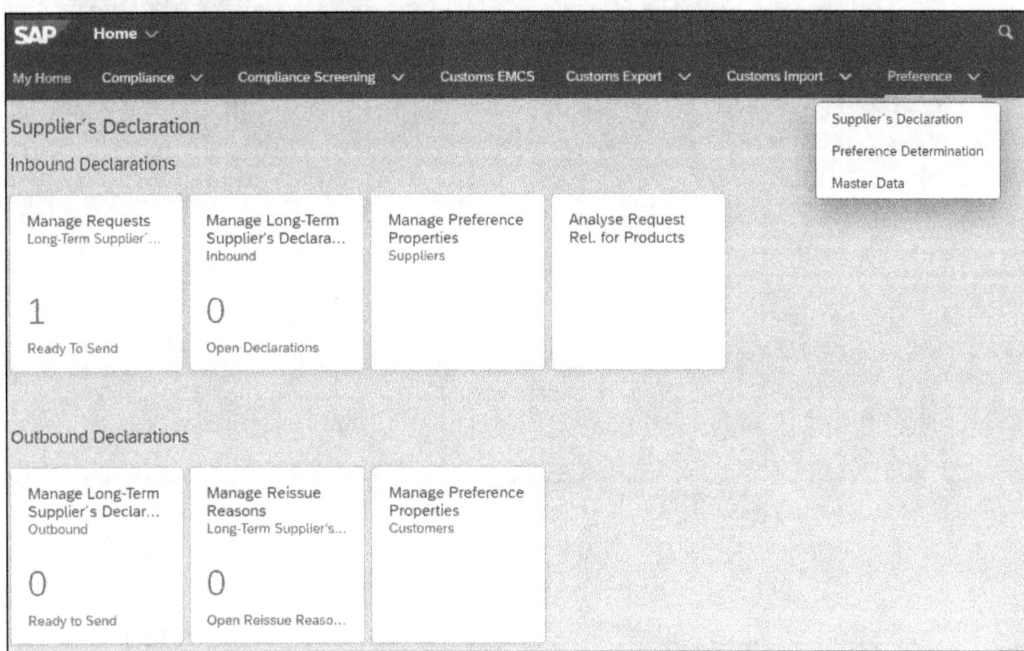

Figure 2.11 Preference Management: Related Spaces, Pages, and Tiles

The activities related to the **Preference** space are subdivided into three different pages, each focused on a specific area or task:

- **Supplier's Declaration**
  The **Supplier's Declaration** page groups the SAP Fiori tiles to manage both inbound and outbound long-term supplier's declarations as well as the tiles to manage supplier and customer preference properties.

- **Preference Determination**
  The **Preference Determination** page allows users to utilize apps such as **Perform Preference Determination** and **Display Preference Logs**.
- **Master Data**
  Under the **Master Data** page, users can display bills of material, preference data, and procurement indicators. The page also contains a section to manage business partners and products.

2.3.5 Products and Classification

In the SAP Fiori launchpad, a user can find SAP Fiori tiles related to product and classification, as shown in Figure 2.12.

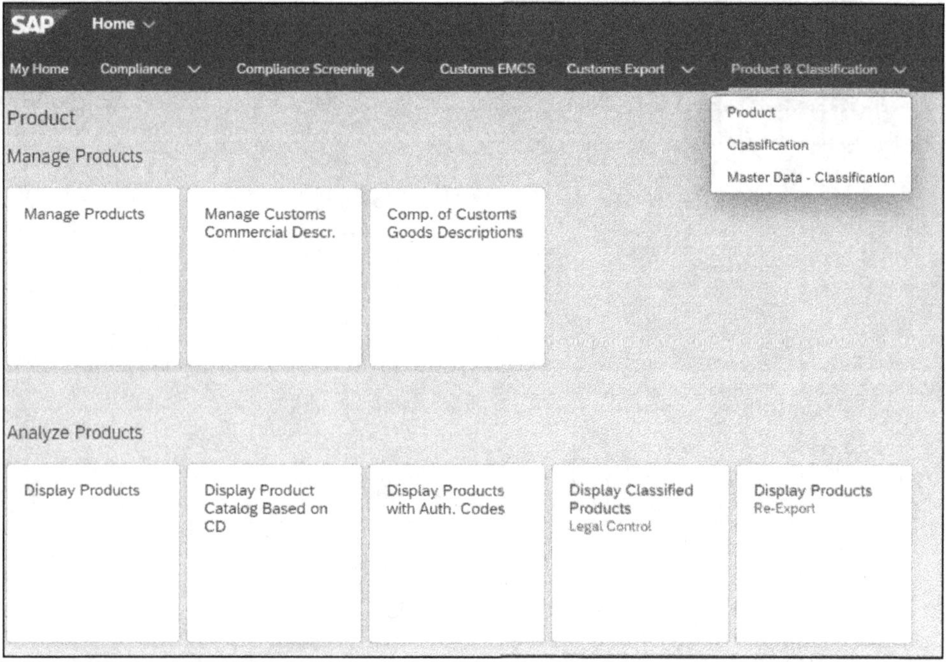

Figure 2.12 Product & Classification: Related Space, Pages, and Tiles

The activities related to the **Product & Classification** space are subdivided into three different pages, each focused on a specific area or task:

- **Product**
  The **Product** page groups the SAP Fiori tiles to allow users to manage and analyze products.
- **Classification**
  The **Classification** page allows users to classify and reclassify products both for Customs Management and Compliance Management.

- **Master Data – Classification**
  The **Master Data – Classification** page allows users to utilize SAP Fiori apps to upload commodity codes and tariff codes as well as to display the upload logs. Users can also navigate to the SAP Fiori tiles to upload duty rates, exchange rates, and more.

These examples are a few of the available product and classification functionalities in the SAP Fiori launchpad used in SAP GTS.

2.3.6 Intrastat

In the SAP Fiori launchpad, a user can find all the SAP Fiori tiles related to **Intrastat**, as shown in Figure 2.13.

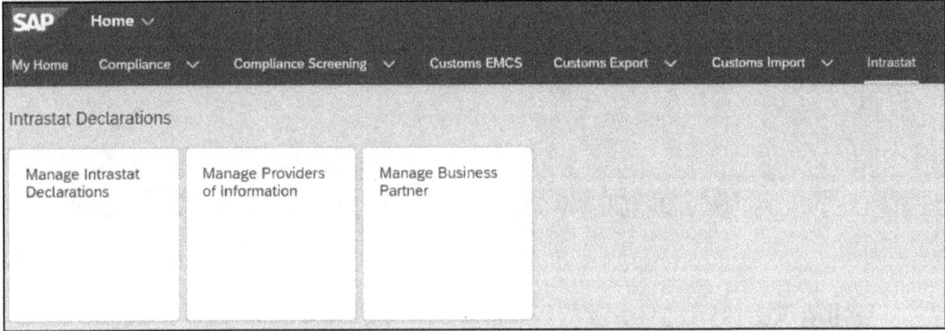

Figure 2.13 Intrastat Declarations: Related Space, Page, and Tiles

The activities related to the **Intrastat Declarations** space are limited to one page. Therefore, the space for **Intrastat** doesn't have a dropdown to display different pages. The **Intrastat Declarations** page contains three different SAP Fiori tiles (i.e., **Manage Intrastat Declarations**, **Manage Providers of Information**, and **Manage Business Partner**). These are designed for users to manage Intrastat within SAP GTS.

2.3.7 Administrative Tasks

Administrative transactions (e.g., transactions background processing, default data, etc.) aren't available in the SAP Fiori launchpad, but are still accessible through the SAP GUI, as shown in Figure 2.14.

The administrative menu allows access to the SAP GUI transactions for administrative and maintenance activities such as **Archiving**, **Data Upload**, **Background Processing**, **Default Data**, and more. In addition, certain core functionalities in SAP GTS such as **System Connection** are still available through this menu in the SAP GUI. The administrative menu becomes available with the authorization `/SAPSLL/GTS_ADMINISTRATOR`, which is explained in Section 2.4.

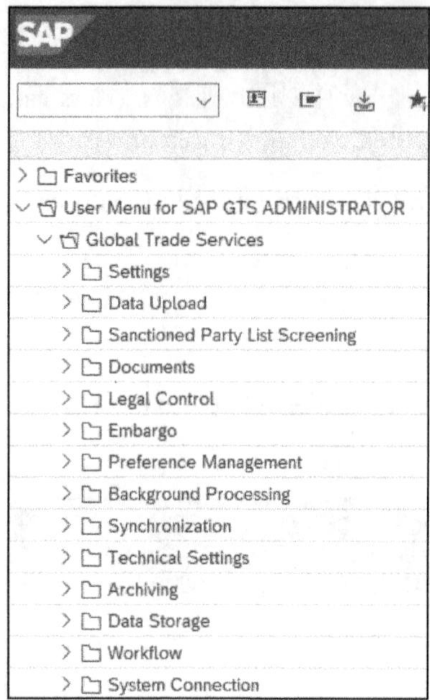

Figure 2.14 Administrator Menu in SAP GUI

2.4 Using Authorizations

The SAP GTS authorization capability allows the segregation of business roles. The SAP GTS authorization capability is the same as the one used in SAP S/4HANA. Therefore, similar techniques can be applied to set up the necessary user authorizations. This section introduces the authorization concept and gives some basic information on how to define user authorizations.

To use SAP GTS, all users are required to log on to the system using a unique user ID. For each user, user master data containing the password and other relevant details need to be created. SAP programs use authorization checks to protect the system from unauthorized access based on roles.

A role contains a set of transactions and specific access to these transactions, such as create, change, and display, which can be configured in the role. Restrictions to these roles can be given based on organizational units such as foreign trade organizations (FTOs), administrative units (AUs), and other functional objects (numbering schemes, etc.). Finally, a role can be assigned to certain users via the user master data.

A business process consists typically of several activities that are performed by users, and these activities are included in a role. Such subsets of activities usually relate to the job function (or position/designation) a user has in the organization.

In the SAP Fiori authorization landscape, these business processes and activities are mapped into catalogs that contain various groups of transactions in the form of SAP

Fiori tiles to perform certain tasks. A role can have multiple catalogs and groups assigned to it. SAP provides a set of standard business roles for different business flows and processes, as shown in Table 2.3.

Business Role Description	Technical Role (PFCG)
Trade Classification Specialist	SAP_BR_CLSFCTN_SPCLST_LLS
Trade Compliance Manager	SAP_BR_CMPLNC_MANAGER_LLS
Trade Compliance Specialist	SAP_BR_CMPLNC_SPCLST_LLS
Trade Compliance Specialist – Screening	SAP_BR_SCRNG_SPCLST_LLS
Preference Specialist	SAP_BR_PREF_SPCLST_LLS
Customs Export Specialist	SAP_BR_EXPORT_SPCLST_LLS
Customs Import Specialist	SAP_BR_IMPORT_SPCLST_LLS
Customs Special Procedure Specialist	SAP_BR_SP_PROCED_SPCLST_LLS
EMCS Specialist	SAP_BR_EMCS_SPCLST_LLS
Intrastat Specialist	SAP_BR_INTRASTAT_SPCLST_LLS

Table 2.3 Business Roles for SAP GTS

For all administrative tasks in SAP GTS, backend role /SAPSLL/GTS_ADMINISTRATOR is provided as a template.

To create roles in SAP GTS, Transaction PFCG (Role Maintenance) is used, as shown in Figure 2.15. This profile generator is organized as follows:

- **Menu**
  Allows the user to select individual transactions or SAP Fiori launchpad groups, spaces, and catalogs, which group a predefine set of transactions.
- **Applications**
  Consists of all available transactions.
- **Authorizations**
  Provides the access rights given to different business functions and data.
- **User**
  Contains a list of all the user IDs that have access to this role.

SAP roles can be single or derived roles. Further, these single or derived roles can be combined to form composite roles, as well. Single roles are the template roles that can be used to derive regional roles. The single role passes on its authorizations to the derived role as default values that can be changed afterward. This inheritance concept reduces the workload for creating new roles as derived roles are a way of maintaining roles that don't differ in their functionality (identical menus and identical transactions) but have different characteristics related to the organizational level. A composite role, on the other hand, is simply a group of single roles.

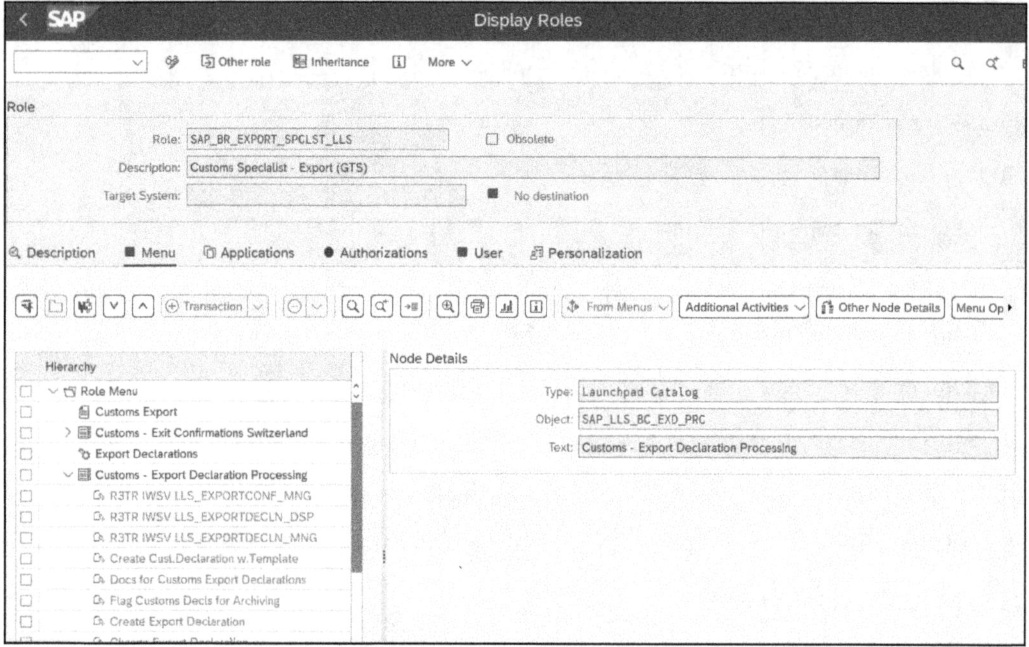

Figure 2.15 Role Maintenance (Transaction PFCG)

When you use the profile generator via the **Authorizations** tab, the authorization objects of all the transactions are automatically determined, and when you generate the role, the authorizations are created. Typically, a role is created using the profile generator. You can use predefined roles, catalogs, and groups provided by SAP as a basis for creating customer-specific roles.

2.5 Activation of SAP Fiori Apps

Since the launch of the Rapid Content Activation task lists, the activation effort has been greatly reduced. Instead of activating application by application, we can activate the delivered SAP business roles (see Table 2.3) as a consolidated unit.

There are two tasks to run through Transaction STC01 (Task Manager):

1. Run the task SAP_FIORI_FOUNDATION_S4,which will perform the basic configuration for the SAP Fiori launchpad, SAP GUI for HTML, and Web Dynpro ABAP common services, as well as provide generic user roles for the administrator and end user.
2. Run the task SAP_FIORI_CONTENT_ACTIVATION, which will activate all SAP Fiori apps and Web Dynpro for ABAP applications based on the business roles selected, as shown in

Figure 2.16. This will generate the business roles with the default authorizations as described previously in Section 2.4.

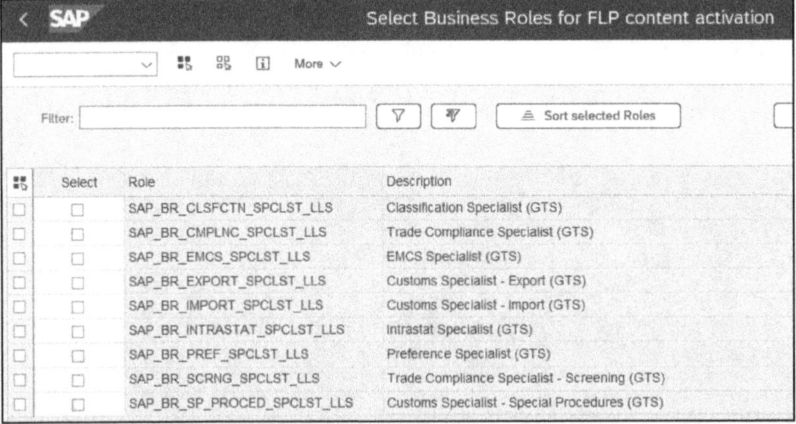

Figure 2.16 Select SAP Business Roles for SAP Fiori Launchpad Content Activation

After assigning the created business roles to a test user, the setup can be tested by accessing the SAP Fiori launchpad via Transaction /UI2/FLP. After a short initialization period, access to the various groups and applications based on the assigned roles is provided.

As a final step, it's possible to activate spaces for the SAP Fiori launchpad instead of using the group-based UI that is activated by default. To enable this feature, execute Transaction /UI2/FLP_CUS_CONF to open the **Launchpad Configuration**. Here, you can enable or disable the parameters **SPACES** and/or **SPACES_ENABLE_USER** per your requirements. With the space-based layout, you'll have a process-oriented approach that is more closely related to the daily tasks connected to each business role.

2.6 Summary

SAP GTS requires a connection with a transactional or feeder system to perform the different SAP GTS functionalities. SAP S/4HANA, including embedded TM and EWM, are the most-used feeder systems for SAP GTS, but you can also use other feeder systems, such as non-SAP systems.

This chapter also described the basic navigation available in SAP GTS and the different modules. We described the user authorizations available in SAP GTS and how to set them up. We also provided an overview of the necessary steps to activate SAP Fiori apps.

The next chapter will cover the baseline settings for SAP GTS. These baseline settings are general, cross-modular settings that need to be configured at the start of any SAP GTS implementation.

Chapter 3
Baseline Settings

This chapter explains the baseline settings that need to be in place across all SAP GTS functionalities.

Baseline settings are activities related to master data and IMG Customizing. The baseline settings in SAP GTS cover areas such as the organizational structure, partner roles, document types, and so on, as well as some specific global trade concepts, such as the legal regulation. The baseline settings require extensive mapping activities that connect feeder systems and SAP GTS.

This chapter will first cover the settings for the SAP GTS organizational structure (Section 3.1), followed by the plug-in settings (Section 3.2) and the general settings in SAP GTS (Section 3.3).

3.1 Organizational Structure in SAP GTS

Just as in the feeder system, a company's organizational structure must be set up in SAP GTS. The correct setup of the organizational structure is crucial because it will drive the tax ID number determination for relevant transactions, applicable export licenses, or customs authorizations, among other functions.

The organizational structure is the backbone of every implementation. It's a necessary step to ensure communication between SAP GTS and feeder systems. This section covers how the organizational elements in SAP GTS will be mapped to the organizational structure of the feeder system, and it walks through the Customizing steps and leading practices to consider.

3.1.1 Foreign Trade Organization

The foreign trade organization (FTO) in SAP GTS represents the legal entity of the company. In SAP GTS, the FTO acts as the authorization owner and applicant for import/export licenses. Logistics processes from the feeder system are integrated with foreign trade processes in SAP GTS.

You define the FTO within business partner maintenance by creating a new business partner and assigning the business partner role SLLFTO. Each FTO has its own address, which represents the official address of the legal entity. Therefore, only one country assignment per FTO is feasible. The country will determine the legal regulations that apply to global trade activities.

The FTO corresponds to the company code in the feeder system. In SAP GTS, each company code can be assigned to only one FTO, but one FTO can be assigned several company codes (including company codes from different systems). Note, however, that it's a leading practice to assign no more than *one* feeder system company code per FTO in SAP GTS.

3.1.2 Legal Unit

Legal units (LUs) correspond and are linked to the plants in the feeder system. As with the company code, each LU can be assigned to only one FTO, but the same FTO can be assigned to multiple LUs. The LUs are set up as business partners with the business partner role SLLSIT.

There are many different options to map LUs with plants in the feeder systems. Several plants can be assigned to the same LU, but each plant can be assigned to only one LU. SAP S/4HANA storage locations can also be assigned to an LU, even if the physical locations of the storage locations are different. This is because the LU determines the legal regulation for compliance.

Figure 3.1 maps the company codes, FTOs, plants, storage locations, and LUs.

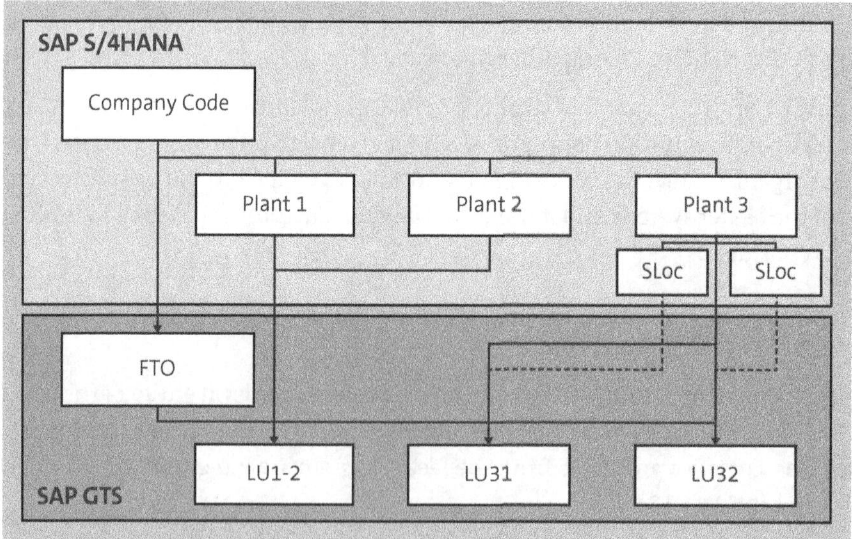

Figure 3.1 Organizational Structure of SAP GTS

Note

The user should assign LUs to FTOs only within the same country. If there is an LU in one specific country, but the actual FTO resides in another country, then additional enhancements may be necessary.

3.1.3 Administrative Unit

Administrative units (AUs) are used in the Preference Management functionality of SAP GTS, which will be covered more thoroughly in Chapter 8. They are used to set up the organizational structure for trade preference determination. AUs are used to group all the enterprise entities (e.g., plants or plant groups) needed to request, issue, and revoke vendor declarations. They are set up as business partners with business partner role SLLMGR.

3.1.4 Customs ID

A *customs ID* is a unique combination of plant and storage location that represents the physical location where special customs procedures can take place. Special customs procedures, such as bonded warehouses (BWHs), inward processing relief (IPR), or outward processing relief (OPR), will be further explained in Chapter 7 on Customs Management. Customs IDs represent an authorized reporting unit for special customs procedures. This means that one administration can be retained per customs ID for reporting purposes. Because a customs ID consists of a combination of plant and storage location, this combination needs to be set up in the feeder system and SAP GTS.

3.1.5 Customizing

You need to set up these organizational structures in SAP GTS using Customizing steps, or via Transaction SPRO by choosing **SAP Global Trade Services, edition for SAP HANA • General Settings**.

First, define the FTO as a business partner in Transaction BP, as shown in Figure 3.2. In our example, we've entered "FTO_G_US" in the **Business Partner** field, and we've assigned the business partner role **SLLFTO (New)** in the **Create in BP role** field.

Next, assign the newly created FTO to a company code in the feeder system, a group of logical systems, or a group of feeder systems, as shown in Figure 3.3. This can be accessed via Transaction SPRO: **SAP Reference IMG • SAP Global Trade Services, edition for SAP HANA • General Settings • Organizational Structures • Assignment of Organizational Units from Feeder System to Foreign Trade Org. • Assign Company Code at Feeder System (Group) Level**. Then, you can repeat the same process for the creation and assignment of LUs, which you can assign to a single feeder system or a group of feeder systems. You can assign a combination of plant and storage location from the feeder system to the LUs.

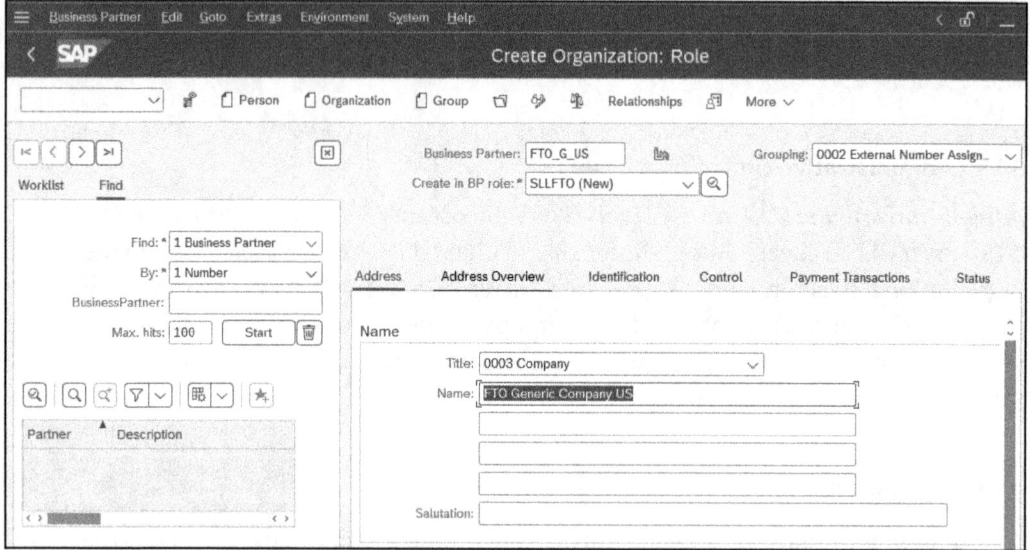

Figure 3.2 Creating an FTO in Transaction BP

Display View "Mapping: Company Code Group VS to Organizat..": Overview

Mapping: Company Code Group VS to Organizational Unit

LS Group	Description	FS Org.	FT Org	Na.
EH4FEEDGRP	S/4 HANA Feeder Logical System Group	1010	FTO_1010	Delo...
EH4FEEDGRP	S/4 HANA Feeder Logical System Group	1710	FTO_1710	Com...
EH4FEEDGRP	S/4 HANA Feeder Logical System Group	2510	FTO_2510	Delo...
EH4FEEDGRP	S/4 HANA Feeder Logical System Group	US00	FTO_US00	

Figure 3.3 Link between FTO and Company Code

After the user creates the FTOs and LUs, the LUs need to be assigned to the relevant FTO. You need to set up these assignments in SAP GTS using Customizing steps or via Transaction SPRO by choosing **SAP Global Trade Services, edition for SAP HANA • General Settings • Assign Legal Unit to the Foreign Trade Organization**. Figure 3.4 shows the assignment of LUs to FTOs.

If the Preference Management functionality is activated, the AUs need to be set up within the SAP GTS organizational structure. See Chapter 8 for more information. When special customs procedures are in scope, customs IDs need to be set up in both the feeder system and SAP GTS.

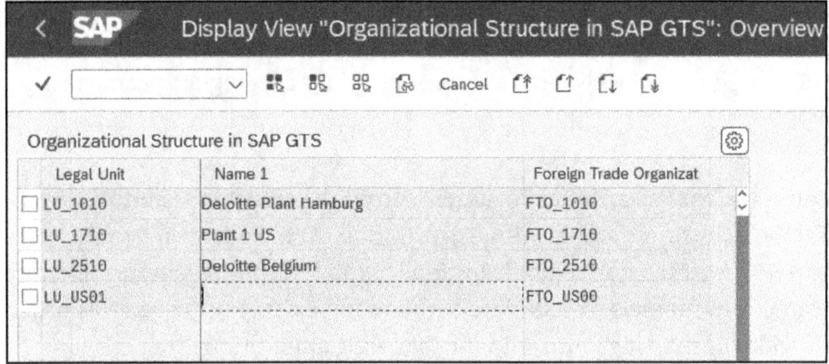

Figure 3.4 Mapping LUs to FTOs

3.1.6 Leading Practices

When you're setting up the organizational structures in SAP GTS, you should take some leading practices and guidelines into consideration:

- Each FTO has a single tax ID number. This means that two company codes with two different tax ID numbers can't belong to a single FTO.
- The FTO is the legal entity in SAP GTS. It's a leading practice to assign one company code to one FTO.
- SAP GTS will look at the country assigned to the FTO to trigger the customs legal regulations. Therefore, two company codes, located in different countries, can't be mapped to one FTO because two separate set of customs legal regulations would need to be triggered.
- SAP GTS will look at the country assigned to the LU to trigger the compliance/trade preference legal regulations. Therefore, two plants located in different countries can't be mapped to one LU because two separate compliance/trade preference legal regulations would need to be triggered.

3.2 Plug-In Settings in SAP S/4HANA

The feeder system and SAP GTS are linked via a plug-in to enable communication between both systems. The plug-in allows the transfer of master data and transactional documents between the feeder system and SAP GTS. You can transfer a transactional document from the feeder system only if the user has created all required master data—that is, the materials and business partners (customers, vendors, etc.) used in the transactional document—in SAP GTS. If the master data isn't available in SAP GTS, then the document gets a technical incomplete status in the Compliance Management functionality or doesn't generate a complete customs declaration in the Customs Management functionality.

Note

The SAP GTS, edition for SAP HANA (release 2023) plug-in is an integrated component of SAP S/4HANA.

Before you can transfer transactional documents from the feeder system to SAP GTS, you have to activate business add-ins (BAdIs). A BAdI is an enhancement to the standard version of SAP. You can define BAdIs by application levels, which in turn identify transaction categories (namely inbound or outbound, purchase order or sales order, etc.). We can divide the existing BAdIs into the following application levels:

- MMOA: Receipt/Import: Purchasing Document (e.g., purchase orders)
- MMOB: Receipt/Import: Inbound Delivery Document (e.g., inbound deliveries)
- MMOC: Receipt/Import: Material Document (e.g., goods receipt material documents)
- SDOA: Dispatch/Export: Sales Document (e.g., sales orders)
- SDOB: Dispatch/Export: Outbound Delivery Document (e.g., outbound deliveries)
- SDOC: Dispatch/Export: Billing Document (e.g., pro forma invoices)

Figure 3.5 shows the application levels, which are also used during the configuration of the document structure in SAP GTS. These settings can be retrieved by navigating the following path in the Customizing settings of SAP S/4HANA accessed via Transaction SPRO: **SAP Reference IMG • Integration with Other SAP Components • Integration with Governance, Risk and Compliance • SAP Global Trade Services • Control Data for Transfer to SAP Global Trade Services • Configure Control Settings for Document Transfer.**

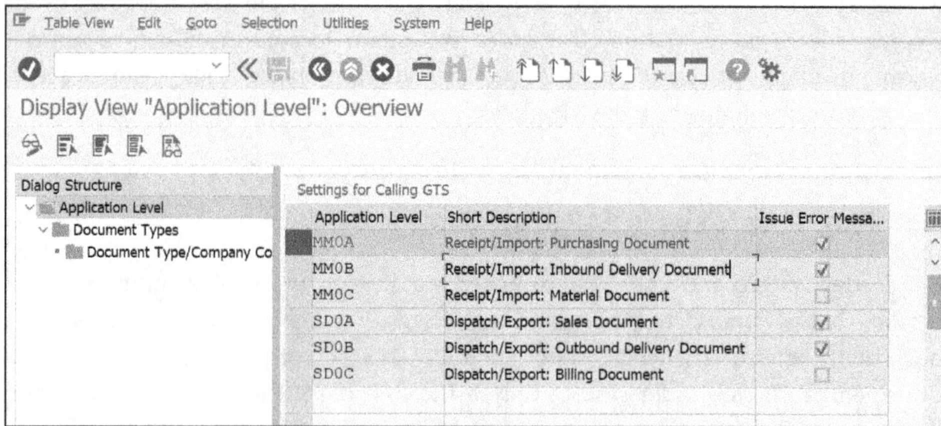

Figure 3.5 Application Levels

After you've activated the relevant BAdIs, you can specify transactional documents for each application level via Transaction SPRO: **Integration with Other SAP Components Integration with Governance, Risk and Compliance • SAP Global Trade Services • Control Data for Transfer to SAP Global Trade Services • Configure Control Settings for Document Transfer.** In this transaction, you choose which document types need to be transferred for every application level; for example, you might transfer the purchase orders with document type NB. In this scenario, an entry is added for application level MMOA with document type FNB. For document type FNB, F represents the document category (purchasing), and NB represents the purchase order document type.

As shown in Figure 3.6, you can specify the services (e.g., **Compliance**, **Customs**, **Risk**, etc.) for which the document must be transferred and whether an output dialog should be shown in the feeder system (e.g., stating that the document is blocked for compliance).

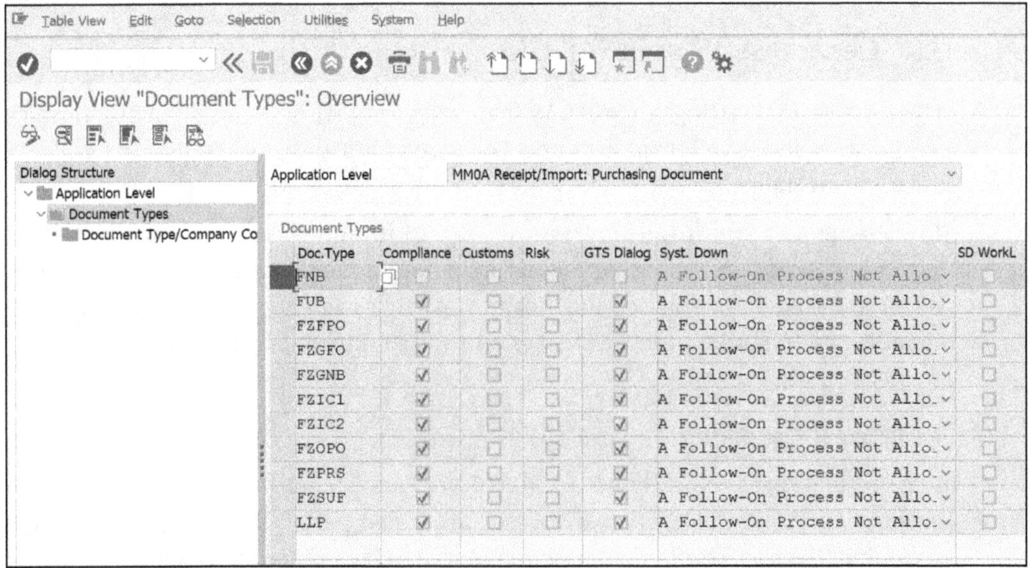

Figure 3.6 Document Application MMOA

In the SAP S/4HANA feeder system, as of version 2021, you can restrict document transfer using the company code in addition to the document type. You specify the relevant company codes in the **Document Type/Company Codes** subfolder, as shown in Figure 3.7.

In some cases, you may need to execute additional configuration. The available configuration settings depend on the relevant application level.

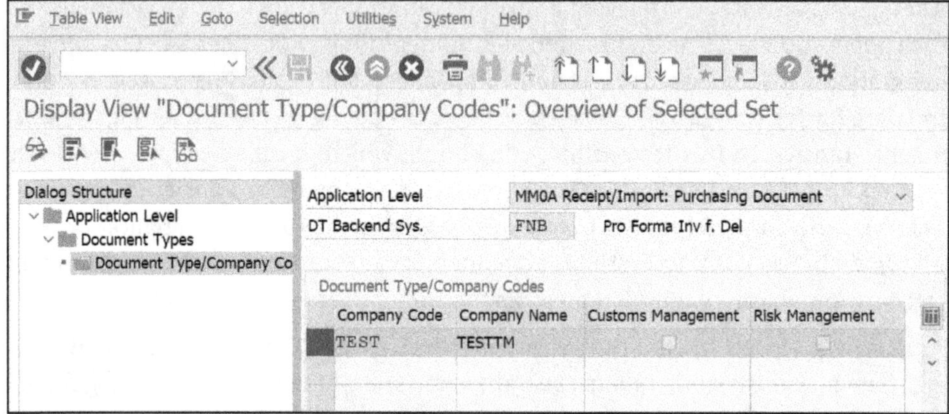

Figure 3.7 Document Type FNB and Company Codes

3.3 General Settings in SAP GTS

There are several settings in SAP GTS that apply to most of its functionalities, which we'll discuss in the following sections. They are grouped in a section of the SAP GTS Customizing menu.

The settings also apply to all processes in SAP GTS, including defining and maintaining legal regulations, document types, and mapping to corresponding document types within the feeder system; defining business partner roles and mapping to the corresponding partner roles of the feeder system; and setting up number ranges.

3.3.1 Legal Regulation

Legal regulations are largely used in the Compliance Management functionality for import/export control checks, including sanctioned party lists (SPLs) and embargo screening. The legal regulation is also referenced in Preference Management flows as a representative of the free trade agreement and for customs processes.

The bottom of Figure 3.8 gives an overview of the **General Settings** menu for legal regulations in Customizing.

Legal regulations are assigned to a country and/or country group. As a result, it's necessary to first define the appropriate values for the countries and country groups and then define and assign the legal regulation to a specific process such as customs within SAP GTS. The legal regulation could apply to only the import, only the export, or both. You can set this within the activation of the legal regulation.

Next, assign the legal regulation a determination procedure that includes a determination strategy. This setting regulates at which country level the legal regulation determination is executed or whether the determination is completed at the country or country group level for both import and export functions. It's possible to define different determination strategies by partner group (a partner group is a grouping of business partners).

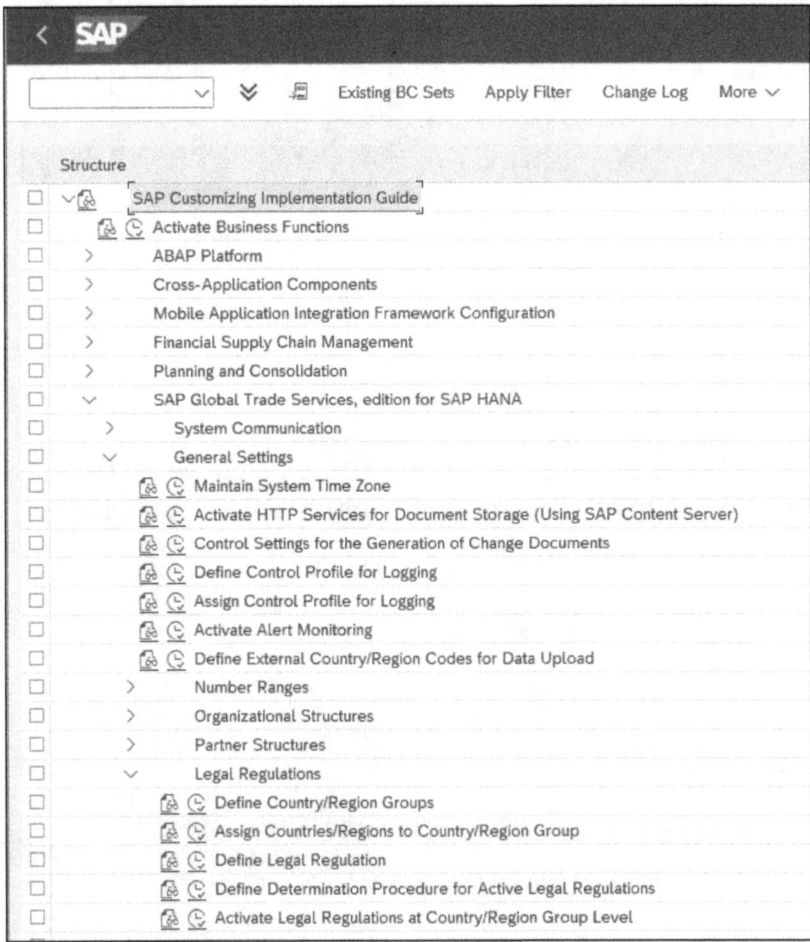

Figure 3.8 General Settings Menu for Legal Regulations

You need to set up these determination procedures for legal regulations in SAP GTS using Customizing steps or via Transaction SPRO, as shown in Figure 3.9, by choosing **SAP Global Trade Services, edition for SAP HANA • General Settings• Legal Regulations • Define Determination Procedure for Legal Regulation.**

Two approaches are possible in setting a determination strategy:

- The determination executes all strategies to determine all the relevant legal regulations.
- The determination stops after the first successful strategy applies a legal regulation as shown in Figure 3.10.

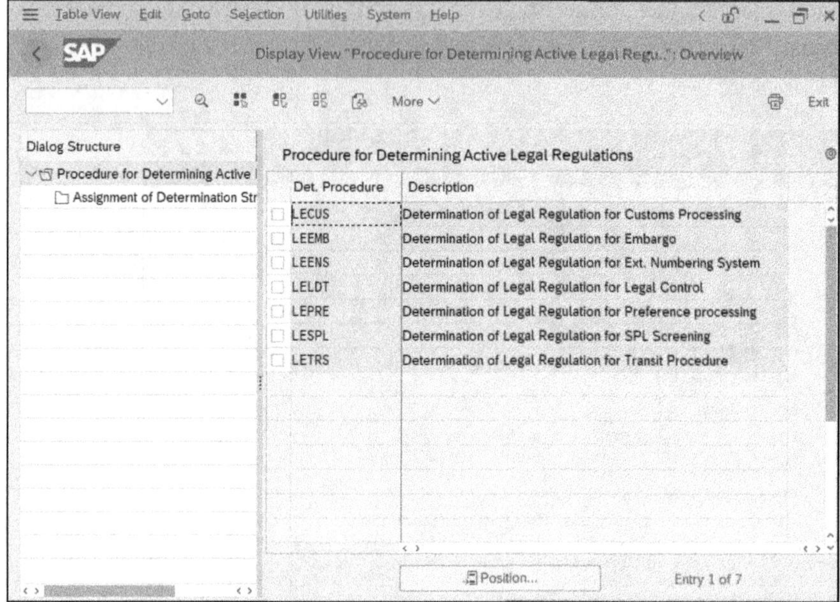

Figure 3.9 Define Determination Procedure to Automatically Determine Legal Regulations

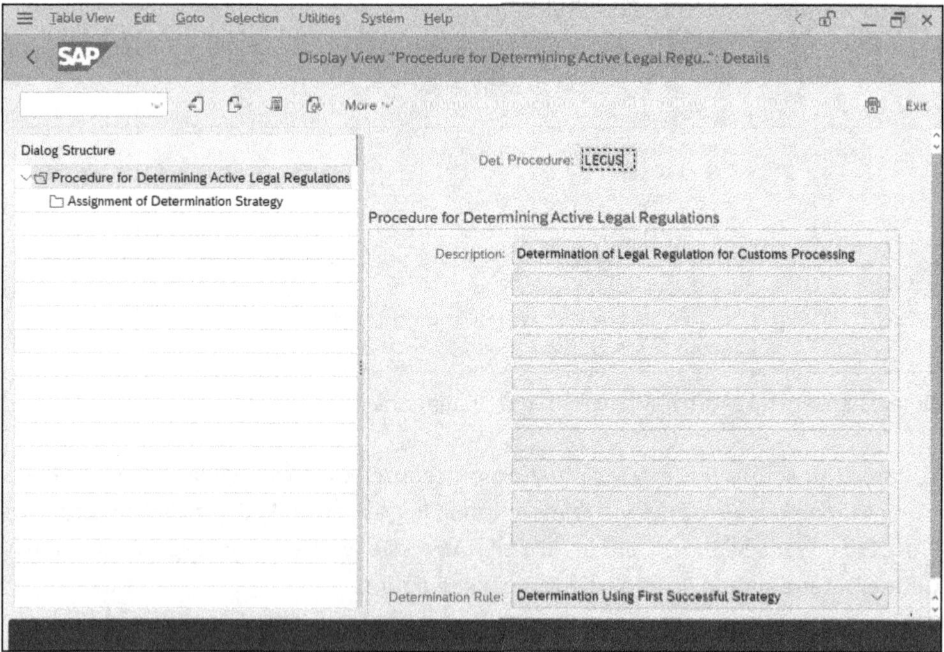

Figure 3.10 Assignment of Determination Strategy to Determination Procedure

Subsequent settings for legal regulations are done at a functionality-specific level and will be covered in more detail in the following chapters.

3.3.2 Document Types and Mapping

You can transfer transactional documents from SAP S/4HANA or non-SAP feeder systems to SAP GTS. Through this transfer, the document types and item categories used in the feeder system are also transferred. You need to map these feeder system categories with those in SAP GTS; otherwise, the document has a technical incomplete status in the Compliance Management functionality, or the customs document isn't created in the Customs Management functionality.

To set up the document structure in SAP GTS, follow these steps, shown at the bottom of Figure 3.11:

1. Define the SAP GTS-specific document type/item category.
2. Activate the SAP GTS-specific document type/item category for the required functionalities.
3. Map the SAP GTS-specific document type/item category to the corresponding feeder system-specific document type/item category.

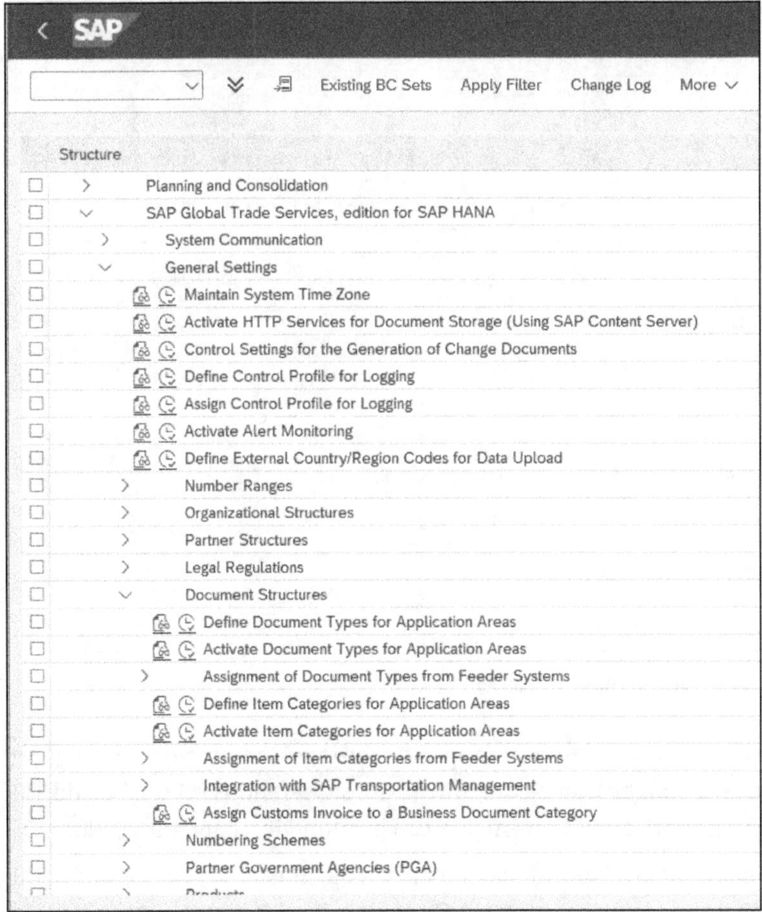

Figure 3.11 General Settings Menu for Document Structures

In SAP GTS, you can define specific document types via Transaction SPRO by choosing **SAP Global Trade Services, edition for SAP HANA • General Settings • Document Structure • Define Document Types for Application Areas.** Each document type belongs to a specific document category. The document category defines the functionalities for which the document can be used.

SAP provides a number of standard document types that can be used for standard SAP GTS processes, as shown in Figure 3.12. However, sometimes you have to set up additional specific document types, such as if your company creates a sales order for the sales of physical goods. For these physical goods movements, legal control checks are required. If the sales orders are also used for nonphysical shipments, then they might not require legal control checks. To make this scenario feasible, you'd need to customize an additional document type for the sales flow that doesn't require legal control checks.

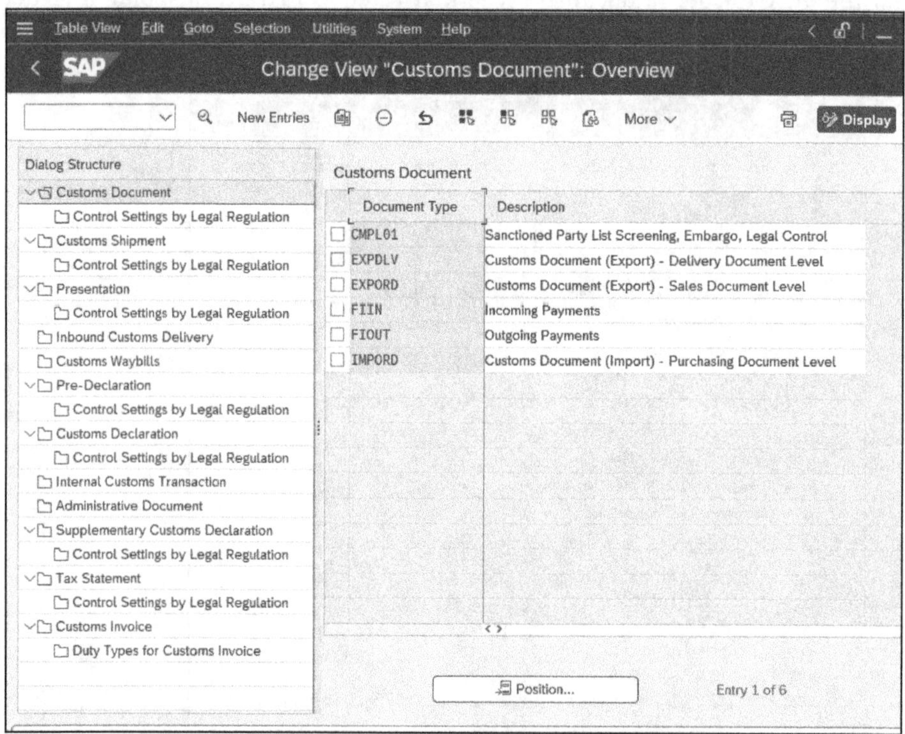

Figure 3.12 Document Types for Application Areas

You can control each defined document via specific settings. Among these settings are the number range used and the transfer for SAP Business Warehouse (SAP BW) and SAP BW/4HANA reporting. More details on SAP BW can be found in Chapter 9, Section 9.3.

After you've defined SAP GTS-specific documents, you must activate them for the applicable functionalities via Transaction SPRO by choosing **SAP Global Trade Services,** edition for SAP HANA • General Settings • Document Structure • Activate Document Types for Application Areas**. All functionalities are mentioned on the left-side menu, as shown in Figure 3.13. An entry needs to be added for a specific document type to activate the specific document type for that functionality.

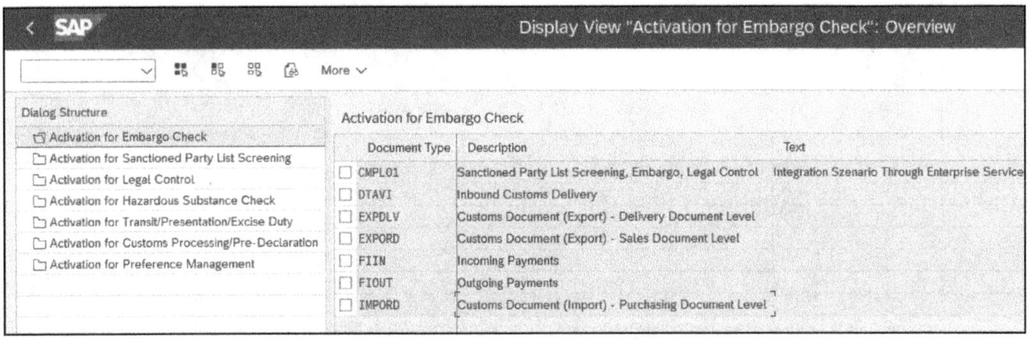

Display View "Activation for Embargo Check": Overview

More

Dialog Structure
- Activation for Embargo Check
- Activation for Sanctioned Party List Screening
- Activation for Legal Control
- Activation for Hazardous Substance Check
- Activation for Transit/Presentation/Excise Duty
- Activation for Customs Processing/Pre-Declaration
- Activation for Preference Management

Activation for Embargo Check

Document Type	Description	Text
CMPL01	Sanctioned Party List Screening, Embargo, Legal Control	Integration Szenario Through Enterprise Service
DTAVI	Inbound Customs Delivery	
EXPDLV	Customs Document (Export) - Delivery Document Level	
EXPORD	Customs Document (Export) - Sales Document Level	
FIIN	Incoming Payments	
FIOUT	Outgoing Payments	
IMPORD	Customs Document (Import) - Purchasing Document Level	

Figure 3.13 Activating Document Types for Application Areas: Functionalities

Finally, you need to map feeder system document types to their corresponding SAP GTS document types in Transaction SPRO by choosing **SAP Global Trade Services, edition for SAP HANA • General Settings • Document Structure • Assign Document Type at Feeder System Group Level**. If the mapping is done at the logical system level (i.e., assigned to a specific system), then it must be redone in case the configuration is transported across systems (e.g., from a development to a quality system). If the mapping is performed at the logical system group level (i.e., logical system group feeder system), no additional Customizing is needed. The assignment of a logical system to a logical system group ensures that the existing Customizing entries are still correct (e.g., by changing the assigned logical system to the logical system group from the development system to the quality system).

Figure 3.14 shows document types mapped at the logical system group level. You have to perform this mapping for each relevant application area. For example, the first line shows the potential mapping for the previously mentioned purchase order NB:

- **Application Level**: MMOA
- **LS Group** (logical system group): Feeder system
- **Document Type**: IMPORD

After you've completed the mapping process, SAP GTS determines the appropriate functionality on the transferred purchase order NB.

After you've set up the document types, you can configure item categories. The setup here is similar. First, set up SAP GTS-specific item categories, and activate them for specific functionalities. Then, map the corresponding item categories between the feeder system and SAP GTS, as shown in Figure 3.15.

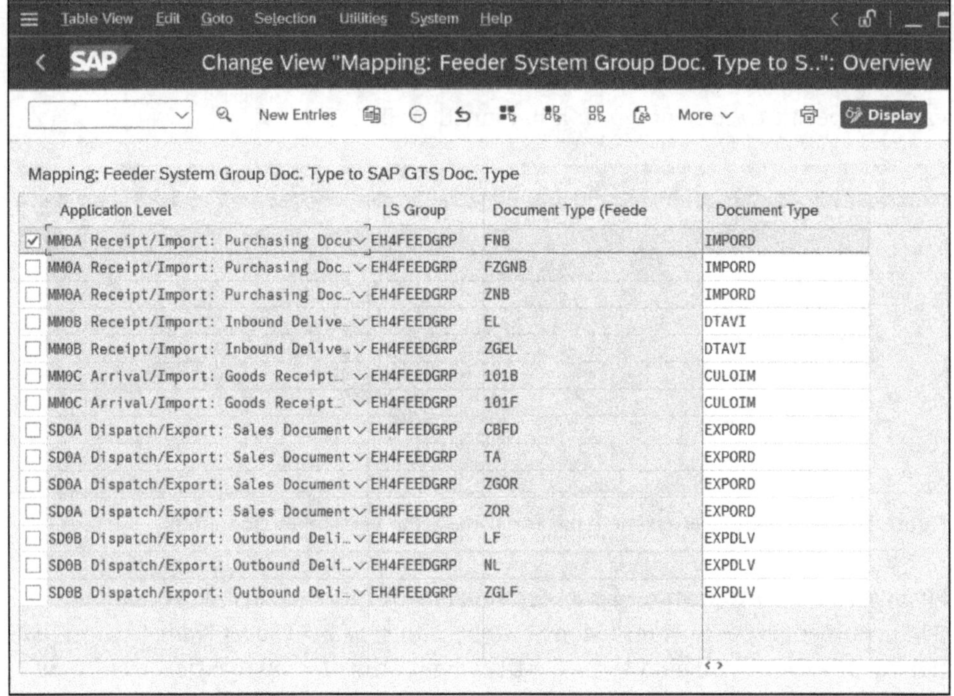

Change View "Mapping: Feeder System Group Doc. Type to S..": Overview

Mapping: Feeder System Group Doc. Type to SAP GTS Doc. Type

Application Level	LS Group	Document Type (Feede	Document Type
MM0A Receipt/Import: Purchasing Docu	EH4FEEDGRP	FNB	IMPORD
MM0A Receipt/Import: Purchasing Doc...	EH4FEEDGRP	FZGNB	IMPORD
MM0A Receipt/Import: Purchasing Doc...	EH4FEEDGRP	ZNB	IMPORD
MM0B Receipt/Import: Inbound Delive...	EH4FEEDGRP	EL	DTAVI
MM0B Receipt/Import: Inbound Delive...	EH4FEEDGRP	ZGEL	DTAVI
MM0C Arrival/Import: Goods Receipt...	EH4FEEDGRP	101B	CULOIM
MM0C Arrival/Import: Goods Receipt...	EH4FEEDGRP	101F	CULOIM
SD0A Dispatch/Export: Sales Document	EH4FEEDGRP	CBFD	EXPORD
SD0A Dispatch/Export: Sales Document	EH4FEEDGRP	TA	EXPORD
SD0A Dispatch/Export: Sales Document	EH4FEEDGRP	ZGOR	EXPORD
SD0A Dispatch/Export: Sales Document	EH4FEEDGRP	ZOR	EXPORD
SD0B Dispatch/Export: Outbound Deli...	EH4FEEDGRP	LF	EXPDLV
SD0B Dispatch/Export: Outbound Deli...	EH4FEEDGRP	NL	EXPDLV
SD0B Dispatch/Export: Outbound Deli...	EH4FEEDGRP	ZGLF	EXPDLV

Figure 3.14 Document Types Mapped at the Logical System Group Level

Display View "Mapping: FS Item Cat. Group to SAP GTS Ite..": Overview

Mapping: FS Item Cat. Group to SAP GTS Item Cat. Group

Application Level	LS Group	Doc. Item Category (	Item Category
MM0A Receipt/Import: Purchasing Document	EH4FEEDGRP	0	IMORD1
MM0A Receipt/Import: Purchasing Document	EH4FEEDGRP	1	IMORD1
MM0A Receipt/Import: Purchasing Document	EH4FEEDGRP	2	IMORD1
MM0A Receipt/Import: Purchasing Document	EH4FEEDGRP	3	IMORD1
MM0A Receipt/Import: Purchasing Document	EH4FEEDGRP	4	IMORD1
MM0A Receipt/Import: Purchasing Document	EH4FEEDGRP	5	IMORD1
MM0A Receipt/Import: Purchasing Document	EH4FEEDGRP	6	IMORD1
MM0A Receipt/Import: Purchasing Document	EH4FEEDGRP	7	IMORD1
MM0A Receipt/Import: Purchasing Document	EH4FEEDGRP	8	IMORD1
MM0A Receipt/Import: Purchasing Document	EH4FEEDGRP	9	IMORD1
MM0A Receipt/Import: Purchasing Document	EH4FEEDGRP	A	IMORD1
MM0A Receipt/Import: Purchasing Document	EH4FEEDGRP	C	IMORD1
MM0A Receipt/Import: Purchasing Document	EH4FEEDGRP	P	IMORD1
SD0A Dispatch/Export: Sales Document	EH4FEEDGRP	TAN	EXORD1
SD0B Dispatch/Export: Outbound Delivery	EH4FEEDGRP	TAN	EXDLV1

Figure 3.15 Mapping Item Categories

3.3.3 Mapping Partner Roles

Business partners are persons or organizations that perform a role within a company. These roles can include customers, vendors, FTOs, LUs, customs offices, data providers, and so on. It's necessary to link business partner roles of the feeder system to the partner roles in SAP GTS. Business partners are grouped in partner groups, which are relevant to several SAP GTS processes. For example, business partner groups assigned to a combination of legal regulation and document types determine which partner functions appear in a customs document. You need to define and assign a partner structure in Transaction SPRO by choosing **SAP Global Trade Services, edition for SAP HANA • General Settings • Partner Structure**.

You must create, group, and assign business partner functions according to the following steps:

1. Create the business partner functions (e.g., CUSEXP as exporter).
2. Define groups of partner functions. Partner groups provided by SAP group partner functions for specific functionalities (e.g., customs), specific transaction flows (e.g., import), and specific countries (e.g., Belgium).
3. The partner function groups allow you to define specific system behaviors for a specific group of partners (i.e., certain determination procedures by legal regulation).

Assign the partner function to a partner function in the feeder system. Then, set up a one-on-one link between the SAP GTS partner function and the feeder system partner function, as shown in Figure 3.16. The **Logical System Group** field refers to the feeder system in which the partner role will be mapped. If different feeder systems are connected to SAP GTS, multiple logical system groups need to be specified. The **BS Partner Role** field contains the partner roles as maintained in the feeder system. The **Partner-Function** field is respective partner mapping in SAP GTS.

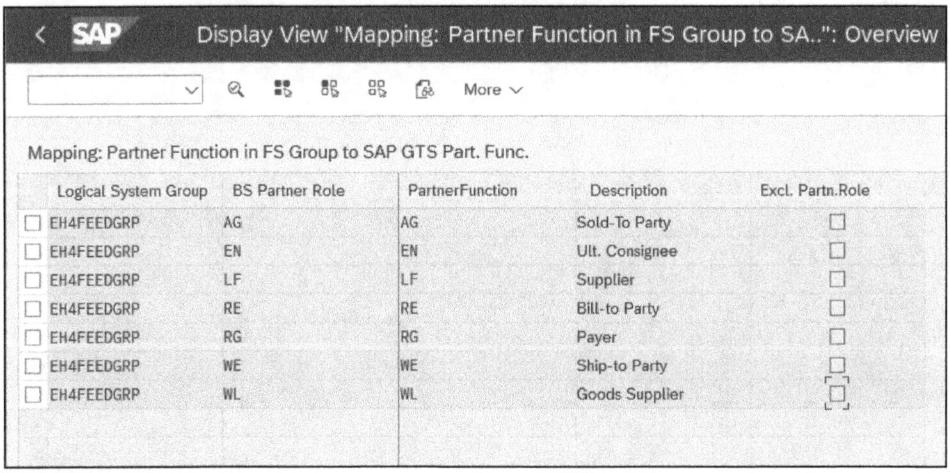

< SAP Display View "Mapping: Partner Function in FS Group to SA..": Overview

More

Mapping: Partner Function in FS Group to SAP GTS Part. Func.

Logical System Group	BS Partner Role	PartnerFunction	Description	Excl. Partn.Role
EH4FEEDGRP	AG	AG	Sold-To Party	☐
EH4FEEDGRP	EN	EN	Ult. Consignee	☐
EH4FEEDGRP	LF	LF	Supplier	☐
EH4FEEDGRP	RE	RE	Bill-to Party	☐
EH4FEEDGRP	RG	RG	Payer	☐
EH4FEEDGRP	WE	WE	Ship-to Party	☐
EH4FEEDGRP	WL	WL	Goods Supplier	☐

Figure 3.16 Mapping of Partner Functions

3.3.4 Number Ranges

Before you can execute any transactions in SAP GTS, you need to define number ranges that determine how SAP GTS internal numbers are being generated for transactions and master data. Number ranges can contain letters, numbers, and the validity year for which they have been set up. For example, consider the number range 0200000000–0299999999.

Some other examples are number ranges for business partners, customs products, licenses, vendor declarations, and so on. These number ranges have to be set up per type of document, meaning that different types of documents trigger different number ranges. You can perform the setup of these number ranges in Customizing by following this transaction path: **SAP Global Trade Services, edition for SAP HANA • General Settings • Number Ranges.**

3.4 Summary

The baseline settings are the foundation of the SAP GTS implementation that allows the subsequent configuration of the SAP GTS functionalities. You need to have a detailed understanding of the global trade requirements and business flows to set up a stable and reliable baseline settings structure that can support all the processes of a company.

Now that these main settings have been set up, it's time to set up the relevant master data in SAP GTS, including master data transferred from the feeder system and master data specific to SAP GTS. These topics are discussed in the following chapter.

Chapter 4
Master Data

Master data is critical for all SAP GTS functionalities. Some master data must be transferred from the feeder system to SAP GTS, while other master data can be maintained directly in SAP GTS. This chapter provides an overview of the main types of master data, the source systems, and how the data is transferred.

Master data is the single source of basic business data that is used across multiple systems, applications, and/or processes. Creating master data once and sharing it across systems, rather than creating the same data in multiple different systems, increases efficiency. SAP GTS operates on this principle. Some master data is maintained in the feeder system and replicated to SAP GTS, while other master data isn't available in the feeder system and is directly maintained in SAP GTS.

This chapter describes the business processes involving master data (Section 4.1) and discusses the different types of master data: business partner master data (Section 4.2), product master data (Section 4.3), bill of materials (BOM) master data (Section 4.4), licenses and authorizations (Section 4.5), and other types of master data (Section 4.6). The chapter also outlines the Customizing considerations (Section 4.7) to set up master data maintenance in SAP GTS.

4.1 Business Processes

Master data in SAP GTS represents all nontransactional data. Master data is the foundation for all processes supported by SAP GTS. Master data is created once in a system and can be reused in each transaction, though sometimes it must be updated where necessary. The data is either transferred from the feeder system (e.g., business partners) or wholly maintained in SAP GTS (e.g., customs offices). The master data that is transferred from the feeder system can be enriched in SAP GTS with trade-specific data elements such as tax ID numbers for business partners or commodity codes for products.

This section will discuss master data source systems and the initial transfer of the master data from the feeder system to SAP GTS.

4.1.1 Source System

The source system for the master data is where the data is set up and maintained. As shown in Table 4.1, either the feeder system or SAP GTS can act as the source system. Commonly, the business partners and materials are maintained in the feeder system and transferred to SAP GTS, while business partner exceptions and material attribute exceptions are maintained in SAP GTS.

Source System of Master Data	Feeder System	SAP GTS
Business partners	X	
Business partner exceptions (e.g., customs offices)		X
Materials	X	
Material attribute exceptions (e.g., classification codes)		X

Table 4.1 Source System of Master Data

When the source system is a feeder system, an initial transfer of master data (e.g., business partners and materials) is performed to SAP GTS, which avoids the need to transfer master data each time via transaction and thus makes the system operate more efficiently.

For example, information that is updated infrequently can be directly copied into customs documents in SAP GTS. Transferring master data changes enables compliance checks such as sanctioned party list (SPL) screening or influence trade preference calculations for the Preference Management functionality.

Note

Multiple feeder systems can be connected to one SAP GTS system. When there are multiple feeder systems, it's still important to maintain business partners and materials in the feeder system as the source system.

4.1.2 Initial Transfer from the Feeder System to SAP GTS

Maintaining the feeder system's master data in SAP GTS requires an initial transfer of data. Master data from the feeder system is transferred asynchronously (i.e., on a periodical basis) to SAP GTS via batch jobs. Business partners (e.g., customers and vendors) and material master records are usually transferred this way because these data elements aren't frequently updated. However, change pointers can be set up in the feeder system to ensure that critical data elements (e.g., customer addresses) are automatically retransferred to SAP GTS when changed in the feeder system.

Before transferring master data from the feeder system to SAP GTS, the communication between the two systems should be defined. The technical setup for communication between the feeder and SAP GTS systems is covered in Chapter 2.

Figure 4.1 shows how the feeder system and SAP GTS communicate by transferring data through the plug-in (and through advanced shipping and receiving [ASR] and web services in the case of embedded extended warehouse management [EWM] and transportation management [TM]). Note that Figure 4.1 distinguishes between master data and transactional data. Master data (double arrow) is transferred once to SAP GTS, with updates via change pointers, and only relevant data is transferred back to the feeder system. Transactional data is transferred to SAP GTS in a one-way transfer (single arrow), which means that documents (e.g., billing documents) are transferred to SAP GTS on a transactional basis and not transferred back to the feeder system.

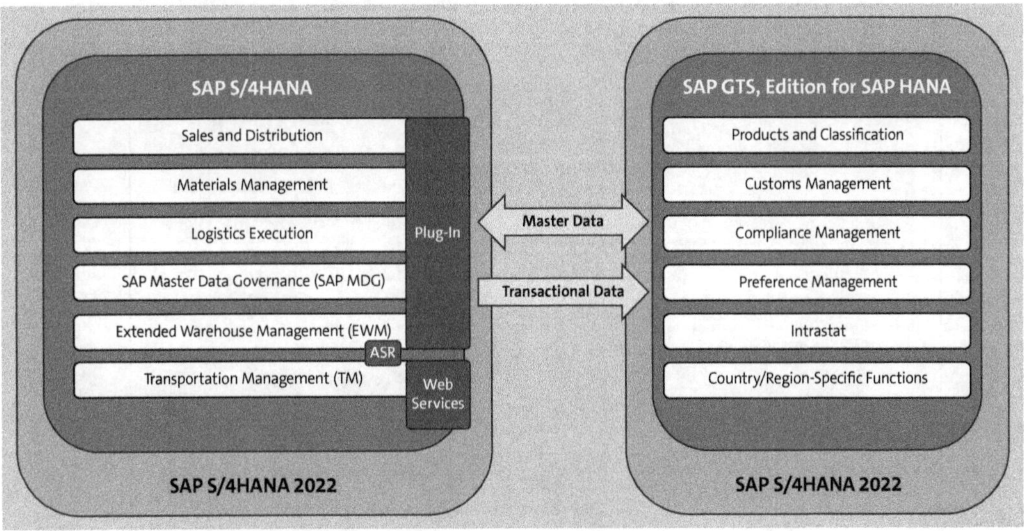

Figure 4.1 Communication between Feeder System and SAP GTS

The initial master data transfer from the feeder system to SAP GTS must happen before transactions can occur in SAP GTS. Standard SAP functionality can enable data transfer from the feeder system to SAP GTS; therefore, no additional Customizing settings or interfaces are required. Figure 4.2 shows the screen in SAP S/4HANA where the initial transfer of master data can be executed via Transaction /SAPSLL/MENU_LEGALR3.

The following initial transfers of master data can be launched via the plug-in:

- **Transfer Vendors**
- **Transfer Customers**
- **Transfer Contact Persons**
- **Transfer Material Masters**
- **Transfer Bills of Material**
- **Transfer Banks**

Most master data is cross-area, which means it can be used for the Compliance Management, Customs Management, and Preference Management functionalities.

Note

The Restitution Management functional area and the Letter of Credit Processing functional area, which previously existed in SAP GTS 11.0, aren't available in SAP GTS, edition for SAP HANA. The Risk Management functionality is now replaced with Preference Management as only this functionality exists in SAP GTS. However, the feeder system plug-in menu within SAP S/4HANA still reflects the old module name **Risk Management**, as illustrated by Figure 4.2.

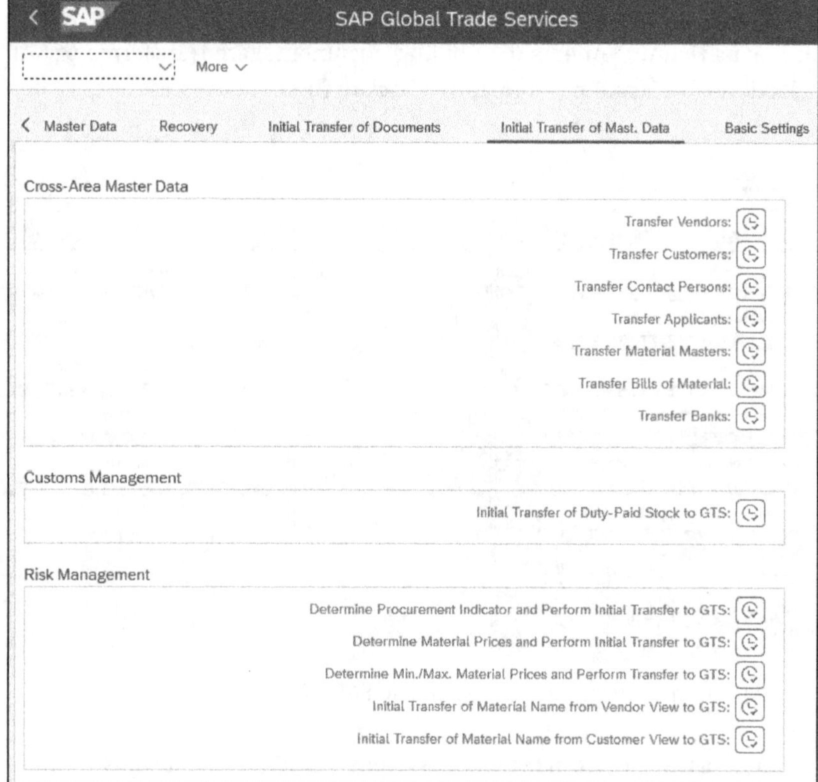

Figure 4.2 Initial Transfer of Master Data from SAP S/4HANA to SAP GTS

There are two area-specific transfers:

- Transfer of specific data for Preference Management, such as material prices
- Transfer of specific data for Customs Management, such as duty-paid stock when implementing a special customs procedure

After the initial transfer is executed, users can activate change pointers to ensure the feeder system syncs with SAP GTS. Change pointers track the changes made to specific master data objects (e.g., materials) and flag the new/changed entries. A batch job fetches all the changed entries and transfers all the information from the feeder system to SAP GTS. The change pointers are set up in the plug-in menu within SAP S/4HANA, accessible via Transaction /SAPSLL/MENU_LEGALR3, on the **Basic Settings** tab, as shown in Figure 4.3.

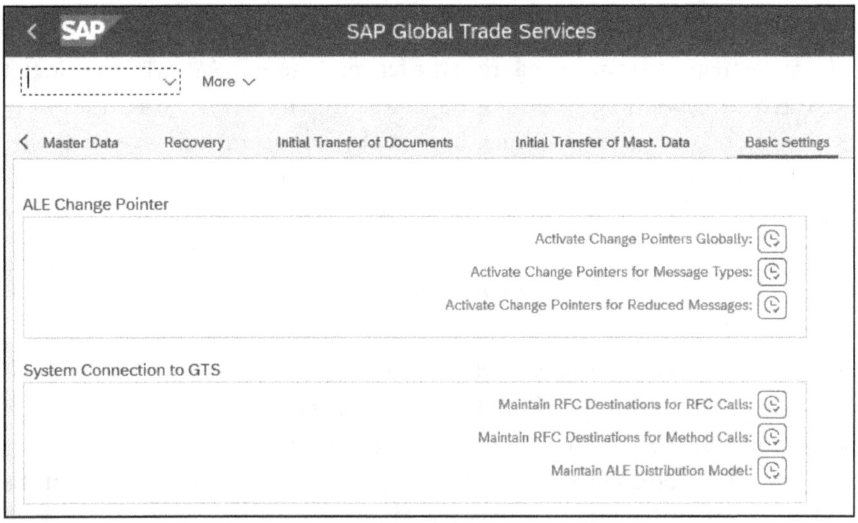

Figure 4.3 Basic Settings for Change Pointers Setup

Several master data elements are maintained in SAP GTS. As shown in Figure 4.4, a distinction is made between business partners, products, BOMs, and licenses/authorizations. Recall that some of this data uses the feeder system as the source system, and other data uses SAP GTS as the source system.

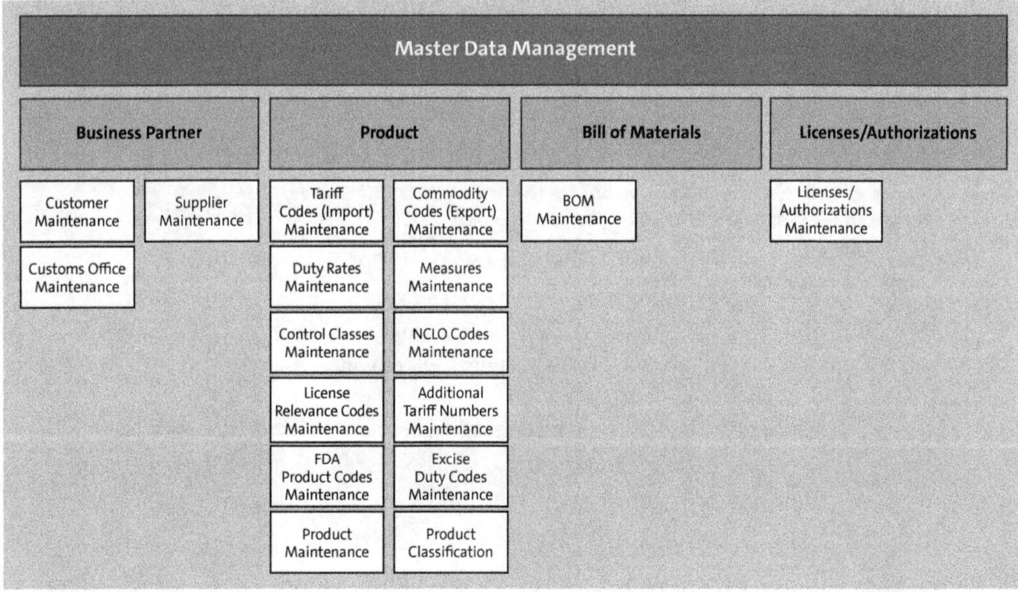

Figure 4.4 Master Data Types

4.2 Business Partner Master Data

Business partners that can be transferred from the feeder system are customers, vendors, contact persons, applicants, and banks. Customs office master data is maintained directly in SAP GTS.

When business partners are transferred from the feeder system to SAP GTS, they aren't assigned to their corresponding account groups (e.g., ship-to party). As mentioned in Chapter 3, this is done on a transactional basis and requires an upstream mapping between partner functions in the feeder system and the business partner functions in SAP GTS. Because a customer can exist in SAP GTS without a business partner function, and the assignment is done on a transactional basis, a customer can, for example, be assigned to a ship-to partner function in one transaction and a payer partner function in another transaction.

When transferred to SAP GTS, business partners are created as different types: persons, organizations, or groups. Each type has different roles assigned to it. For example, a person can be given the role of contact person, and an organization can be given the role of customer or vendor. Within the different partner types, several roles can be selected. Table 4.2 lists available partner roles in SAP GTS.

Person	Organization	Group
■ 000000 – Business Partner (Gen) ■ BUP001 – Contact Person ■ BUP002 – Prospect ■ BUP005 – Internet User ■ BUP008 – Response. for Preference ■ BUP012 – Collaboration user ■ FS0000 – Financial Services BP ■ SLLSTL – Person Responsible	■ 000000 – Business Partner (Gen) ■ BUP002 – Prospect ■ BUP004 – Organization Unit ■ FS0000 – Financial Services BP ■ SLLAUT – Government Agency ■ SLLBNK – Bank ■ SLLCOF – Customs Office ■ SLLCPC – Customs Bus. Partner (Cus) ■ SLLCPS – Customs Bus. Partner (Sup) ■ SLLDAP – Data Provider ■ SLLFTO – Foreign Trade Org. ■ SLLFTX – FTO for Re-Export ■ SLLLCB – Customs Broker ■ SLLMGR – Admin. Unit of Vend. Decl. ■ SLLPOI – Provider of Information ■ SLLSIT – Legal Unit ■ SLLTPD – Third Party Declarant	■ 000000 – Business Partner (Gen) ■ BUP002 – Prospect ■ FS0000 – Financial Services BP

Table 4.2 Business Partner Types and Roles

Note

Before you can transfer any business partners to SAP GTS, the following Customizing activities must be performed (see Chapter 3, Section 3.3):

- Define number ranges for business partners.
- Create and assign partner functions.

We'll provide an overview of the key business partner master data elements in the following sections.

4.2.1 Customer Maintenance

Customer data (e.g., customer name, customer number, customer address, Incoterms, shipping conditions, partner functions) is managed in the customer master record in the feeder system. As not all customer data is relevant for SAP GTS, only certain data elements are transferred from the feeder system to SAP GTS. To maintain the master data of customers in SAP GTS, the following relevant master data must be transferred from the feeder system to SAP GTS:

- Name
- Address
- VAT number (if relevant)
- Military use flag (relevant for the legal control functionality where the civilian or military flag could potentially impact the license determination from a compliance perspective)

When customers are transferred from the feeder system to SAP GTS, they aren't assigned to a partner function or account group. This means that a customer is set up as a customer and not as a sold-to or ship-to partner. All customer account groups in the feeder system (e.g., sold-to, ship-to, bill-to) are mapped to a single business partner role in SAP GTS (e.g., customer) because the different partner functions aren't distinguished in SAP GTS. These partner functions are mapped on a transactional basis (e.g., via sales order documents and billing documents). This enables one customer to be a sold-to party in one transaction and a ship-to party in another transaction. To assign these different partner functions, upstream mapping between the partner functions in the feeder system and the partner functions in SAP GTS must be completed.

After customer master data is transferred, additional information specific to global trade is added in SAP GTS. For example, whenever a business partner is screened against SPL, the result is maintained in the business partner master data in SAP GTS and is represented by the **Sanction List Block** checkbox. Customer information can be updated by navigating to one of the spaces capturing the main SAP GTS functionalities (such as **Compliance** or **Customs Import**), followed by selecting the **Master Data** page and then opening the **Manage Business Partner** SAP Fiori tile under the **Business Partner and Products** section. The customer can be searched using different criteria (e.g., the external number, which is the business partner reference number from the feeder system, or the business partner number in SAP GTS).

Note that customers receive an internal number within SAP GTS and don't use feeder system numbering. This is due to the possibility that SAP GTS could be connected to multiple feeder systems with conflicting numbering structures. Both the feeder system and SAP GTS numbers can be displayed in the business partner master record and can be used in business partner searches. Figure 4.5 shows both SAP GTS and feeder system numbers as **Business Partner** and **External BP Number**, respectively. Before transferring customers, the number ranges must be set up in SAP GTS to enable this number assignment (see Chapter 3, Section 3.3.4).

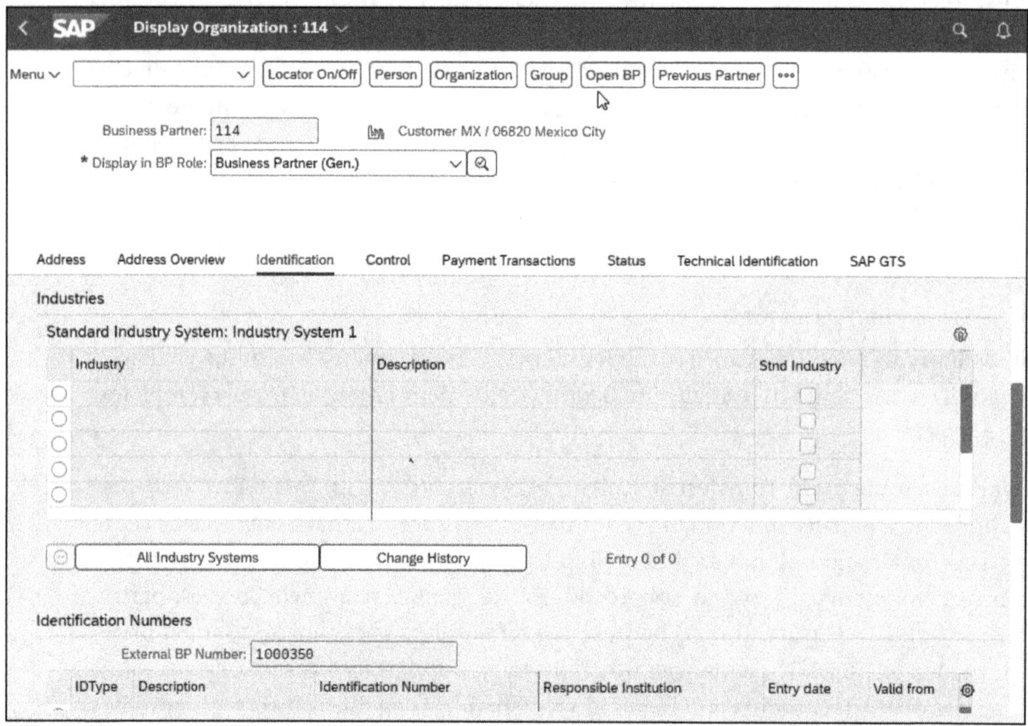

Figure 4.5 Customer Master Data in SAP GTS

4.2.2 Vendor Maintenance

As with customers, vendors can be maintained in the feeder system. Some vendor master data elements (e.g., vendor number, name, address) are relevant for SAP GTS, while others (e.g., currency used, terms of payment, tax information) aren't. When vendor data is transferred to SAP GTS, only the information relevant to global trade is captured, including the following:

- Name
- Address
- VAT number (if relevant)
- Military use indicator (relevant for the legal control functionality where the civilian or military indicator could potentially impact the license determination from a compliance perspective)

Similar to how customers are maintained in SAP GTS, vendors are enriched with additional information and given their own internal numbering in SAP GTS after the initial transfer from the feeder system. This enables companies to connect multiple feeder systems to one SAP GTS client. Vendor information can be updated similar to customer information by navigating to the **Customs Import** or **Customs Export** spaces in the SAP Fiori launchpad, followed by selecting the **Master Data** page and then opening the **Manage Business Partner** tile under the **Business Partner and Products** section. The vendor can be searched using different criteria (e.g., the external number, which is the business partner reference number from the feeder system, or the business partner number in SAP GTS).

4.2.3 Customs Offices Maintenance

Customs offices are branches of central customs agencies that monitor customs-relevant activities. As customs offices are specifically used in customs management, they are maintained as business partners directly in SAP GTS rather than in feeder systems. A customs declaration in SAP GTS may require the user to specify the customs office of departure, exit, destination, transit, entry, or the main customs office. For example, in the context of an export from the EU, the office of departure is the office where the goods are declared for export, whereas the office of exit is the office where the goods physically depart the EU. Depending on the transaction, a given customs office may fulfill different roles (office of departure, office of exit, etc.); therefore, the mapping between each business partner (i.e., office) and role is performed on a transactional basis.

Customs offices can be created manually or uploaded using a file provided by government agencies. Customs offices can be created manually by navigating to one of the spaces capturing the main SAP GTS functionalities (such as **Compliance** or **Customs Import**) and then going to the **Master Data** page and opening the **Manage Business Partner** tile under the **Business Partner and Products** section, as shown in Figure 4.6. First, the customs offices are created as business partners with the **Customs Office** role. Then, these business partners are assigned an office number by selecting the **Identification** tab and creating an entry with **SLLCID** as **IDType**. The office number represents an external ID number provided by the relevant customs authority so that each customs office can be uniquely identified.

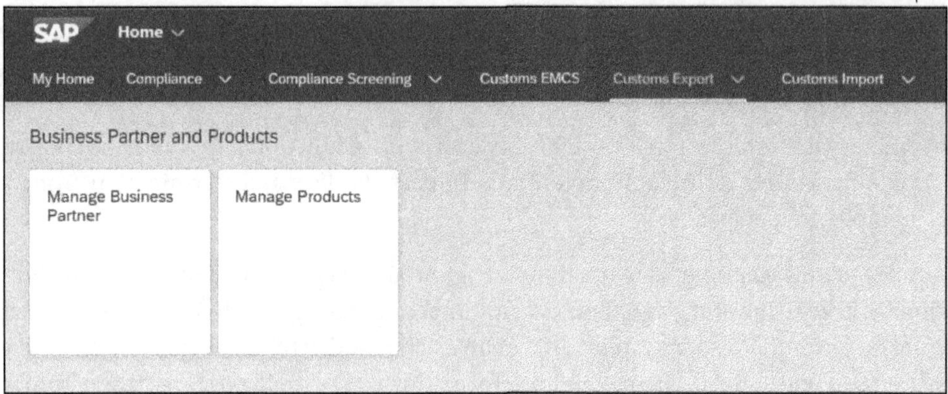

Figure 4.6 Business Partner and Products Related SAP Fiori Tiles under the Customs Export Space

Alternatively, customs offices can be uploaded through the administrator role by executing Transaction /SAPSLL/CUSTOFF_UPL, as shown in Figure 4.7. You need to perform the following configuration activities:

- **Upload Customs Offices**
- **Display Upload Logs Customs Offices**
- **Upload EU Customs Offices**

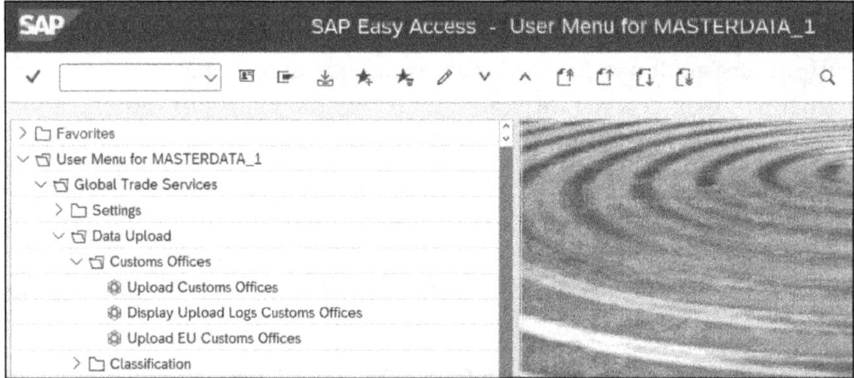

Figure 4.7 Administrator Menu with Related Transactions for Customs Offices Upload

4.3 Product Master Data

Product master data is required to populate customs documents, perform compliance checks, and determine preferential statuses in SAP GTS. Product master data is transferred to SAP GTS from the feeder system or, for some data elements, maintained directly in SAP GTS.

Product master data includes the definition of a product itself and the product's various classifications for trade purposes. The most common classifications are tariff codes (along with associated duty rates) and commodity codes. Some countries may require additional classification codes for specific tariff or commodity codes, such as measures in Germany, non-customs law obligations (NCLO) codes and license relevance (Authorization Requirement Code [ARC]) codes in Switzerland, or additional tariff codes and Partner Government Agency (PGA) codes in the US. SAP GTS also covers legal control classification via control classes, excise duty codes, and Food and Drug Administration (FDA) product codes.

We'll take a closer look at the product master data components in the next sections.

4.3.1 Tariff and Commodity Codes

Tariff codes are used for import processing, whereas *commodity codes* are used for export processing. These codes are used to determine import duties and import/export taxes, track trade flows, manage quota restrictions, and more.

The process of assigning a specific code to a product is referred to as *classification*. From a global trade perspective, classification for both commodity and tariff codes can be maintained in SAP GTS (see Chapter 5, Section 5.1.3). For some regions, commodity codes may also be required in the feeder system (e.g., Intrastat in the EU). Therefore, SAP GTS provides a standard transaction to transfer commodity codes from SAP GTS to the feeder system. The transfer of tariff codes isn't available within the standard functionality but can be implemented via an enhancement.

When an export declaration is created in SAP GTS, the classification code can come from two sources:

- **When SAP GTS is the leading system**
  Products are classified in SAP GTS, and the classification from the SAP GTS is used in the export declaration.
- **When the feeder system is the leading system**
  Products are classified in the feeder system, and commodity codes are transferred to SAP GTS on a transactional basis. In SAP S/4HANA, the material commodity code is maintained on the numbering scheme level, which is linked to one or multiple countries.

4.3.2 Control Classes

Control classes represent codes that are assigned to different products. They are used to check products against import and export legal control regulations. The classification numbers determine product license requirements, reporting requirements, record-keeping requirements, and other regulatory obligations. They are commonly used for dual-use or military products, and—in the case of dual-use products—are referred to in the US as export control classification numbers (ECCNs). The classification functionality is further detailed in Chapter 5.

4.3.3 Products

Although products can be created in the feeder system or in SAP GTS, they are typically created in the feeder system. When products are set up and maintained in the feeder system, the feeder system is treated as the leading or source system for product information. Even though the feeder system is the source system, some information (e.g., customs commercial description) is maintained directly in SAP GTS.

Unlike business partners, products have the same naming in SAP GTS as in the feeder system. As multiple feeder systems can use the same SAP GTS instance, the combination of the logical system group (covering the systems where the product originates from) and the product number uniquely defines a product in SAP GTS. A product can be displayed by entering the combination of its logical system group and its external product number.

Only relevant product master data, as follows, is transferred from the feeder system to SAP GTS and copied into the **Basic Data** tab:

- Product number
- Product description
- Base unit of measurement (BUoM)
- Alternative unit of measure (AUoM)
- Gross weight
- Net weight

After the product is transferred from the feeder system to SAP GTS, additional views are created in SAP GTS.

The product information can be managed by navigating to the **Product & Classification** space, followed by selecting the **Product** page and then opening the **Manage Products** tile under the **Manage Products** section. The following tabs display extended product master data and are shown in Figure 4.8:

- **Commercial Description**
  Maintains the commercial description of the product, including in several languages when necessary for global trade purposes.
- **Classification**
  Manages the assignment of commodity and tariff codes (see Chapter 5).

- **Legal Control**
  Includes product master settings related to specific legal regulations, such as reexport for the US (see Chapter 6). Control classes are maintained in this tab as well.
- **Preference Statements**
  Includes settings that support the Preference Management functionality (see Chapter 8). This tab also shows the result of trade preference calculations performed for a specific product.
- **Preference Properties**
  Includes properties (e.g., relevancy for request) for supplier declarations from suppliers as well as agreement-specific properties for preferential trade agreements (see Chapter 8).
- **Prices and Values**
  Displays product prices and values transferred from the feeder system to SAP GTS to support the preference determination process (see Chapter 8).
- **Procurement**
  Displays procurement indicators transferred from the feeder system to SAP GTS to support the Preference Management functionality (see Chapter 8).
- **Partner-Specific Data**
  Displays product names and descriptions used by business partners in their own systems to facilitate the supplier declaration request process within the Preference Management functionality (see Chapter 8).
- **Special Customs Procedures**
  Indicates use of special customs procedures such as bonded warehouses (BWHs), inward processing relief (IPR), and outward processing relief (OPR) (see Chapter 7). Indicators of relevance for special customs procedures need to be transferred back from SAP GTS to the feeder system.
- **Tax Classification**
  Defines the tax liability of materials used in SAP GTS and contains information for calculating import turnover tax (within the EU).
- **Authorities**
  Maintains product-specific documents and evidence that the system uses for customs shipments (to process transit procedures with the customs authorities in EU and EFTA member states) and customs declarations.
- **PGA**
  Maintains the PGA codes that regulate the import and export of products in the US (see Chapter 5).
- **Bill of Material**
  Maintains data regarding BOMs, which are explained in Section 4.4.

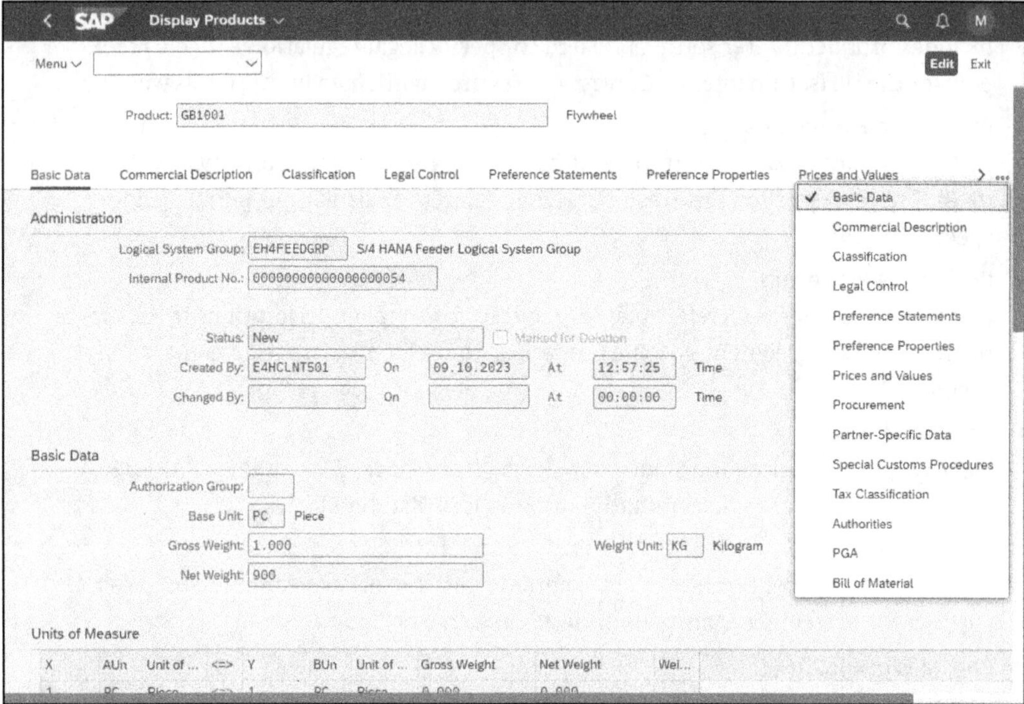

Figure 4.8 Product Master Data

> **Note**
>
> Country of origin isn't part of the SAP GTS product master. Country of origin data is retrieved for each transaction (outbound delivery, material document, and invoice) from the feeder system and transferred to SAP GTS on a transactional basis.

4.3.4 Product Classification

Product classification is the process of assigning a code to a product. For codes to be assigned to products, they should be set up first in SAP GTS via one of two options:

- Upload tariff, commodity, measures, NCLO, PGA, additional tariff, license relevance, FDA product codes, and excise codes, along with control classes, using a file from an external data provider (XML file format). The external data provider must be defined as an external business partner in SAP GTS.
- Maintain classification codes manually.

Before uploading or creating codes, numbering schemes and numbering scheme content need to be defined in SAP GTS. Schemes represent the nomenclatures for the various types of codes. SAP has predefined templates for some countries (e.g., Germany).

There are two ways to assign a code to a product:

- **Individual classification**
  Classify products individually. This method allows users to review and classify products individually.
- **Classification based on a worklist**
  Classify products for a single numbering scheme at a time, such as for compliance export classification or Customs Management Harmonized System (HS) numbers. Products can be filtered based on various criteria (e.g., products newly transferred to SAP GTS or products created by a specific user).

Individual classification can be performed by navigating to the **Product & Classification** space, then to the **Product** page, and finally to the **Manage Products** tile under the **Product** section, as shown in Figure 4.9.

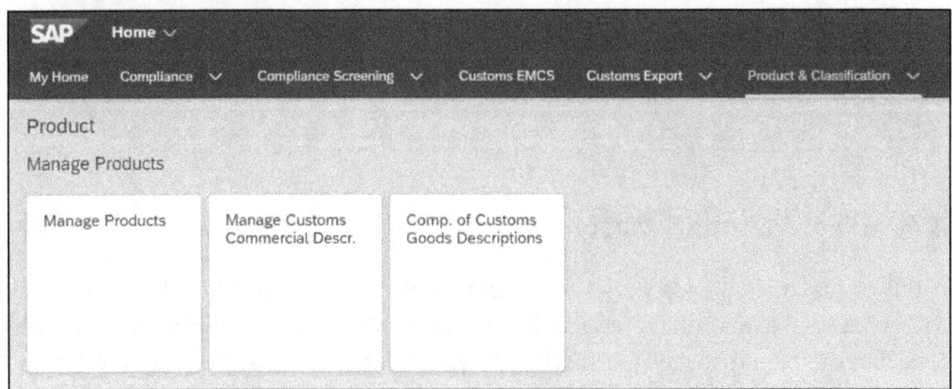

Figure 4.9 Manage Products Section and Related SAP Fiori Tiles under the Product & Classification Space

Classification based on a worklist can be performed by navigating to the **Product & Classification** space, then to the **Classification** page, and finally to the **Classify Products** tile under the **Customs** or **Legal Control** sections, as shown in Figure 4.10. Separate from the initial product classification, there are additional tiles that allow users to reclassify products en masse (manually or via an XML file), delete existing classifications en masse, and perform product classification checks.

The classification process is described in further detail in Chapter 5.

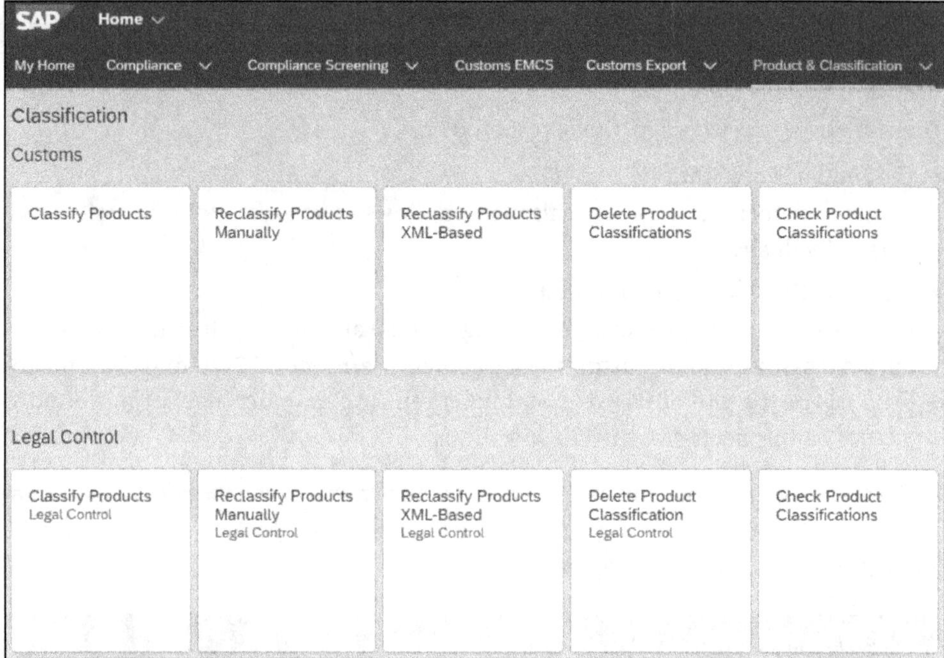

Figure 4.10 Classification Page with Customs and Legal Control Sections and Related SAP Fiori Tiles

4.4 Bill of Materials Master Data

A bill of materials (BOM) shows the components (e.g., raw materials) and quantity thereof needed to manufacture a finished product. BOMs are typically in the feeder systems but can be transferred to SAP GTS. BOMs are maintained at the plant level in the feeder system.

BOMs are used throughout different SAP GTS functionalities, including Compliance Management, Customs Management, and Preference Management. This section distinguishes between BOMs that are manually maintained in SAP GTS and those that are transferred from the feeder system to SAP GTS.

4.4.1 BOMs Manually Maintained in SAP GTS

BOMs can be manually maintained in SAP GTS for two functionalities: Compliance Management and Customs Management. These functionalities relate to requirements to export controls (e.g., International Traffic in Arms Regulations [ITAR]), IPR, and OPR. Though the process of creating and maintaining BOMs is the same for both functionalities, the BOMs are used in different ways. Because not all BOM components in the feeder system are relevant for these functionalities, it's considered more efficient to maintain BOMs manually in SAP GTS with the minimum information required rather than transferring them from the feeder system in their entirety.

Let's consider an example. To maintain BOMs in SAP GTS, go to the **Manage Products** tile by navigating through the **Product & Classification** space, **Product** page, and **Manage Products** section. On the selection screen, select the product, click on the **Change** button, and go to the **Bill of Material** tab to view the **BOM ID** and components. As shown in Figure 4.11, the BOM with **BOM Number ID001** has five components (**Flywheel**, **Wires**, **Radiator Return Hose**, **Fuel Line**, and **Aircraft Engine Parts**) that are relevant for Compliance Management or Customs Management. Other components aren't maintained in SAP GTS.

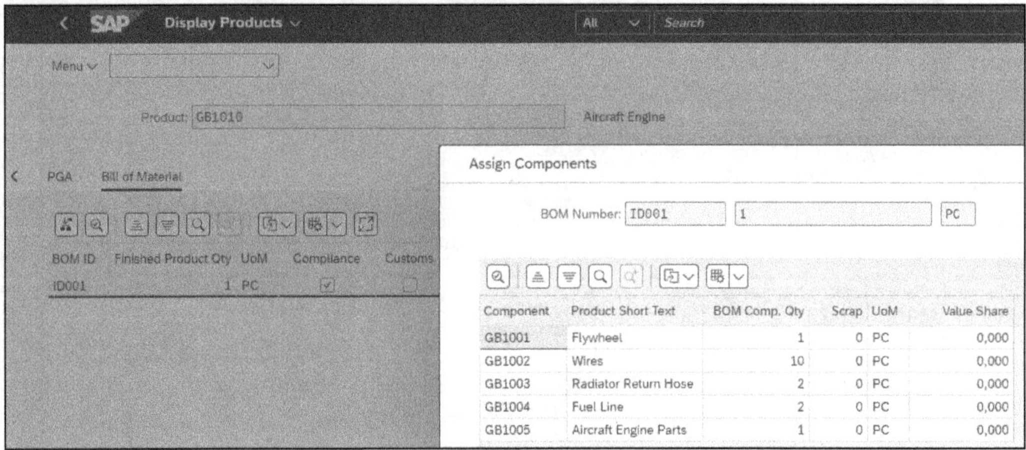

Figure 4.11 Manually Maintained BOM in SAP GTS

In Compliance Management, BOMs are used to help determine the export classification of items. For example, when a product has a component in its BOM that is classified for ITAR, an ITAR authorization is required to export that product.

BOMs in the Customs Management functionality are also used for IPR and OPR. These two special customs procedures are further explained in Chapter 7. The functionality uses the BOM to identify the duty suspended components integrated in the finished or semifinished product.

4.4.2 BOMs Transferred from SAP S/4HANA to SAP GTS

Certain processes in Compliance Management and Preference Management require complete BOMs with all components. Therefore, for these processes, BOMs are transferred in their entirety from SAP S/4HANA to SAP GTS.

BOMs are transferred from SAP S/4HANA via Transaction /SAPSLL/BOMMAT_DIRR3, which can also be reached through the plug-in menu within SAP S/4HANA. This itself is accessible via Transaction /SAPSLL/MENU_LEGALR3 by navigating to the **Initial Transfer of Master Data** tab and executing the **Transfer Bills of Material** application under **Cross-Area Master Data**. The transferred BOM from SAP S/4HANA can be accessed in SAP GTS via the **Preference** space, **Master Data** page, **Master Data – Preference** section, and **Display Bills of Material** tile.

Figure 4.12 provides an example of a BOM transferred from a feeder system. The product (**Casual Leather Shoe**) is displayed along with all of its components (**Leather**, **Wool**, etc.).

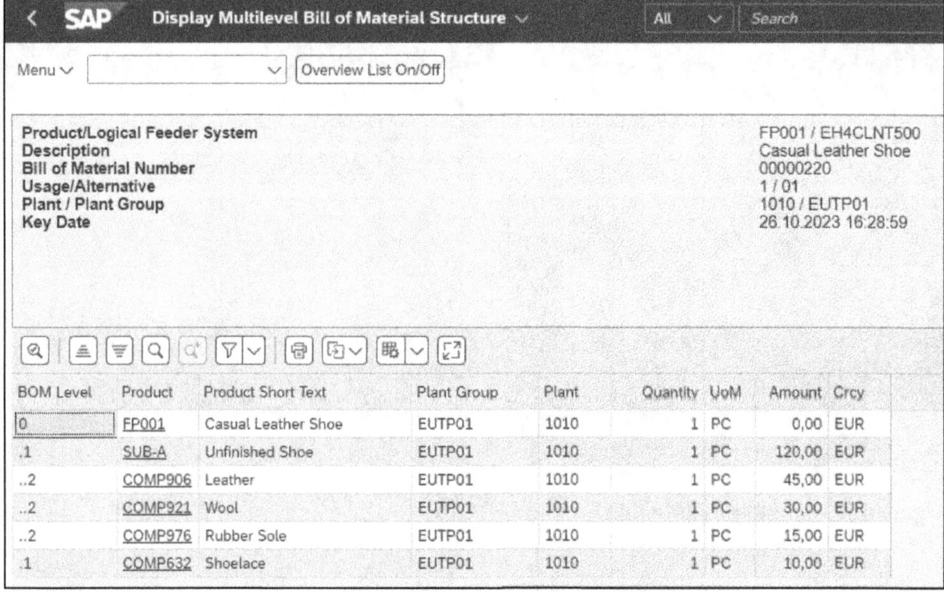

Figure 4.12 Transferred BOM from SAP S/4HANA to SAP GTS

From compliance perspective, BOMs can also be transferred from the feeder system for EAR de minimis rule calculations. These calculations are performed on goods manufactured outside of the US to determine whether the US-origin content of those goods exceeds a defined threshold. If the threshold is exceeded, the goods will be subject to US dual-use export regulations and may require a US license to export.

Several rules for this calculation are set up in SAP GTS. To execute the calculations, all components of a BOM need to be available in SAP GTS, and, therefore, the whole BOM is transferred from the feeder system to SAP GTS. More information can be found in Chapter 6.

BOMs are also used for trade preference calculation in the Preference Management functionality to analyze whether a product meets the rules of origin for preferential status under relevant free trade agreements. This calculation also uses every component of a finished product, so it requires that a complete BOM is transferred from the feeder system to SAP GTS.

4.5 License and Authorization Master Data

Licenses and authorizations are used in SAP GTS for two functionalities:

- For Compliance Management, licenses are maintained for import and export control. Figure 4.13 shows an example of this for routers classified with control class **5A002** under the US numbering scheme. Licenses can be maintained in SAP GTS by navigating to the **Compliance** space, **Legal Control** page, and **License Management** section, and then selecting the **Manage Licenses** tile.
- For Customs Management, authorizations are used when operating under a simplified or special customs procedure. For example, authorization is required when a BWH or a simplified export procedure is set up. Authorizations can be maintained in SAP GTS by navigating to the **Customs Import** or **Customs Export** space and the **Master Data** page, and then selecting the **Authorizations** section for **Import** or **Authorizations and Securities** section for **Export**. Under these sections, the **Manage Transit Authorizations** and **Manage Simplified Procedures Authorization** tiles can be found. The **Manage Storage Authorizations** tile can only be found under the **Customs Import** space.

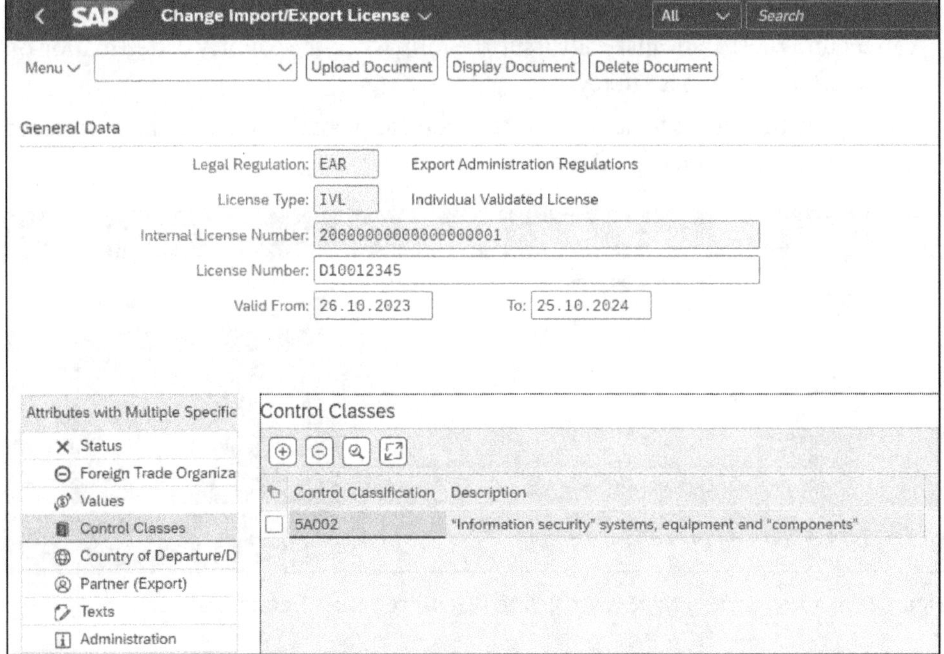

Figure 4.13 License for Compliance Management

4.6 Additional Master Data

Additional master data can be maintained in SAP GTS. Some of this data is readily available, whereas other data requires some setup.

One example of additional master data is the sanctioned party list (SPL). SPLs are maintained manually using the **Manage Sanctioned Party List Master** tile under the **Compliance Screening** space, **Sanctioned Party Lists** page, **Master Data – Sanction List** section, or are uploaded in SAP GTS via an XML file purchased from a content provider (using Transaction /SAPSLL/SPL_UL01). SPLs are used for screening business partners to identify those that are subject to sanctions at the time of business transactions (see Chapter 6). Whenever adjustments are made to the lists, updated versions can be uploaded, and business partners can be rechecked.

An example of master data setup is the default data for customs purposes and determination strategies for compliance purposes. Default data consists of fields that can be populated automatically in customs declarations according to a predefined logic (defaulting rule), outputs that need to be created each time a specific scenario takes place, and automatic email notifications that are triggered by predefined criteria. These fields can be defined in SAP GTS via the administrator menu (see Figure 4.14) to facilitate the completion of customs documents in an automated fashion. For example:

- A company that has shipments to China always by plane may elect to default the mode of transportation on an export declaration to air when the country of destination is China.
- A company that requires a specific form to be issued for shipments bound to Turkey can set up a rule to automatically issue the form when the country of destination on an export declaration is Turkey.
- A company that needs to deliver goods as soon as possible can set up an automated email notification when goods are released by customs.

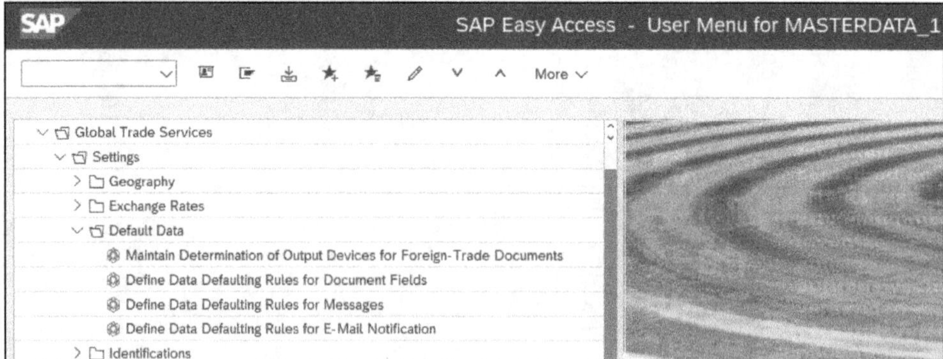

Figure 4.14 Administrator Menu with Related Transactions for Default Data Setup

Another example of master data setup is updating the determination strategy for licenses maintained in SAP GTS. The determination strategy allows SAP GTS to determine whether a license is required based on the parameters of a given transaction. This strategy can be maintained manually via the **Manage Determination Strategies** tile via the **Compliance** space, **Legal Control** page, and **License Management** section, or it can be uploaded via an XML file via Transaction /SAPSLL/CD_UPLOAD.

4.7 Customizing

Customizing for master data is limited because the transfer of master data from the feeder system to SAP GTS is part of the standard functionality offered by SAP. However, some cases require customization:

- When transferring business partners from the feeder system to SAP GTS, the country and region codes in the feeder system must exist in SAP GTS as well. To enable the transfer of this master data, the country list must be aligned between the feeder system and SAP GTS.
- When transferring products to SAP GTS, the units of measurement in the feeder system must already exist in SAP GTS and must be mapped correctly.

In addition, users need to map partner functions in the feeder system with those in SAP GTS to enable relevant functions, for example, customer-partner functions to be determined when sales documents (i.e., transactional data) are transferred to SAP GTS.

4.8 Summary

This chapter covered how SAP GTS handles master data used across multiple systems, applications, and/or processes. The master data set up in the feeder system is partially transferred to SAP GTS and is supplemented by data directly maintained in SAP GTS.

This chapter discussed the business processes related to master data and provided an overview of business partner master data, product master data, BOM master data, authorization master data, and related Customizing steps.

Later chapters will discuss the business processes for Compliance Management (Chapter 6), Customs Management (Chapter 7), and Preference Management (Chapter 8) in detail using specific business scenarios to illustrate the applicable flows or processes. Before exploring these main SAP GTS functionalities, Chapter 5 details the classification capabilities in SAP GTS.

Chapter 5
Classification

Customs authorities require product classification codes for each imported and exported product. This chapter explains various classification codes and the fundamental business and system processes associated with them.

5

In SAP GTS, classification connects product master data with importing and exporting requirements. It primarily involves assigning precise legal codes to products, and these codes, along with other factors, play a crucial role in making key determinations, such as the following:

- The relevant duties applicable when importing and/or exporting goods into/out of a particular country (e.g., calculating duties for importing pharmaceutical products into the European Union, a task handled through Customs Management)
- The necessary permits for importing and/or exporting certain goods into/out of a specific country (e.g., exporting pharmaceutical, dual-use, military, or International Traffic in Arms Regulations [ITAR]-restricted products from the United States, managed through Compliance Management)
- The applicable regulations for ascertaining the preferential origin of exported goods (e.g., determining which specific rules under the Mercosur agreement should be applied to establish if a product can be exported under a preferential status, which is a function carried out through Preference Management)

Commodity codes follow an internationally recognized standard of names and numbers known as the Harmonized System (HS). More than 210 countries and economies base their classification schedules on this system, developed and maintained by the World Customs Organization (WCO). Member countries use universal headings and nomenclature for products; however, they can independently determine the country-specific rates of duty. Harmonized codes are broadest at the four- and six-digit levels, with country-specific descriptions from the eight-digit level.

Classifications are generally country specific, depending on the importing and exporting country requirements. Classification codes are sometimes determined on a country group/customs union basis (e.g., the commodity codes for the EU member states).

This chapter details the purpose of product classification and how to set it up in SAP GTS. We first explain the business and system processes of classification in Section 5.1, including uploading classification codes, classification in both Compliance Management and Customs Management, and reclassification. We then consider system setup based on the most important Customizing steps related to product classification in Section 5.2.

5.1 Business and System Processes

When preparing and submitting documentation for customs clearance, it's imperative to accurately specify each product's official customs code. This information is crucial for regulatory compliance and ensuring smooth customs processing. SAP GTS plays a vital role in simplifying and optimizing this classification process.

It's essential to manage and maintain the master data linked to specific classification numbers to assign the appropriate classification code to each product. This process involves uploading the classification numbers to SAP GTS before initiating the actual classification process, as illustrated in Figure 5.1. After setting up the master data for classification numbers, it's important to update and maintain this information. SAP GTS facilitates the uploading of reclassification files, allowing for efficient updates to master data and the associated products linked with updated classification codes. When changes occur in customs regulations or when product classifications are revised, businesses need to adapt swiftly to comply with the updated requirements.

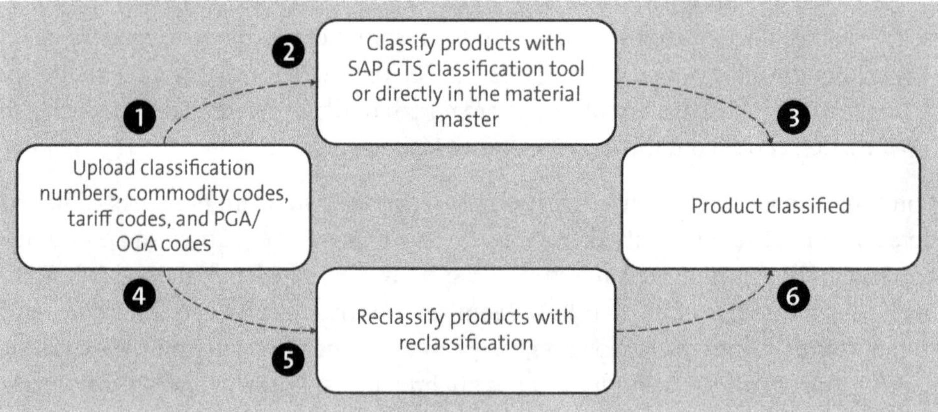

Figure 5.1 Classification Process in SAP GTS

SAP GTS comes equipped with a predefined set of SAP Fiori tiles aimed at facilitating classification tasks for users. These tiles can be activated using the designated standard user role (trade classification specialist) outlined in Chapter 2, Section 2.4.

Under **Product & Classification**, as shown in Figure 5.2, the functionality is organized into three pages: **Product**, **Classification**, and **Master Data – Classification**.

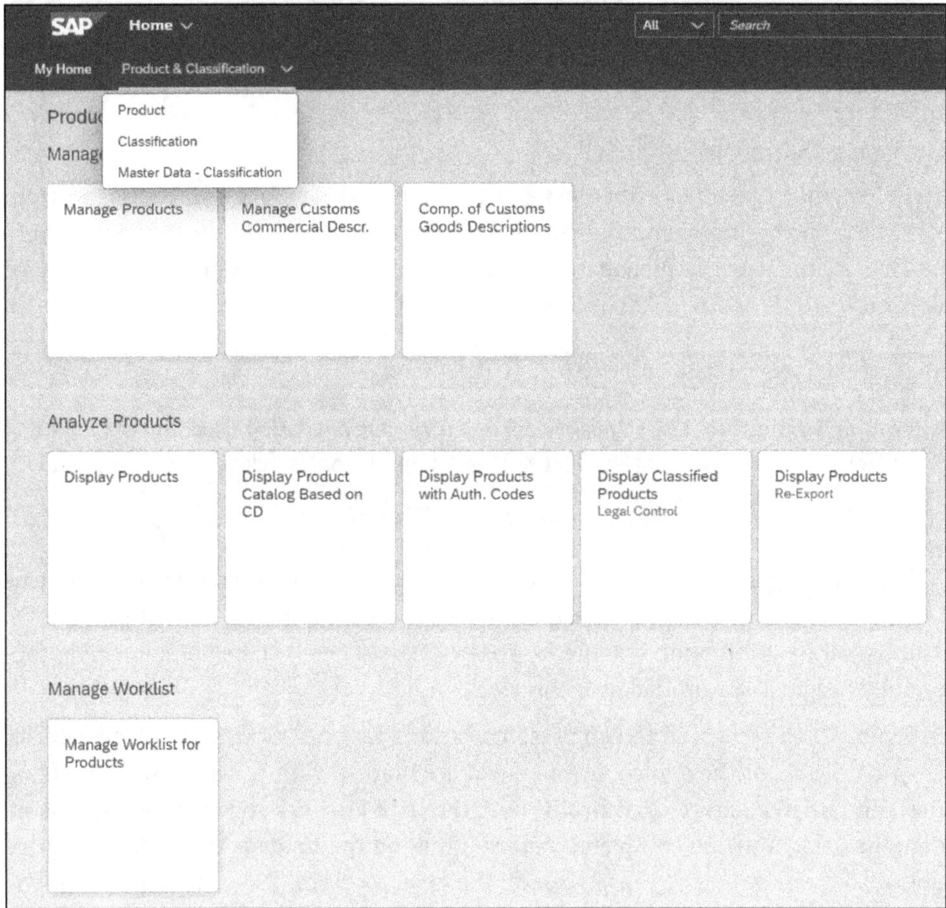

Figure 5.2 SAP Fiori Launchpad for Trade Classification Specialist

We'll explain the key processes that are enabled by these functionalities in the following sections.

5.1.1 Uploading Classification Codes

Users have the option to upload classification numbers and codes using XML format files provided by a third-party content provider. Leveraging this approach eliminates the need for manual upkeep of classification codes, automating a significant portion of the classification process.

The XML files obtained from a third-party content provider can be conveniently uploaded using a dedicated SAP Fiori tile that aligns with the specific content within the file. From the **Master Data – Classification** page, users have access to tiles designed for uploading various essential components, such as import and export codes, duty rates, measures, control classes, and more.

The following sections explain in detail how to upload and maintain control classes, tariff numbers, commodity codes, and relevance codes.

Uploading Control Classes from File

The tile named **Upload Control Classes** is dedicated to uploading control classes from XML files. Furthermore, another tile is available to showcase summary logs encompassing all previously uploaded codes, along with detailed insights into a specific upload.

> **Note**
>
> In SAP GTS, edition for SAP HANA, the term *control classes* is being used for both export control classification numbers (ECCNs) and import control classification numbers (ICCNs).

Before initiating the classification code upload process, certain prerequisites need to be met. It's necessary to define the data provider as a business partner within SAP GTS. Additionally, a numbering scheme for control classes needs to be defined and appropriately assigned, as outlined in Section 5.2.1.

In Figure 5.3, the selection screen presents various parameters for the upload process. Essential fields for the control class upload are marked with red asterisks, signifying that they are mandatory. Specifically, both the **Data Provider** and **Numbering Scheme Content** fields must be completed before initiating the upload. The XML file can be uploaded directly either from your local file explorer (**Local Data Medium**) or via the **Application Servers**.

Within the upload process, the **Select Languages for Upload** field offers the flexibility to choose the preferred language for text display and input, tailoring the experience to the user's language preference.

In the **General Control Parameters** section, there are two important options available. The **Update Schema Definition** checkbox determines whether the schema definitions should be transferred from the upload file, consequently updating the description of the numbering scheme content. This update includes modifications to the stage, length, processing type, and name of the individual structure stages of a tariff number if it doesn't exist yet.

On the other hand, the **Simulation Mode** checkbox serves as a useful tool for testing the program. This allows users to conduct trial runs and test the functionality of the upload process without making actual changes, enhancing the safety and precision of the upload procedure.

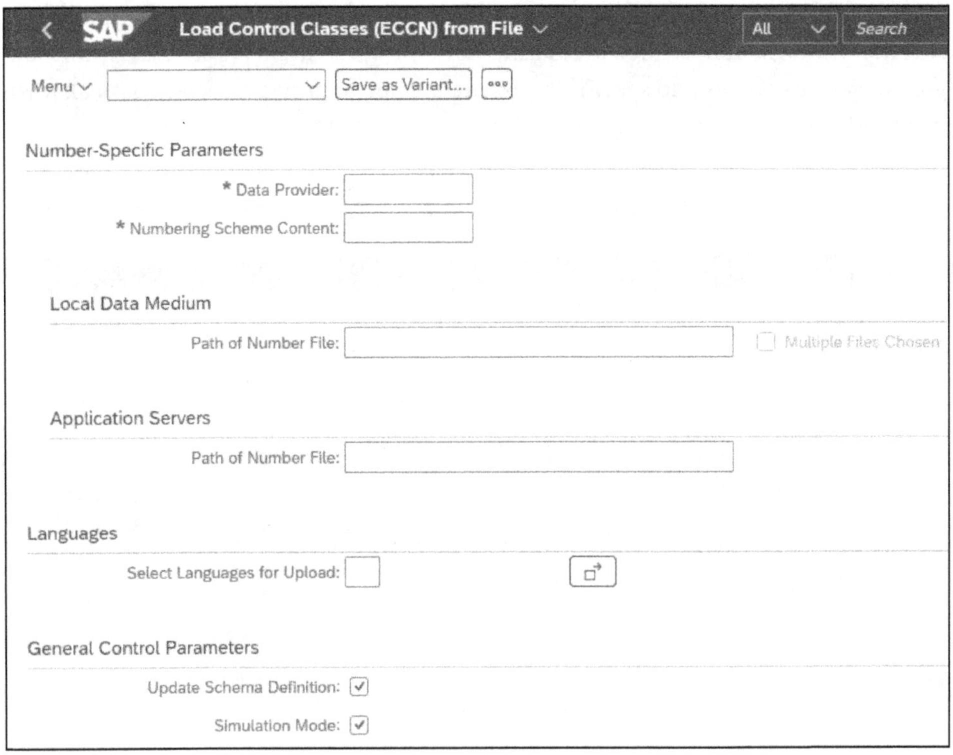

Figure 5.3 Load Control Classes (ECCN) from File Tile

Maintain Import/Export Control Classification Numbers

In addition to uploading classification numbers through XML files, SAP GTS offers the capability to manually maintain and display all classification numbers. This manual management allows users to enter or modify classification codes based on the respective numbering scheme, incorporating relevant categories and classes. However, it's important to note that once a classification number is assigned to one or more products within a numbering scheme, it can't be deleted from that scheme.

A key consideration is that manual maintenance of classification numbers can't be combined with the process of uploading XML files from a content provider within the same numbering scheme. These approaches need to be managed separately to maintain data integrity and avoid conflicts.

To perform manual maintenance of the classification numbers, you navigate to the **Manage Control Classes** tile, where you fill in the parameters **Numbering Scheme Content** and **Language for Text Types**. Consequently, click **Maintain Numbers**, and you'll arrive at the screen shown in Figure 5.4.

The screen is divided into three sections. The upper part features the **Category/Class** section, where category **0** encompasses **Nuclear Materials, Facilities and Equipment** falling under the Export Administration Regulations (EAR) using the numbering scheme Uniform Standard for Evaluating Content Competency (USECC). Within category 0, various classes and specific ECCNs are applied to provide a detailed breakdown of export control requirements for distinct product groups.

The lower part of the screen is divided into a list displaying all ECCNs for a particular class on the left side with the details of each classification number on the right side.

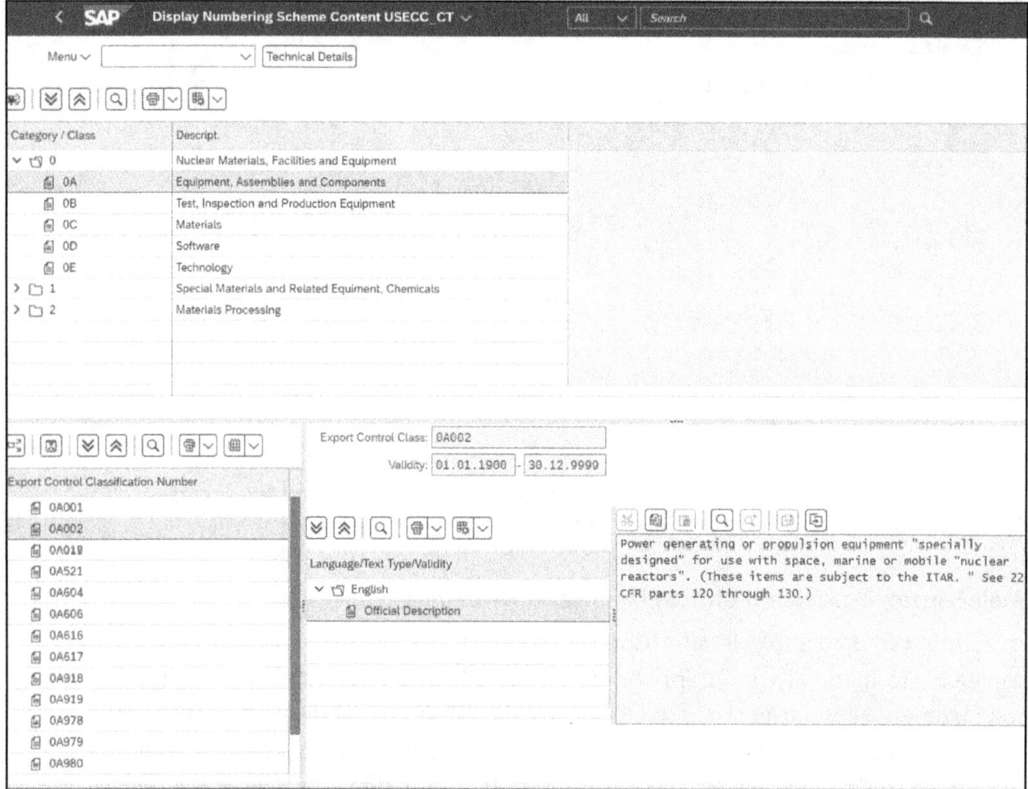

Figure 5.4 Maintaining Control Class Numbers

Uploading Tariff Code Numbers and Commodity Codes

Tariff codes and commodity codes can be uploaded via a similar process to control classes. The **Upload Commodity Codes** and **Upload Tariff Numbers** tiles are activated for the trade classification specialist role, as shown in Figure 5.5.

A correct upload of classification master data depends heavily on the information provided by the content provider. The files allow for the addition of footnotes, notes, and keywords to the uploaded codes and numbers. If no files are provided, you can manually maintain the classification codes and numbers.

Like with control classes, one key consideration is that manual maintenance of classification numbers can't be combined with the process of uploading XML files from a content provider within the same numbering scheme.

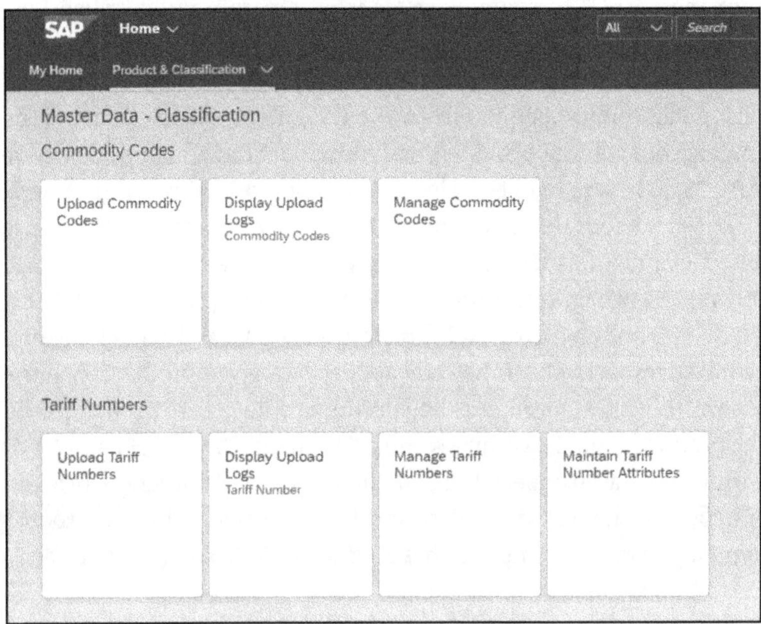

Figure 5.5 Tiles for the Trade Classification Specialist

For tariff and commodity codes, users have the option to perform a package check in addition to the same general control parameters available for control classes. When selected, the package check only loads the next version if all packages or files were loaded successfully for the previous version. It's strongly advised to activate this check because if you're loading a particular version of the numbers, without having loaded the previous version completely, its content may become inconsistent. As a consequence, no more versions can be loaded.

Assigning US Partner Government Agency Codes

Partner Government Agencies (PGAs) are divisions within the US government that regulate products imported into the country and exported from the country. PGAs have historically been referred to as Other Government Agencies (OGAs). From a customs reporting perspective, each of these agencies works with the US Customs and Border Protection (CBP) to obtain additional information on imported or exported products from importers or exporters. This information is collected by CBP under a single-window concept whereby CBP enforces the reporting, inspection, and other requirements on behalf of the other agencies for goods sent to or received from foreign countries. In SAP GTS, these reporting requirements are covered by two concepts: (1) relevance codes/PGA flags and (2) additional PGA data elements. If an importer or exporter doesn't provide the relevance code(s) and the corresponding additional PGA data elements to CBP, CBP may block their import from entering the US or export from leaving the US.

Relevance codes are the SAP GTS terminology for what are commonly called PGA codes. These PGA codes indicate whether the product is potentially relevant to declare additional information upon import/export under a specific PGA and identified based on the product's classification number. Based on the HS data provided as part of the tariff schedule content from content providers and individual tariff/commodity code assignments in SAP GTS, items are marked to indicate that a product classified under a given HS may be subject to additional PGA reporting requirements, and the user is forced to maintain relevant information accordingly. Some examples of relevance codes are DT2, FD1, and EP3, which correspond to the National Highway Traffic Safety Administration (NHTSA), Food and Drug Administration (FDA), and Environmental Protection Agency (EPA), respectively. If the code applies to a given product, then the company must "claim" the code and report the additional data elements; if the code doesn't apply to the product, then the company must "disclaim" the code. Examples of the additional information that may need to be maintained include program, process, and claim/disclaim codes to help PGAs understand the imported/exported product declared by the company and determine whether additional PGA data elements (and/or paperwork) need to be provided to the respective PGA.

PGA data is structured, and each agency has its own implementation guide for managing PGA information based on product classifications. Configurable PGA record maintenance enables editing of necessary PGA information for products based on their classifications, ensuring compliance with relevant authorities.

PGA reporting requirements in the US and other jurisdictions are unique, affect particular product categories, and require additional reporting requirements associated with customs declaration filings. Similar concepts exist in other jurisdictions, and SAP GTS provides functionalities at the master data level to maintain the necessary information to support the filings. At the time of declaration, this master data flows through as part of the transactional data sets and can be communicated to the customs authorities. See Chapter 7, Section 7.1.4, for more information on how SAP GTS handles customs filing processes.

For example, Figure 5.6 shows a product in SAP GTS for which a record has been created for one of the PGA programs (i.e., **FDA - Food and Drug Administration**). The **<default>** value indicates that the record isn't supplier specific.

Before a PGA record can be assigned to a product, a PGA relevance code needs to be assigned to the product's classification. This allows individual products to be classified via the **Manage Products** tile. The assignment of the PGA relevance code is automated once a product is classified. This mapping is provided in the classification upload files from the content provider.

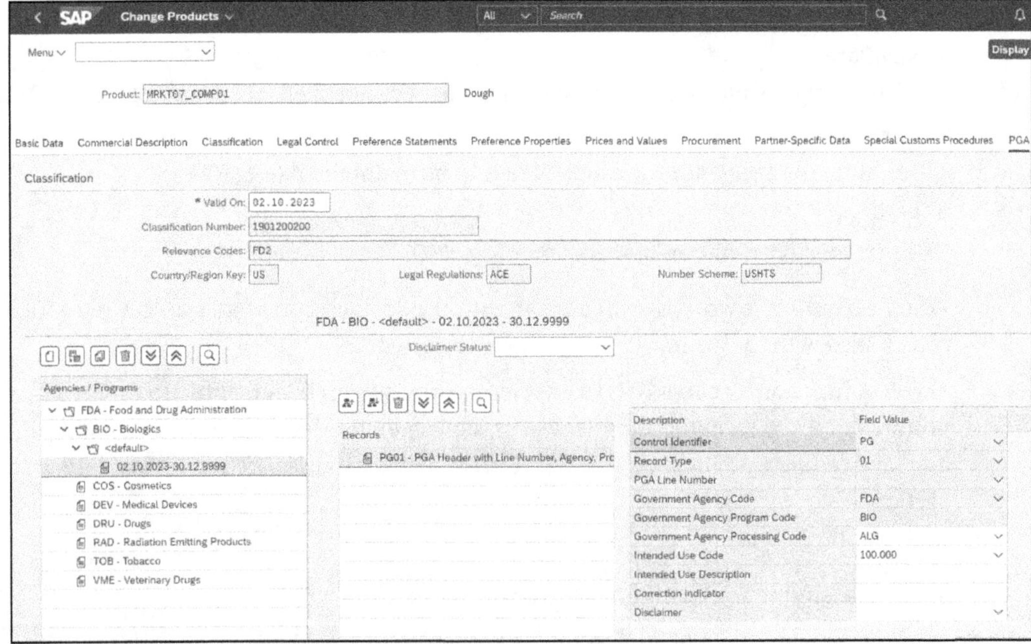

Figure 5.6 PGA Tab in the Manage Products Tile

Alternatively, you can maintain PGA records via a worklist through the **Create PGA Records for Products** tile, as shown in Figure 5.7, which can be found under the **Product & Classification** space.

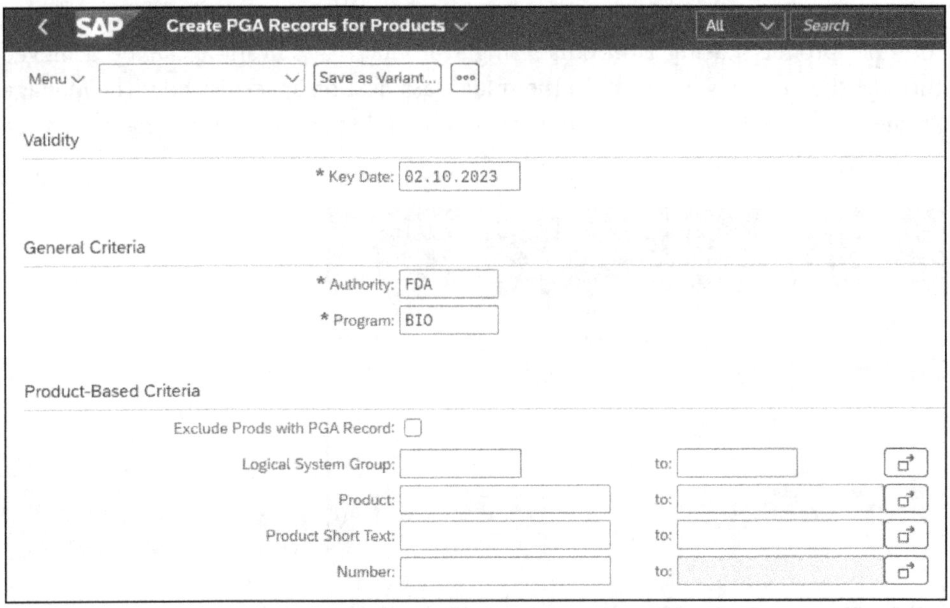

Figure 5.7 Selection Screen Fiori Tile: Create PGA Records for Products

To use the worklist, populate the following fields:

- **Key Date**
  Dictates from which date onward the PGA data can be filed to the authorities.
- **Authority**
  Selection of the agency for which data is to be maintained (e.g., **FDA**).
- **Program**
  The respective program of the agency (e.g., **BIO**).

When executing, a worklist will appear with products assigned and a PGA relevance code linked with the authority and program selected.

The PGA functionality in SAP GTS enables users to assign PGA records to product and product/supplier combinations. However, setting up this legal requirement isn't standard and requires a PGA specialist. The Customizing activities are further explained in Section 5.2.

5.1.2 Classifying for Compliance Management

Exporters are legally obligated to report control classes to the appropriate authorities when engaging in the trade of goods. These classification numbers ascertain a company's eligibility to trade particular products and whether a license is necessary. After uploading the respective codes, it's imperative to assign these codes to the relevant products within the SAP GTS product master.

SAP GTS provides diverse product classification options accessible through the trade classification specialist role. Users have the ability to classify individual products or multiple products using a worklist. Single product classification can be achieved through the products page within the trade classification specialist role. The **Manage Products** tile (see Figure 5.8) enables users to maintain various product characteristics, including classification.

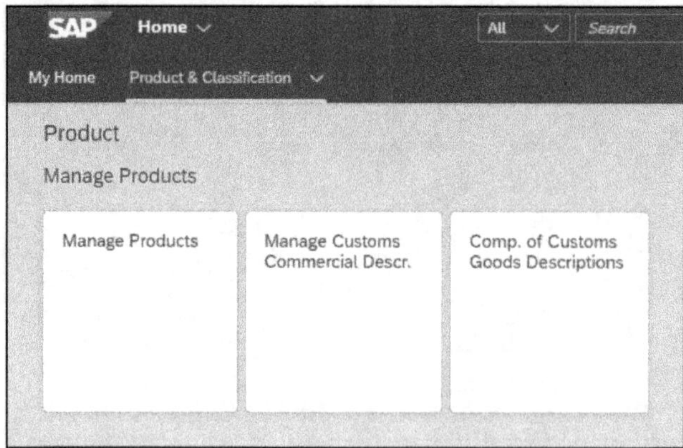

Figure 5.8 Product Page: Manage Products Tiles

For mass classification, you can navigate to the **Classification** page within the trade classification specialist role. Under **Legal Control**, the **Classify Products** tile (see Figure 5.9) permits you to classify multiple products through a worklist but also allows for single product classification.

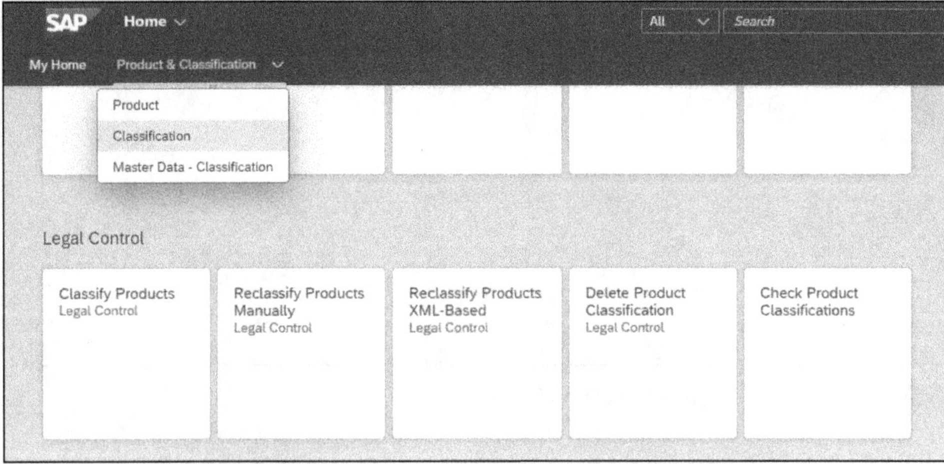

Figure 5.9 Classification Page: Classify Products Tile

The following sections further elaborate on the two product classification options: individual product classification and classification using a worklist.

Individual Product Classification

To classify products individually, access the **Manage Products** tile, as shown in Figure 5.10. This tile allows you to search for products based on different filtering criteria (e.g., product number, created on/by, etc.). After entering those parameters, you click **Change** to classify the product(s).

To access a product's master data for classification, you need to select the specific product intended for classification. Within the product master's **Legal Control** tab (see Figure 5.11), choose the relevant legal regulation for classification. Use the magnifying glass icon found in the **Number** area to initiate the classification process. This action generates a list of classification numbers based on either the upload or manual maintenance methods previously described. You can then review the available numbers within the system and select the appropriate one. Upon saving the product master, the product is officially assigned the chosen classification number for the specified legal regulation. When assigning a new classification code to the product, you can enter a **Classification Period**. If you leave this time period empty, the classification period is defaulted with start date today and ends with the expiration date of the number.

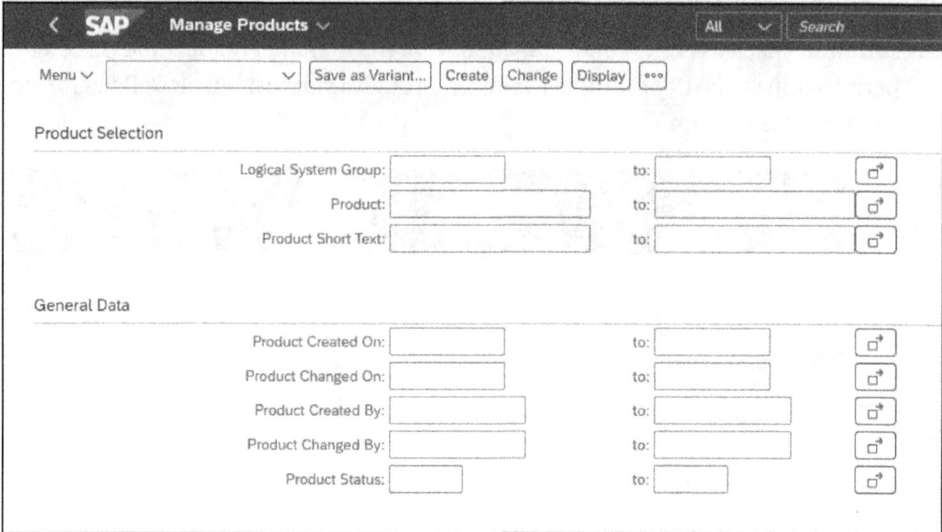

Figure 5.10 Manage Products Selection Screen

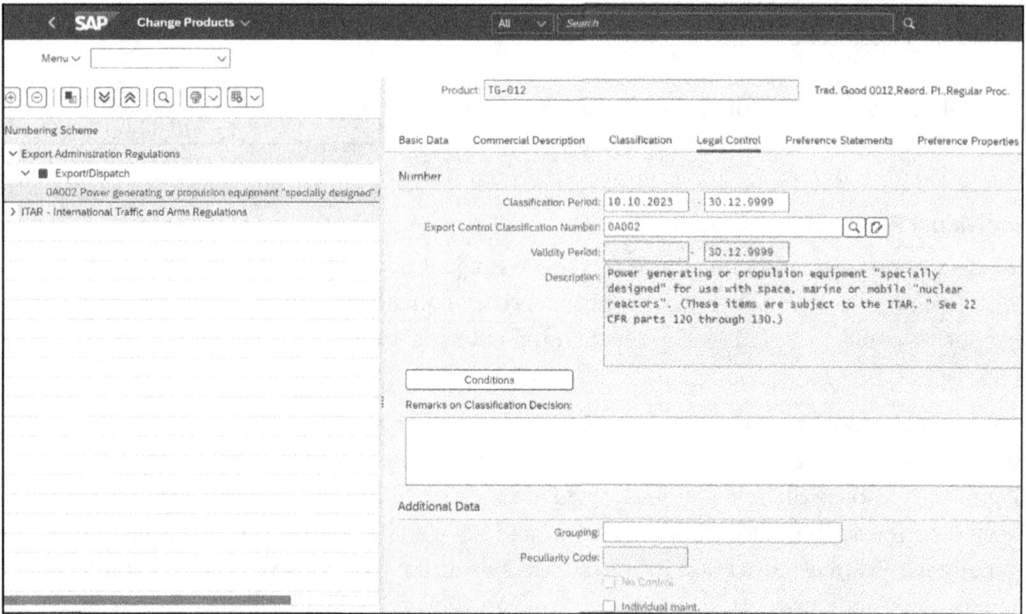

Figure 5.11 Change Products: Legal Control

If you don't need a classification, you can select the **No Control** checkbox to indicate that the product-legal regulation match doesn't require a specific classification code, and a search for further details during the screening process is ignored. If you do want the product to be taken into account during the legal control check, you select the **Individual maint.** checkbox.

Classification Using the Worklist

Classifying via worklist facilitates the creation of a list with all products that still have to be classified for a legal regulation within SAP GTS. In the **Classify Products (Legal Control)** tile, you can specify **General Criteria**, such as **Import/Export** and the **Legal Regulation**, as well as **Product-Specific Criteria**, shown in Figure 5.12.

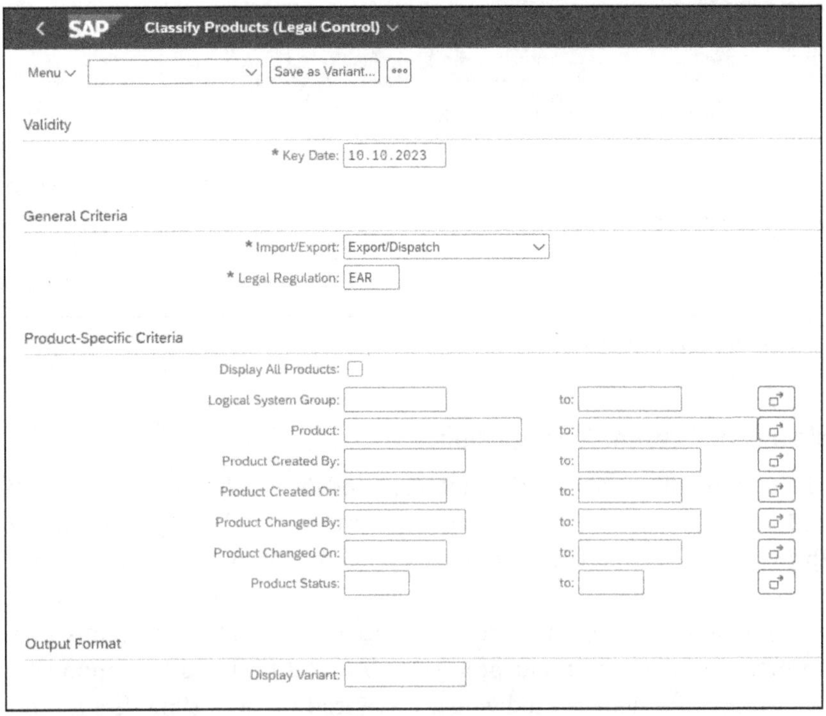

Figure 5.12 Classify Products (Legal Control) Worklist

After generating the worklist (displayed in Figure 5.13), there are two options to further classify the products:

- Classify each product individually by selecting them one by one.
- Opt for the mass classification feature, which classifies all selected products under a single chosen classification code within the same legal regulation. This is referred to as the mass classification process. For this purpose, you need to select all products you want to classify, followed by clicking the **Classify Products** button, after which you enter the **Mass Classification** screen.

Next to the **Classify Products (Legal Control)** tile, there is a dedicated application under **Legal Control** named **Check Product Classifications** that can be used to validate the assigned classification codes. This app allows you to perform a consistency check on classified products for specific numbering schemes. If consistency issues are detected, you can use the **Remediate** button to immediately resolve issues.

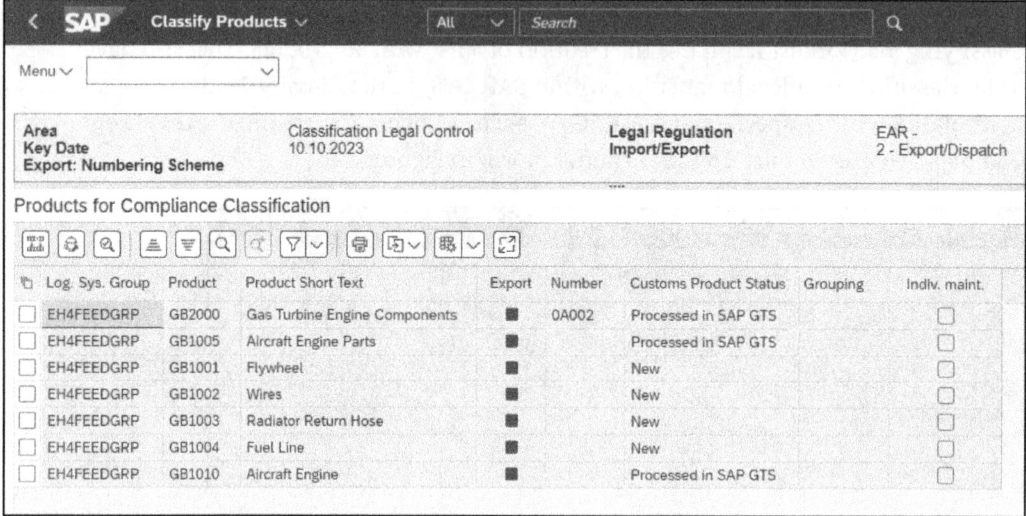

Figure 5.13 Classify Products Worklist

5.1.3 Classifying for Customs Management

Specific legal requirements, trade security measures, and duty tariffs apply to products as they cross international borders. The legal implications and duty applications are facilitated through a comprehensive system that governs import and export procedures. Products are associated with classification codes designed for customs-related activities, including statistical reporting, customs declarations, duty calculations, preferential treatment, and other relevant processes. SAP GTS provides two options to classify for Customs Management: individual classification (accessible through the **Manage Products** tile illustrated previously in Figure 5.8) or mass classification using a worklist. The latter can be accomplished by using the **Classify Products** tile underneath **Classification • Customs**, as depicted in Figure 5.14. We'll explore these options further in the following sections.

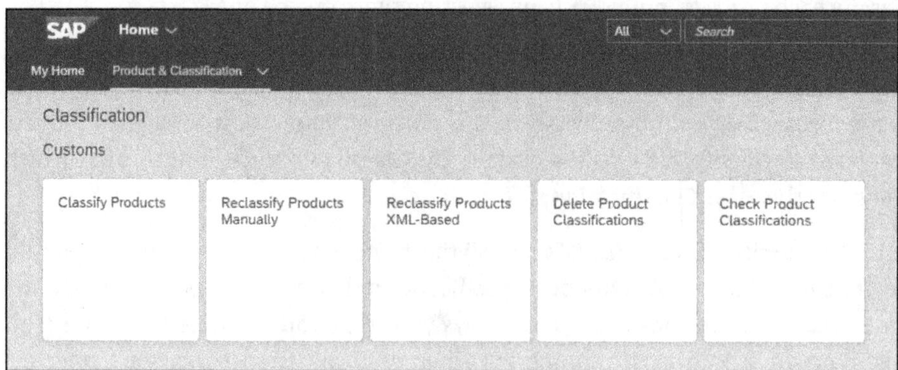

Figure 5.14 Classification Customs Section

Individual Product Classification

Product classification can be performed on an individual basis, mirroring the process described in Section 5.1.2. By accessing the **Manage Products** tile, you can choose specific products for subsequent classification (see Figure 5.15). A thorough classification can be achieved by specifying the appropriate **Numbering Scheme** to accurately classify products. In Figure 5.15, we've selected the EU tariff scheme, represented as **TARIC**, and the product is classified with **TARIC-Code 8544111000** for period **10.10.2023 – 01.01.4000**.

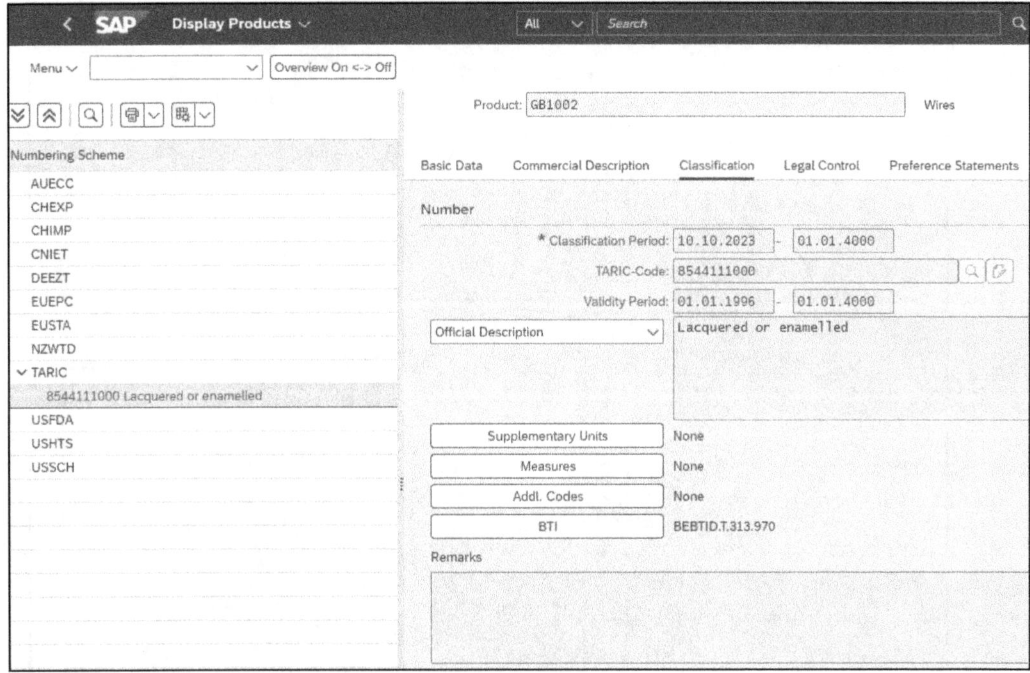

Figure 5.15 Manage Product Tile for Customs Classification

To attain legal certainty regarding the tariff classification of goods, companies can apply for a written tariff classification decision, which national customs authorities can issue. If a classification decision has been provided by the customs authorities for a specific product, it's important to reference this decision in the customs declaration. This legal reference can be stored in the SAP GTS product master (under the Binding Tariff Information [**BTI**] button, as shown in Figure 5.15), allowing for the automatic inclusion of a document with the decision number in the SAP GTS customs document when the corresponding product is imported or exported.

Classification Using the Worklist

The classification process via a worklist closely mirrors the process outlined in Compliance Management in Section 5.1.2. SAP GTS in Customs Management offers similar functionalities, including mass classification. Moreover, the tile specific to customs classification features an expanded selection screen, enabling you to refine product searches more comprehensively (see Figure 5.16). You can filter specific classification codes under the **Classifications** section, allowing for reclassification under a particular code or validation of all products assigned to the same classification code.

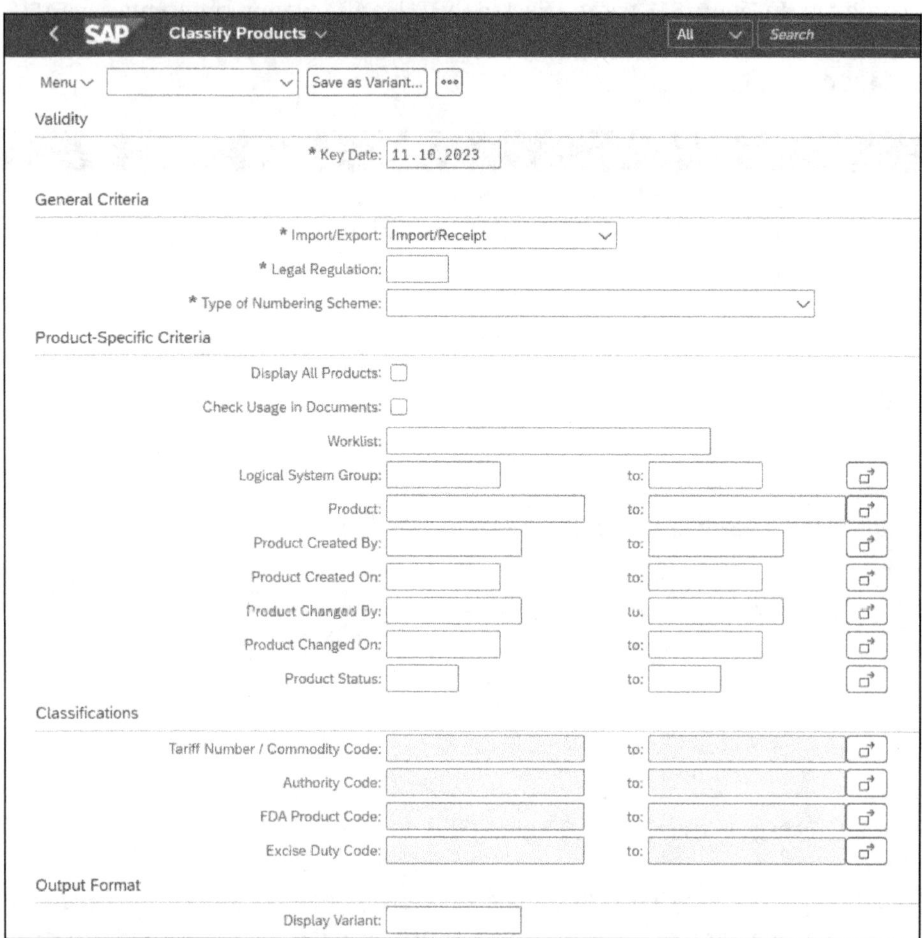

Figure 5.16 Classify Products Application for Customs Classification

5.1.4 Reclassifying Products

Product reclassification assists users in managing classification numbers already recorded in SAP GTS as they undergo changes. This feature allows users to reclassify products for various reasons, including correcting an initially incorrect product classification or conducting periodic reclassifications due to alterations in regulatory requirements. SAP GTS offers two SAP Fiori apps per functionality that facilitate this process: Reclassify Products Manually and Reclassify Products XML-Based, the latter through a reclassification file received from a content provider.

When a reclassification transaction is initiated using either of these apps, SAP GTS automatically updates all product classifications associated with the selected classification numbers within the respective numbering scheme. Content providers are proactive in supplying updated XML files whenever regulatory changes are instituted by the relevant authorities.

An advantage of using SAP GTS is the retention of historical classification data for a user-defined validity period. This feature ensures that users can access and reference previous classification data, enhancing compliance and record-keeping capabilities.

5.1.5 Tariff Number Attributes

SAP GTS can preserve tariff number attributes specified by the Chinese customs authorities. These attributes contain supplementary information and specifications associated with each tariff number. The relevant attribute data is typically included in the files provided by the content provider. Once these files are uploaded into the system, the attributes become visible during the product classification process (see Figure 5.17).

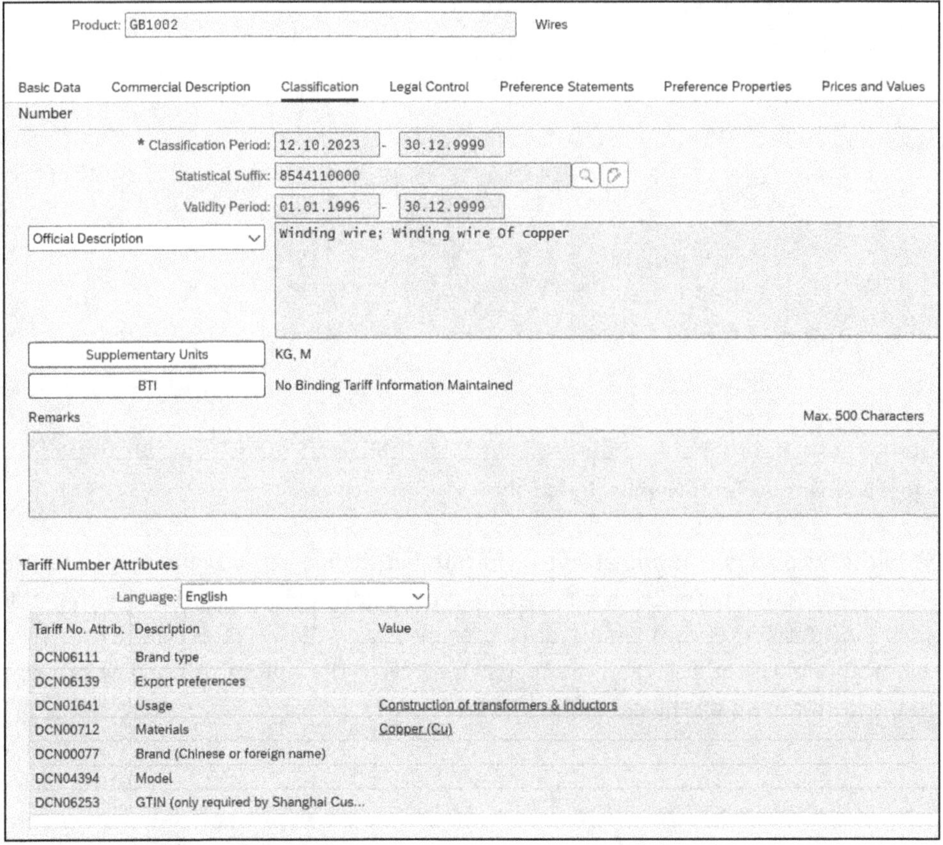

Figure 5.17 Tariff Number Attributes

These attributes vary based on the HS code. When creating a customs declaration, these attributes are presented in an additional tab at the item level, enhancing the completeness and accuracy of the declaration.

5.1.6 Analysis of Customs Management Classification

For a specific feature within classification, the standard SAP GUI transaction is still required to be used and thus to be entered in the command field (Transaction /SAPSLL/CDS_CUS). In this transaction, called **Display Tariff Number Data Sheet** (see Figure 5.18), users can run reports for specific classification codes. This feature enables users to access detailed information related to tariff numbers.

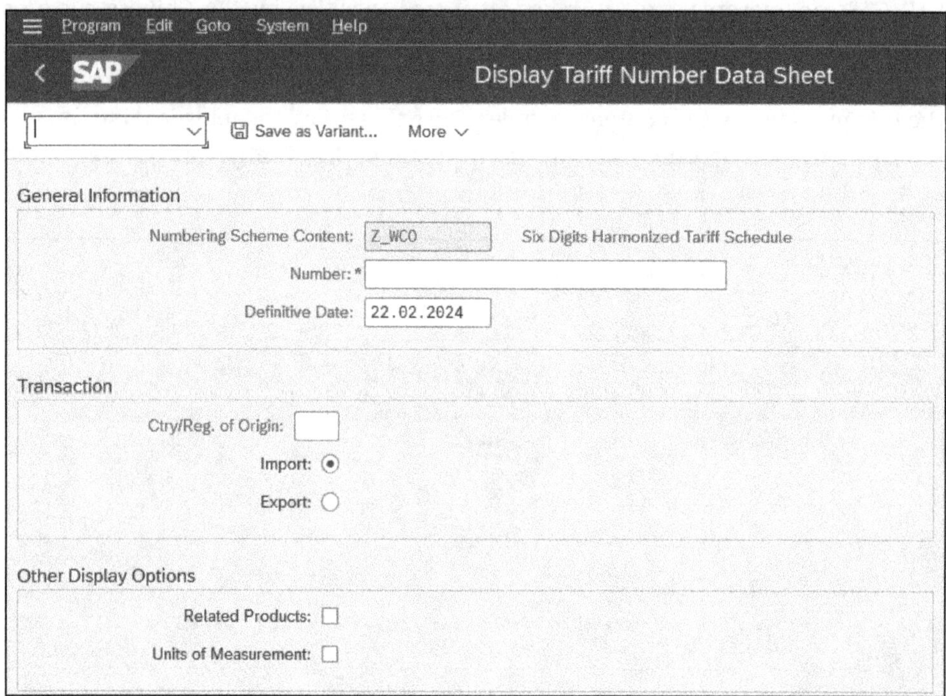

Figure 5.18 Display Tariff Number Data Sheet

Per tariff number, a comprehensive tariff number data sheet is available (see Figure 5.19). This data sheet provides a summary of crucial information associated with a classification number, encompassing attributes such as validity date, numbering text, technical details, and assigned products. Users have the option to preview or print these informative data sheets.

Tariff number data sheet for 85441110 Number valid: Jan 1, 1996 to Jan 1, 4000

Numbering text	
85	CHAPTER 85 - ELECTRICAL MACHINERY AND EQUIPMENT AND PARTS TH
85	EREOF; SOUND RECORDERS AND REPRODUCERS, TELEVISION IMAGE AND
85	SOUND RECORDERS AND REPRODUCERS, AND PARTS AND ACCESSORIES
85	OF SUCH ARTICLES
8544	Insulated (including enamelled or anodised) wire, cable (inc
8544	luding coaxial cable) and other insulated electric conductor
8544	s, whether or not fitted with connectors; optical fibre cabl
8544	es, made up of individually sheathed fibres, whether or not
8544	assembled with electric conductors or fitted with connectors
854411	Winding wire; Of copper
85441110	Lacquered or enamelled

Technical information	
Numbering scheme	EUSTA EU Commodity Codes
Legal regulations that use the scheme	CUSAT, CUSBE, CUSDK, CUSES, CUSFR, CUSNL, CUSNZ, ATLAS

Figure 5.19 Example of Tariff Number Data Sheet

5.2 Customizing

Specific system configurations are often needed to upload control class codes or classify products. The system configuration in SAP GTS revolves around establishing numbering schemes.

5.2.1 Setting Up Numbering Schemes

Numbering schemes define the behavior of classification numbers. Numbering schemes are used for all classification-related master data: control class codes, commodity codes, tariff code numbers, PGA/OGA codes, and excise duty codes.

The configuration for these numbering schemes can be found in the **SAP Global Trade Services, edition for SAP HANA** node under the Implementation Guide (IMG). The essential transactions for this setup are located under **General Settings • Numbering Schemes**.

The setup process involves two main steps: setting up a numbering scheme and assigning numbering scheme content. Afterward, the numbering scheme needs to be assigned to the related legal regulations. This last assignment is merged into a single step for all numbering schemes, simplifying the configuration.

Considering that most numbering schemes, aside from a few subsettings, follow a similar definition pattern, examples will be provided using tariff code numbers and PGA codes, which typically present the most complexity during setup. Let's dive into the configuration activities:

1. **Define Unit of Measurement System**
   In this customization task, measure systems are defined, and units of measure (UoMs) are assigned to them. The system validates UoMs entered in the application and during tariff number uploads by using the units from the designated UoM system. Note that a UoM system can be used across multiple numbering scheme contents.
2. **Define Numbering Scheme for Tariff Numbers**
   In this customization task, a numbering scheme for tariff numbers is established through defined steps. The process commences by providing a name and description for the numbering scheme, followed by specifying the required length of the classification numbers within the scheme (the length of a classification code can vary based on country-specific regulations). Additional settings are available to determine if the numbering scheme supports end use. Furthermore, there is an option to select a reference scheme if the need arises to cascade the classified number between different numbering schemes.
3. **Define the Numbering Scheme Content for Tariff Numbers**
   In this customization task, the primary objective is to define the structure of the tariff numbers for customs processing. The following settings need to be made:
   - **Numbering Scheme Content for Tariff Numbers**: After defining a name for the **Numbering Scheme Content**, a key consideration is whether the data input will be manually sourced or sourced from an external provider. If data is externally sourced, adjustments can't be made manually. Additionally, the option to make the numbering scheme content time-dependent is provided, allowing for a dynamic setup. Furthermore, an important step involves assigning the defined UoM system. This assignment determines the specific UoMs communicated to the customs authorities.
   - **Define Structure Stages**: The next step involves defining the hierarchy levels of tariff numbers, as shown in Figure 5.20. A numbering scheme must consist of at least two levels. For each level, detailing the total length of the numbers (**Length**) and the processing (**Processing Mode**), as well as assigning a description (**Name**), are essential.
   - **Assign Dependent Procedures**: Additionally, there is a provision to assign dependent numbering schemes to the selected numbering scheme content, such as the scheme for additional customs tariff numbers.
   - **Define Chapter Restrictions**: Lastly, chapter restrictions can be defined to specify the chapters to be transferred during a number upload, allowing for a more selective data transfer process. If no chapters are specified, all chapters are uploaded by default.

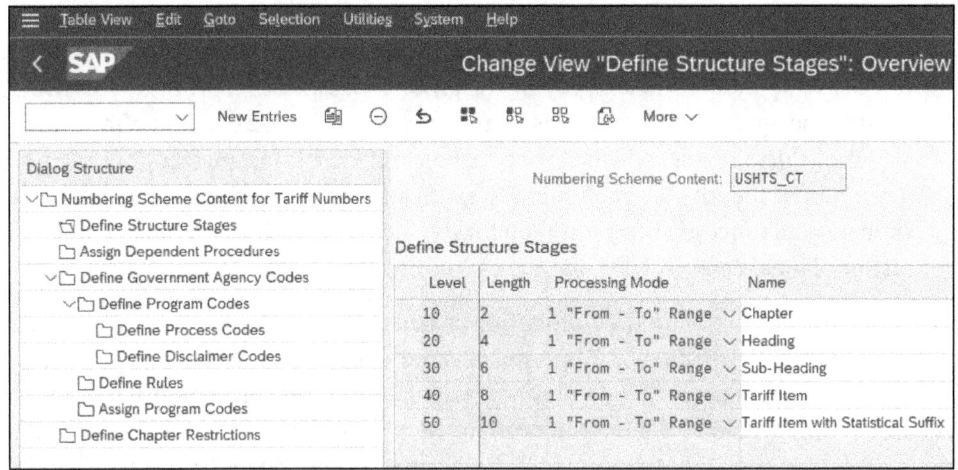

Figure 5.20 Numbering Scheme Structure for USHTS_CT Numbering Scheme Content

4. **Activate Numbering Scheme Content for Tariff Numbers**
   In this customization task, the objective is to enable and activate the designated numbering scheme content. The process involves selecting the relevant numbering scheme and entering the applicable validity period and associated numbering scheme content.
5. **Assign Numbering Schemes to Legal Regulations**
   In this customization process, the objective is to link numbering schemes to their corresponding legal regulations within specific application domains, such as legal control, import and export processing, excise duty monitoring, and so on.

   Figure 5.21 shows that the standard settings in the Customizing already encompass predefined numbering schemes and their respective assignments to legal regulations across different application areas.

SAP Change View "Assign Scheme for Tariff Numbers": Overview

New Entries More Display Exit

Dialog Structure
- Assign Scheme for Control Classes
- Assign Scheme for Numbers in Legal Control
- Assign Scheme for Commodity Codes
- Assign Scheme for Tariff Numbers
- Assign Scheme for FDA Classification
- Assign Scheme for Excise Duty Codes

Assign Scheme for Tariff Numbers

Legal Regulation	Valid From	Schema for Import	Schema for Export	Measure Scheme
ACE Customs Processing - US Automated Customs Environment (A...	01.01.1900	USHTS	USHTS	
ATLAS Customs Processing - Germany (ATLAS)	01.01.1900	DEEZT	TARIC	DE
CUSAT Customs Processing - Austria (eZoll)	01.01.1900	TARIC	TARIC	AT
CUSBE Customs Processing - Belgium	01.01.1900	TARIC	TARIC	BE
CUSCH Customs Processing - Switzerland	01.01.1900	CHIMP	CHEXP	
CUSCN Customs Processing - China	01.01.1900	CNIET	CNIET	
CUSDK Customs Processing - Denmark	01.01.1900	TARIC	TARIC	DK
CUSES Customs Processing - Spain	01.01.1900	TARIC	TARIC	ES
CUSFR Customs Processing - France	01.01.1900	TARIC	TARIC	FR
CUSNL Customs Processing - Netherlands	01.01.1900	TARIC	TARIC	NL
CUSNZ Customs Processing - New Zealand	01.01.1900	NZWTD	NZWTD	
CUSSE Customs Processing - Sweden	01.01.1900	TARIC	TARIC	SE

Figure 5.21 Assigning Numbering Schemes for Tariff Code Numbers to Relevant Legal Regulations

6. **Define Number Set**
   In this customization task, the objective is to define a number set within the IMG. Number sets play a role in simplifying the product classification process across both country and application levels. This feature allows for the use of a predefined number set in Customizing, which can be further refined during master data maintenance. Consequently, it enables the classification of a product across multiple countries and functionalities simultaneously, such as Compliance Management and Customs Management.

To initiate this task, ensure the numbering schemes for customs code numbers (e.g., Harmonized Tariff Schedule [HTS] numbers, commodity codes) have been previously maintained. These customs code numbers should already be present either through manual input or upload. Define the number set by providing a distinct name and a detailed description. Then, associate all the relevant numbering schemes that should be part of this number set.

Select a reference scheme, which serves as the definitive numbering scheme used to determine the number set during classification. A number set can only have one reference scheme.

5.2.2 Setting Up Partner Government Agencies

The PGA functionality needs to be activated per numbering scheme. The agency and program-specific format definition of records and fields isn't available out of the box but must be maintained by PGA experts.

As detailed in Section 5.1.1, a PGA is a governmental entity collaborating with CBP to regulate commodities being imported or exported into the US. SAP GTS offers a tailored solution to meet the specific requirements set forth by PGAs. To configure the system, the PGA functionality must be activated for the required numbering scheme.

The precise format definition for records and fields specific to agencies and programs isn't readily available as a standard configuration. This level of customization requires expertise from PGA specialists, who are responsible for maintaining and defining these essential formats. These necessary customizations can be accessed via the IMG path **SAP Global Trade Services, edition for SAP HANA • General Settings • Partner Government Agencies (PGA)**. The configuration activities are as follows:

1. **Define PGA Agencies and Programs**
   The initial step involves defining the agencies and their respective programs. An illustrative example of such an agency is the FDA, accompanied by programs like Biologics (BIO), Cosmetics (COS), and others.
2. **Define and Assign PGA Relevance Codes**
   The next step entails defining and assigning relevance codes to the relevant programs. This allows the system to associate the PGA relevance codes, which are assigned once a product is classified, to the correct program codes.

3. **Assign PGA Structures to Programs**
The third step involves defining the structures, that is, PGA records, for the respective agency programs. This process involves specifying which records are necessary for each program. Numerous PGA records exist, and every agency defines which records are required, or optional, to satisfy the requirements when filing PGA information.

4. **Assign PGA Fields to the Structures**
After assigning the structures to the relevant programs, the fields per structure are assigned, as shown in Figure 5.22. All fields are provided by SAP; however, it's up to the PGA expert to identify per agency and its respective program which fields are mandatory, conditional, or optional.

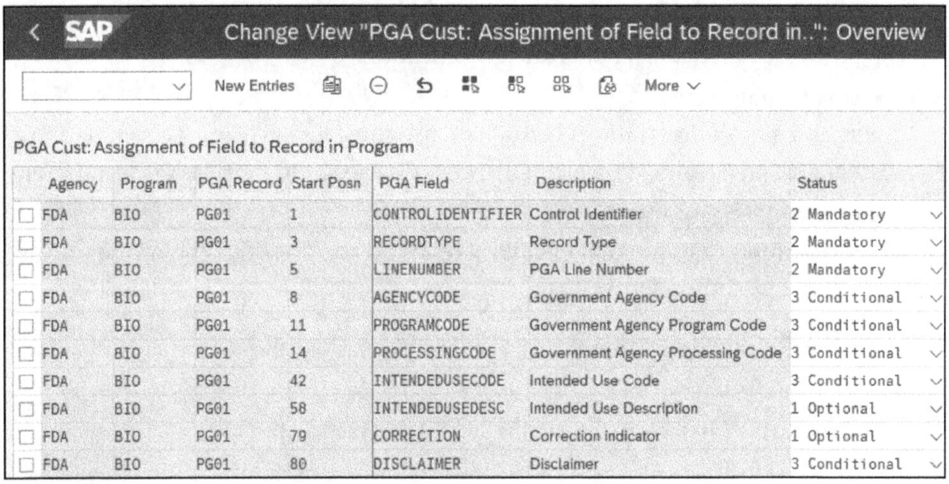

Change View "PGA Cust: Assignment of Field to Record in..": Overview

New Entries More

PGA Cust: Assignment of Field to Record in Program

Agency	Program	PGA Record	Start Posn	PGA Field	Description	Status
FDA	BIO	PG01	1	CONTROLIDENTIFIER	Control Identifier	2 Mandatory
FDA	BIO	PG01	3	RECORDTYPE	Record Type	2 Mandatory
FDA	BIO	PG01	5	LINENUMBER	PGA Line Number	2 Mandatory
FDA	BIO	PG01	8	AGENCYCODE	Government Agency Code	3 Conditional
FDA	BIO	PG01	11	PROGRAMCODE	Government Agency Program Code	3 Conditional
FDA	BIO	PG01	14	PROCESSINGCODE	Government Agency Processing Code	3 Conditional
FDA	BIO	PG01	42	INTENDEDUSECODE	Intended Use Code	3 Conditional
FDA	BIO	PG01	58	INTENDEDUSEDESC	Intended Use Description	1 Optional
FDA	BIO	PG01	79	CORRECTION	Correction indicator	1 Optional
FDA	BIO	PG01	80	DISCLAIMER	Disclaimer	3 Conditional

Figure 5.22 Assignment of PGA Fields to PGA Records

5. **Define PGA Field Values**
To conclude the setup, the permissible values for each field corresponding to the PGA must be defined. The agencies set out strict rules in how the data should be maintained in the records. For some data fields, only predefined values are allowed. As an example, the FDA only allows codes "A" and "F" to be filed as disclaim codes—any other code would be inadmissible while maintaining the field in the product master data.

5.3 Summary

This chapter emphasized the critical task of classification, which involves assigning specific legal codes to products. Proper classification is essential for customs documentation to indicate applicable duties on imported and/or exported goods, necessary permits and licenses for import or export, and any relevant rules determining the preferential origin of exported goods.

SAP GTS addresses these legal requirements by allowing the upload and configuration of classification codes. This functionality enables classification within both Compliance Management and Customs Management, facilitating reclassification as needed. The chapter provides a detailed account of Customizing steps to configure classification settings within SAP GTS.

The subsequent chapters delve into the three main modules of SAP GTS, each strongly influenced by the concepts elucidated in Chapter 4 on master data and this current chapter on classification:

- **Compliance Management**
  Transactions relevant for import and export control will be screened by SAP GTS using the associated master data and classification.
- **Customs Management**
  Import and export declarations in SAP GTS are constructed using a combination of master data and classification elements, ensuring accurate transaction declarations.
- **Preference Management**
  The calculation of preferential eligibility relies on the classification of the finished product and its components.

We'll begin with Compliance Management in the next chapter.

Chapter 6
Compliance Management

Export and import control regulations are constantly evolving, and violating these regulations can result in major fines and penalties. The Compliance Management functionality in SAP GTS helps companies comply with complex international trade regulations.

International regulations have evolved as a result of increased activities in global trade. One of the major trends is the increase in rules and export controls imposed by governments to control and monitor the movements and transfers of certain goods, software, and technology. These export controls prevent the movement of goods without the proper authorizations or to prohibited destinations. Governments implement these controls as a way to prevent the proliferation of weapons of mass destruction, combat terrorism, protect their countries' interests, prevent the unwanted use of sensitive goods, and more. Other controls can be very industry-specific in nature, such as food safety requirements, pharmaceutical licenses, and so on.

Throughout this chapter, we'll elaborate on the different functionalities of Compliance Management (sanctioned party list [SPL] screening, embargo, and legal control). For each functionality, we'll distinguish between the business process and the system process. First, the business process explains the steps and events a company generally goes through in the applicable compliance process. Subsequently, the system process links the steps and events from the business case to the activities performed in SAP GTS to support the process. We'll explain both processes using an applied business case scenario. Finally, we'll cover the relevant master data and Customizing steps required for each compliance functionality.

The Compliance Management functionalities can be accessed from the **Compliance** space in the SAP Fiori launchpad (see Figure 6.1).

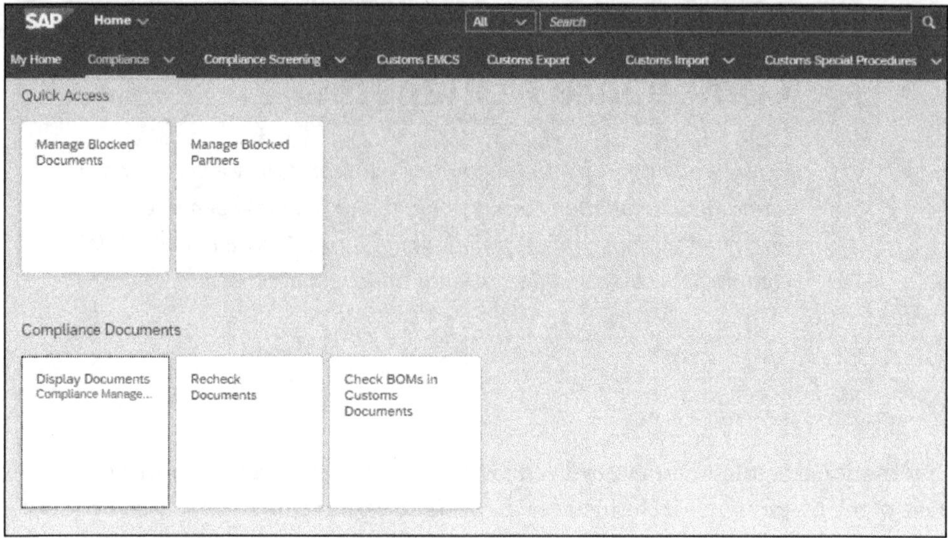

Figure 6.1 Compliance Management SAP Fiori Access

After a discussion of customs documents for Compliance Management in Section 6.1, this chapter explains the functionality of the following main processes:

- **SPL screening (Section 6.2)**
  Users can manage the screening of business partners against government-sanctioned parties using this functionality. The screening process covers customers, logistics partners, and financial institutions. Processing options, such as the blocking and releasing of business partners, are available, as well as detailed information to analyze the reasons for a block or release.
- **Embargo (Section 6.3)**
  Using the embargo functionality in SAP GTS, companies can manage possible embargo situations. Users can monitor, release, and analyze shipments to business partners situated in embargoed countries.
- **Legal control (Section 6.4)**
  The legal control aspects applicable to inbound and outbound processes can be managed using the legal control functionality in SAP GTS, which includes all transactions to manage and automate the required license assignments.

  Additionally, the legal control outbound process includes US reexports. To track and manage licenses for US-controlled products, a product's US content needs to be calculated, and, if applicable, the appropriate license must be included.

SAP GTS also provides a menu of options for classification and master data, which is where components may be classified for both import and export purposes. Every material that is relevant for legal control processes needs to be classified. Master data relevant for SPL screening is included in a designated space accessed via **SAP Global Trade**

Services, edition for SAP HANA • Compliance Management • Sanctioned Party Lists • Master Data – Sanction List.

Note

We won't elaborate extensively on classification and master data for Compliance Management in this chapter. For a more detailed overview, refer to Chapter 4 and Chapter 5.

Beta's Global Trade Requirements

Using a fictitious company, Beta, this chapter will take you through the main business processes supported by Compliance Management in SAP GTS by demonstrating how the relevant government authorities control imports and exports of certain sensitive items in most jurisdictions.

Beta, a company in the aerospace and defense industry, is an independent subsidiary of an international manufacturing company. Beta produces several types of turbine engines used in multiple aircrafts and tanks. A significant supplier for several aircrafts has its European headquarters in Belgium, where various production activities take place.

Figure 6.2 illustrates the business case, which serves as the backbone of this chapter and captures the following import/export transactions (discussed further in Section 6.4):

1. Beta receives an order for one of its products from Customer Y based in Malaysia. This flow illustrates a restriction on a specific organization in Malaysia, imposed by the national government (Section 6.2.1) because the screening of the Malaysian business partner against SPLs results in a block imposed on that customer.

   When attempting to sell products to Customer Z in Syria, the sales representative in Belgium encounters an embargo hit (Section 6.3.1), illustrating a destination restriction on Syria.
2. The Belgian subsidiary temporarily exports aerospace military products (aircraft engine parts) and gas turbine engine components to a US vendor (Vendor A) and reimports them to perform manufacturing operations on them. The business process in Section 6.4.1 approaches the legal control checks from the US vendor's point of view (import into the US).
3. Subsequently, both products are incorporated in the manufacturing process, and an aircraft engine is produced in Belgium. This engine is exported to Customer W in Qatar. Section 6.4.2 explains the export business process from Belgium to Qatar.
4. Assemblies (with US content) and integrated circuits are similarly exported to Customer X in Brazil, as addressed in Section 6.4.5.

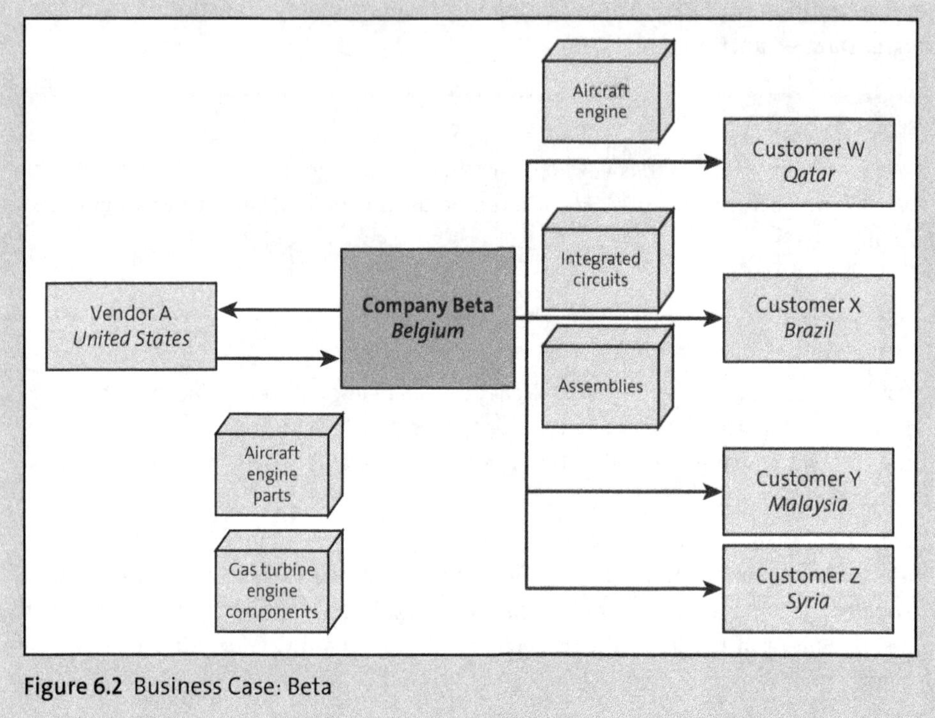

Figure 6.2 Business Case: Beta

6.1 Customs Documents

Upon creation of certain feeder system documents, mirror documents called *customs documents* are created in SAP GTS, and checks are immediately performed upon the creation and transfer of feeder system documents to SAP GTS.

Depending on the feeder system document, there may be a distinction between inbound and outbound documents. The SAP GTS customs documents for both inbound and outbound processes (e.g., purchase orders, sales orders) can be retrieved under the **Compliance Documents** space.

Customs documents in SAP GTS contain information at the header and item levels, as displayed in Figure 6.3. The header level applies to the entire document and includes the organizational data, business partners, values, and so on. Information regarding the specific items in the document is found on the item level and includes general data, classification information, item values, and more.

The SAP GTS customs document contains an overall processing status, as well as an individual status of each screening process for which the document may be blocked. Figure 6.4 displays the customs document **Status** tab on the header level. A status check is available for the following Compliance Management processes:

- SPL screening
- Embargo
- Legal control
- Hazardous substance check

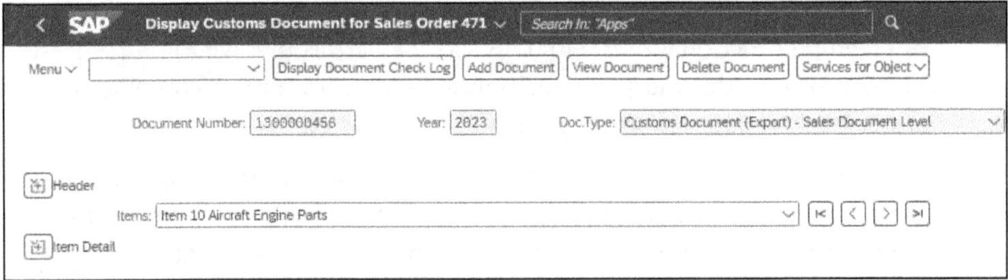

Figure 6.3 SAP GTS Customs Document

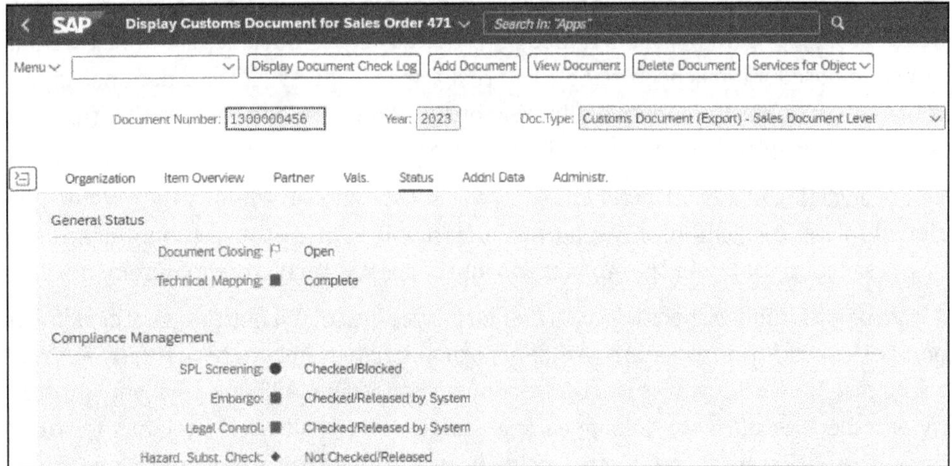

Figure 6.4 Customs Document Status Overview

To reflect the latest information, the customs document is updated instantly each time the corresponding SAP feeder system document undergoes a change and is saved. These feeder system updates result in the Compliance Management functionalities being automatically rechecked.

In the next three sections, we'll elaborate on each Compliance Management functionality. Section 6.2 covers the SPL screening functionality, which helps prevent companies from dealing with sanctioned business partners.

6.2 Sanctioned Party List Screening

In today's global environment, it takes just a few clicks to create sales orders and fulfill shipments across international borders. As a result, the risk of dealing with parties and entities named on government agency or country-specific denied party lists has increased steadily over the past decade.

The SPL screening functionality in SAP GTS screens parties involved in a company's export, import, domestic, and payment transactions to minimize compliance risks with the regulatory requirements across applicable jurisdictions.

There may be strict prohibitions or export authorization requirements when engaging with entities on a restricted or denied party list, which requires additional due diligence. For example, the United Nations and supranational government agencies issue denied party lists (or boycott lists). These lists are generally provided by, and may be procured from, external content service providers.

Similarly, national governments may restrict persons, organizations, groups, and entities by designating them as sanctioned or restricted parties. Such denials or restrictions can be imposed if it has been established that a specific party violated national export control regulations or is involved in weapons proliferation or activities that threaten national security and/or foreign policy concerns.

Before engaging in any business, the company must establish whether the potential recipients of items, including any intermediaries involved in the transaction, are included in any sanction list under the applicable country-specific regulatory requirements.

Some nations publish specific lists of denied parties, restricted entities, or other similar persons to whom transactions are denied or restricted, but many nations don't. In either case, it's the responsibility of the company or other applicable legal entity to verify whether the ultimate consignee (e.g., customer or end user), freight forwarder, banks issuing letters of credit or financing transactions, or any other entity involved in a potential transaction, appear on any denied parties or restricted parties lists through SPL screening. If so, it may be necessary to cancel the shipment of items or the transfer of technology to that person or entity.

The SPL functionality in SAP GTS can do the following:

- Automate SPL checks in international or domestic trade, reducing manual review efforts
- Increase transparency of supply chain transactions and associated parties to prospective transactions for audit purposes
- Minimize compliance risks with international trade regulations
- Potentially serve as a mitigating factor in penalty scenarios

Let's apply SPL screening to our business case, before exploring the system processes, Customizing, and master data.

6.2.1 Business Process

Belgian company Beta receives an order for one of its products from a customer, Freelance Aviation Corporation, based in Malaysia. At the time of order placement, Beta should screen all parties involved in the transaction against the applicable jurisdictions' denied party lists.

Figure 6.5 indicates what Beta's export process looks like. Multiple business partners are involved, including Freelance Aviation Corporation, the carrier who will transport the shipment overseas, and a forwarding agent who will ship the products from the port to the end customer. All partners involved in the transaction should be screened. Furthermore, if any of the partners change (e.g., there is a change of carrier), the new partner needs to be checked as well.

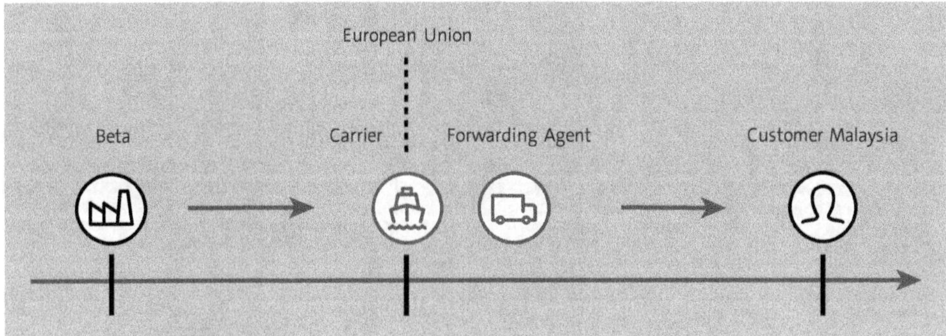

Figure 6.5 Export Screening Business Process

Beta can either perform the screening process or outsource the screening process to a third party. Regardless of the option it chooses, Beta remains ultimately responsible for not transacting with denied parties.

Manually screening business partners is a highly time-consuming, error-prone task, so Beta decides to use SAP GTS to automate the screening process.

6.2.2 System Process

From a system perspective, SPL screening in SAP GTS can occur at several different points:

- Creation of a business partner (e.g., vendor, customer, bank)
- Changes to a business partner
- Changes to government agency-specific lists when changes are uploaded (delta files)
- Transactions (e.g., sales orders, purchase orders, quotes, or deliveries)

All relevant business partner types can be subjected to SPL screening in SAP GTS. Domestic business transactions can be included or excluded from SPL screening.

Business partners are screened either synchronously when they are created or transferred from the feeder system to SAP GTS or asynchronously via a manual trigger or batch job. Names and addresses of business partners in documents are screened only if they have been manually modified. Otherwise, the result of the previous screening for each business partner (status check) is used to determine whether a document should be blocked.

Figure 6.6 links the SPL screening business to the system process applied in SAP GTS. Beta chooses to screen all business partners in feeder system documents, including at time of quotation, sales order, outbound delivery, and billing document creation.

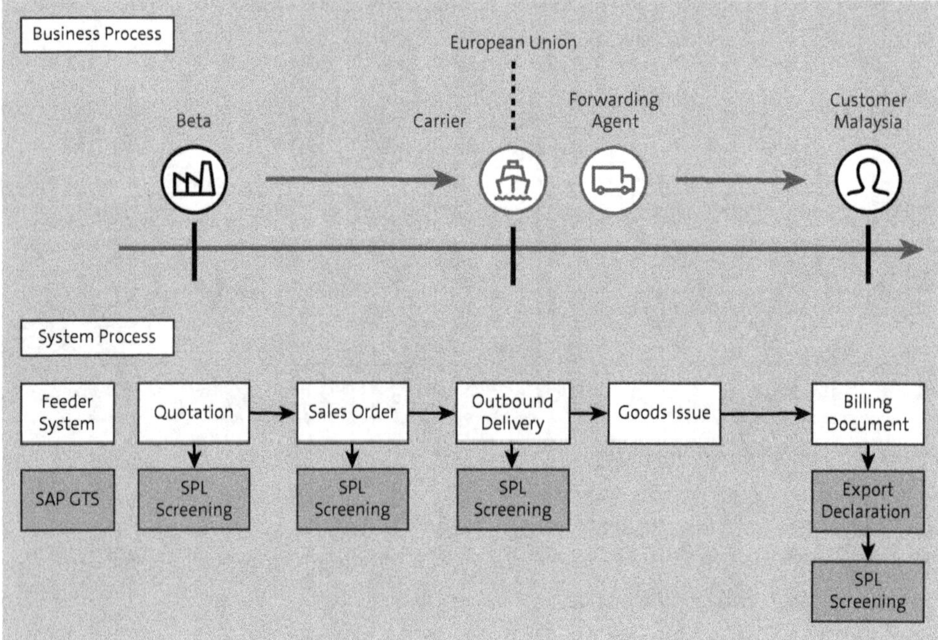

Figure 6.6 Export Screening System Process

It's worth noting that the SPL screening may take place for import scenarios as well (e.g., against the supplier in a purchase order).

SAP GTS performs an automatic check of all the business partners in the feeder system document. The system screens the business partners one by one against the SPL entries present in SAP GTS, as displayed in Figure 6.7. You can upload these lists as master data, which is covered in Section 6.2.4.

If a match is detected, meaning that the business partner appears on a denied party list, SAP GTS blocks the business partner. Users can configure the system to block the feeder system document and prevent any further transactions until action is taken in SAP GTS.

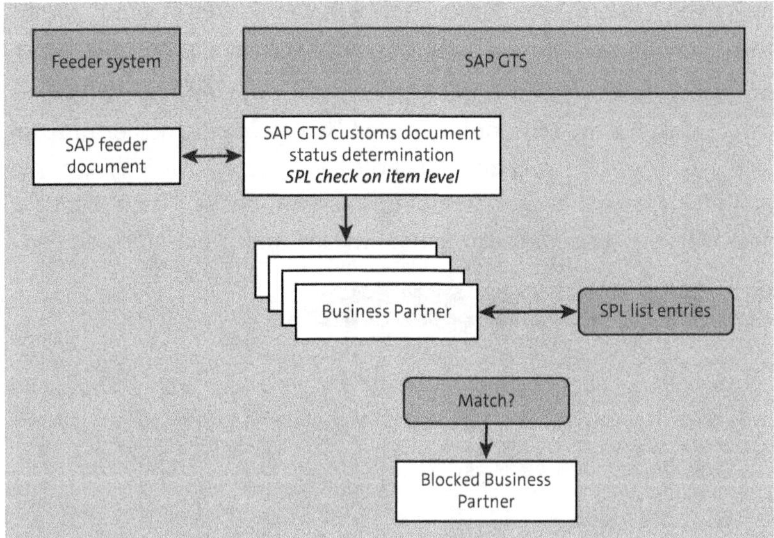

Figure 6.7 Logic for SPL Screening

Prior to the transactional screening, users can transfer the business partner master data from the feeder system to SAP GTS for screening. The SPL status is assigned to every SAP GTS business partner, and the partners maintained in a document only undergo a new screening (versus a status check) if there is a change in the document.

As soon as the sales order is created in Beta's feeder system, it's transferred to SAP GTS. Then the sales order is checked against the legal regulation for SPL screening. In this instance, Freelance Aviation Corporation results in a hit on the US SPLs, and the popup window in Figure 6.8 appears.

Reference Number of a Document from Back Error

Item Service	Status
471	Blocked / Checked
Customs Document	Blocked / Checked
10 Embargo Check	Released / Checked
10 Sanctioned Party List Screening	Blocked / Checked
10 Legal Control	Released / Checked

Figure 6.8 SPL Block in the Feeder System

As a result of the screening and displayed in the popup message, the order can't be processed due to an SPL compliance hit, and a block is placed on the subsequent activities (e.g., creating the outbound delivery). The shipping operator attempting to create the outbound delivery receives an error message stating that the order can't be delivered due to the SAP GTS block.

The trade compliance team at Beta monitors the status of the blocked business partner and relevant blocked documents in SAP GTS. Trade compliance personnel may display the SAP GTS log and process (i.e., analyze, release, or confirm) the block in SAP GTS.

Figure 6.9 displays the log for the sales order created by Beta. Each partner function (e.g., sold-to, ship-to, payer, etc.) is screened against the SPL lists. Because Freelance Aviation Corporation (business partner number **661** in this example) has been identified as an entry on these lists, the customs document is blocked for further processing.

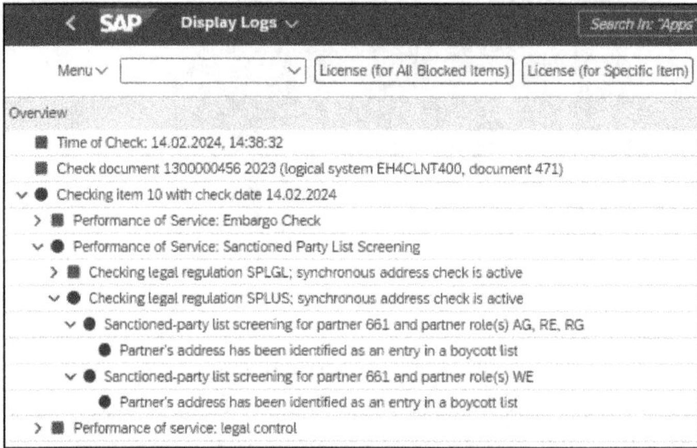

Figure 6.9 Customs Document Log

Users can configure SAP GTS to notify the responsible department via automatic email when an SPL block occurs.

Figure 6.10 displays the blocked Malaysian business partner as a result of the SPL screening, accessed via the **Manage Blocked Partners** app under the **Partner Screening** section.

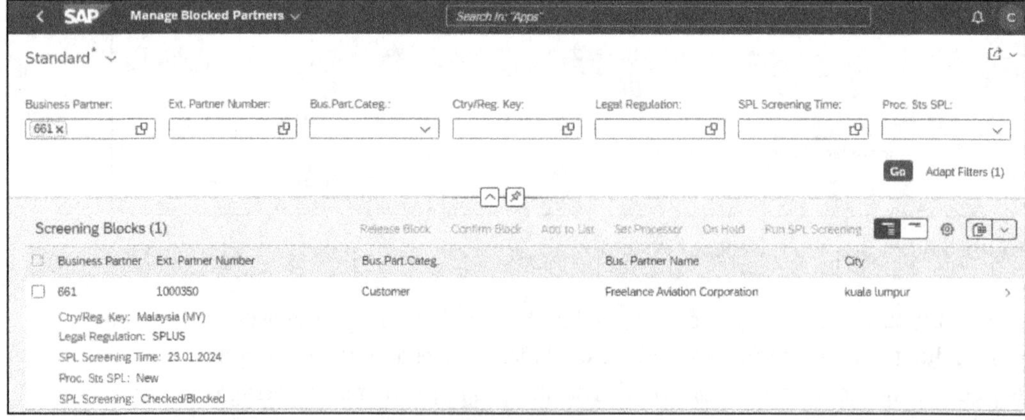

Figure 6.10 Business Partner Blocked for SPL Screening

The middle of the screen shows the different options available to process the blocked business partner:

- **Release Block**
  When a business partner is released, SAP GTS won't block this partner again in future transactions, unless there is a change to the business partner or a relevant change to the SPL triggers a new block on that business partner.
- **Confirm Block**
  This option may be used if a hit can be confirmed or identified as a true SPL match. Further processing of the document in the feeder system isn't possible, and the partner remains blocked.
- **Add to List**
  A business partner can be put on a positive or negative list. These partners are excluded from further checks and will keep that same status until a rescreening of that list is performed. The positive list indicates that the business partner is always released; the negative list indicates that the business partner is always blocked.
- **Set Processor**
  Here, users can forward the task to review the business partner to another individual, or users can adopt the block to add it to their list of blocks to analyze.
- **On Hold**
  If the person handling the release needs more time or further details to decide whether to release or confirm the block status, the business partner may be temporarily put on hold.
- **Run SPL Screening**
  This option is used to perform a new check or a rescreening of the business partner.

To further analyze the actual SPL block, navigate to the **Display Audit Trail – Partners** tile under the **Sanctioned Party Lists** space. Figure 6.11 displays the entry on the government's SPL that **Freelance Aviation Corporation** was identified against.

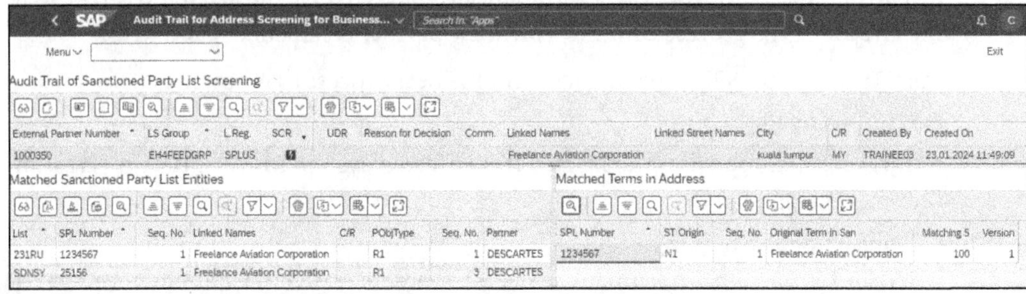

Figure 6.11 Analyzing SPL Screening

Additional processing options are available, such as adding comments to detail the analysis results or pass along information to the group handling the worklist of a blocked business partner.

Further processing options may be available for the customs document as well. Beta may view its blocked documents and release a particular document while postponing the decision about the blocked business partner by selecting the **Manage Blocked Documents** tile under the **Compliance** space. The release procedure can be managed in different ways. The business can detail a workflow and define specific release reasons (predefined or free-text options).

In the different business areas, SAP GTS provides functionalities to conduct asynchronous or periodic screenings and simulate screenings, along with different monitoring options for business partners and documents.

The transfer of the relevant master data to SAP GTS takes place via change pointers. Each time the user creates or changes a business partner name or relevant address field in the feeder system, a change pointer is updated, and a report selects the relevant partner address. The updated business partner is then sent to SAP GTS via the remote function call (RFC) connection. Depending on the SPL setup, users can perform one of the following two kinds of screening:

- **Asynchronous**
  With asynchronous screening, users will need to schedule batch jobs to trigger the screening. Otherwise, the screening must be initiated manually. In other words, the screening doesn't happen at the same time as the data is updated.
- **Synchronous**
  The system automatically triggers synchronous screening each time a business partner is created or changed in SAP GTS via the feeder system change pointer mechanism.

Users may select the type of the SPL block when using asynchronous screening. There are two options: (1) the process is automatically blocked until the partner is screened in SAP GTS, or (2) the process continues until the partner is screened in SAP GTS.

SAP GTS integrates SPL screening into the business process as follows:

- **Master data screening**
  Check business partners transferred from a feeder system to SAP GTS (after creation or modification).
- **Transactional screening**
  Screen transactional documents (e.g., sales orders, deliveries, purchase orders, or payments).
- **Simulate screening**
  Check new business partners prior to creation in feeder system environment.

Furthermore, the SPL screening functionality provides the following advantages:

- Provides flexible audit and monitoring mechanisms
- Screens against changes to government agency-specific lists when delta files are uploaded

Note

In this chapter, the term *business partner* represents all business area partners (e.g., vendors, account holders, and banks), and the term *document* represents the relevant business area documents (e.g., sales orders, purchase orders, deliveries, and payment documents).

The SPL screening can be performed on both the business partner level and document level, as follows:

- **Business partner or master data screening**
  The business partner or master data screening displayed in Figure 6.12 is triggered on newly created or changed partners. The screening result is saved in the SAP GTS SPL status information and subsequently retrieved for comparison in transactional screening.

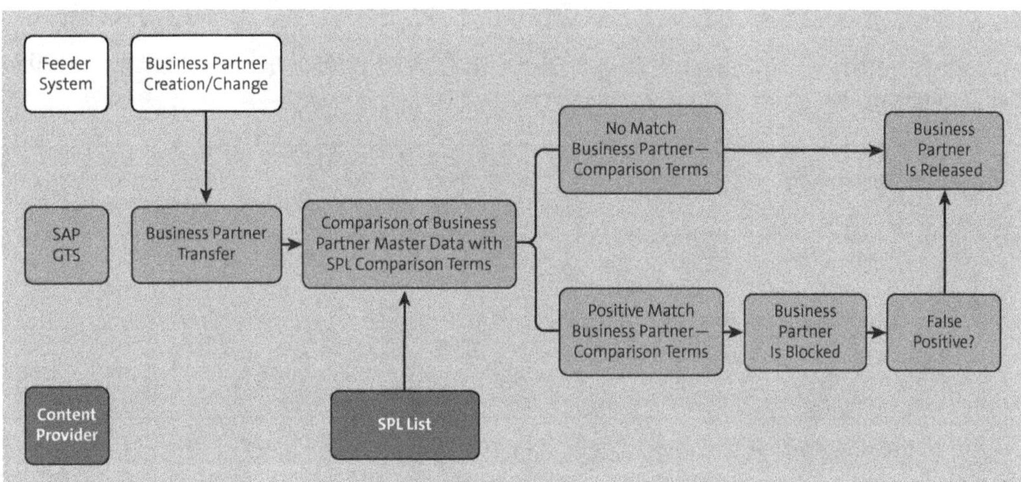

Figure 6.12 Business Partner SPL Screening

- **Document or transactional screening**
  In transactional screening, document partner addresses are compared with SPL entries. If the address hasn't been manually changed, a status is retrieved based on the latest master data screening result. If the address was manually changed since the last screening performed, a new screening takes place to determine whether the

address matches an entity on an SPL. The SAP GTS process for this scenario is displayed in Figure 6.13.

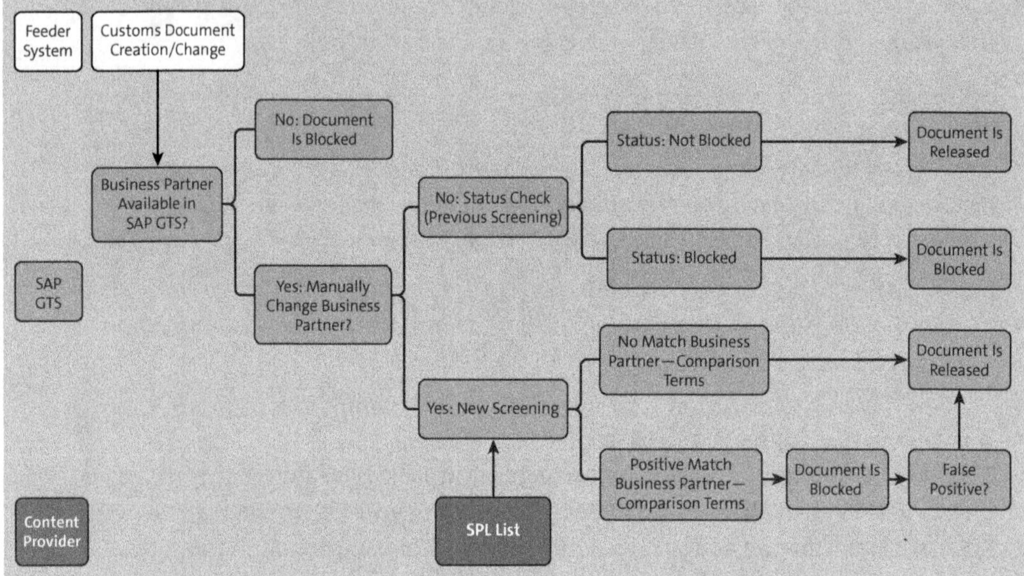

Figure 6.13 Document SPL Screening

The different SPL-related app tiles within the **Partner Screening** section are displayed in Figure 6.14.

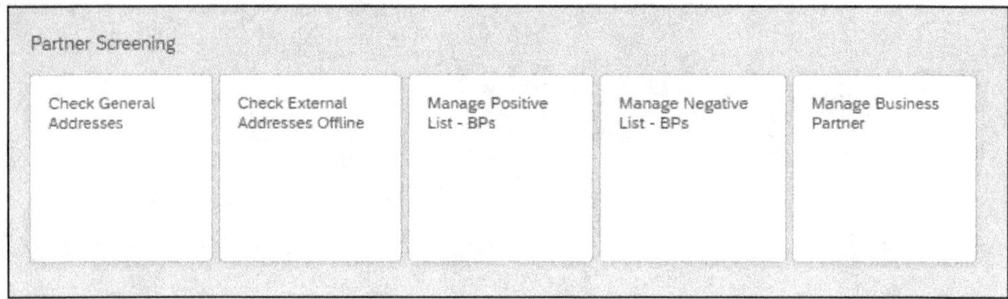

Figure 6.14 Partner Screening Tiles in SAP GTS

The compliance manager in our Belgian-Malaysian (Beta-Freelance Aviation Corporation) SPL entity scenario monitors the process from this area.

The type of scenario is based on two parameters:

- Screening of business partner address and/or document partner address
- Synchronous and/or asynchronous screening

Table 6.1 lists SAP GTS predefined scenarios. Understanding these scenarios is crucial to correctly screen business partners.

Scenario	Name	Data to Be Screened
A1	Business Partner Address Check	New/changed addresses in business partner master data
A2	Document Partner Address Check	New/changed addresses in document partner
B1	Periodic BP Address Check	Unchanged and already-screened business partner
B1(P), B1(N)	Positive/Negative BP Address Check	Business partners placed on positive/ negative list
B2	Periodic Document Partner Check	Unchanged and already-screened documents
C1	SPL Data Basis Change – Partner	Unchanged and already-screened partner (against new SPL entries—delta files)
C2	SPL Data Basis Change – Document	Unchanged and already-screened documents (against new SPL entries—delta files)
O3	Offline, External SPL Check	External XML file containing address data
S1	Simulation BP Master Data	One business partner master data
S2	Simulation Document Check	One document partner address
S3	Simulation Independent Address	Free address input

Table 6.1 SAP GTS Predefined Scenarios

It's considered a leading practice to use asynchronous screening during the initial transfer of master data (or other high-volume transfer activity) and to use synchronous screening for the ongoing transfer of new or changed business partners. Next, we'll look at each SAP GTS scenario in more detail:

- **A1/A2**

  Scenarios A1 and A2 are used in the initial transfer and changes of business partner master data. If a new business partner is created or changed (or an address of a business partner in the document is created or changed), these two scenarios are triggered, and the data is compared to the complete SPL. Partners that have been placed on positive or negative lists aren't considered in these screening scenarios.

  If users choose synchronous screening, this scenario is started in the background during transfer, and they don't have to set up batch jobs to initiate the screening. For asynchronous screenings, users have to trigger the screening scenarios via a batch job or trigger them manually.

- **B1/B2**
  The periodic SPL check allows users to rescreen all business partners and documents against the entire SPL database at regular intervals (e.g., on a quarterly basis). The partners that users have put on a positive/negative list are only considered in the specific B1 scenarios for these partners.
- **C1/C2**
  Business partners and documents are checked against the SPL delta files (i.e., only against changed or new SPL records). Existing blocked business partners or those placed on a positive or negative list aren't taken into account.
- **O3**
  Offline screening allows you to screen external master data files (in XML format) against the SPLs. The purpose of this scenario is to validate business partners from an external system. The result is a target file with the status of the business partner; there is no status change in the system.
- **S1/S2**
  The simulation functionality allows screening against only one business partner master data or one document. Users can use this scenario for testing purposes to check whether the address will result in a block.
- **S3**
  In this scenario, users can screen a manually entered address (without master data) against the SPL database.

The overall screening process should follow a specific sequence that users can partially run using batch jobs. The exact sequence depends on the selected screening algorithm. Consider the following sequence when the SAP GTS search screening algorithm is used:

1. The initial volume master data (technical go-live before business go-live) is uploaded, which includes the following steps:
   - Master data transfer, no document transfer: Documents aren't screened on transactional basis yet. Only the business partner master data itself will be screened against the SPLs in the SAP GTS environment (preparation activity during technical go-live).
   - Asynchronous screening: The master data is screened all together (not at business partner creation or change).
   - Manual: Execution of scenario A1 is performed manually.

   The action to take after these items have been executed is to evaluate the results and to release the relevant business partners, to confirm the SPL blocks, or to put the business partner on the positive/negative list (based on the different results of the evaluation).

2. Transactional screening takes place (business go-live), which includes the following steps:
   - Activate document transfer: At this point, the relevant documents are transferred from the feeder system to SAP GTS.
   - Synchronous screening: Via the change pointer mechanism and RFCs, automatic screening is performed on master data changes and document addresses.
   - Regular delta file upload: The changes in SPLs need to be uploaded directly into the SAP GTS environment.

As soon as these preparation steps are completed, the following scenarios can be executed to make the environment ready for further processing:

- Scenario C1: Business Partner Delta Screening
- Scenario C2: Document Delta Screening

When following this sequence, the screening should be in line with the latest published SPLs. In addition, a periodic screening process may be performed against all SPL entries.

6.2.3 Customizing

In this section, the Customizing settings for SPL screening will be covered. Note that all general settings discussed in Chapter 3, Section 3.3 must be correctly set up (document transfer, legal regulations) before this Customizing can be done.

We'll cover the following settings:

- **Define Reasons for Releasing Blocked Documents and Business Partners**
- **Sanctioned Party List Screenings**
- **Define Comparison Procedure**
- **Defining Identity Card Types for Ad Hoc Sanctioned Party List Screening**
- **Control Settings for Sanctioned Party List Screening**

Define Reasons for Releasing Blocked Documents and Business Partners

This Customizing step, found via **SAP Global Trade Services, edition for SAP HANA • Compliance Management • Define Reasons for Releasing Blocked Documents and Business Partners**, defines the reasons for releasing blocked documents and blocked business partners. Having predefined reasons for release expedites the release process and can be used in the audit trail. Different reasons can be chosen for SPL and embargo.

General Sanctioned Parties List Customizing Settings

The following sections cover the general SPL Customizing settings located in Transaction SPRO by choosing **SAP Global Trade Services, edition for SAP HANA • Compliance Management • Sanctioned Party List Screenings.**

Activate Business Partner at the Business Partner Function Level

Through this setting, users can define which business partner types will be screened, the timing and sequence of screening, the block type that is triggered, and the business area (i.e., logistics, finance) for which the screening takes place.

Users define the application object (e.g., BUPA – SAP business partner, or FICO – financial conditions) and the business partner role (e.g., SLLBNK – bank, or SLLCPC – customer) that is screened.

The timing of screening is a crucial decision. Choosing between the following two settings defines when SPL screening is performed if a new address is created or if an address has been changed:

- **0 Asynchronous (by calling the function/service)**
  Users schedule batch jobs to run the SPL screening or manually trigger the screening using scenario A1 (business partner address) and/or A2 (document address). The system sends the new or changed address data to SAP GTS and assigns a status, depending on the block type selection.
- **1 Synchronous (at update)**
  The system sends the address information to SAP GTS and screens it at runtime. This means the status is updated directly and available in the feeder system for further processing.

Assign Determination Procedure for Active Legal Regulations

Users must configure a procedure for the determination of active legal regulations in SPL screening. To configure this, users first define a determination procedure for active legal regulations under **General Settings** (see Chapter 3, Section 3.3.1). LESPL is the standard SPL determination procedure, predefined by SAP, and is based on the country level. Keeping this setting means the legal regulation used for SPL screening uses the country of the legal unit and the business partner.

Activate Legal Regulation

Table 6.2 shows options for the import and export side to activate the SPL legal regulations for multiple countries or country groups.

SPL Import Active	SPL Export Active
1 – Check: Receipt (exclusively)	1 – Check: Dispatch (exclusively)
2 – Check: Import (exclusively)	2 – Check: Export (exclusively)
3 – Check: Receipt/Import (excluding domestic)	3 – Check: Dispatch/Export (excluding domestic)

Table 6.2 Activation Options for SPL Screening

SPL Import Active	SPL Export Active
4 – Check: Receipt/Import (including domestic)	4 – Check: Dispatch/Export (including domestic)
Check: Not Activated	Check: Not Activated

Table 6.2 Activation Options for SPL Screening (Cont.)

Because SPL screening is typically intended for all transactions, option **4** is often deployed because it checks all import and export scenarios and domestic transactions. This means that all inbound and outbound flows for a specified country or country group are subject to SPL screening.

Define Comparison Procedure

In this step, accessed via **SAP Global Trade Services, edition for SAP HANA • Compliance Management • Define Comparison Procedure**, the comparison procedure can be customized to specific master data requirements. The comparison terms generated in the SPL screening are built based on the settings defined under this Customizing step.

There are two possible screening algorithms to choose from, each with its strengths and limitations: SAP GTS search and SAP HANA search. Note that TREX search is no longer available in SAP GTS, edition for SAP HANA.

SAP GTS Search

The main benefit of this algorithm is its flexibility. It provides the user complex but extensive configuration options, from which the algorithm can be customized to specific master data requirements. An important issue is the "word-in-word" limitation, which requires the algorithm to find the full business partner name in the sanctioned party entity or vice versa, leaving limited error margin for manual/spelling mistakes when creating the master data.

SAP HANA Search

This is the most recent screening algorithm that was made available. The control structure of the SAP HANA search requires a slightly different approach during configuration but has proven to increase performance. SAP HANA search must be run on an SAP HANA database.

Compared to the standard SAP GTS search, the SAP HANA search is less flexible for customization. The algorithm allows users to select a fixed set of predefined screening fields, which makes configuration simpler but limited. The SAP HANA search also allows hits on solely name or address elements, which isn't possible with previous search algorithms.

Defining Identity Card Types for Ad Hoc Sanctioned Party List Screening

This setting, accessed via **SAP Global Trade Services, edition for SAP HANA • Compliance Management • Defining Identity Card Types for Ad Hoc Sanctioned Party List Screening**, defines the types of identity documents in a scheme. Users can define these documents as optional or mandatory information during a simulated screening of general addresses (scenario S3). Users must activate the ID entries in the control settings of SPL screening. Users can also define a scheme for the ID card types and then assign the desired types to use in that scheme, which groups the ID types for a specific legal regulation.

Control Settings for Sanctioned Party List Screening

This Customizing activity sets forth the control settings for the legal regulations intended for use in SPL screening. The first step to complete the **Control Settings for Sanctioned Party List Screening** transaction is to define the control settings for the legal regulations. This step can be found by navigating to **SAP Global Trade Services, edition for SAP HANA • Compliance Management • Control Settings for Sanctioned Party List Screening**.

General Control Settings for Legal Regulation

Figure 6.15 displays the following different control settings to be completed for the legal regulations:

- **Control Settings for SPL Screening**
  Define procedures that determine how SPL screening compares address data.
- **Document Control**
  Assign a control procedure and partner groups for each legal regulation used in SPL screening, both for import and export activities.

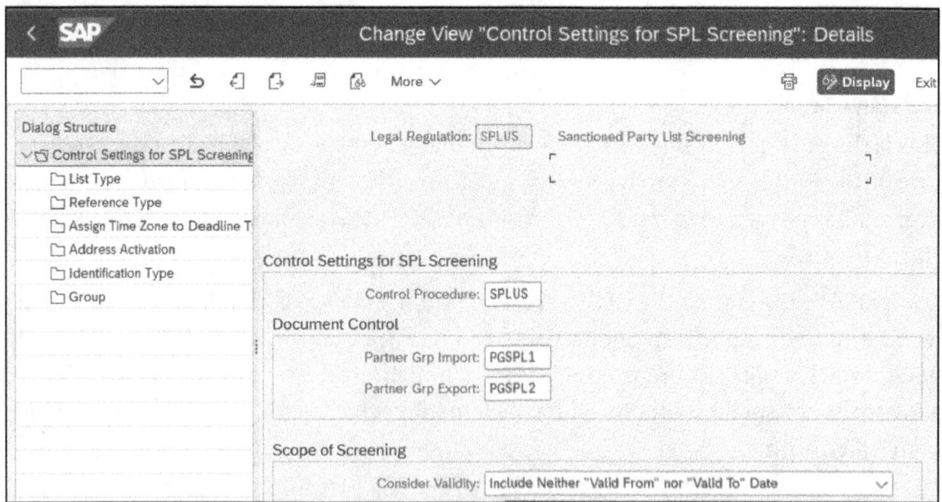

Figure 6.15 Control Settings for SPL Screening for Legal Regulation (Part 1)

- **Scope of Screening**
  Consider the validity periods for SPL master data during screening, and if the validity period has expired, flag the corresponding SPL entry for deletion. Users can also include SPL entries flagged for deletion in the generated comparison index and consider the address validity for screening.
- **Audit**
  Set the usage of the audit trail, which is an important function in SPL screening for both external and internal audit/reporting purposes.

The **Scope of Screening** section shown in Figure 6.15 contains the following important checks:

- **Compare Address Elements with Each Other**
  This cross-check indicator has a significant influence on the overall performance of the SPL functionality. If this indicator is active, address elements are compared with one another regardless of the address section (e.g., comparing street data in an SPL entry with the name entry in a business partner address).
- **Take Past Screening Result into Account**
  Considering past SPL screening results may improve the performance of your SPL configuration. The SAP GTS functionality defines whether the SPL screening is rerun with similar circumstances or if the results from the last screening run are used. All criteria for the business partner master data and the SPL master data must be the same as the preexisting screening parameters; otherwise, new checks will be performed.
- **Multiple Address Screening Active**
  This setting enables users to screen business partners with several address types in the business partner master data. Users must define the address activation level at which the address should be checked in the SPL screening.
- **Screening Logic Docs**
  The check logic for documents provides two options:
  - **Status Check of Unchanged Partner Addresses (Partner Address)**
  - **New Check of Changed Addresses**

  By choosing the first option, a manual change of the document partner address will trigger a block in SAP GTS SPL screening, stating that the **Partner's address was manually changed**. The system performs checks of the business partner master data via a status query for unchanged document addresses of the business partner.

  By choosing the second option, the document addresses are screened synchronously when changes from the master business partner are detected.

The selected setting influences the logic behind the check and the resulting block of the screening types as follows:

- Status query of business partner screening causes a block.
- Changed document address causes a block.

- **Search Strategy**
  The search strategy provides two options:
  - **SPL Entry Searches in Address Entry**
  - **SPL Entry Searches in Address Entry and Vice Versa**

  The search strategy setting in the control procedure Customizing steps provides a choice of logic to compare addresses. The search strategy setting provides a solution to the constraint that the target word must be longer than the source word.

  Selecting the second option removes the constraint, meaning it's no longer important if one word is longer or shorter than the other. This setting is adjusted in the first screen, **Define Control Procedure for Address Comparison**, where it determines which terms are matched and in which direction.
- **Scr. Pkg. Sze**
  SPL screening is performance intensive; dividing the data into packages can provide the following performance improvements:
  - Minimizes runtime for screening and update
  - Optimizes use of buffer mechanisms
  - Optimizes use of the queue server
  - Prevents memory overflows

  SPL screening is performed for each package and the result is updated in the database. Due to buffering mechanisms, the more partners included in one package, the shorter the runtime for each object. Other processes may be affected due to locking partners and documents, so there's an increased risk of memory overflow with the package size. The default setting is 1,000, but the system and master data properties determine and affect which value should be applied.

Next, we'll discuss the relevant settings in the **Audit** section:

- **Business Partner Screening**
  This setting confirms the audit trail is updated during SPL screening for business partners.
- **Document Screening**
  This setting confirms the audit trail is updated during SPL screening for documents.
- **Simulation Screening**
  This setting confirms that an entry is logged for business partners that are screened in simulation modes (S1 or S2). Users can also log an entry in the audit trail for an ad

hoc screening simulation of general address data (S3) or through a screening performed on external addresses from an XML file (O3).

There are a few remaining control settings that must be completed, as shown in Figure 6.16:

- **System Control**
  Users can improve SPL check runtimes by storing the comparison index in the application server buffer. This way, running an SPL check, users can read the SPL comparison index quickly in the application server rather than from the database. Frequently run SPL checks can significantly improve response times. Note that response times may be slow if the application server buffer isn't large enough.

  Another way to improve the performance of SPL checks is to store the SPL comparison index in summarized form in a cluster table. The system can access this cluster table more quickly than a traditional transparent database table.

  The **Time Initialized** setting defines when the generated comparison terms are initialized, that is, when they are reset to allow new comparison words to be created. This is necessary if the SPL logic changes or new terms are generated.
- **Notification**
  This setting activates the email notification workflow. The system automatically sends an email if a business partner or document obtains a blocked status during an SPL check.

  There are two available options:
  - **Send Mail When Document Is Blocked**
  - **Send Mail When Bus. Partner Is Blocked**

  Users can configure the email settings in the communication settings of SAP GTS and further define user groups emailed per FTO and legal regulation.
- **Incompleteness Check for Transit and Customs Processing**
  The incompleteness check for transit and customs processing is linked to the Customs Management functionality in SAP GTS. When the mirrored feeder system document is blocked after a compliance check in SAP GTS (in this case, an SPL block), a message is categorized in the incompleteness log of the customs declaration. If the message is added to the Customs Management functionality incompleteness check as an error message, users can't send the customs declaration to the customs authorities until the underlying document is released in Compliance Management.
- **Ad Hoc SPL Screening**
  Finally, users can also activate IDs, specified in an earlier step. The ID settings for SPL ad hoc screening are **Optional**, **Mandatory**, or **Inactive**.

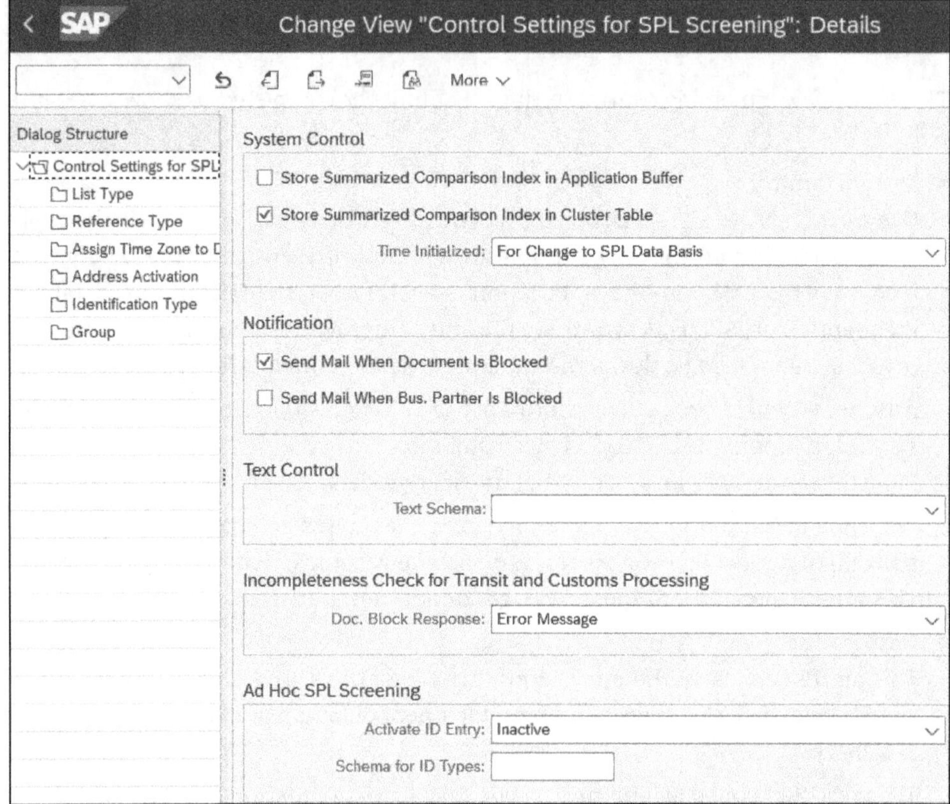

Figure 6.16 Control Settings for SPL Screening for Legal Regulation (Part 2)

List Type

This setting defines the list types (e.g., specially designated nationals [SDNs]) relevant for a company's legal requirements. Certain list types are predelivered in the standard legal regulation SPLUS, but other list types would be defined here as well.

For each list type, users must define several parameters that influence the comparison term generation. Defining the license type description and assigning a government agency (defined as a specific business partner type in SAP GTS, e.g., SLLAUT Government Agency) is optional.

The **Maximum Number of Index Entries per Object** setting defines how many terms are generated for each name/address of a master data entry. Though the quality of SPL screening results will improve with the number of search terms generated, the increased load on the system can present performance issues. Both the quality of screening and system load pressure depend on the master data in use and the system sizing, so users should consider performing several tests to identify the optimal settings.

There are also text fields where users have the option to add the legal regulation behind the SPL and additional comments. These fields don't influence the SPL screening logic and are only informational in nature.

Reference Type

Users can define reference types for each legal regulation and each name in an SPL master data record as a reference name. To use this functionality, define the reference types in the master data (e.g., "also known as" becomes "AKA" and "formerly known as" becomes "FKA").

Assign Time Zone to Deadline Type

Time zones are defined per legal regulation and allow a time zone adaptation per country or region. Time zones affect validity periods and inclusion dates for compliance functions.

Address Activation

Users can maintain possible usage types for addresses. When maintaining business partners, the address usage function can assign business partner addresses to address types (e.g., correspondence address and delivery address). The address types differ from the address versions in that the address version is connected to the language setting, and the type is connected to the type of address.

6.2.4 Master Data

SPL-related master data is maintained in a dedicated set of SAP Fiori tiles. Users can access the master data via **Master Data – Sanction List** under the **Sanctioned Party Lists** space, as shown in Figure 6.17. The following tiles are available:

- **Manage Sanctioned Party List Master**
- **Display Sanctioned Parties**
- **Display Expiring Sanctioned Parties**
- **Display SPL Change History**

To effectively use the SPL logic defined in Customizing, users need to define certain master data elements in the **Master Data** apps in SAP GTS.

We'll look at two key master data settings that need to be maintained: the SPLs upload functionality and control settings for comparison term generation. Note that within SAP GTS, these functionalities remain within the SAP GUI. They can be found by navigating to **Compliance Management: Sanctioned Party List Screening • Master Data**.

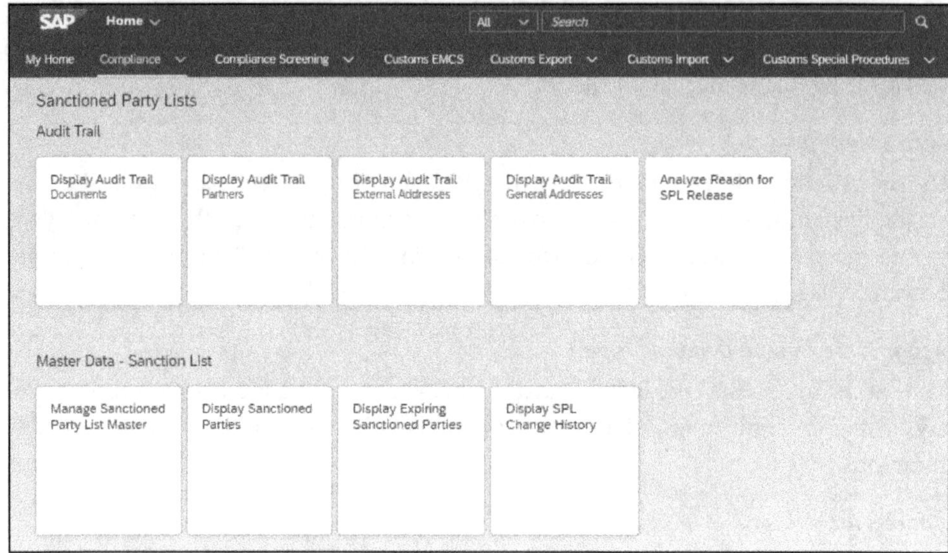

Figure 6.17 Master Data – Sanction List Tiles in SAP GTS

Sanctioned Party Lists

Typically, content providers publish a master file including the SPL entries that exist at that point in time. Users then update the master data file using delta files, which include only new and changed entries. Users can automate or manually upload the purchased XML lists to create unique, internally generated restricted party lists by defining a specific list type for the relevant legal regulation.

Note that users can't manually add or update entries in the previously uploaded lists when uploading an SPL. Users must maintain a separate list for these entries in which the address and name details to screen against can be manually added. The master data purchased from various content providers is uploaded via XML files. Most of the data providers provide specific user guides on how to handle the initial upload and subsequent update files.

This transaction can be accessed via **Compliance Management: Sanctioned Party List Screening • Master Data • Sanctioned Party Lists**. To verify that uploads were performed correctly, users can search the different monitor logs.

Once the SPL content has been loaded in SAP GTS, users can navigate to the **Master Data – Sanction List** section (refer to Figure 6.17) to manage and display the loaded content.

Control Settings for Comparison Terms for SAP GTS Search

The general control settings include the following different details users should consider defining per comparison procedure. They can be accessed via **Compliance**

Management: Sanctioned Party List Screening • Master Data • Comparison Terms. Let's walk through the main settings:

- **Delimiters**
  With delimiters, users can define characters and/or numbers to be used as separators in the SPL screening logic. If a separator is used, the algorithm splits the name into separate words, and matches can be recognized.

  Most companies start with standard delimiters, such as (, : ; / \ ( ) ¨ [ ] {} ? . -), which are adapted to each specific master data requirement. Take, for example, mother-in-law. If the hyphen (-) is defined as a separator in the delimiter, the algorithm splits the word into three words for SPL match determination.
- **Excluded texts**
  Excluded texts are ignored during the SPL check. The words used depend on the master data of each company, industry sector, and relevant business partners. Some examples of possible exclusion words are AG, AVE, GROUP, LLP, ORG, RD, AND, INC, LT, THE, LP, AB, ON, BANK, OF, MBH, PTE, CORP, ASSOCIATION, FOR, INDUSTRIES, PLC, PT, and STREET.
- **Alias terms**
  SAP GTS allows the replacement of terms to defined aliases before performing SPL screening. The alias basis defined in this setup is replaced by the term defined in the **Alias** field in the SPL screening (e.g., *Dan* for *Daniel* and *Ted* for *Theodore*).

Depending on the language used, users should consider replacing letters or strings of letters with other letters or combinations. In Germany, for example, the letter *ö* could be replaced with *oe*. The same applies for French words, where *é* and *è* could be replaced with *e*.

Control Settings for SAP HANA Search

The control settings for SAP HANA search contain the following two configurable steps. Both can be defined per comparison procedure.

- **Term mapping**
  This step is comparable to the alias terms that can be defined for SAP GTS search.
- **Stop words**
  Stop words can be compared to exclude texts for the SAP GTS search. The impact heavily depends on the business partner master data. If a stop word is maintained in the settings and is present during the screening, the impact of that word on the minimum score will be lower than the other matching words.

  Delimiters as shown for SAP GTS search are now standardly set up for SAP HANA search.

Note

SAP GTS can screen non-Latin characters. In this way, for example, Chinese character and Cyrillic alphabet screening is supported.

6.3 Embargo

Destination restrictions, such as embargoes, refer to restrictions on trade with certain countries ranging from a comprehensive embargo, targeted financial sanctions, or more limited arms (i.e., weapons) embargo. Government agencies administer and enforce comprehensive embargoes and sanctions, such as UN Security Council (UNSC) resolutions against certain countries. Because embargoes and sanctions are imposed to address a specific political situation or foreign policy objective, each program can vary extensively.

Embargo and sanction programs may cover imports, exports, and/or financial transactions. They may block assets or target specific individuals, entities, or commodities. They may apply to foreign countries, as well as nationals and legal entities from those countries, and may extend to imports or exports of services. They may prohibit or restrict exports and reexports, financial transactions, and travel to certain countries.

The terms *embargo* and *sanction* technically have different meanings, but they are often used interchangeably to refer to controls prohibiting trade and transactions with certain countries. A comprehensive embargo prohibits the export of *all* goods and services to certain countries with a few limited exceptions, such as trade in certain eligible medical or agricultural goods to certain end users. A sanction (or partial embargo) has a narrower scope than an embargo and restricts the sale of specific materials and services to particular countries or parties within a country.

Embargoes and sanctions can take on a variety of forms, although the most common forms include the following, ordered from least to most restrictive:

- **Prohibitions on financial transactions**
  Prohibitions may be put in place for transactions with certain persons related to certain countries, including restrictions on investment, payments, or capital movements; the freezing of assets and economic resources; or the withdrawal of tariff preferences that are designed to target specific persons, groups, or entities by freezing their funds and economic resources.
- **Arms embargoes**
  Arms embargoes are applied to keep arms and military equipment from reaching conflict areas or regimes that are likely to use them for internal repression or aggression against a foreign country. Arms embargoes normally involve the prohibition of two things:

  - Selling, supplying, transferring, or exporting arms or related material of all types, including weapons and ammunition, military vehicles or equipment, paramilitary equipment, or specific spare parts
  - Providing financing or financial assistance, technical assistance, brokering services, or other services related to military activities, or the provision, manufacture, maintenance, or use of arms or related material of all types

- **Restrictions**
  Restrictions on the export of strategic goods may be applicable, including prohibitions or restrictions on the export of controlled items to sensitive countries.
- **Comprehensive trade embargoes**
  Finally, trade embargoes may include the prohibition of all exports to and specific imports from a particular foreign country.

As we did with SPL, let's walk through the business and system processes for embargoes, before explaining the Customizing and master data.

6.3.1 Business Process

Belgian company Beta receives an order for one of its products from a new customer based in Syria. At the time of the order, the sales representative in Belgium faces an immediate issue: Syria is part of the embargoed country list.

Figure 6.18 indicates what Beta's export process looks like. In this scenario, the export shipment can't occur because Beta isn't allowed to export products to any customer in Syria.

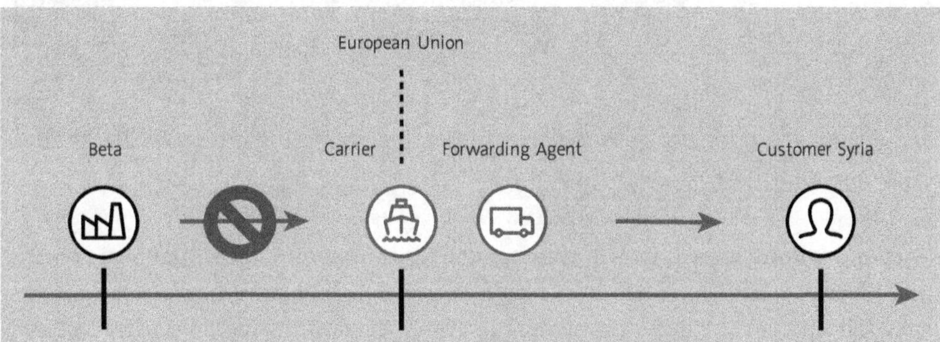

Figure 6.18 Embargo Export Business Process

6.3.2 System Process

Embargo checking in SAP GTS is used to confirm the system blocks all documents containing business partners from embargoed countries.

Embargo screening therefore works on the basis that the check is performed in relation to the business partner's country/location and in relation to the relevant prospective transaction. As such, the embargo functionality in SAP GTS only considers the country/location of the customer or supplier, and not the material being transacted. For embargo screening that includes product information, the license determination functionality can be used in SAP GTS. This functionality will be described in Section 6.4.

Figure 6.19 shows the feeder system documents and SAP GTS checks performed for an outbound process flow. Similar checks are applicable on the inbound process flow. Because Beta is trying to sell a product to a Syrian customer, SAP GTS will produce an embargo block.

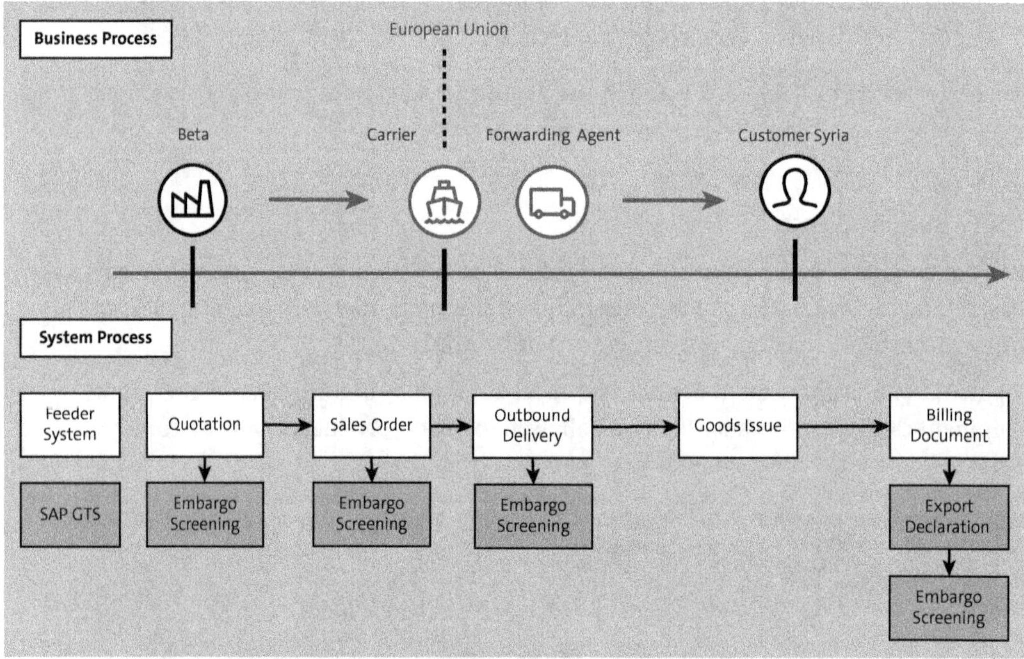

Figure 6.19 Embargo Export System Process

When quotations, sales orders, outbound deliveries, and other documents are created in the feeder system, they are transferred to SAP GTS, and the embargo check is executed. This process includes the screening of the relevant business partners' country, which is maintained in the system (master data). If a business partner's location is on the SAP GTS embargo list, SAP GTS blocks the feeder system document, which also prevents the creation of any subsequent documents in the feeder system.

Returning to our case study, a sales employee at Beta is trying to sell a non-export-controlled kit to a customer in Syria. The employee receives a popup in the sales document identifying the block. The message in Figure 6.20 notifies the feeder system user that the document is blocked by SAP GTS. Unless a trade compliance officer approves the sale by releasing the document in SAP GTS, it won't be possible to create the outbound delivery in the feeder system and ship the non-export-controlled kit to Syria.

Reference Number of a Document from Back Error

Item Service	Status
472	Blocked / Checked
Customs Document	Blocked / Checked
10 Embargo Check	Blocked / Checked
10 Sanctioned Party List Screening	Released / Checked
10 Legal Control	Released / Checked

Figure 6.20 Embargo Block in the Feeder System

As soon as one of the business partners in a feeder system document (e.g., the customer, payer, freight forwarder, etc.) is identified as located in an embargoed country, that identification causes a block on the customs document.

This section explains in detail how the embargo status for all items in the SAP GTS document is determined based on the business partners used in the document and the existing master data settings for embargo in SAP GTS. As with SPL screening, users can't perform further processing steps if documents are blocked for embargo reasons, but they can take steps to analyze and possibly release the block imposed on the document.

Users can access a worklist of blocked documents via the **Manage Blocked Documents** tile under the **Compliance** space. Note that the selection screen for this transaction in Figure 6.21 is used across the various Compliance Management functionalities; users can apply filters on documents blocked for SPL, embargo, or legal control reasons.

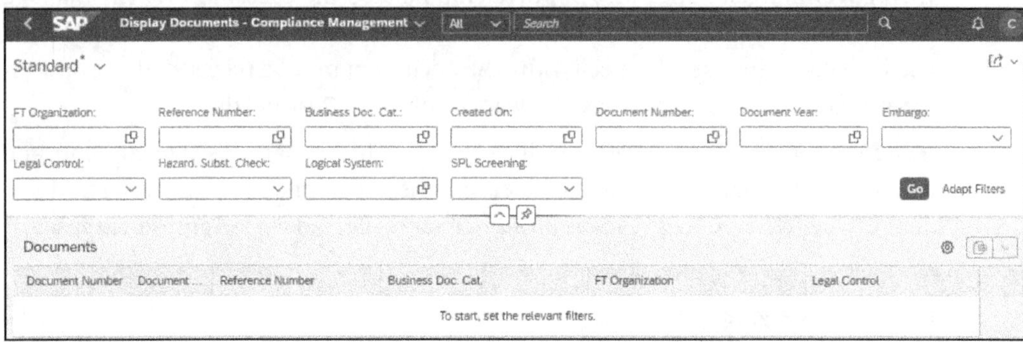

Figure 6.21 Selection Screen for Displaying Blocked Export Documents

For example, using the feeder system's document reference number as a search criterion in the selection screen, Beta can easily retrieve the blocked document. Figure 6.22 shows the result of executing the transaction and depicts how the sales order is cleared for all but the embargo check.

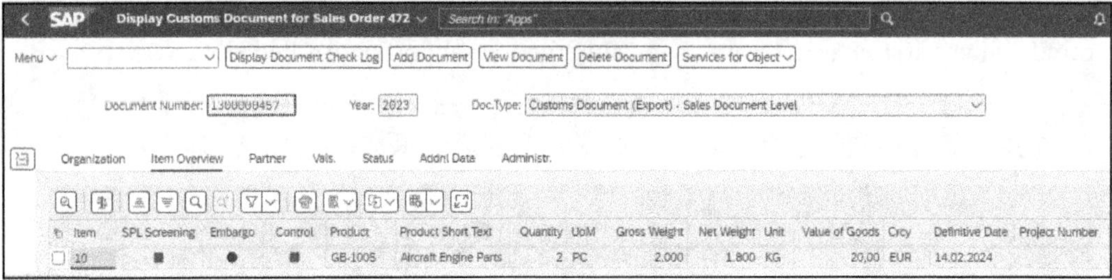

Figure 6.22 Customs Document Blocked for Embargo Reasons

Users can access the document's log to further analyze the reason for the imposed block. Figure 6.23 shows the corresponding log for Beta's sales order.

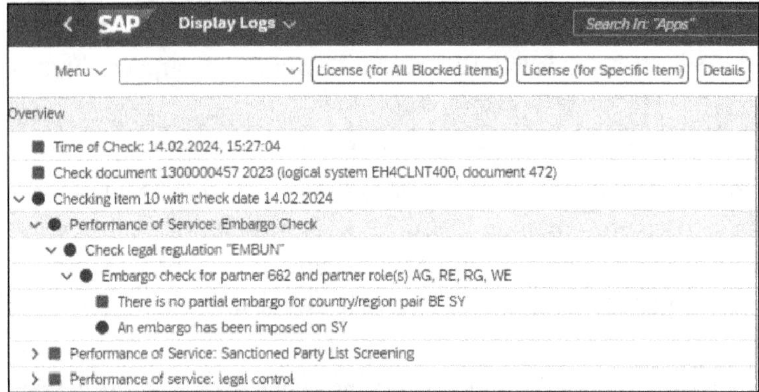

Figure 6.23 SAP GTS Customs Document Log Describing the Embargo Block

Users are allowed to release blocked documents in specific circumstances, such as when governmental authorities approve shipment to the customer, via the **Manage Blocked Documents** tile under **Compliance Management**. A dropdown list and/or text field is available for users to specify why the document may be released; these options correspond to acceptable reasons of release outlined in Customizing.

As soon as the document is released, it's placed on the list of released documents and included in the **Analyze Reasons for Release** transaction for audit purposes. The document can be retrieved from the **Display Documents** tile under the **Compliance** space.

6.3.3 Customizing

After users have taken the baseline Customizing steps, identified the general settings in Compliance Management (including defining the applicable legal regulations for embargo purposes), and defined the release reasons as explained previously in Section 6.2, they may apply the three settings in **SAP Global Trade Services, edition for SAP HANA • Compliance Management • Embargo Checks**:

- **Assign Determination Procedure for Active Legal Regulations**
- **Activate Legal Regulations**
- **Control Settings for Embargo Check**

This assumes that the relevant document types have also been flagged for compliance in the feeder system plug-in. Next, we'll look at each of these settings.

Assign Determination Procedure for Active Legal Regulations

In this activity, users specify which determination procedure to use to determine the active legal regulations for embargoes. The relevant legal regulations are determined based on the destination and/or departure country.

Standard SAP GTS defines the default determination procedure as LEEMB, which consists of the following determination strategies:

- Determine legal regulation at the country level
- Determine legal regulation at the country-group level

Depending on the rule option selected, the system will run both determination strategies or stop after the first strategy that results in a successful legal regulation.

In standard SAP GTS, the **Determination Using First Successful Strategy** option is selected, which means that when a legal regulation can be found using the country level, the country-group level isn't triggered.

Activate Legal Regulation

Activating legal regulations for embargo purposes is done similarly to the corresponding Customizing step for SPL screening, described in Section 6.2.3. This step defines whether the legal regulation is active for the various combinations of directions based on the countries or country groups defined in the baseline settings.

By adding countries to this table, users can specify whether the selected legal regulation is active on the country or country-group level. Depending on those settings, the legal regulation is triggered for both import and export, as well as from and to the specified countries or country groups.

Control Settings for Embargo Check

After the embargo regulations for which a check should be performed have been defined, users define the business partners to be checked by assigning partner groups to the legal regulations used in import and export transactions, as noted in Figure 6.24.

SAP GTS has predefined default partner groups that can be changed based on specialized business requirements. For more details on how to define partner groups, refer to Chapter 3, Section 3.3.3.

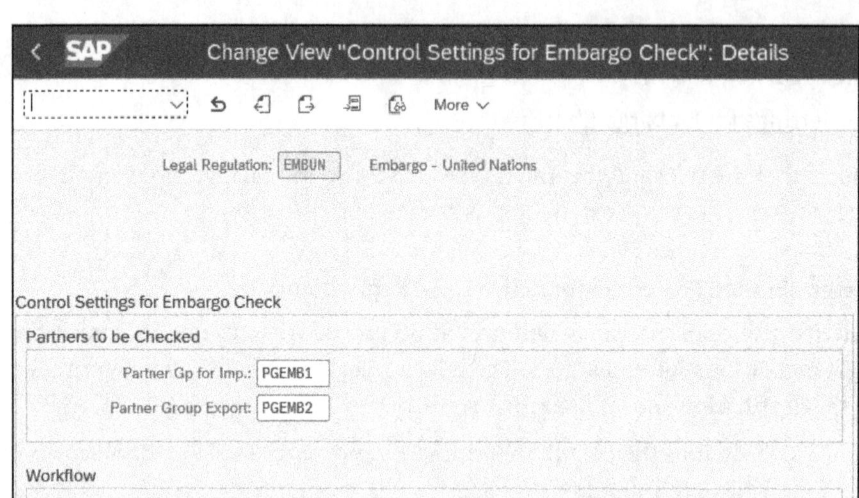

Figure 6.24 Embargo Check Control Settings

In the **Workflow** step, users can define whether to receive an email notification when a document is blocked with the **Mail for Doc. Block** setting.

6.3.4 Master Data

The master data related to embargo checks is limited to the combinations of countries and legal regulations for which an embargo block must be imposed.

There are three possibilities for entering master data to impose embargoes, which can all be accessed via the **Manage Embargo Situations** tile under the **Embargo Checks** section:

- **Embargo on country level**
  Set an embargo on a particular country that is valid for all possible cases. This implies that transactions to or from this country are blocked, regardless of the departure or destination location of the business process flow. Note that users can set the embargo flag for a specified validity period, as necessary.
- **Embargo on legal regulation and country of destination level**
  Set an embargo on a country that is valid for only a particular legal regulation. Users can specify different embargoed countries for each legal regulation. Standard SAP GTS setup includes one legal regulation for embargo, but various business requirements may dictate otherwise.

- **Embargo on legal regulation and country of departure/destination**
  Set an embargo on a country that is valid for only a legal regulation and a country of departure and/or destination. This option allows the flexibility to define embargoed countries at the greatest level of detail.

Figure 6.25 gives an example of the third scenario. The entry implies, for example, that a shipment from Belgium (**BE**) to Syria (**SY**) includes an embargo block if the legal regulation **EMBUN** is triggered.

6

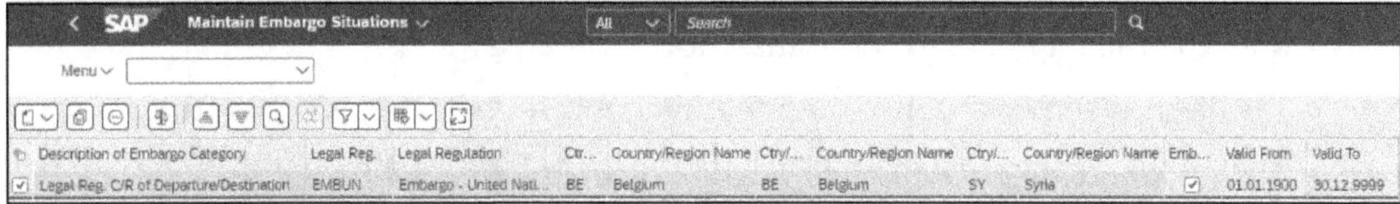

Figure 6.25 Embargo Set by Legal Regulation and Countries

6.4 Legal Control

In most jurisdictions, importing and exporting certain sensitive items and technology is controlled by the relevant government authorities. The export control classification of such items and technologies, their end use, the intended end user, as well as the destination and departure country determine whether an import or export authorization is required.

Licenses/authorizations are granted on a case-by-case basis and can cover a single or multiple transactions. License/authorization determination requires an understanding of all the facts related to a prospective transaction, including an item's technical characteristics.

Prior to exporting (or in certain cases, importing) items and technology in accordance with national import and export control legislation, the exporter or importer must establish whether a license/authorization is required for the prospective shipment or technology transfer. The following five elements should be examined to establish whether an export license/authorization is required:

- **What is the item being exported?**
  Before an item can be exported or reexported to a permissible end user, the item must have a proper *export control classification number* (ECCN; export control code) assigned to it, which identifies what the item is and determines the related controls.
- **Where is the item going to?**
  The country of ultimate destination, transshipment locations, export, or reexport are important factors in determining export licensing requirements. Exports to some countries are more restricted than exports to other countries.

- **Who will receive the item?**
  The ultimate end user of the exported item can't be a restricted or prohibited end user (in some cases, a specific authorization can be obtained depending on the citation-specific requirements).
- **What will the end user do with the item?**
  The ultimate end use of the exported item can't be a restricted end use.

The inbound and outbound license/authorization determination process in SAP GTS is the compliance process that checks the existence of the appropriate license/authorization when an import or export transaction is initiated.

The license/authorization determination process validates whether the required license/authorization is available for the specified product and transaction. There are numerous options for configuring SAP GTS to determine and assign the applicable license/authorization to the product and associated transaction. One option is to automatically assign transactions to the appropriate license/authorization when available. Alternatively, SAP GTS can be configured to allow for a user to manually assign the appropriate license/authorization to a blocked line item on an order or delivery.

The determination is performed when the applicable feeder document type—such as a sales order, purchase order, or outbound and inbound delivery—is created and transferred to SAP GTS. If a line item on a sales order or purchase order is blocked for a missing license/authorization, SAP GTS can be set up to block the creation of the subsequent document in the feeder system document flow. For example, when the user creates a sales order with two items (e.g., one with the relevant license/authorization in place and one without), the user can create a delivery note for only the released/unblocked line item.

Given the broad number of configuration settings available in SAP GTS, a user can define various rulesets to authorize or block prospective goods movements. For example, some country-specific encryption controls require a license/authorization when sending encryption items to government end users. A user can also configure SAP GTS to address end-use controls and determine export license/authorization requirements for entities that may be involved in restricted military- or weapons-based programs.

Many companies also use SAP GTS to determine license requirements related for other regulatory obligations, such as drug precursors or government agency requirements related to medical, chemical, food and drug agency authorizations. Thus, when considering an automated process to manage a specific regulatory requirement that requires specific transactional permits, authorizations, licenses, and other related requirements, user should review SAP GTS across all its capabilities and build it accordingly.

This section presents the multiple benefits of the SAP GTS legal control functionality. We'll begin with an overview of the business and system process flows for the standard licensing functionality, including the automatic bill of materials (BOM) explosion, and then describe the reexport functionality and intercompany sales screening. We'll close with details on the Customizing and master data.

6.4.1 Import Licensing

An import license/authorization may be required prior to an import transaction being executed. This section will use the example of the Beta company to introduce import licensing. Via this example, the business process will be linked to the system process.

Business Process

In this business process scenario, Beta exports defense articles to a US vendor and reimports them after necessary modifications have taken place. The import licenses/authorization discussed in this section are required by the US vendor to import the products shipped by Beta. This implies that the US vendor also uses SAP GTS to manage its import process. This process is depicted in Figure 6.26.

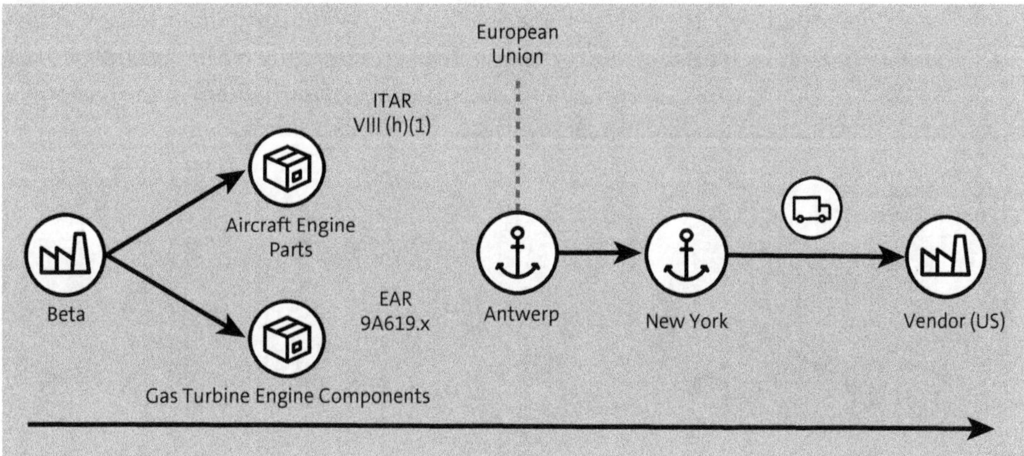

Figure 6.26 Import Business Process of Defense Articles into the United States

The imported articles include aircraft engine parts and specially designed gas turbine engine components. Because the aircraft engine parts are for military applications, they are subject to the jurisdiction of the US Department of State's International Traffic in Arms Regulation (ITAR). The US Munitions List (USML) is part of the ITAR legal regulation, containing all restricted military products that are heavily monitored and subject to license requirements. More specifically, the aircraft engine parts are classified in the USML list as VIII(h)(1) products.

Furthermore, the gas turbine engine components temporarily exported by Beta aren't considered military products but are considered dual-use products. In our case, the gas turbine engine components are a direct result of US technology and are therefore subject to the Export Administration Regulations (EAR), which covers the control of dual-use products from a US export controls purposes. To classify products in the EAR regulation, products are assigned a specific ECCN (export control code). The gas turbine engine components are classified with export control code 9A619.x.

When importing goods from third countries, a customs import declaration must be filed with the relevant customs authorities (see Chapter 7 for further information on customs processes). In addition, certain items also require an import authorization. In our example, the import would be a temporary import into the US of a military item. Therefore, Vendor A (the US vendor) decides to apply for a DSP61 license to cover the temporary import of the aircraft engine parts.

After Vendor A receives the applicable license from the US Directorate of Defense Trade Controls authorizing the import of the aircraft engine parts, it notifies Beta to continue with the export process.

System Process

The US vendor temporarily imports two types of products from Beta in Belgium: aircraft engine parts and specially designed gas turbine engine components. The SAP GTS system process for handling this scenario is displayed in Figure 6.27. For simplicity reasons, and to avoid the complexity of setting up subcontracting flows in the feeder system, we're using a standard purchase order to cover this scenario.

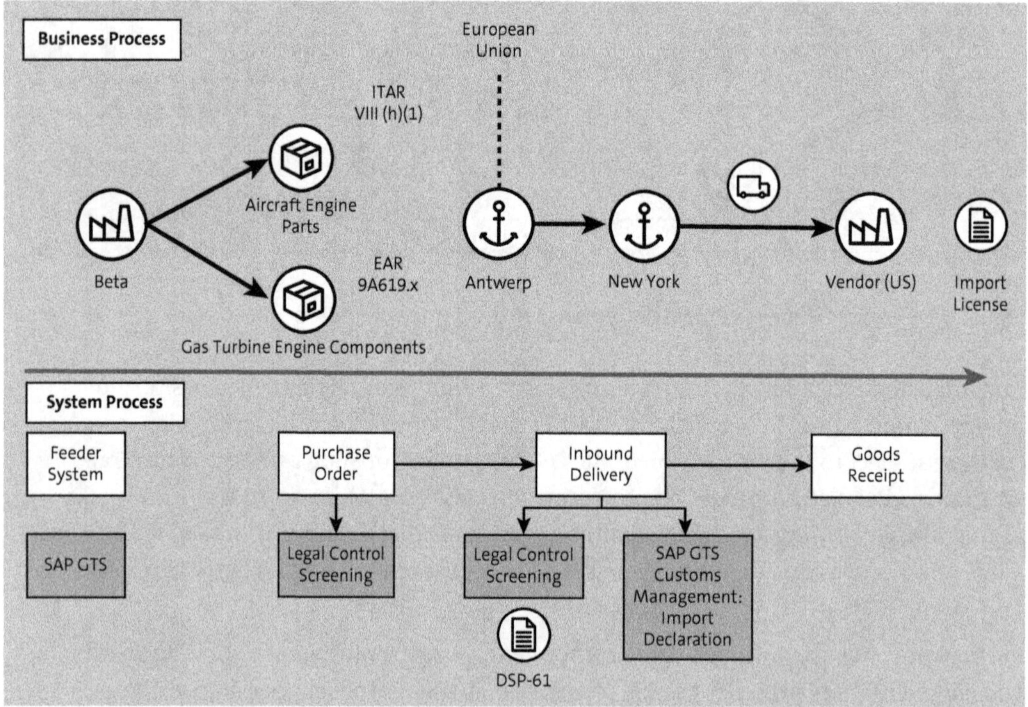

Figure 6.27 Import System Process of Importing Defense Articles into the US

When creating the purchase order at the US vendor, users get the popup notification shown in Figure 6.28 about the SAP GTS block stating that item 10—that is, the aircraft engine parts, ITAR part classified as VIII(h)(1)—needs a temporary US import authorization. The import of item 20—the gas turbine engine components, EAR part classified as 9A619.x—doesn't result in a legal control block.

Reference Number of a Document from Back Error

Item Service	Status
4500000765 Blocked / Checked	
Customs Document	Blocked / Checked
10 Embargo Check	Released / Checked
10 Sanctioned Party List Screening	Released / Checked
10 Legal Control	Blocked (License)
20 Embargo Check	Released / Checked
20 Sanctioned Party List Screening	Released / Checked
20 Legal Control	Released / Checked

Figure 6.28 Legal Control Block in the Feeder System

The inbound license determination takes place based on the Customizing setup, determination strategy, product classification, and departure and destination countries.

In this process flow, a license is required to import the aircraft engine parts because they are classified as defense articles subject to ITAR. A legal control check is also performed on the second item, and this item will get screened and immediately released as it doesn't require an import license.

To analyze the block, users can display the blocked document in SAP GTS via the **Manage Blocked Documents** tile under the **Compliance** space using the purchase order number as the reference number. Note that this is the same app used to display the documents blocked for embargo reasons and SPL.

Figure 6.29 displays the blocked document in SAP GTS. The US vendor's purchase order is blocked for legal control, specifically for the aircraft engine parts (product number **GB1005**).

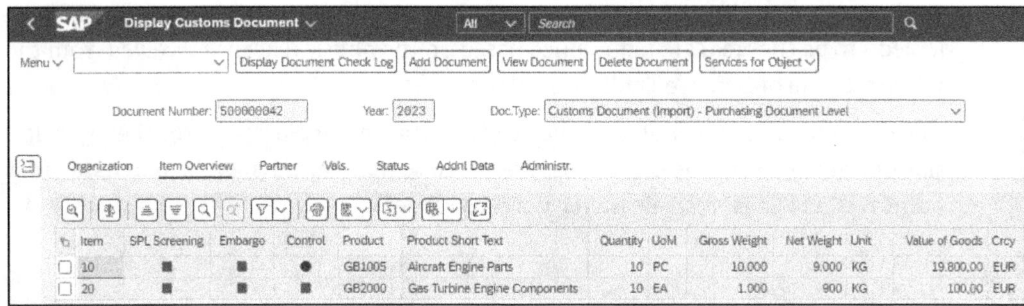

Item	SPL Screening	Embargo	Control	Product	Product Short Text	Quantity	UoM	Gross Weight	Net Weight	Unit	Value of Goods	Crcy
10	■	■	●	GB1005	Aircraft Engine Parts	10	PC	10.000	9.000	KG	19.800,00	EUR
20	■	■	■	GB2000	Gas Turbine Engine Components	10	EA	1.000	900	KG	100,00	EUR

Figure 6.29 Import Document Blocked Due to Legal Control Check

To find out why the document is blocked, display the SAP GTS check log, as shown in Figure 6.30. The log contains the two items included in the purchase order and provides information on the checks performed on that document.

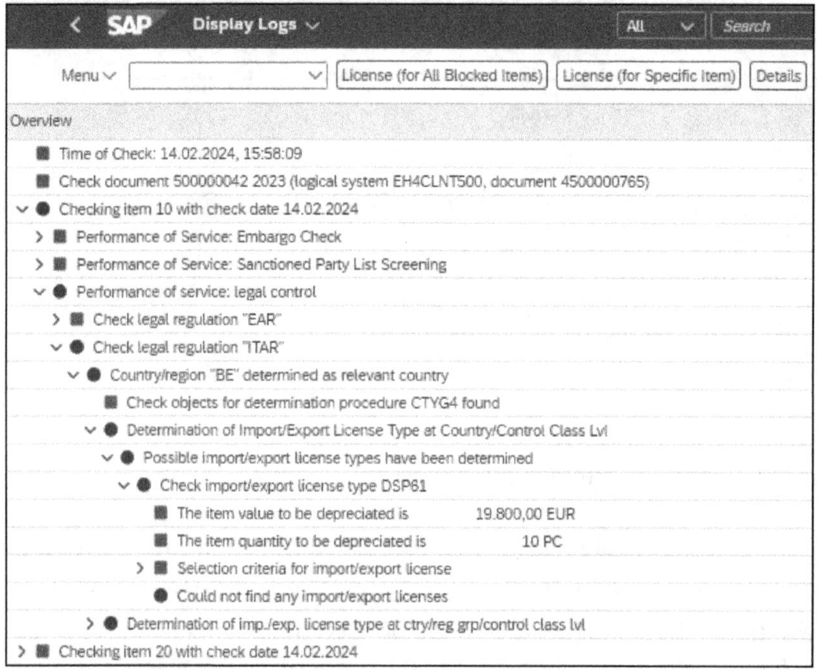

Figure 6.30 SAP GTS Log for the Blocked Import Document

A license is required to temporarily import these parts into the US from Belgium (material GB1005). The US vendor doesn't yet have an applicable license to cover the temporary import of that product, creating a block under the ITAR regulation. Furthermore, item 20—gas turbine engine components (material GB2000)—is an EAR-classified material and doesn't require an import license. Thus, it's cleared for the legal control check.

The US vendor decides to apply for a DSP61 license, which is specifically designed for the temporary import of military items. As soon as the vendor has received the license, the license master is maintained in SAP GTS, as displayed in Figure 6.31. All attributes (objects to be checked) are maintained based on the physical license received from the US Directorate of Defense Trade Controls.

One of the required attributes on the license is **Document Numbers** (see the lower-left side of Figure 6.31), so the US vendor is required to manually assign the purchase order number to the license to release the purchase order.

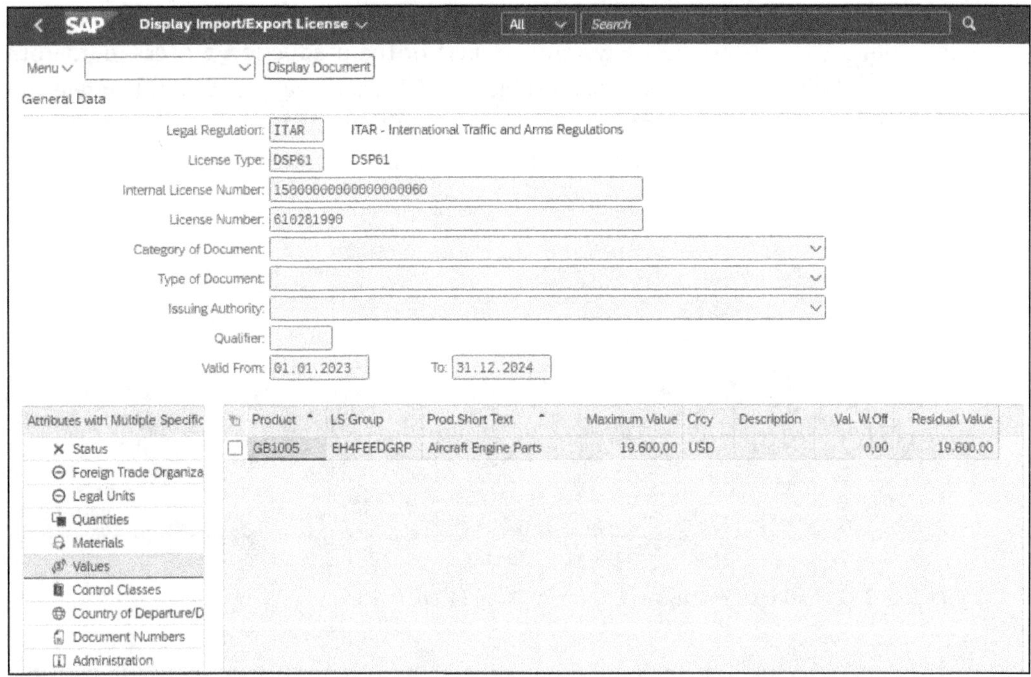

Figure 6.31 DSP61 License and License Depreciation

In the license, the value is written off for the depreciation group at the purchase order level, as shown in Figure 6.31.

The next section will focus on export licensing. Although there are a lot of similarities between import and export licensing, there are some key differences that should be highlighted.

6.4.2 Export Licensing

An export license/authorization may be required prior to an export transaction being executed. This section will build further on the example of the Beta company to introduce export licensing. Via this example, the business process will be linked to the system process.

Business Process

As soon as the US vendor performs the necessary operations on the aircraft engine parts and gas turbine engine components, both products need to be reexported from the US to Beta in Belgium using the same DSP61 license (for the aircraft engine parts) and an individual validated license (IVL) 748QV export license (to cover the gas turbine engine components).

Beta manufactures the aircraft engine according to the technical routing sheets (usually not set up in the feeder system). As part of the production process, numerous items, including the defense articles subject to ITAR, are assembled. The ITAR relevant aircraft engine parts are incorporated in an aircraft engine for final export to an end user in Qatar. The latter export transaction is the basis for this sequence, as shown in Figure 6.32.

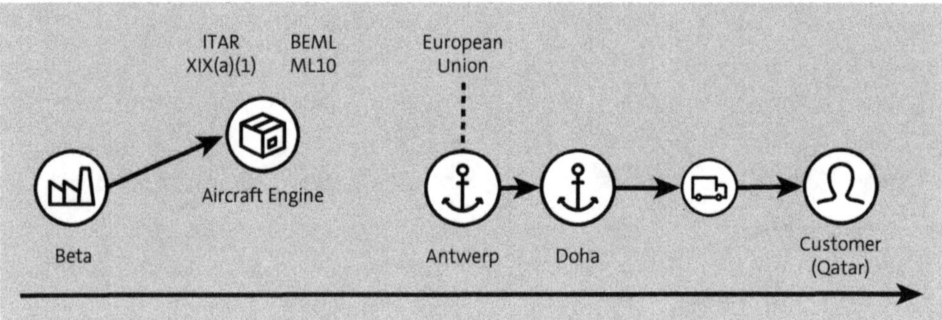

Figure 6.32 Business Process of Exporting Regulated Articles

The regulatory and business requirement is to perform license determination on the ITAR components embedded in the aircraft engine. Additionally, the shipment of the aircraft engine is also subject to the Belgian regulatory framework for defense articles. To comply with the imposed regulations, Beta must apply for licenses to cover the export of a product with an ITAR component, as well as a Belgian military license (ML).

System Process

Beta is planning to export controlled items to its customer in Qatar. Besides analyzing the finished product to determine the required licenses/authorizations, Beta also needs to analyze the finished product's components.

In SAP GTS, the necessary licenses assigned at the transaction level (order, outbound delivery) in the legal control functionality can be automatically included in the customs export declaration on an item level. Therefore, they can be contained within the electronic declaration, and the license information can be printed on the customs shipping documentation generated in SAP GTS. Figure 6.33 displays the SAP GTS system process in which the company determines the required licenses for both the finished product and its components by analyzing the finished product's BOM.

Beta manufactures the aircraft engine according to specific technical routing sheets. During production, the flywheel, wires, radiator return hose, two fuel lines, and imported aircraft engine parts are assembled. Note that the BOM composed in this example is for illustration purposes only. As a prerequisite to perform automatic BOM explosion on the export transactions using SAP GTS, the BOM should be transferred to SAP GTS as a master data element. Alternatively, BOMs can be created in the SAP GTS product master by navigating to the **Manage Products** tile under the **Business Partner and Products** section of the **Master Data** space directly, as illustrated in Figure 6.34.

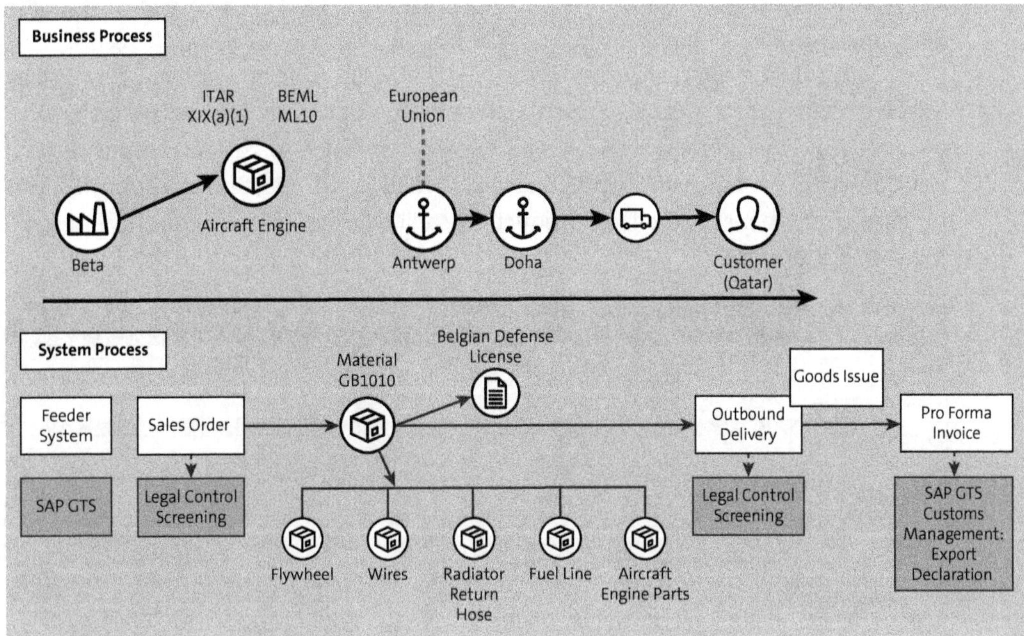

Figure 6.33 System Process of Exporting Regulated Materials

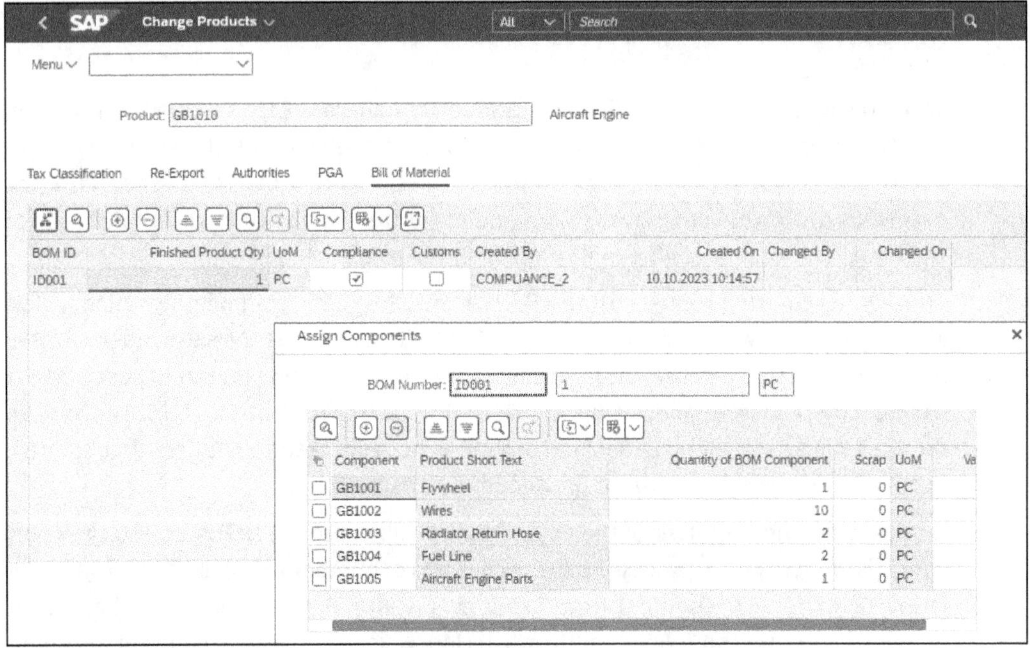

Figure 6.34 Bill of Materials of the Aircraft Engine

First, a license/authorization determination is performed on the ITAR components embedded in the aircraft engine. Second, the export of the aircraft engine is subject to the Belgian military regulations.

Beta creates a sales order for the sale of the engine (material GB1010). Using SAP GTS, the customs document recognizes that a BOM is embedded in material master GB1010. Initially, the customs document contains only one line item, representing the final product. However, as soon as the BOM is exploded in SAP GTS, the customs document also contains all components of the final product. In Figure 6.35, the legal control status for each component for the final product is provided (item numbers on the left, starting with **B000000001**).

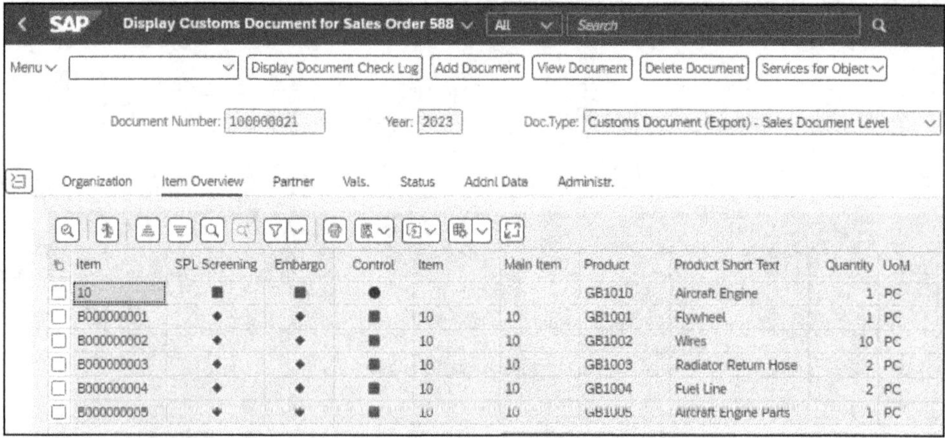

Figure 6.35 Customs Document of Beta's Export Sales Order

When recognizing that Beta's finished product contains an ITAR-classified component, SAP GTS proposes a DSP5 license (alongside other possible license types). Beta is notified of this license requirement upon saving the sales order in the feeder system (the parties to this prospective transaction and other item-specific details are defined for the sale of the aircraft engine to the customer in Qatar).

The BOM explosion proposes using a DSP5 license for the aircraft engine, which is classified in SAP GTS with USML classification XIX(a)(1). As soon as the sales order number is added to the DSP5 license, the legal control status for this item is cleared because Beta has already requested and set up the DSP5 license up front in SAP GTS. The system requires the sales order customs document to be rechecked to confirm that all attributes match before the status of the document changes.

Additionally, the shipment of the engine is also subject to the Belgian military export controls. In the system process flow, the customs document recognized the legal regulation BEML for the shipment from Belgium to Qatar. Beta has classified the aircraft engine as ML10 for the Belgian military list. Hence, users must also maintain the determination strategy for the Belgian Military Legal Regulation for classification code ML10 for shipments "from" Belgium "to" Qatar in SAP GTS to assign the Belgian ML in legal regulation BEML. Because the sales order remains blocked and the shipment can't occur, Beta decides to apply for the Belgian defense license via the official request form issued by the Belgian government. In this form, Beta indicates all details about the customer in Qatar, including the classification and description of the products to be sent, among other details, and attaches the necessary export control authorization details.

As soon as authorities provide the license, the license master is set up in SAP GTS, the sales order is rechecked, and the outbound delivery can be created, as long as all attributes of the Belgian license master are in line with the characteristics of the customs document in the sales order.

6.4.3 Legal Control Fundamentals

After explaining Beta's import and export licensing both from a business and a system point of view, this section will go deeper into the fundamentals of the legal control functionality by explaining the key fundamentals in determining license requirements for both inbound and outbound flows. We'll provide a detailed overview of how license determination is performed for specific flows. We'll take a closer look at the license determination strategy and authorizations master data setup.

To determine whether an import or export license/authorization is required for a material (i.e., for an item in an SAP feeder system document), SAP GTS performs an automatic validity check of the license/authorization criteria. Before the actual license check is performed in the customs document, SAP GTS determines the legal regulations that apply to the document. For each legal regulation, available licenses in SAP GTS are assigned when applicable.

The legal control functionality is shown in Figure 6.36 and will be further explained in the following sections.

We'll explain in detail how the legal control status for all items in SAP GTS documents are determined based on the business transaction (business partner countries), product classification of the items included in the shipments, determination strategy setup, and existing license master data in SAP GTS.

The following import and export legal regulations are used in the Beta business case:

- Export Administration Regulations (EAR) and US Reexport (USREX)
- International Traffic in Arms Regulations (ITAR)
- EU Dual-Use (EUDU)
- Belgian regulatory framework for military licenses (BEML)

As mentioned previously, the SAP GTS customs document determines the legal control status at the item level. To determine the status for legal control, SAP GTS first determines the applicable legal regulations to be checked based on the country/country group activation. The country/country group activation is performed in Customizing settings (Section 6.4.7).

For example, to import aircraft engine parts and gas turbine engine components (import business process flow), the SAP GTS customs document retrieves all active legal regulations in the Customizing "from" Belgium "to" the US. In another example, for the SAP GTS check on the sales order to export the aircraft engine (export business flow) to a customer located in Qatar, SAP GTS processes all applicable legal regulations active "from" Belgium "to" Qatar. Therefore, for each item in the SAP GTS customs document, multiple legal regulations can be applicable.

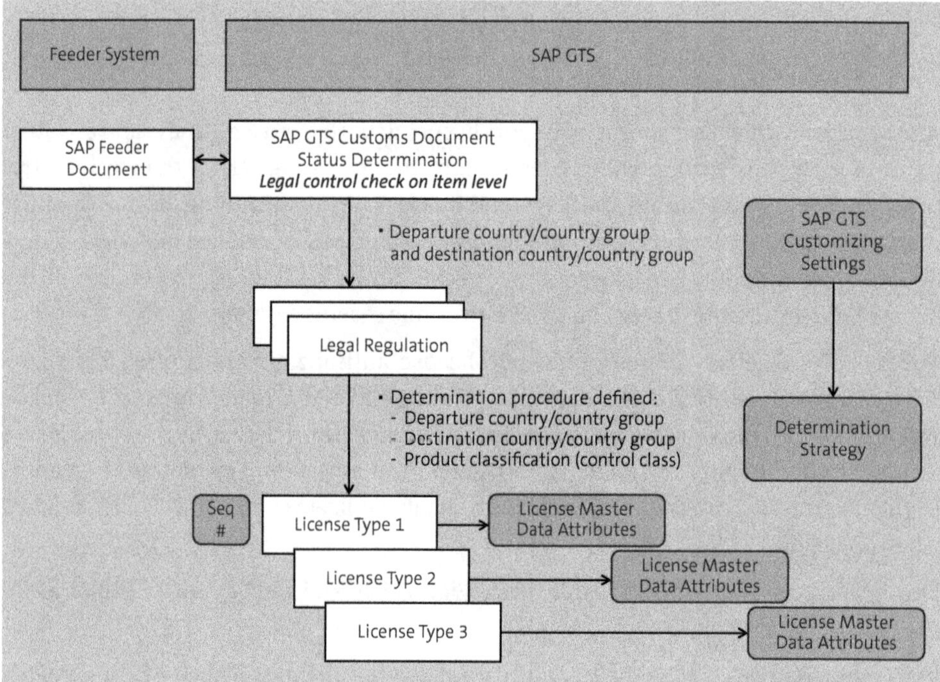

Figure 6.36 Legal Control Structure

Determination Strategy

Regulatory requirements architected in SAP GTS are set forth by the applicable authorities. Specific import and export transactions of controlled items require explicit approval. As described earlier, Beta's US vendor is receiving (importing)—and Beta will be shipping (exporting)—materials for which a license is also required.

In the example discussed, Beta has to apply for the required licenses/authorizations with the applicable regulatory authorities. Once the relevant license/authorization is received, the authorized company can maintain the approved license as master data in

SAP GTS. Relevant licenses are assigned to the customs document items representing the feeder system document items. For this to happen, SAP GTS customs document system logic goes through two additional steps:

- Identifying the license/authorization types (e.g., exceptions or exemptions) that could be applicable for this transaction based on the regulatory jurisdiction, product classification, and business transaction
- Identifying whether a license/authorization is already setup as SAP GTS master data, for which all relevant attributes match those on the transaction (see the next section)

As soon as the legal regulations are identified at an item level in the customs document, the determination strategy is triggered.

In addition to the necessary Customizing steps, users can define the determination strategy as master data. Users can maintain the license determination strategy manually (via the **Manage Determination Strategies** tile under the **License Management** section of the **Legal Control** space) or upload it via an XML file. This table can be used to define the applicable license/authorization types for each item in a customs document. Typically, these can include the departure country or country group, the destination country or country group, and a classification code. This classification code should be one of the codes that are maintained/uploaded for the relevant numbering scheme assigned to that legal regulation.

Figure 6.37 displays the determination strategy applicable to the Beta situation.

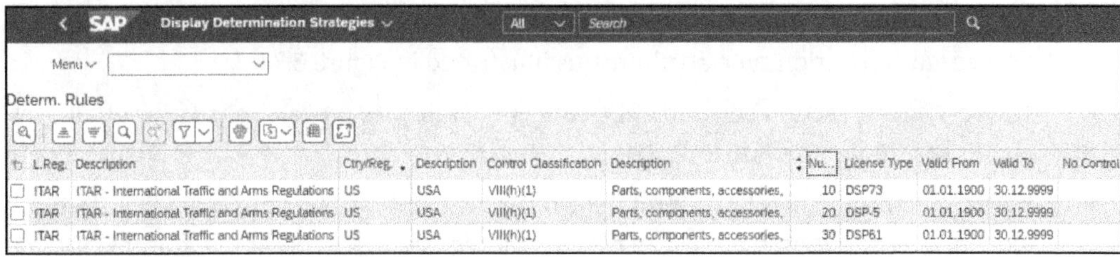

L.Reg.	Description	Ctry/Reg.	Description	Control Classification	Description	Nu...	License Type	Valid From	Valid To	No Control
ITAR	ITAR - International Traffic and Arms Regulations	US	USA	VIII(h)(1)	Parts, components, accessories,	10	DSP73	01.01.1900	30.12.9999	
ITAR	ITAR - International Traffic and Arms Regulations	US	USA	VIII(h)(1)	Parts, components, accessories,	20	DSP-5	01.01.1900	30.12.9999	
ITAR	ITAR - International Traffic and Arms Regulations	US	USA	VIII(h)(1)	Parts, components, accessories,	30	DSP61	01.01.1900	30.12.9999	

Figure 6.37 Determination Strategy

A specific entry in the table is valid only if the customs document check is done between eligible **Valid From** and **Valid To** dates.

Referring to the business import scenario, the US supplier is planning to import aircraft engine parts from Belgium, and thus SAP GTS checks whether a license for import of a material classified as military is available. In the log of the customs document for the inbound delivery (refer to Figure 6.30), SAP GTS first checks all license master records against potentially applicable license types within the ITAR legal regulation. If SAP GTS can't identify an applicable license master, the system searches for a license master with another potentially applicable license type. The sequence number in the determination strategy defines which license types are being checked first.

To summarize, within each legal regulation, the determination strategy defines the applicable import/export license types for the item in the customs document. The only exception occurs when the material master of the item is flagged as **No Control** for the specific legal regulation. In this case, the item isn't subject to the specific regulatory jurisdiction requirement (e.g., ITAR). In other words, the product won't be screened for the specific legal regulation in SAP GTS.

Licenses

Beta receives and ships items that require various country-specific authorizations. Therefore, the capability within SAP GTS to perform automatic legal regulations and authorization determination can support Beta in managing the applicable trade requirements.

The license control functionality requires that the users maintain the import and export license approvals received from the relevant governmental agencies as master data in SAP GTS.

Users may create import and export licenses in SAP GTS as master data via the **Manage Licenses** tile under the **License Management** section.

One of the main advantages of using SAP GTS is the automatic comparison of the item data in the customs document and the data maintained in the license attributes of all license/authorization master data set up for each license type in SAP GTS. The criteria for licenses/authorizations assignments are automatically assessed based on the license/authorization type attributes, as illustrated in Figure 6.38.

The rest of this section sets forth the concepts of license types, license type attributes, and license assignment in SAP GTS.

We've illustrated how the system determines the potentially applicable license types via the determination strategy. By design, it's because of this backbone framework that every license master in SAP GTS can be set up for only one legal regulation and one license type. When creating a license master, users need to specify the legal regulation and license type as part of the initial setup.

Prior to creating the license master, users must define the different license types in Customizing. When defining the license type, users need to determine the license attributes that need to be maintained to enable the automatic license determination for the customs document, as displayed in Figure 6.39.

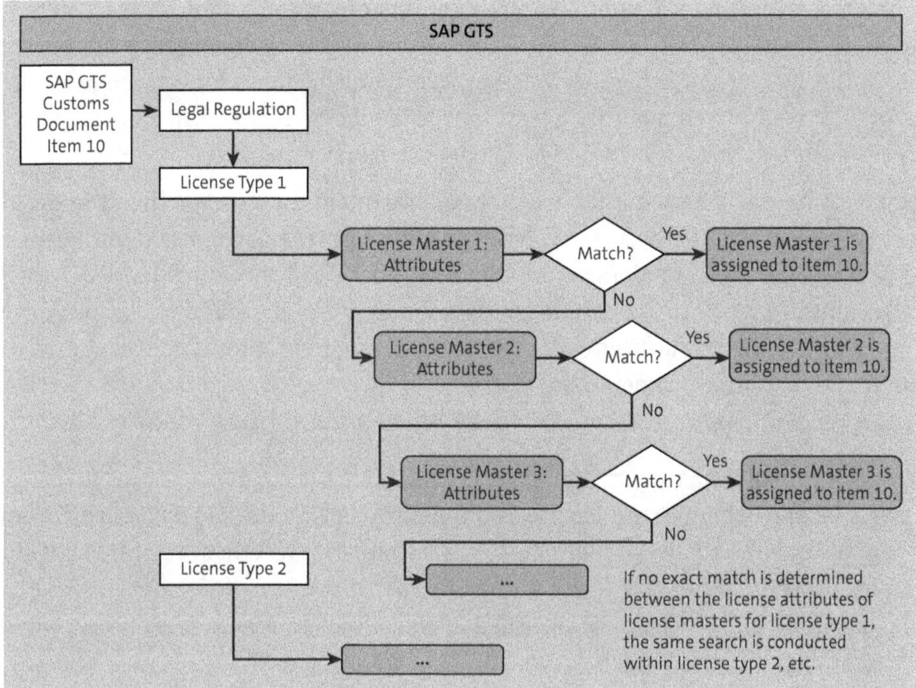

Figure 6.38 SAP GTS License Assignment Structure

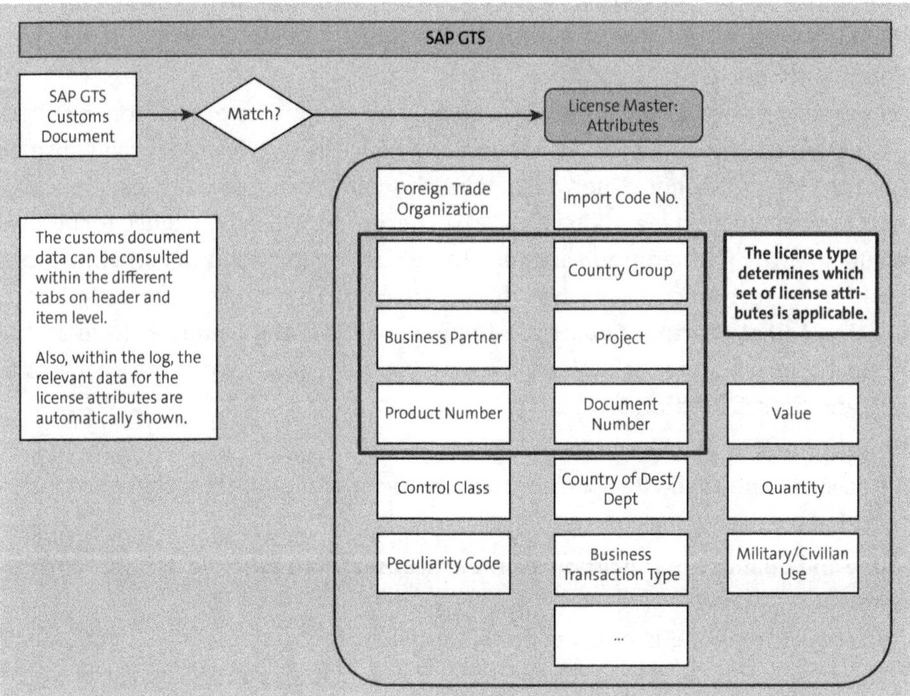

Figure 6.39 License Attributes

The purpose of setting up multiple license types per legal regulation is to accommodate the different types of licenses/authorizations issued by the applicable government authorities. Many export licenses/authorizations include a total approved value and/or a total approved quantity over a specified validity period, while other export licenses/authorizations don't specify a maximum value or quantity.

When creating the license master, users should maintain the data for all attributes in the license based on the license received from the relevant government authorities. The license type setup predefines the attributes that need to be maintained, which clearly indicates when the license master attributes are to be added to complete the license. The license type attributes are key objects to support automatic license assignment to the customs document item line.

The most common license attributes in SAP GTS are the following:

- **Organizational structure attributes: FTO and LU**
  This specifies whether the license is valid for one or multiple FTOs, which are mapped to the company codes in the feeder system, and whether the license is valid for only specific legal units, which are mapped to the plants in the feeder system. For example, some licenses might be issued to Beta for all entities across Europe, while others might be specific to the one plant in Belgium.
- **Control class attribute**
  This determines whether the license is valid for only specific export control classifications such as export control codes. During license assignment, the system compares the classification code of the product to the codes maintained in the license.
- **Country of destination/departure**
  This captures whether the license is valid only for shipments to and from business partners in specific countries. All countries for which a given license is valid must be captured in this attribute. The system takes the country from the first existing business partner function maintained in the partner group, which is assigned to the legal regulation. The partner group assigned to the legal regulation can contain various business partner functions, such as ship-to party, end user, and sold-to party. Selecting the **Multiple Partner Countries** setting entails that the countries from all the business partner types maintained in the partner group assigned to the legal regulation are checked (Section 6.4.7).
- **Product number**
  If a license is only valid for specific products, the product numbers should be maintained in the license master.
- **Value and quantity of the products in the licenses (with possible definition of tolerances)**
  Different settings in Customizing determine whether you should specify the values and quantities on the license or product level. For example, a DSP5 issued by the US

Directorate of Defense Trade Controls usually contains a maximum value and maximum quantity allowed per product. However, Beta might deal with many other licenses for which no maximum value is specified.

- **Business partner**
  This controls whether a license is valid for only a specific business partner or multiple business partners.
- **Partner function**
  When this object is flagged, it indicates that users need to maintain a partner group. The system can check whether all the partner functions maintained in the partner group are also maintained in the corresponding document (e.g., sales order). This option administers the end user to be filled in as an ultimate consignee.
- **Document number**
  Enabling this object forces the compliance users to fill in every relevant document for which the license is valid. When a sales order is created, the order is blocked because the system requires the sales order number to be maintained in the license. The same happens for the delivery and subsequent documents. This setting sets forth a manual license assignment. Users can architect SAP GTS to assign import/export licenses manually or by using an automated process. These are key decisions in the system design. When using the automated export authorization setup, users should consider import/export licenses with potentially overlapping attributes (e.g., a DSP5 could overlap with a DSP73 and have the same part numbers, export control codes, legal entities, and destination countries). Thus, users need to include specific configuration to prevent inadvertent assignment to the incorrect import or export control authorization.
- **Flag to check the first document in the document flow**
  Facilitates some level of automation as the compliance user manually adds the license to the first document in the feeder system document flow (e.g., sales order), and the same license is then automatically assigned to the subsequent documents (e.g., outbound delivery) if all other attributes match.
- **Value and quantity update**
  Activates automatic decrementing of the license. Several options exist to adjust the specifics of the functionality. Users can, for example, specify whether the maximum value is valid for all products or whether the maximum value can be exceeded for specific products.

In summary, the actual import/export license received from the relevant authorities are set forth in SAP GTS via the specific license types that are set up as part of Customizing. The applicability of certain attributes will depend on the license type.

For the license determination functionality to perform effectively, the legal regulations and specific master data needs to be set up properly, namely the license determination strategy and the license master data.

As soon as the system finds a match with a specific license/authorization type, SAP GTS assigns the license/authorization (including license exceptions or exemptions, as applicable) to the customs document, and the document status for legal control is flagged as positive for that item. Furthermore, SAP GTS notifies the user if a *nearly* identical match is identified but not all the attributes match. An example of this is illustrated in the import system process flow in Figure 6.30 shown earlier, where the system notified the SAP GTS user that a DSP61 license could be applicable, but the open value wasn't sufficient.

Agreements

Agreements in SAP GTS allow the company to define and manage export control authorizations that can be linked with licenses as applicable or serve as the sole export authorization. This functionality is primarily set forth to meet applicable ITAR requirements, but other export control jurisdictions have similar requirements. The functionality also allows the grouping of several export authorizations within one agreement, making use of the "nested" license concept.

In SAP GTS, agreements can be customized to manage the value or quantity depreciation.

Note

The following discussion illustrates the use of agreements, separate from the business case import/export transactions introduced in Section 6.1.

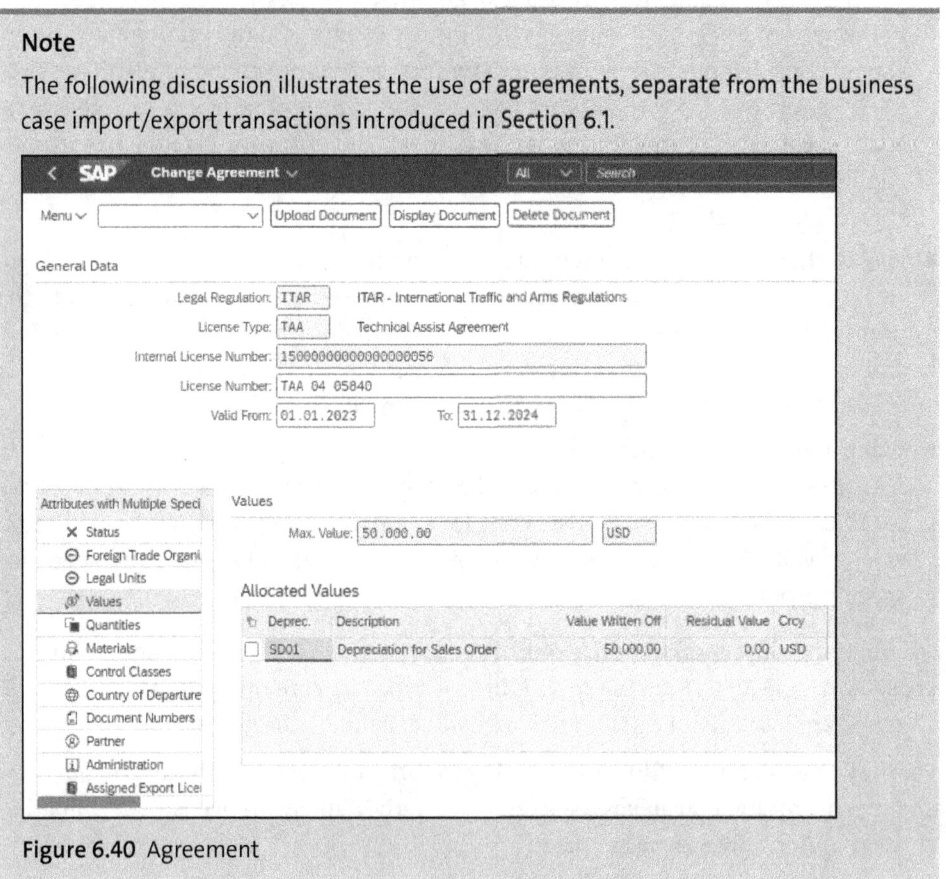

Figure 6.40 Agreement

In this example, Beta has set up an agreement in SAP GTS to represent the Technical Assistance Agreement (TAA). The total quantity of the agreement is 100 PC for a specific aircraft engine part. The total value allowed to be transacted against is $50,000. Furthermore, all additional agreement attributes are maintained similar to license attributes, as shown in Figure 6.40. This can be displayed via the **Display Agreements** tile in the **ITAR** section of the **Compliance** space.

The process in SAP GTS to clear the blocked document can be split into two steps:

1. **Agreement-license assignment**
   After the DSP5 license is assigned to the agreement, the open quantity of 100 PC in the agreement is decremented by **20**, which is the maximum quantity specified in the corresponding DSP5 license (see Figure 6.41).

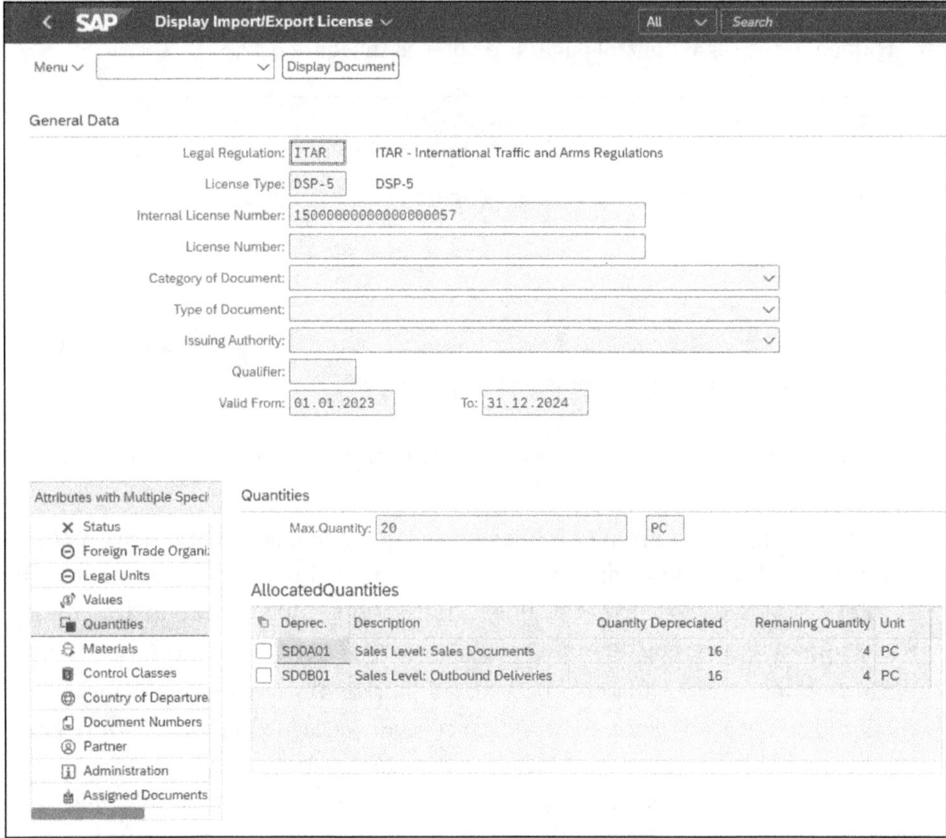

Figure 6.41 DSP5 License with Agreement

2. **License-document assignment**
   After the sales order assignment to the DSP5 license, the license itself has a remaining open quantity of **4** as the feeder system document contains **16** pieces. If the

feeder system document contains 35 parts of the material, this license is insufficient, and an additional DSP5 license or other export control license/authorization needs to be obtained. The amounts can be depreciated on the sales document level and outbound delivery level, depending on the Customizing settings for depreciation groups.

Users can consult both the total and remaining quantity of the agreement, as the total and remaining quantity of the license, in the **Quantities** attribute of the license shown in Figure 6.41.

6.4.4 Applications for Legal Control

In the previous sections, the import and export licensing processes were explained. In this section an overview will be given of the different apps the user can use to manage outbound and/or inbound documents, licenses, and agreements. All the different apps are grouped in the **Legal Control** page, as displayed in Figure 6.42.

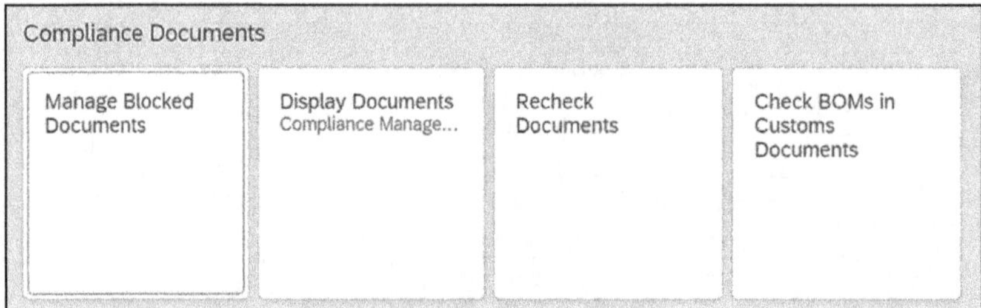

Figure 6.42 Legal Control Export: Compliance Documents

The following functions can be used when dealing with legal control monitoring:

- **Monitoring**

  The monitoring functions in Compliance Management enable Beta to maintain an overview of all business transactions. This function shows which documents the system has blocked for further business transactions and acted upon accordingly, for example, by assigning the appropriate licenses for import or export transactions.

  Within **Compliance Documents**, the **Manage Blocked Documents** worklist can help track and review transactions. Users can set several selection criteria to work effectively. For archiving purposes, all trade compliance information is captured in the SAP GTS customs document for each transactional document in the SAP feeder system. Users can view the available information by going into the document itself (header/item) or the log.

  The overall status of a document determines whether users can execute follow-on activities. If a document is technically incomplete, the log of the document won't be available. The SAP GUI can be used to access technically incomplete documents through Transaction /SAPSLL/CUHD_IC_EXP (Display Technically Incomplete Customs Documents).

In the example in Figure 6.43, the material GB1011 hasn't been transferred yet to SAP GTS. When the feeder system document was transferred to SAP GTS, the system could not identify material GB1011 and thus resulted in a technically incomplete document. The customer and product must be transferred to SAP GTS before the legal control checks can be performed. After making the necessary corrections, users can manually retransfer the document to SAP GTS.

When Beta reports a license assignment that occurred in the past, the SAP GTS reporting functionality allows searching in both directions: from document to license and from license to document. The Display Assigned Documents app allows users to search by document number and then display which licenses have been assigned to each individual item. Users can also report all created, active, and expired licenses in SAP GTS by legal regulation, license type, and FTO through the Display Existing Licenses app. When going into the license master, the assigned documents report is available in the **Assigned Documents** attribute.

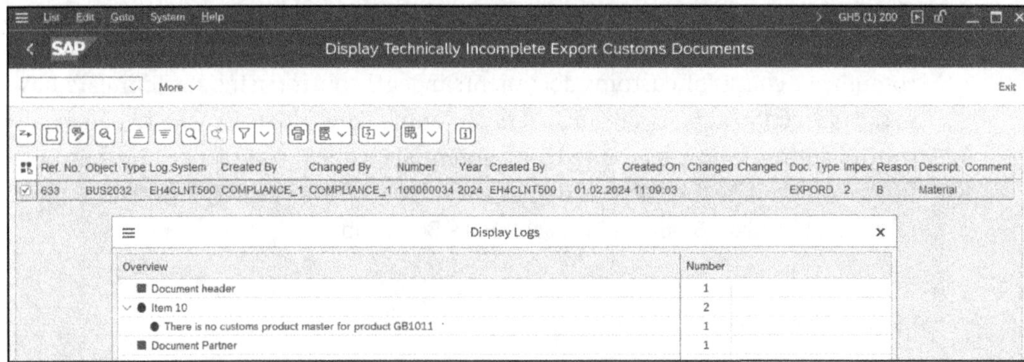

Figure 6.43 Technically Incomplete Documents

- **Recheck**
  The recheck of the customs documents helps determine the latest status of a document when updates have been made in SAP GTS (e.g., a license creation/change). Users can execute the recheck in the specific dedicated Recheck Documents app (not shown) or with the **Recheck** button in the Manage Blocked Documents app.

  This function enables users to recheck all documents in the feeder system regardless of their processing status.

- **Legal control data**
  The Change Control Data for Documents app allows users to manually change the license assignment to a specific document from the SAP feeder system.

- **Agreements**
  Similar to the Display Existing Licenses app, users can generate the Display Agreements report as an overview of all ITAR agreements or other related export-control jurisdiction requirements in the system. Another application provides an overview of the assigned licenses to the existing agreements one level higher in the structure than the overview provided by the Display Assigned Documents app. This assignment is shown in Figure 6.44.

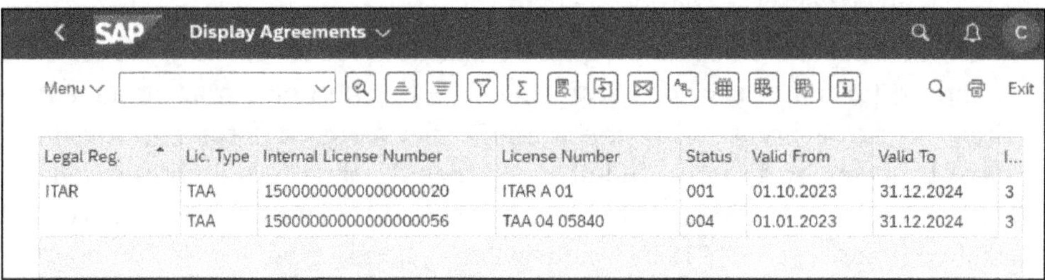

Figure 6.44 License – Agreement Assignment

- **Simulation**
  Beta can also use the simulation functionality in SAP GTS. Users can simulate the compliance check for customs documents in legal control using the Simulate Business Transaction app. As a result, users can immediately identify whether an export control authorization is required for a potential business transaction. Furthermore, users can simulate the check of document-related BOMs in legal control for exports using the Simulate Business Transactions – BOM app.

6.4.5 US Reexport

When goods subject to US controls are reexported to another country, special US export control requirements apply. For example, the reexport might be subject to US authorization if an item is of US origin, contains more than a specified percentage of US controlled content, or is based on certain US technology or software.

This section explores the legal requirements for reexports by demonstrating the SAP GTS de minimis functionality. Calculating the de minimis value makes it possible to determine whether a non-US-produced item that incorporates US controlled content (including technology or software) is subject to the EAR and thus would require a US export authorization even if exported from outside the US. Under the EAR, there are various thresholds for de minimis calculation based on the type of controlled item, the destination of the export, and the entities involved. It's also important to note that certain items aren't eligible for de minimis treatment.

Business Process

As part of outgoing shipments from Belgium, Beta provides specially designed assemblies and integrated circuits to an end user in Brazil. The assemblies contain parts that were previously imported and contain US content and are thus also subject to the EAR controls mentioned earlier in this section.

Beta must perform a de minimis calculation to determine whether the reexported materials require a US authorization. In other words, if the US content in the exported materials exceeds a given threshold determined by the country of destination and classification of the US components, a US license may be required.

Beta's business process for exporting materials containing US content is displayed in Figure 6.45.

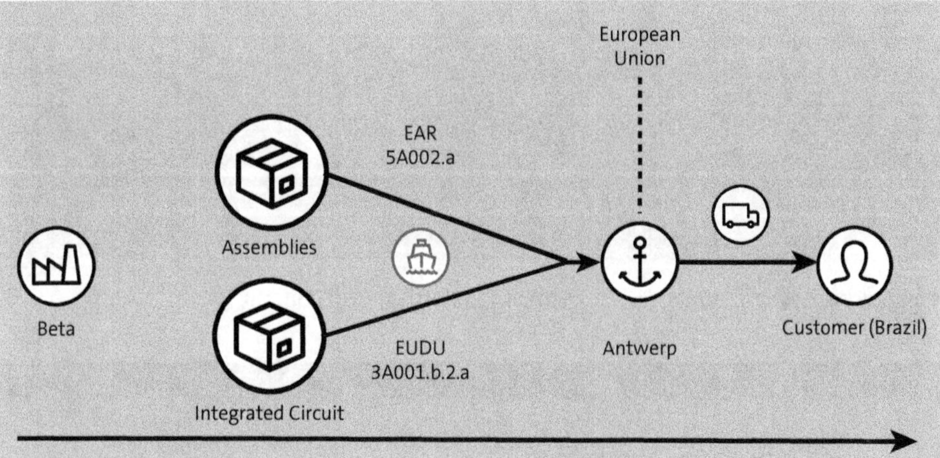

Figure 6.45 Business Process of Exporting Materials with US Content

In addition to the US reexport legal regulations, the integrated circuit is also subject to EU dual-use control (similar to the aircraft engine from Section 6.4.2).

System Process

Figure 6.46 illustrates Beta's system process for exporting materials containing US content to a customer in Brazil. SAP GTS picks up the BOM for both products when creating the sales order in the feeder system and determines the US content of both materials. This amount of US content is compared against a set threshold percentage to determine whether a US license is required.

At the product master data level, as shown in Figure 6.47, the reexport control relevance can be maintained. For finished products, this reexport control relevance and its related threshold value are the result of the de minimis calculation performed in SAP

GTS. The result of the de minimis calculation can also be displayed via the **Display Re-Export Calculation Logs** tile in the **Legal Control Re-Export** section of the **Compliance** space.

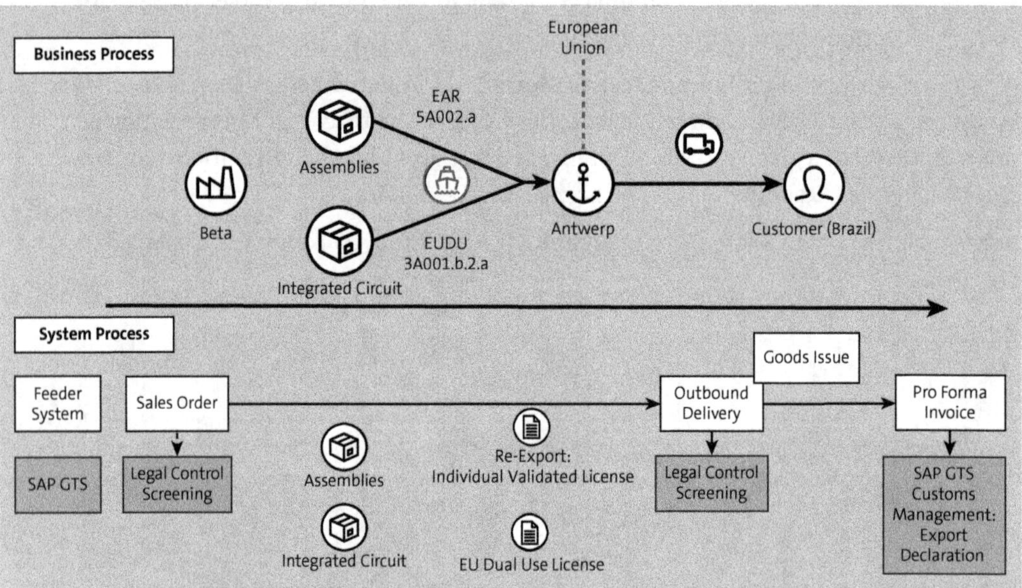

Figure 6.46 System Process for Exporting Materials with US Content

SAP Display Products All Search

Menu Edit Exit

Product: GB9100 Specially designed assemblies

Special Customs Procedures | Tax Classification | Re-Export | Authorities | PGA | Bill of Material

FTO Re-Exp	L.Reg.	Ctry Group	Re-Export Ctrl Relevance	Re-Export Calculated Control Relevance	Calculated Hardware Share	Crcy	% HW	% SW	% Techn.
FTO_2510	USREX	USREX_CRIT	Determine Value-Dependent Control Relevance	Determine Value-Dependent	141,00	EUR	0,00	0,00	0,00
FTO_2510	USREX	USREX_UNCR	Determine Value-Dependent Control Relevance	Determine Value-Dependent	141,00	EUR	0,00	0,00	0,00

Figure 6.47 Re-Export Tab on Finished Material

This data needs to be entered for each FTO using this product (given recent regulatory changes as part of the US Export Control Reform, a 0% de minimis tolerance applies to specific items classified under EAR 600 series to certain destination countries). In this example, two lines have been added to the product master of the finished product: one for the critical country group and one for the uncritical country group. Both values have been set to €141.00 (value for the US controlled component). Figure 6.49, later in this section, shows how SAP GTS has performed the de minimis calculation for the finished products with a total value of €141.00 based on the US reexport shares of the components in the BOM.

Three different statuses apply in the **Re-Export Ctrl Relevance** field; the setting determines the system behavior regarding de minimis for these products:

- **Check/Control**
  The product is always relevant for US reexport.
- **Do Not Check/Control**
  The product is never relevant for US reexport.
- **Determine Value-Dependent Control Relevance**
  A US reexport value is determined for this product.

Before calculation in SAP GTS can take place, the components in the finished product's BOM that are subject to the EAR regulation must be flagged as subject to the EAR. Recall from Chapter 5 that this can be done in various transactions, including the **Products Classification** app accessible from the tile in the **Product & Classification** page.

Furthermore, for all the relevant components within the finished product BOM, users must maintain a value of the US content in SAP GTS. For components that aren't EAR controlled, users aren't required to enter a value. SAP GTS makes it possible to use the values per legal unit/plant that are transferred from the feeder system and stored in the **Prices and Values** section of the product master.

In another example, Beta is shipping the specially designed assemblies and integrated circuits on one shipment to its customer in Brazil as follows:

- The integrated circuits are built with only components produced in house, and the value of US content is equal to zero.
- The specially designed assemblies (product GB9000) have a BOM with three components: one main US-controlled part classified with export control code 5A002.a and two parts of Belgian origin. The value for the US-controlled component (GB10001) is €141, whereas the value for the Belgian item not subject to EAR is €0. Consecutively, SAP GTS calculates the total US reexport share for the specially designed assemblies based on the BOM.

Users can find several calculation transactions in the **Calculate Products for Re-Export** app under the **Legal Control Re-Export** section of the **Compliance** page, as shown in Figure 6.48. With this app, users can perform the calculation for a range of products (**Calculate Multiple Products**), one individual product (**Calculate Individual Product**), or the products for which the BOM changed in a certain period (**Calculate Changed Products**).

If SAP GTS is unable to retrieve the US value for a component (e.g., if value isn't maintained in the product master), the solution considers the finished product as controlled, meaning that an EAR authorization is required regardless of the price of the product. This can be achieved through Customizing and applying the worst-case principle; users can configure SAP GTS so that when no US value is maintained, the component has an infinite US value.

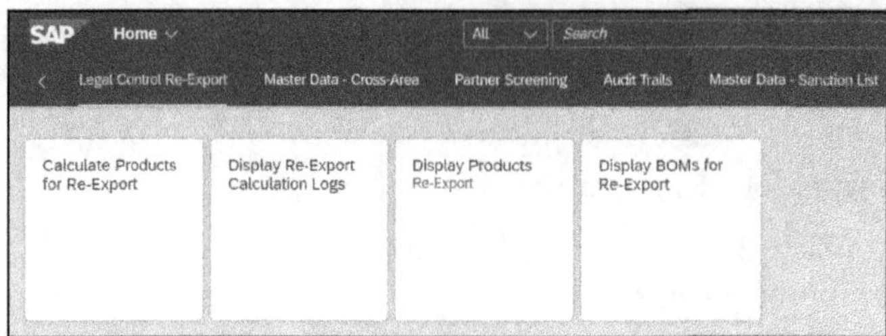

Figure 6.48 SAP Fiori Tiles for Legal Control Re-Export

If users properly maintain the product master, including transferring the value and BOM from the feeder system, SAP GTS calculates the actual US reexport share for the applicable assemblies. Figure 6.49 shows where the de minimis calculation was executed for the finished product (**Calculate Individual Product**), the specially designed assemblies **GB9100**.

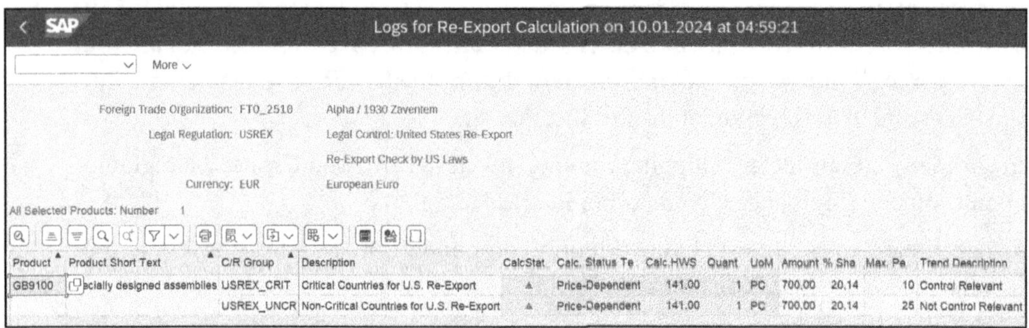

Product	Product Short Text	C/R Group	Description	CalcStat.	Calc. Status Te	Calc.HWS	Quant	UoM	Amount	% Sha	Max. Pe	Trend Description
GB9100	...ecially designed assemblies	USREX_CRIT	Critical Countries for U.S. Re-Export	▲	Price-Dependent	141,00	1	PC	700,00	20,14	10	Control Relevant
		USREX_UNCR	Non-Critical Countries for U.S. Re-Export	▲	Price-Dependent	141,00	1	PC	700,00	20,14	25	Not Control Relevant

Figure 6.49 Logs for Reexport Calculation

SAP GTS determines that the finished good has a total value for the US reexport share of €141. As previously discussed, this value can be extracted from the component's **Re-Export** tab, as shown in Figure 6.50.

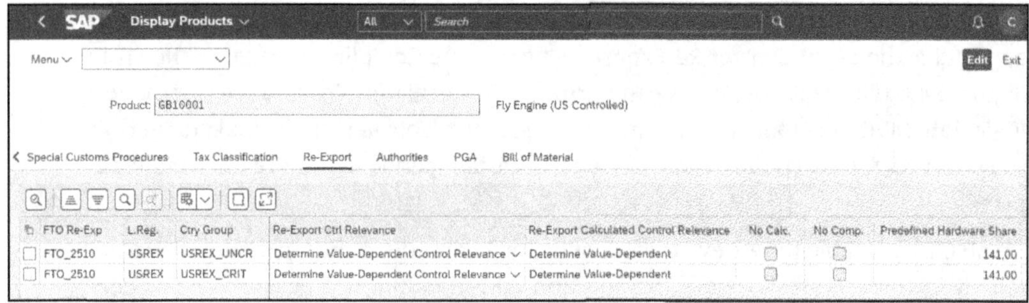

FTO Re-Exp	L.Reg.	Ctry Group	Re-Export Ctrl Relevance	Re-Export Calculated Control Relevance	No Calc.	No Comp.	Predefined Hardware Share
FTO_2510	USREX	USREX_UNCR	Determine Value-Dependent Control Relevance	Determine Value-Dependent	☐	☐	141,00
FTO_2510	USREX	USREX_CRIT	Determine Value-Dependent Control Relevance	Determine Value-Dependent	☐	☐	141,00

Figure 6.50 Re-Export Tab on Component Material

When comparing this with the price of the product as maintained in its material master, the system determines that the 25% threshold isn't surpassed (while the 10% is). It's always possible that the transactional net value of the product is higher or lower than the price of the product. The result gives an indication of a possible EAR license requirement.

After the calculation is executed with this transaction, the value is stored in the product master of the finished product (the **Calc.HWShr** column shown earlier in Figure 6.49). Users can overwrite this value by manually correcting a value to make sure that the value of the US content of the finished product is never lower than a specified amount, even if the sum of values of the components is lower (the **Predefined Hardware Share** column).

De Minimis Calculation

At the transaction level, SAP GTS picks up this US reexport share from the product master of the items that are shipped and compares it with the actual sales price in the document. The log per document in SAP GTS displays the threshold percentage, item value, and reexport share percentage. If the reexport share percentage exceeds the threshold percentage, the finished product is subject to the EAR legal regulation.

Figure 6.51 shows the SAP GTS log for the shipment for Beta going to the customer in Brazil; the specially designed assemblies exceed the 25% de minimis threshold for the "uncritical" country group (which includes Brazil) in the USREX legal regulation because the selling price of the finished product assembly **GB9100** varies from the price indicated in the product master (€700 versus €520.50).

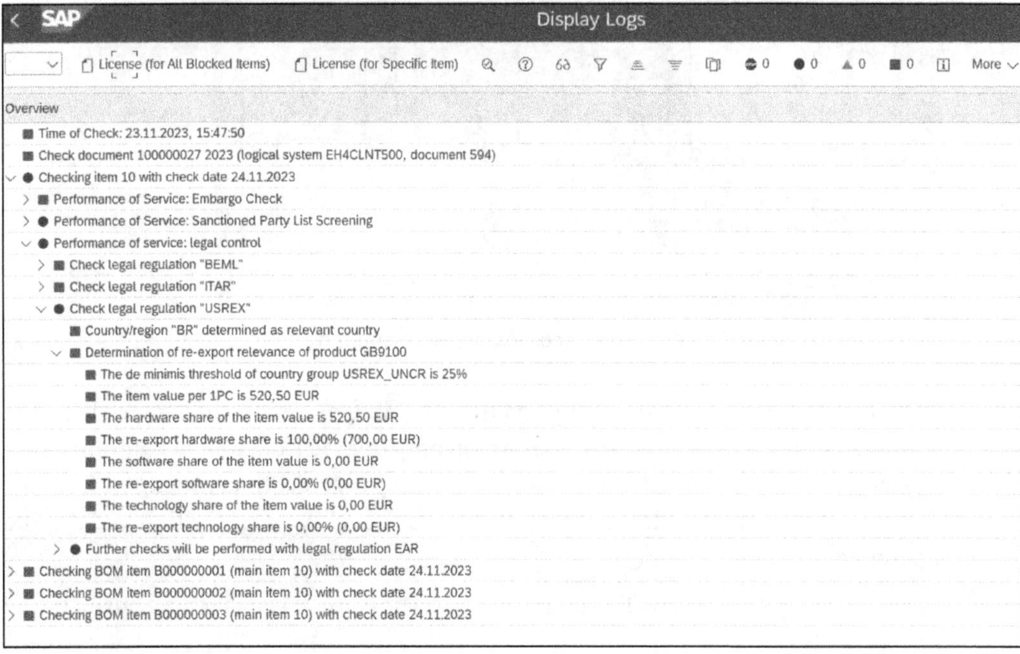

Figure 6.51 Reexport Relevance Determination

An IVL is proposed for the specially designed assemblies (item 10), classified as 5A002.a. This license type is proposed in the EAR legal regulation and triggered via USREX as the threshold is reached and no EAR exception applied.

The de minimis calculation doesn't apply for item 20 because the integrated circuits don't contain US components. However, the integrated circuits are subject to the EU dual-use regulation.

After the assignment of the licenses for the European export controls, as well as the relevant licenses covering the reexport of products subject to US export control requirements, the SAP GTS transaction releases the block. All legal control checks have now been executed, and the transaction is released.

6.4.6 Intercompany Sales Screening

Intercompany transactions are financial transactions between related companies, for example, between a group and a subsidiary or between two subsidiaries of a group. The intercompany sales process involves selling goods or services from one company code to another, as shown in Figure 6.52.

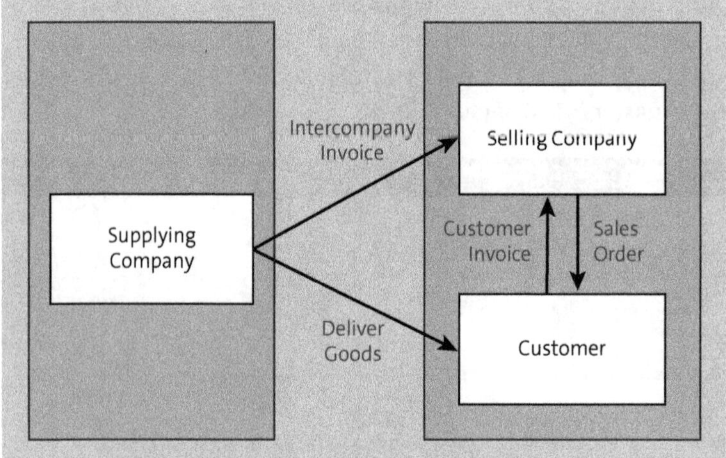

Figure 6.52 Business Process for Intercompany Sales Screening

The legal control functionality supports the screening of import and export licenses/ authorizations for intercompany sales. Intercompany sales documents automatically trigger the system to screen the import to the selling organization, the export from the delivering plant, and the domestic sales between the selling organization and the end customer.

In the following sections, we'll examine the intercompany sales screening business and system process.

6

Business Process

Beta has several entities across the globe, each responsible for their own product groups. Whenever one of these entities receives an order for a product that isn't part of its own product group, another entity can be contacted to deliver the requested product.

Entity A of Beta, for instance, located in Canada, receives an order for an aircraft engine part that should be delivered to its customer, also in Canada. This entity doesn't stock this part, so as a result, it places an order to the Belgian entity of Beta. At that point, the Belgian entity starts with its regular shipping process while the entity in Canada simultaneously continues its procurement process. Both processes are triggered by the same customer order in SAP. In this case, the Belgian entity exports the aircraft engine part to Canada while the entity in Canada imports the same part. Licenses for these movements are required because the aircraft engine part is classified as military (e.g., ITAR in the US, VIII(h)(1) and ML in Belgium). Additionally, after the part arrives in Canada, a domestic license might still be required to ship the goods from Canada to a Canadian end customer. The licenses for these movements need to be present prior to the shipment of the aircraft engine part to avoid any supply chain delays. The shipment will be blocked by the system if any of these licenses are missing or incomplete.

System Process

As shown in Figure 6.53, Beta's Canadian entity generates a sales order indicating the Canadian sales area, combined with the delivery plant of the Belgian entity of Beta. The customer (sold-to) in the sales order is located in the same country as the selling entity, in this case, Canada.

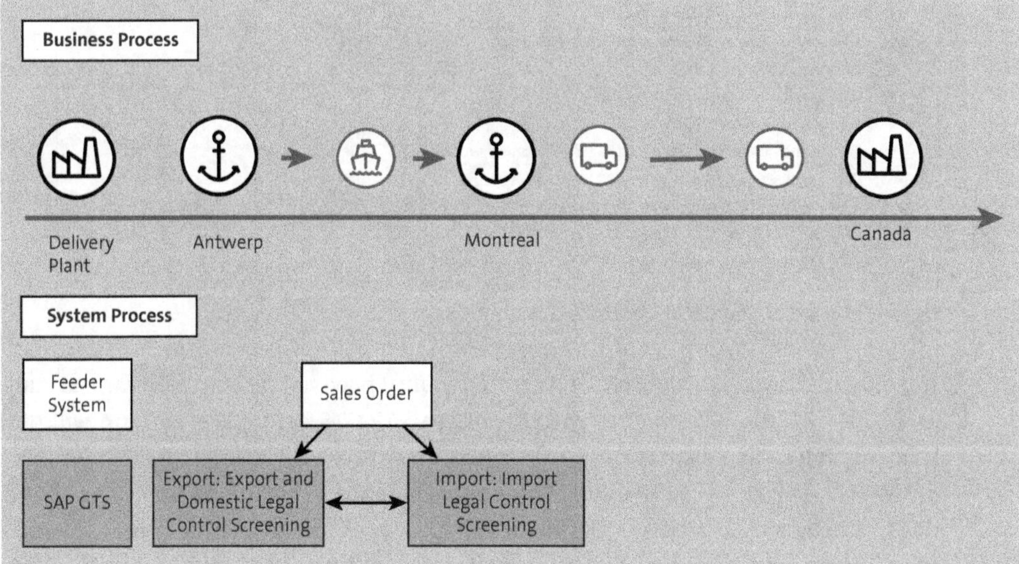

Figure 6.53 System Process for the Intercompany Sales Screening

Upon saving the sales order, SAP GTS simultaneously creates two customs documents in Compliance Management. The system creates a general export customs document and performs an export legal control check to confirm the license requirements between the delivering plant and the customer. In the same export customs document, the system screens for potentially required domestic licenses. The other document generated is an import customs document indicating the screening results for importing the goods into Canada from Belgium.

Users can review the results of the license screenings in the audit trail of the respective customs documents (see Figure 6.54), which can be found using the Display Documents app and selecting the **Display Check Log** after clicking on a document number. The status indicated in this audit trail is sent back to the feeder system. The sales order of the feeder system is then only released if the license requirements for all three licenses—export, import, and domestic—are met. In other words, to send a blocked status from SAP GTS into the feeder system, just one of these checks needs to find a missing license. As with the regular legal control functionality, further processing of the sales order to, for instance, a delivery document, will be prevented if the document is blocked for the license screening.

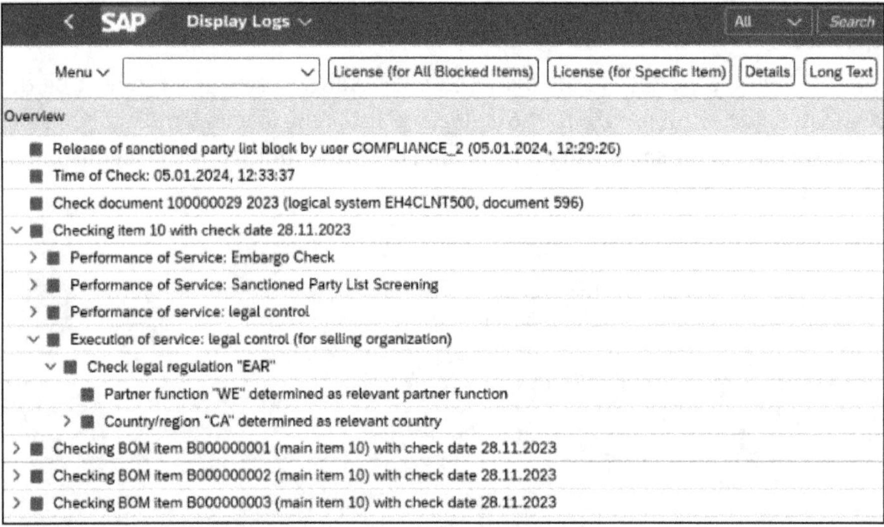

Figure 6.54 Extended Legal Control Screening

The extended legal control screening functionality needs to be activated for the standard SAP GTS sales **Document Type EXPORD** in SAP GTS before the solution will perform any full screening of the intercompany sales order. As shown in the Customizing in SAP GTS, accessed via **SAP Global Trade Services, edition for SAP HANA • General Settings • Document Structures • Define Document Types for Application Areas**, an additional **Advanced Check for Sale Across Foreign Trade Organizations** checkbox has been configured, which users must select to activate this functionality (see Figure 6.55).

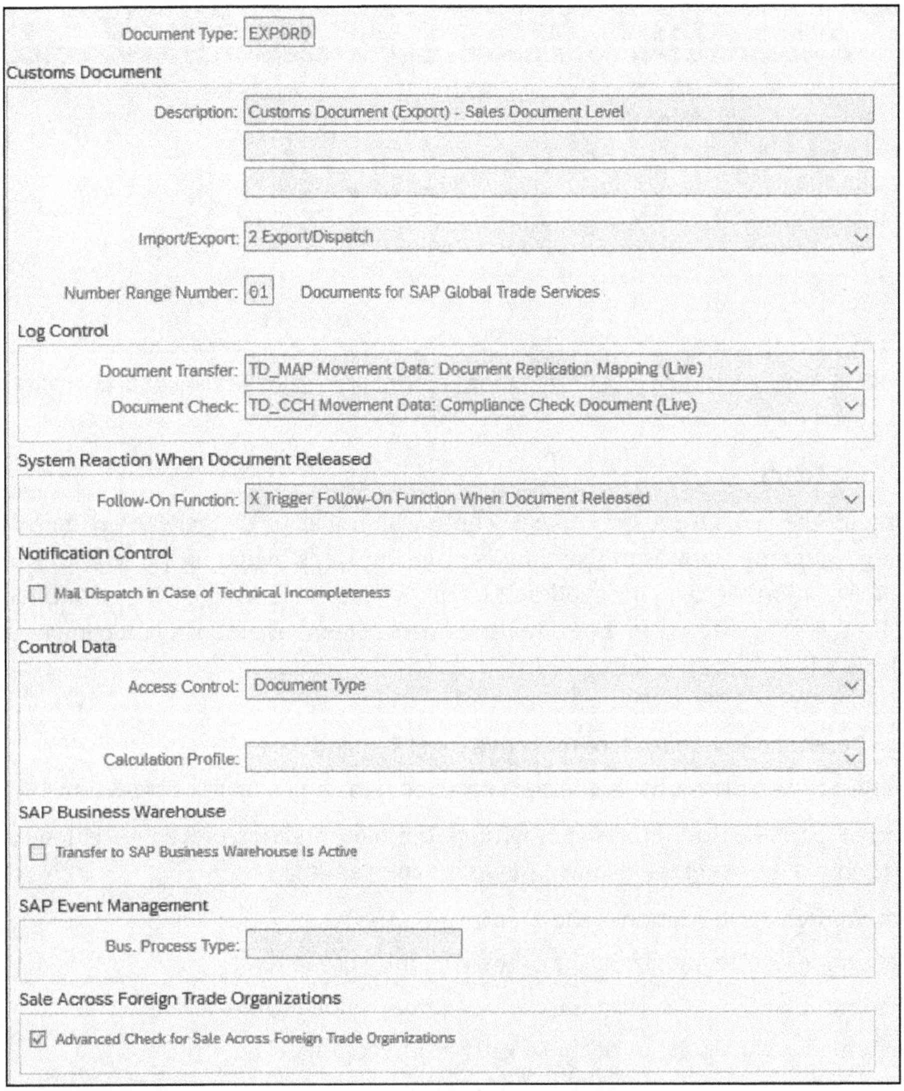

Document Type: EXPORD

Customs Document

Description: Customs Document (Export) - Sales Document Level

Import/Export: 2 Export/Dispatch

Number Range Number: 01 Documents for SAP Global Trade Services

Log Control

Document Transfer: TD_MAP Movement Data: Document Replication Mapping (Live)

Document Check: TD_CCH Movement Data: Compliance Check Document (Live)

System Reaction When Document Released

Follow-On Function: X Trigger Follow-On Function When Document Released

Notification Control

☐ Mail Dispatch in Case of Technical Incompleteness

Control Data

Access Control: Document Type

Calculation Profile:

SAP Business Warehouse

☐ Transfer to SAP Business Warehouse Is Active

SAP Event Management

Bus. Process Type:

Sale Across Foreign Trade Organizations

☑ Advanced Check for Sale Across Foreign Trade Organizations

Figure 6.55 Advanced Check for Sale across Foreign Trade Organizations

To perform the screening, users should indicate the applicable sales order type for the intercompany sales screening in SAP GTS. The relevancy of this document type is specified on both the header and item category levels in the Customizing step **Assignment of Document Types/Item Categories from Feeder Systems** by assigning the newly configured application level **CS01 Import Document for Sales** to the relevant types/categories, as shown in Figure 6.56.

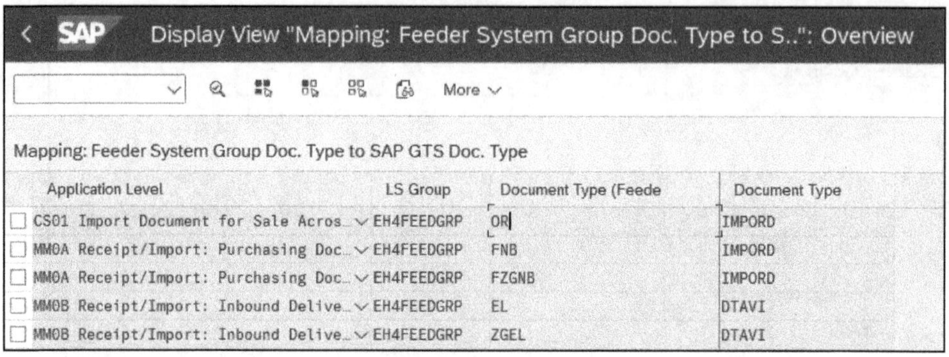

Display View "Mapping: Feeder System Group Doc. Type to S..": Overview

Mapping: Feeder System Group Doc. Type to SAP GTS Doc. Type

Application Level	LS Group	Document Type (Feede	Document Type
CS01 Import Document for Sale Acros...	EH4FEEDGRP	OR	IMPORD
MM0A Receipt/Import: Purchasing Doc...	EH4FEEDGRP	FNB	IMPORD
MM0A Receipt/Import: Purchasing Doc...	EH4FEEDGRP	FZGNB	IMPORD
MM0B Receipt/Import: Inbound Delive...	EH4FEEDGRP	EL	DTAVI
MM0B Receipt/Import: Inbound Delive...	EH4FEEDGRP	ZGEL	DTAVI

Figure 6.56 New Application Level CS01 for the Intercompany Sales Screening Functionality

6.4.7 Customizing

Setting up the general settings of legal control is similar to the general settings of embargo. After maintaining the baseline Customizing steps, including the definition of a legal regulation based on the applicable regulatory requirements, users can continue with the following settings in the Compliance Management legal control functionality. We'll take a look at the key settings next.

Assign Determination Procedure for Active Legal Regulation

This IMG activity is the same as the one explained regarding embargos in Section 6.3.

As part of the standard configuration settings, the determination procedure defined is **LELDT**, which consists of the following determination strategies:

- Determine legal regulation at the country group level
- Determine legal regulation at the country/country group level
- Determine legal regulation at the country group/country group level
- Determine legal regulation at the country group/country level
- Determine legal regulation at the country level

Furthermore, depending on the option selected, the system runs through all five determination strategies or stops after the first strategy that results in a hit (i.e., one or more legal regulations). In standard SAP GTS, the **Determination Using First Successful Strategy** option is selected, which means that when users set a legal regulation using the country group level, the country level isn't triggered.

Users can define whether the legal regulation for those countries and/or country groups is active for the different directions (import/export).

Activate Legal Regulation

This IMG activity is similar to the setup as mentioned in Section 6.2 and Section 6.3 for SPL and embargo, respectively.

The IMG activity to activate legal regulation in legal control requires users to select one or several legal regulations for the legal control service. In this menu, users can activate legal regulations for the countries nominated in **General Settings • Legal Regulations • Activate Legal Regulations at Country/Country Group Level**. This leads to the **Display View "Country": Overview** screen shown in Figure 6.57.

Figure 6.57 Activating Legal Regulations in Customizing

In this IMG activity, users should select the legal regulation that they want to activate for the legal control service. Depending on the previously selected determination procedure, users should activate the legal regulation on different levels.

If the chosen determination procedure runs at only the country level, it's sufficient to select the **Country** folder from the left menu and activate the legal regulation for all the relevant departure countries (see Figure 6.57). However, if the legal regulation runs on the country/country level, users first have to select the departure country by clicking on the **Country** folder on the left and then select the departure country on the right. This enables users to select the destination country by clicking on the **Country/Country** subitem of the **Country** folder.

Define Determination Procedure to Automatically Determine License Types

In this Customizing activity, Beta can define a determination procedure that can be used to automatically determine a license type. The user can access this Customizing activity by navigating to **SAP Global Trade Services, edition for SAP HANA • Compliance Management • Legal Controls • Define Determination Procedure to Automatically Determine License Types.**

Users need to create a determination procedure and assign a determination strategy. The determination strategy can be built by using one or more of the following levels:

- Country/control grouping level
- Country/control class level
- Country group/control grouping level
- Country group/control class level
- Country level
- Country group level
- Control grouping level
- Control class level
- Legal regulation level
- Country/control class/control grouping level
- Country group/control class/control grouping level
- Control class/control grouping level
- Control class/measure/country group level
- Control class/measure/country level
- Country group/country group/control class level
- Country group/country group/control class/control grouping level
- Region/region/control class level
- Region/region/control grouping level

A determination procedure, such as the one in Figure 6.58, can be made up of different strategies that consist of a hierarchy based on a *sequence number*. The system cycles through the different determination strategies based on the sequence number.

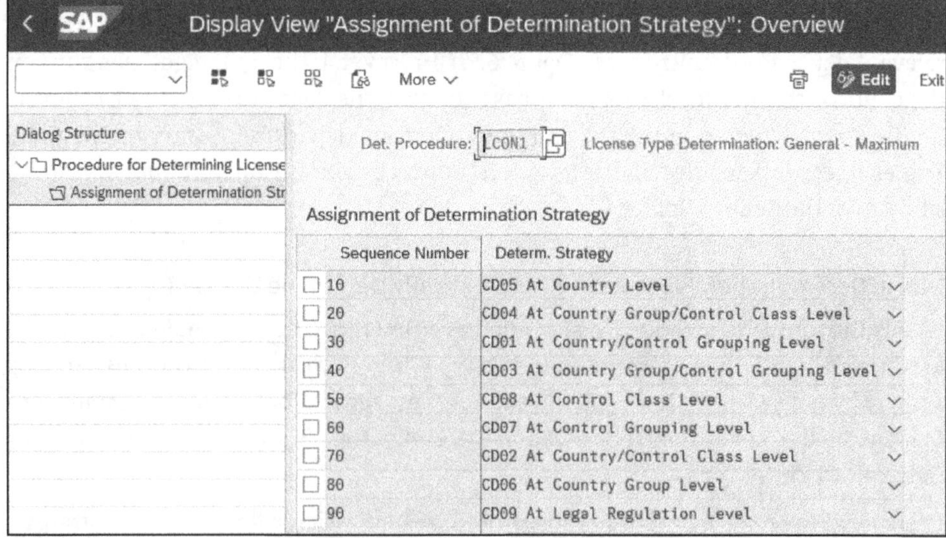

Figure 6.58 Determination Strategy Assignment to Determination Procedure

The system comes with standard determination strategies, though users can also define their own determination strategies.

The determination procedure defined in this Customizing activity is linked to the legal regulation in the **Control Settings for Legal Control** activity, which we'll discuss next.

Control Settings for Legal Control

Users can specify several configuration settings per legal regulation in this Customizing activity. Users can find most of these settings by selecting the relevant legal regulation and clicking on the **Details** icon. Figure 6.59 gives an overview of these settings. In this section, we'll describe partner group setup.

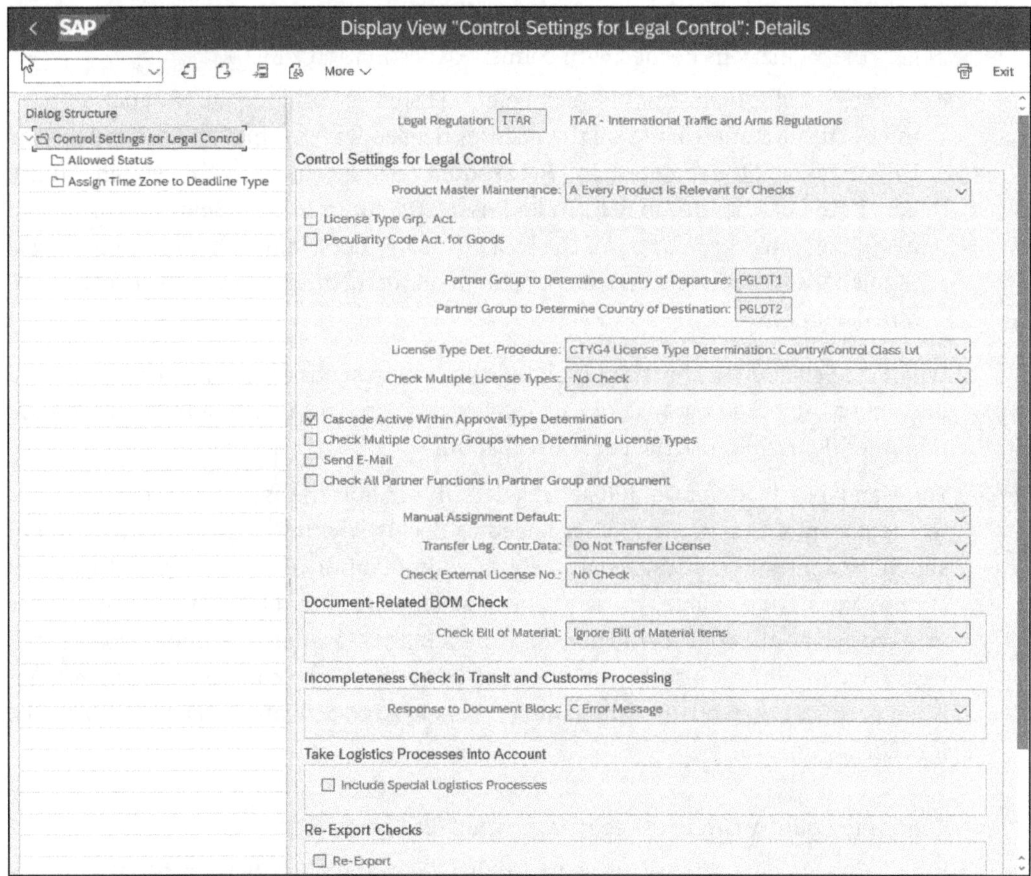

Figure 6.59 Control Settings for Legal Control

For example, for Beta, the partner group in Customizing is set up as in Table 6.3.

Sequence Number	Partner Group
10	Sold to
20	Ship to

Table 6.3 Assignment of Partner Functions to Partner Grouping

The sequence number determines the legal regulations being selected. Following are some examples:

- If there is a shipment from Beta in Belgium listed as "Sold to" the US and "Ship to" the UK, the system triggers all the licenses needed with country of destination "the US". If no "Sold to" partner function is maintained in the document, the system triggers the regulations needed with country of destination "the UK".
- If there is a shipment from Beta in Belgium listed as "Sold to" the UK and "Ship to" the US, the system triggers all the licenses needed with country of destination "the UK". If this isn't the requirement, Beta should consider changing the sequence numbers in the partner group (e.g., 10 **End user**, 20 **Ship to**, and 30 **Sold to**). This would mean that only the country of the "Sold to" party determines the selection of legal regulations if there is no end user or ship-to party maintained in the SAP feeder system document.

When the **Send E-Mail** checkbox is active, the system sends an email if a transactional document is blocked because of this legal regulation. This option also requires additional configuration to activate email capabilities.

The **Manual Assignment Default** setting determines how the assignment of the license to the document takes place. Users need to decide whether the license assignment should be done automatically, manually, or a combination of both. The **Check External License No.** setting enables users to check whether another license master is already using a specific external license number. **Check Bill of Material** determines whether the legal regulation must also be activated for BOM items (the components when a finished product is exploded). For example, users need to activate **Check Bill of Material** for ITAR.

Configure Country Group of Legal Regulation for Reexport

In this Customizing activity, users maintain the reexport settings per country group. Two standard country groups are maintained in Customizing (critical countries and uncritical countries). Here, users can set up the maximum percentage of relevant US content, as well as the calculation rule. This calculation rule defines where the fair market value of the US component is coming from (manually maintained or transferred from the feeder system on the legal unit level) and how the system should respond when no value is maintained.

It's possible to filter out components with specific control codes (e.g., EAR99). This functionality can limit manual deselection of different components.

Define Control Procedures for Address Comparison for License Types

When updating or amending the address information of a business partner in the feeder system, users need to reevaluate the licenses or authorizations that include the specified business partners to confirm validity. This functionality enables users to update the applicable fields.

Define Types of Agreements

Users can set up agreements and create applicable licenses in furtherance of an agreement or with a reference to an agreement. Relevant information is copied over from the agreement to the license, and the quantity and value of the agreement are decremented automatically based on the quantity and value set forth in the license. An agreement is created in the same way as a specific license type.

Define Depreciation Groups (in Documents and Agreements)

Users can depreciate a license in different groups (e.g., on the sales-order level or the delivery level) and specify a maximum amount or quantity to be written off in each deprecation group. The different depreciation groups also allow users to trace the items (e.g., a sale has already been made, but the goods haven't been delivered yet).

Item categories need to be mapped to specific depreciation groups. Users can then define whether the item group needs to add to, deduct from, or neither add to nor deduct from the remaining quantity/value.

Define License Types

Users can maintain different license types per legal regulation. Users are required to map the physical license to relevant SAP GTS license attributes, which can be maintained in the system. Users can define how many values of each attribute can be maintained (one or multiple) and whether the system needs to compare the value of each attribute with the variables in the sales order. If there is a complete match, that license is taken for automatic license assignment.

Return to Section 6.4.3 for more details on the most important attributes.

6.4.8 Master Data

Chapter 4 and Chapter 5 covered the master data elements for Compliance Management in SAP GTS in detail. Following is a brief review:

- **Numbering schemes**
  Users can load the control classes (previously referred to as ICCN and ECCN codes) from a content provider or maintain them manually in SAP GTS, as explained in Chapter 4, Section 4.3.2, and Chapter 5, Section 5.1.1.
- **Product classification**
  As soon as the classification codes exist, users can classify products in SAP GTS with the control codes (previously referred to as ICCN and ECCN codes), as explained in Chapter 4, Section 4.3.3 and Section 4.3.4, and Chapter 5, Section 5.1.2.
- **BOM**
  BOMs are required in SAP GTS to perform the de minimis calculation. For more information on the setup of BOMs in SAP GTS, refer to Chapter 4, Section 4.4.
- **Licenses and agreements**
  Licenses have the same structure for import and export, and users can customize their different attributes (Section 6.4.3 and Chapter 4, Section 4.5). Users can create licenses in reference to an agreement.
- **Determination strategy**
  Users set up the determination strategy as master data.

6.5 Summary

As discussed in this chapter, users perform SAP GTS export control checks at specific times in the order lifecycle. We've outlined and discussed the specific checks related to compliance for import and export flows.

After performing the necessary compliance checks, the next step is for users to confirm that the correct customs declarations are created and filed with the appropriate customs authorities. SAP GTS allows users to assign classification or license information directly into the customs document.

The following chapter deals with Customs Management and elaborates on how SAP GTS supports companies in declaring the import or export of goods to the customs authorities.

Chapter 7
Customs Management

Compliant customs management is critical to any global company. This chapter provides an overview of the customs processes and related SAP GTS capabilities.

In accordance with local rules and regulations, a specific set of documents must be presented to the *customs authorities*. The process for providing this information is made possible through a customs declaration. A customs declaration is a formal statement provided by the importer or exporter to customs authorities detailing the nature, value, and origin of the goods being transported. This can be transmitted either in physical print or digitally through data exchange. Most governing bodies now require the adoption of Electronic Data Interchange (EDI).

Globalization has caused a substantial surge in trade volumes, necessitating an automated environment that enables customs authorities to effectively monitor cross-border movements. Pre-arrival or pre-departure information is essential for the customs authorities to proactively conduct risk assessments using safety- and security-related data.

Warehousing and manufacturing also play a vital role in the supply chain. Governments encourage inward investments into warehousing and manufacturing for companies to remain competitive and create a positive cycle of prosperity by providing tax-efficient structures and incentives. In this respect, various customs authorities across the globe authorize special customs procedures, allowing the economic operator to apply for duty exemption schemes based on the applicable conditions. In this chapter, we'll describe the economic customs procedures applicable in the EU (bonded warehouse [BWH] or customs warehouse [CWH], inward processing relief [IPR], and outward processing relief [OPR]), the US (foreign trade zone [FTZ]), and China (processing trade), while accounting for the principles used by various authorities to implement their special customs procedures.

Note

Since the introduction of the Union Customs Code (UCC) in the EU in 2016, the IPR and processing under customs control (PUCC) functionalities have been merged into the inward processing procedure. From an SAP GTS technical and functional point of view, both functionalities are still separately available.

This chapter draws a clear distinction between business and system processes, as previously delineated in Chapter 6. The business process entails a description of the typical steps and events that a company undergoes, while the system process exclusively outlines the activities that users must navigate within the system. Note that certain steps within both the system and business processes can operate simultaneously.

Figure 7.1 showcases the SAP GTS spaces in SAP Fiori, revealing the four core dimensions of the solution's Customs Management functionality: import, export, special customs procedures, and the Excise Movement Control System (EMCS). Each of these customs spaces houses multiple pages that offer comprehensive insights into their respective functionalities. For example, within the **Customs Export** space, three distinct pages are dedicated to **Declarations**, **Customs Brokers**, and **Master Data**.

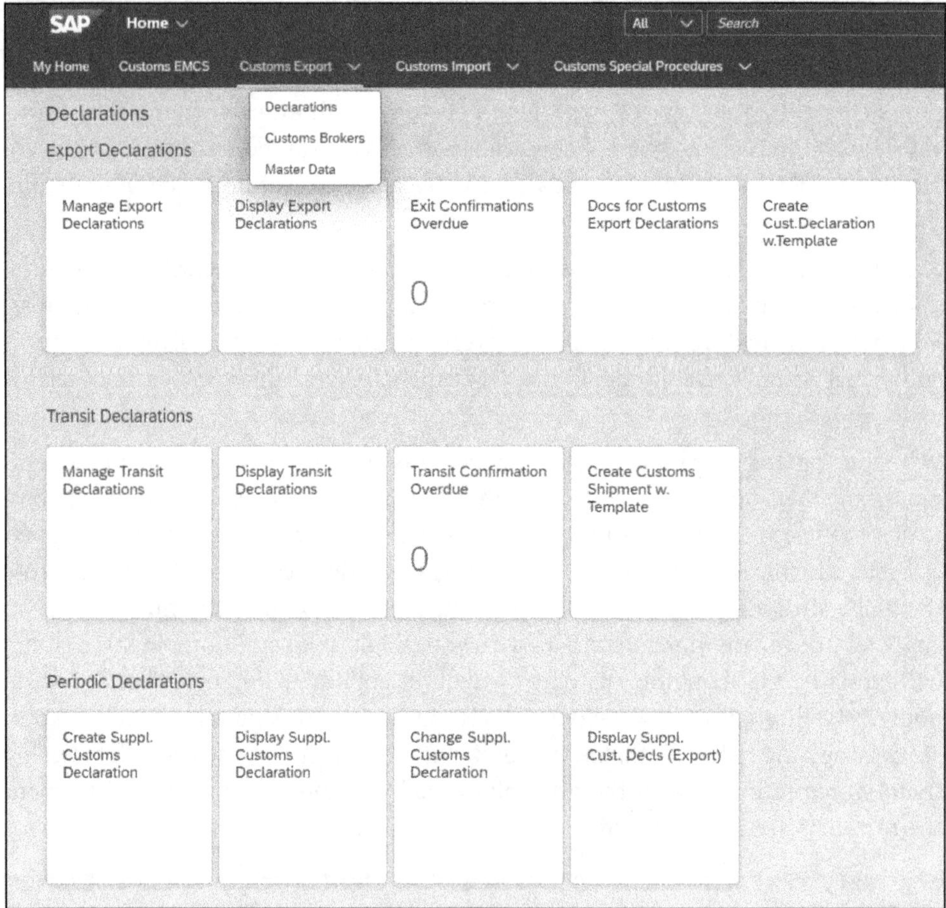

Figure 7.1 Customs Management: SAP Fiori Spaces, Pages, and Sections

After an overview of customs processing through the SAP Fiori launchpad in Section 7.1, this chapter details these processes, while highlighting the complexities of product classification. (See Chapter 5 for a more detailed look at product classification.)

A high-level overview of the functionality underpinning each of these key processes is as follows:

- **Customs import (Section 7.2.1)**
  The **Customs Import** space enables users to oversee general inbound processes, end-of-transit operations, periodic declarations, and pre-declarations. It provides essential information concerning tax statements, waybills, and customs invoices. Additionally, it features dedicated tiles designed to monitor customs brokers and track documents and quantities. This space also includes a dedicated page for master data that houses authorizations and security tiles.
- **Customs export (Section 7.2.2)**
  The **Customs Export** space supports all outbound movements, encompassing the departure of goods from a country or, in the context of EU countries, exiting the customs union. It also enables users to monitor outbound transit activities and manage periodic declarations. The functionality incorporates specialized tiles for customs broker monitoring and offers a dedicated page for master data. This master data page includes authorizations and security tiles, as well as tiles for managing business partners and products.
- **Customs special procedures (Section 7.3)**
  In Customs Management, SAP GTS supports various special customs procedures, such as CWH, IPR, OPR, prescribed end-use, processing trade, and FTZs. Users can manage and monitor these procedures using the tiles on the respective pages for **Customs Warehouse**, **Inward-** and **Outward Processing**, **Foreign Trade Zone**, and **Processing Trade** in the SAP Fiori launchpad.
- **Customs EMCS (Section 7.4)**
  Excise duties may apply depending on the nature of the traded goods (e.g., alcohol, tobacco, and energy). SAP GTS doesn't support keeping excise inventory for trading such goods; specific industry solutions can be used for such purposes. Nonetheless, for the movement of excise goods under excise duty suspension (called the EMCS process in the EU), the SAP Fiori launchpad in SAP GTS offers the **Customs EMCS** space to oversee the associated excise movement and control procedures. These declarations are referred to as administrative documents.

The SAP Fiori launchpad for SAP GTS provides spaces for **Classification** and **Master Data**. In the **Master Data – Cross-Area** space, users can access tiles that are relevant to all business processes: **Products** and **Business Partners**. Users can maintain **Organizational Data** in the tile for **Business Partners** as well. **Authorizations and Securities** need to be maintained in the respective spaces for **Authorizations and Securities for Imports**, **Authorizations and Securities for Exports**, or **Authorizations and Securities for Special Customs Procedures**. Every product that is relevant for Customs Management processes should be classified. This classification is most likely also relevant for Compliance Management and Preference Management.

This chapter also includes guidance on the US reconciliation process (Section 7.5). Reconciliation is a program that allows an importer to file entry summaries with the applicable government authorities.

Section 7.6 is dedicated to electronic communication and provides insight into SAP GTS's capabilities for establishing interfaces between a company and customs authorities or external entities, such as customs brokers—professionals licensed to assist importers and exporters in navigating customs procedures and regulations. This section also outlines the relevant pages and apps that prove invaluable for monitoring the exchange of electronic messages between these parties.

This chapter concludes with Section 7.7, which offers an overview of the countries that receive standard support from SAP GTS.

Throughout this chapter, examples are illustrated using a fictitious company named Alpha. It serves as a practical guide, walking through the primary business processes that are facilitated by Customs Management in SAP GTS. Figure 7.2 visually represents the underlying business case.

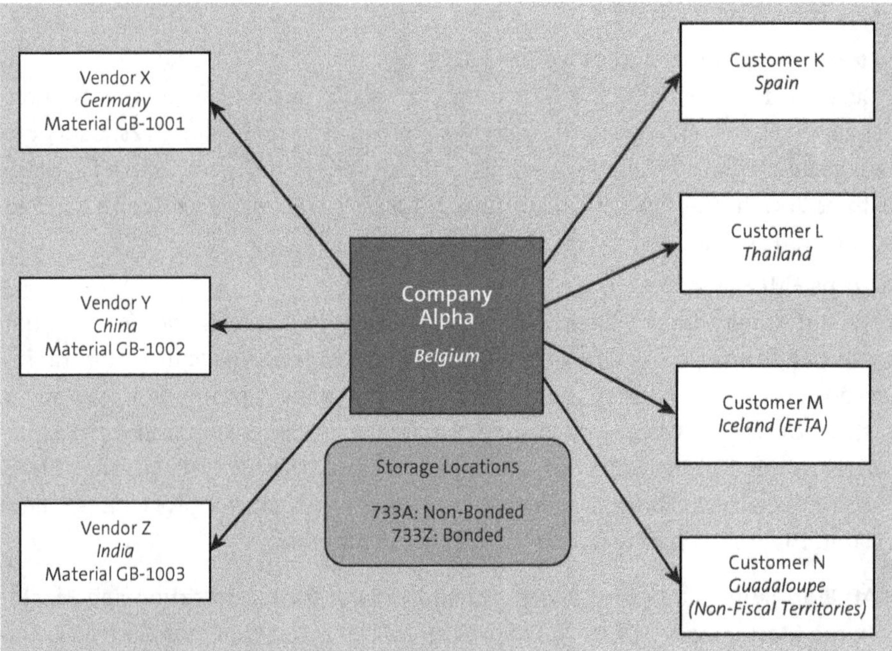

Figure 7.2 Business Case: Alpha

Alpha's Global Trade Requirements

Alpha, a company selling electronic devices (phones, mobile devices, and digital radios) on the business-to-consumer (B2C) market, has a growing volume of export sales. Alpha is an independent subsidiary of an international company. The European headquarters is in Belgium. All orders from Europe, the Middle East, and Africa (EMEA) are shipped from the European Distribution Center (EDC), which is also located in Belgium.

No production activities take place in Belgium. Alpha buys goods from foreign vendors in Asia and an affiliated company in Germany.

Alpha sells a limited number of finished products:

- **Material GB-1001**
  This product is procured exclusively from a vendor located in Germany (Vendor X). This product is designated for both the domestic market and export to Thailand (Customer L).
- **Material GB-1002**
  This product is procured from a vendor in China (Vendor Y), and, after importation, it's stored in a CWH. The product can be exported to Thailand (Customer L) or sold in the Spanish domestic market (Customer K).
- **Material GB-1003**
  This product is imported from a vendor in India (Vendor Z) and is mainly sold in the European market (Customer K). The products are imported and will be stored in a non-CWH. Goods stored in this warehouse are duty-paid (free circulation goods).

7.1 Customs Processing

This section addresses the SAP Fiori launchpad, which provides a daily worklist for customs declarations, as well as the fields that can be maintained in these customs declarations and ways to communicate with customs authorities.

7.1.1 Manage Customs Declarations

SAP GTS offers SAP Fiori apps that provide users with the right tools to administer customs declarations. Users can set up the structure of the report depending on their company's business. SAP GTS can generate separate import, export, inbound, and outbound transit worklist reports through the corresponding tiles.

The clear split in business processes enables users to distinguish different business processes and their corresponding tasks. SAP GTS offers different SAP Fiori apps to manage the various business processes for imports, exports, and transits. Furthermore, users can define separate tasks for each of these processes, making it easy for companies with large trade compliance teams to divide the workload based on these processes. In this way, team members can work on a specific shipment in their domain without interfering with their colleagues.

The **Manage Export Declarations** app (found under **Export Declarations** in the **Customs Export** space) has undergone changes in how it presents documents: instead of grouping documents based on the respective activities they were in previous SAP GTS editions, the new SAP Fiori app provides the user with a **Processing Status** and suggests a follow-up action (**Processing Proposal**) to finalize the activity (illustrated in Figure 7.3). After users finish all relevant, mandatory tasks within a certain activity, the document will automatically transfer to the next required activity.

There is no dedicated tile available to manually create an export declaration in SAP GTS. Instead, declarations and actions are centrally administered under one application, **Manage Export Declarations**. To manually create a customs declaration, users can simply click on the **Create** option within the application.

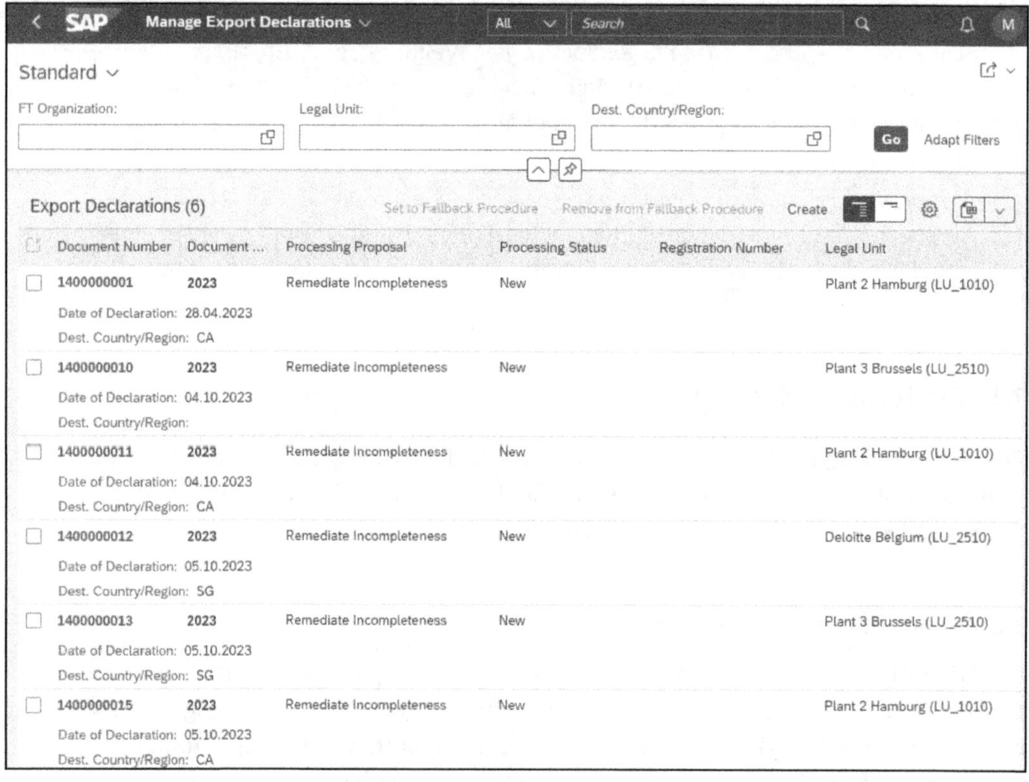

Figure 7.3 SAP Fiori App: Manage Export Declarations

From a business point of view, each main process can be divided into multiple subprocesses. In SAP GTS, each subprocess is called an *activity*. Users must perform one or more tasks in each activity to complete it and move to the next activity.

After all activities are completed, the document will disappear from the worklist under the **Manage Export Declarations** tile. However, users can display it through the monitoring functionality, which can be accessed via the **Export Declarations** space in the **Display Export Declarations** tile.

Alpha's Customs Processing Activities

When designing the SAP GTS solution implemented at Alpha, the business identified the following activities that needed to be performed to provide a correct and complete export declaration. It's crucial to follow the sequence of these activities:

1. Complete the export declaration to ensure that all legally required information is available.
2. Send the information from the declaration to the customs authorities.
3. Anticipate a technical response from the customs authorities, confirming the receipt of the information.
4. Anticipate the validation of the export declaration from the customs authorities. This message confirms the approval of the export declaration.
5. Print the export accompanying document (EAD) for the transporter and internal bookkeeping.

Alpha decided to create the following three sequential activities in SAP GTS:

1. **Export declaration**
   Users complete the declaration and send the information to the customs authorities via an IDoc. This IDoc can't be sent if the document is incomplete (see Section 7.2.3 for more information on an incompleteness check). After the IDoc is sent out (called an M-message in SAP GTS), the document moves to the next status.
2. **Export accepted**
   Users need to wait for the two reply EDI messages (called an I-message in SAP GTS) from the customs authorities to return. The first EDI message confirms that the customs authorities received the information. This message doesn't release the shipment for export. The exportation of the goods is only approved after the company receives the second EDI message. After the second EDI message has been received, the document automatically moves to the next activity.
3. **Print export document**
   Users aren't able to print the export declaration from the system until the document reaches this activity. This system setup prints the most up-to-date information onto the paper documents that accompany the goods during transportation. By restricting the ability to print the document earlier in the process, Alpha ensures the elimination of inconsistencies between the printed export declaration and what has been declared to the customs authorities through the EDI message.

Depending on the legal regulation and the system version, there may be additional activities in the e-filing procedure. For example, an optional activity can be added to the sequence to cancel export declarations.

7.1.2 Display Customs Declarations

The monitoring functionality within SAP GTS provides an alternative method of displaying an overview of all customs declarations via the **Display Export Declarations** app. This tile complements the declarations available through **Manage Export Declarations** app, which specifically retrieves declarations in an open status.

In the **Display Export Declarations, Display Import Declarations**, and **Display Transit Declarations** apps, all customs declarations are listed, irrespective of the activity they are in. This enables users to find back documents that are fully processed, or canceled, in addition to currently active documents.

The report is split into three sections, as shown in Figure 7.4:

- The first section's selection screen enables users to apply certain filters to refine their search for existing declarations, such as per **FT Organization, Legal Unit**, and **Dest. Country/Region**.

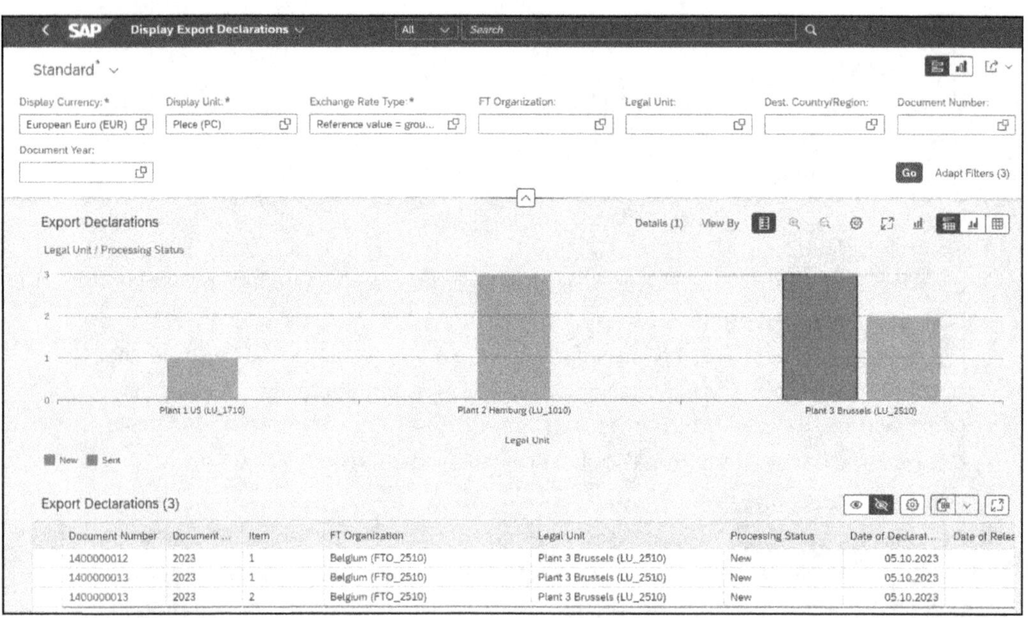

Figure 7.4 SAP Fiori App: Display Export Declarations

- In the second section, SAP GTS launched embedded analytics in the form of customizable *charts* to provide business insights. For instance, standard charts are available to exhibit the number of export declarations per **Processing Status**. The embedded analytics include a variety of reporting charts to choose from, such as pie charts, column charts, or line charts. Before these charts are populated, users need to specify the applicable category and axis. For example, users can create a chart based on the commodity code (category) and the number of items (axis). The charts operate with intuitive and interactive features, allowing users to select specific section of the chart and prefilter the corresponding declarations. For more information on embedded analytics, see Chapter 9, Section 9.2.

- The third section in the application provides the overview of existing declarations, according to the filters activated in the selection screen.

Because the documents aren't grouped into activities, it can be challenging to determine which actions to take to get a certain declaration approved by the customs authorities and to drive it to completion. To tackle this, users can filter documents by **Processing Status** to display the documents associated to a certain status activity.

7.1.3 Customs Declaration

SAP GTS allows users to create a customs declaration for the movement of goods relevant for customs purposes. Each customs declaration created in SAP GTS is structured in the same way and looks similar. All documents consist of a header level and an item level, regardless of whether the customs declaration is an import, export, transit, or EMCS. In SAP GTS, information that is valid for the entire document is available at the header level, whereas item-specific information is available at the item level.

A set of organized header level tabs offers a detailed breakdown of crucial information. Each tab serves a specific purpose to help navigate through the complexities of the declaration process. For instance, **Item Overview** gives a summary of all declared items, as shown in Figure 7.5, while **Communication** handles messages needed for (electronic) communication with customs authorities. **Partners** details the involved entities in the declaration. **Geography** focuses on essential location data such as departure and destination countries. **Transportation** deals with trade specifics such as dates, transport modes, and customs status. **Packaging Data** covers details about packaging and product weights. **Documents** and **Logistics Costs** manage paperwork and duty-related values. **Invoice** refers to invoice details, while **Status** provides updates on declaration progress. **Document Flow** shows documents linked within SAP GTS, **References** manage document numbers from the feeder system, **Texts** allows additional editable inputs, and **Administration** compiles administrative data. These tabs ensure a clear and organized approach to handling customs declarations.

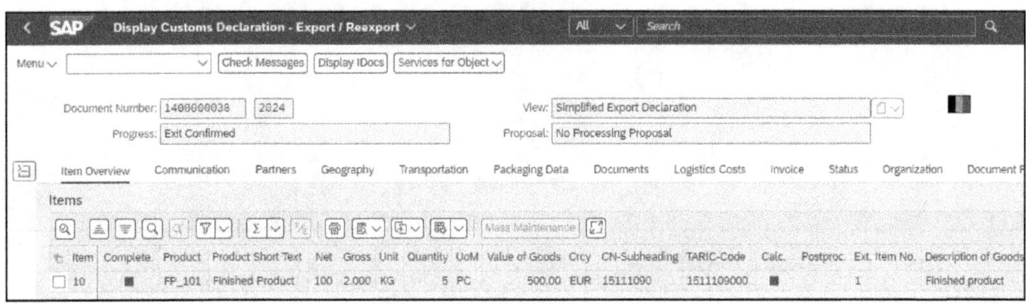

Figure 7.5 Customs Declaration: Export Header

The customs declaration item level comprises several tabs, each capturing crucial details for efficient management. The **General Data** tab encompasses fundamental item information such as product description, weight, origin, and commodity code, as shown in Figure 7.6. The **Placement** tab records invoice specifics, customs status placement, and identification for special procedures. **Packaging Data** logs container numbers and packaging specifics per item, while the **Documents** tab lists necessary additional documents required by customs, including management of the Binding Tariff Information (BTI) reference number. A structured framework within the **Customs Value** tab calculates values, deductions, customs values, import duties, and taxes. The **Status** tab tracks declaration status, while **Preference** indicates eligibility for trade preference (if Preference Management is active). A **Closing Portion** exists for items under special customs procedures, and the **Texts** tab allows for editable additional information specific to the item. Furthermore, there are sections for **Control** involving legal regulations, license types, numbers, and compliance integration, as well as **Administration** to store additional administrative data.

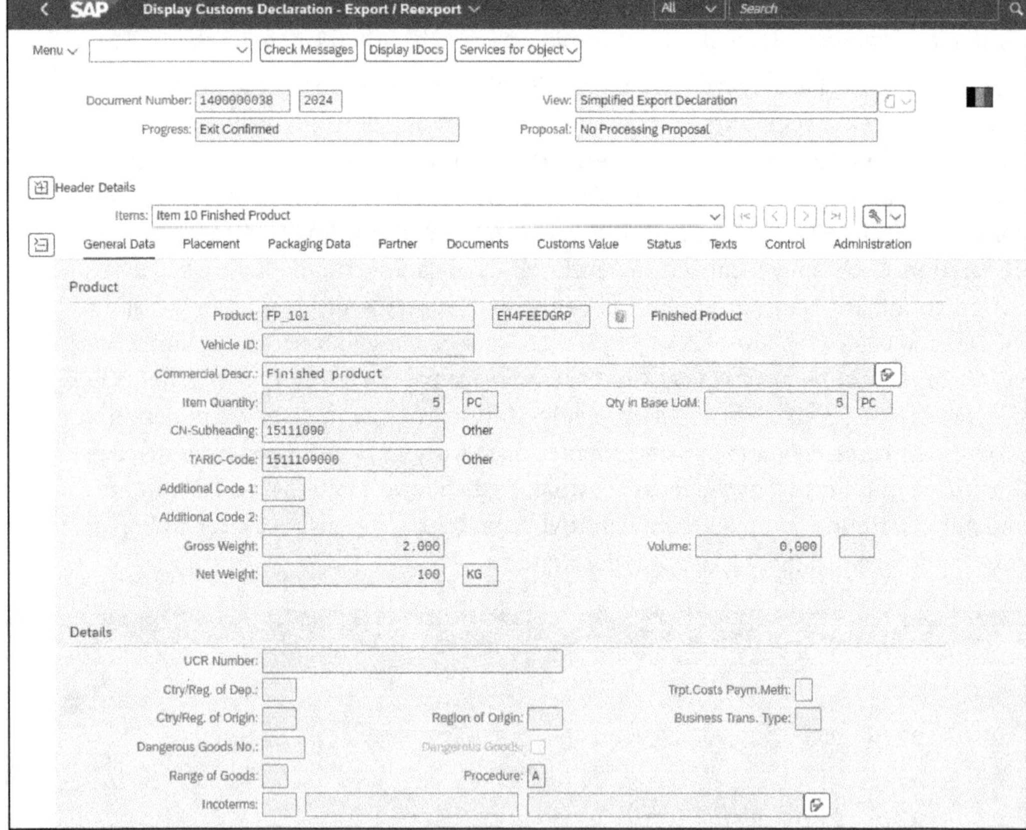

Figure 7.6 Customs Declaration: Item Level

An established standard interface streamlines the transfer of declaration data from the feeder system to SAP GTS, aiming to minimize manual input. Typically, shipping information is readily accessible in the feeder system and automatically transmitted to SAP GTS, effectively cutting down on manual tasks. Users have the flexibility to modify the data transferred from the feeder system. Additionally, SAP GTS permits the manual completion of any missing data. This feature enables users to adjust and supplement information as needed, ensuring accuracy and completeness within the system even after the initial transfer process.

Certain fields within SAP GTS necessitate manual input, while others can be automatically filled using defaulting rules integrated into the system. The determination of which fields are directly transferred from the feeder system to SAP GTS and those populated via defaulting rules depends on various factors. These factors include the nature of the data itself and specific requirements tailored to each company's needs.

For more information on defaulting data into a customs declaration, refer to Section 7.2.3 and Section 7.2.4.

7.1.4 Communication with Customs Authorities

There are three primary methods for declaring goods to the customs authorities: verbal declaration, paper-based declaration, and electronic declaration. However, in international trade scenarios, paper-based and electronic communication methods are used most frequently. The subsequent sections will delve into the specifics of both paper-based and electronic declaration options, outlining their respective processes, advantages, and considerations within the context of international trade.

Paper-Based Declaration

The conventional way of declaring goods to the customs authorities is a paper-based declaration. The forms used for this type of declaration depend on the country's specific rules and regulations.

SAP GTS enables the use of a paper-based declaration because it can incorporate all information from the customs document into a form. This form can be printed directly from the SAP GTS customs document from the **Communication** tab. For countries that are standardly supported by SAP GTS, standard forms are available in the system. However, if a company operates in a country not standardly supported by SAP GTS, additional forms or broker instruction sheets need to be developed.

Figure 7.7 displays a standard EAD for a country in the EU. An exact copy is created in SAP GTS so that users can print a fully automated version of the form from the system.

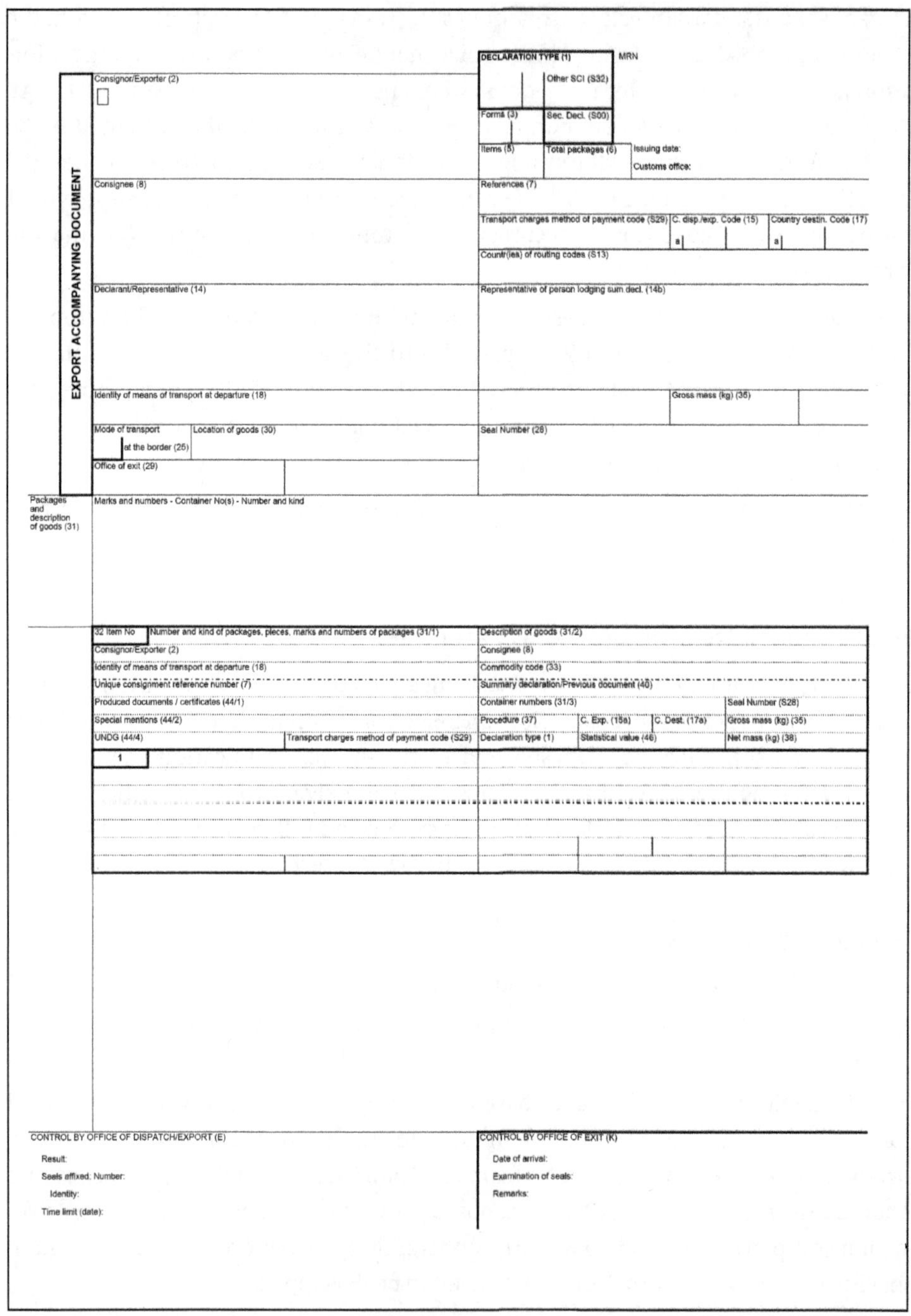

EXPORT ACCOMPANYING DOCUMENT

DECLARATION TYPE (1)
MRN
Consignor/Exporter (2)
Other SCI (S32)
Forms (3)
Sec. Decl. (S00)
Items (5)
Total packages (6)
Issuing date:
Customs office:
Consignee (8)
References (7)
Transport charges method of payment code (S29)
C. disp./exp. Code (15)
Country destin. Code (17)
a
a
Countr(ies) of routing codes (S13)
Declarant/Representative (14)
Representative of person lodging sum decl. (14b)
Identity of means of transport at departure (18)
Gross mass (kg) (35)
Mode of transport at the border (25)
Location of goods (30)
Seal Number (28)
Office of exit (29)
Packages and description of goods (31)
Marks and numbers - Container No(s) - Number and kind

32 Item No
Number and kind of packages, pieces, marks and numbers of packages (31/1)
Description of goods (31/2)
Consignor/Exporter (2)
Consignee (8)
Identity of means of transport at departure (18)
Commodity code (33)
Unique consignment reference number (7)
Summary declaration/Previous document (40)
Produced documents / certificates (44/1)
Container numbers (31/3)
Seal Number (S28)
Special mentions (44/2)
Procedure (37)
C. Exp. (15a)
C. Dest. (17a)
Gross mass (kg) (35)
UNDG (44/4)
Transport charges method of payment code (S29)
Declaration type (1)
Statistical value (46)
Net mass (kg) (38)
1

CONTROL BY OFFICE OF DISPATCH/EXPORT (E)
Result:
Seals affixed: Number:
Identity:
Time limit (date):

CONTROL BY OFFICE OF EXIT (K)
Date of arrival:
Examination of seals:
Remarks:

Figure 7.7 Printed EAD Form

Note

Section 7.7 gives an overview of the countries that are supported by standard SAP GTS.

Electronic Communication (E-Filing)

In addition to paper-based declarations, SAP GTS provides the option to communicate electronically via a customs broker or directly with the customs authorities for declaring imports, exports, EMCS, and transits. This electronic method of communicating with the customs authorities is known as *e-filing*.

E-filing is only possible if the country in which the company is located enables this procedure and the company has an IT system that can handle the processing of electronic communication.

The major benefit of electronic communication is reducing processing time. In some countries, e-filing is compulsory, and paper-based declarations are generally not accepted by the customs authorities (unless, e.g., a temporary fallback procedure is active). Every country that enables e-filing has defined a framework of outgoing messages (M-messages) and incoming messages (I-messages). In this framework, the customs authorities have defined a sequence in which they need to receive and send these messages.

In countries that aren't standardly supported by SAP GTS, companies can set up e-filing through a broker. This requires an interface to facilitate electronic communication (see Section 7.6 for more information about broker interfaces). When permitted by operational business processes, the standard XML structure can be advantageous for external communication. This approach minimizes the developments required to support electronic communication. In the system, a decent design of the customs declaration management page can encourage the correct application of sending and receiving these messages.

Figure 7.8 provides an example of a simplified communication framework for declaring an export shipment. Initially, the company declares the goods to the customs authorities, sending the information in an IDoc called an M-message, providing all the required information.

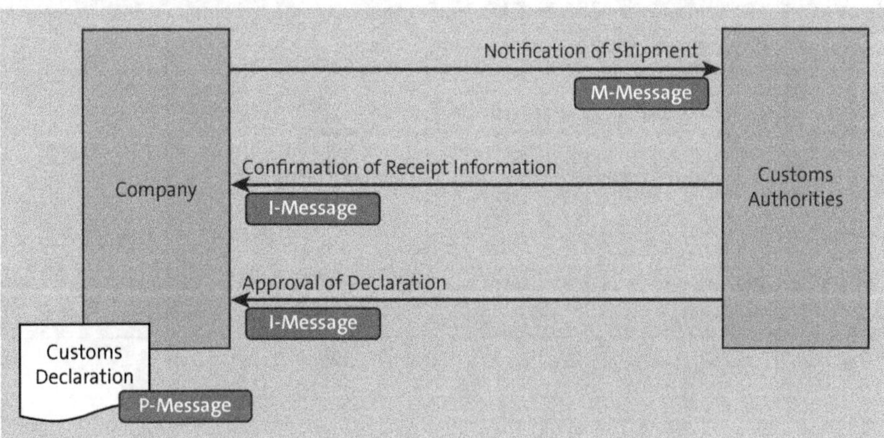

Figure 7.8 E-Filing Communication Scheme

The **Communication** tab in the customs declaration (see Figure 7.9) displays which message can be executed by the system. Note that the SAP GTS proposal for this declaration is to execute the message. Customs Management in SAP GTS has a processing **Proposal** for each declaration. This informational field indicates the next action required for a specific declaration. The possible **Proposal** statements are as follows:

1. **Execute Message**
   In the customs declaration, users can send a message to the customs office.
2. **Remediate Incompleteness**
   In the customs declaration, the next executable message is incomplete. To be able to send the message, remediate the incompleteness.
3. **Check Error Response**
   The customs office has replied to the customs declaration with an error response. Check the error response.
4. **Inspection**
   The customs office requires an inspection of the goods.
5. **Follow-Up**
   The customs office requires a follow-up on the location of the goods. Check the location of the goods.
6. **Document Verification**
   The customs office requires verification of the accompanying documentation for completeness and veracity. Check the documents for completeness and veracity.
7. **Check Duty Statement**
   Check the inbound statement for the taxes and customs duties to be levied.
8. **Request Cancellation**
   The customs declarant wants to request a cancellation of the customs declaration by the customs office. The cancellation message can now be sent.

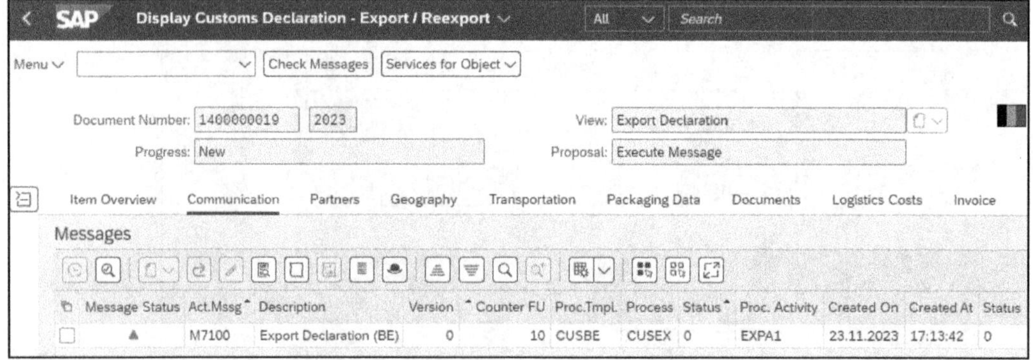

Figure 7.9 Initial Situation before E-Filing

Once the initial message is sent, the system freezes any modification to the declaration. Therefore, the document automatically switches to display mode, meaning users can display the document but can't edit the declaration.

The customs authorities initially validate that they have successfully received the information. They acknowledge this receipt by sending a first message back; as mentioned earlier, in SAP GTS, these are called incoming or I-messages. Users can monitor I-messages from the **Communication** tab of the customs declaration, as shown in Figure 7.10.

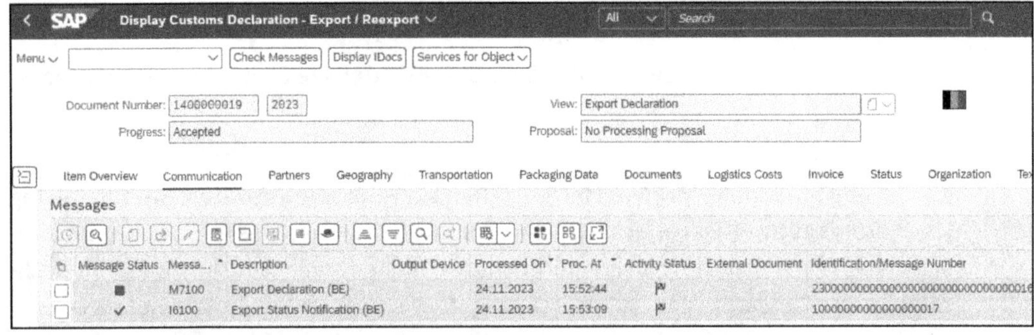

Figure 7.10 Receipt of First Incoming Message from Customs Authorities

Receiving this message doesn't immediately allow the company to start the export process; this indicates that the customs authorities confirm the receipt of the information but haven't yet granted the official approval.

After the customs authorities have validated the data, the declaration may be accepted or rejected. Communication regarding this confirmation from the customs authorities is again organized via IDocs—more specifically, using an I-message. Figure 7.11 displays the receipt of the second incoming message, which informs the company of the approval of the export shipment.

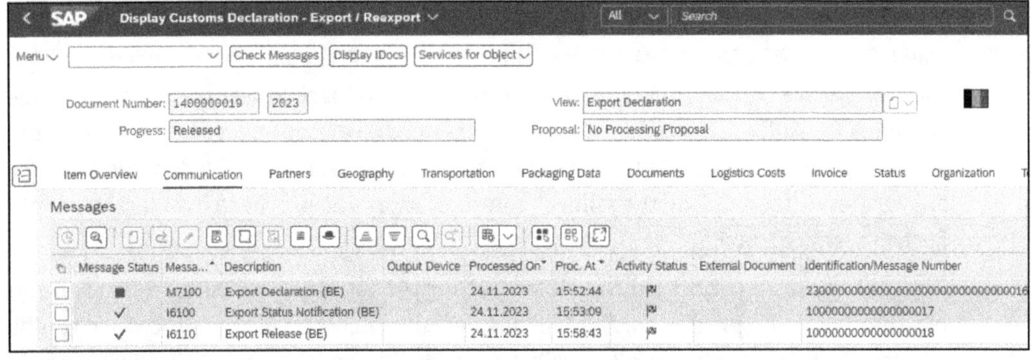

Figure 7.11 Receipt of Second Confirmation Message from Customs Authorities

After all approvals from the customs authorities are gathered in the system, supporting documents (e.g., documents accompanying the goods during transportation, certificates of origin, etc.) can be printed. The activity sequence can be customized in SAP GTS, allowing the document to seamlessly proceed to the subsequent activity, after the execution of a mandatory M-message, I-message, or P-message (print).

Communication (printed forms and/or electronic communication) with the customs authorities takes place in the customs document; all communication activities are grouped in the **Communication** tab on the document header. The system can receive standard error messages from the customs authorities, depending on how the communication platform of the customs authorities is set up.

Section 7.6 will discuss a few useful standard transactions related to the EDI messages.

Fallback Procedure

The e-filing functionality offered by SAP GTS makes the administrative tasks involving the customs authorities easier and faster than the paper-based tasks. However, a downside to e-filing is that the customs authorities' IT system can face downtime from technical issues, updates, and so on. During these periods of downtime, electronic communication with the customs authorities isn't possible, which can cause delays in a company's shipping operations. To prevent logistic interruptions, company Alpha can apply a *fallback procedure*. In SAP GTS, users can allocate a fallback activity sequence to every customs activity sequence. Note that not every customs authority accepts fallback procedures. The fallback sequence contains all the activities necessary to execute and complete fallback procedures related to transits and customs. When the fallback procedure is activated, the system uses the fallback activity sequence as an alternative to the standard activity sequence.

When the authorities grant a company-specific or general fallback procedure, users can enable a functionality in SAP GTS that allows the printing of export documents without prior electronic communication with the customs authorities.

To activate the fallback procedure, users switch the process on via the administrator role (/SAPSLL/GTS_ADMINISTRATOR) and then choose **SAP Global Trade Services, edition for SAP HANA • Documents • Customs Processing • Fallback Procedure.** In the **Manage Numbers for Fallback Procedure** activity, users need to define a fallback number before activating the procedure. This number is the official number provided by the customs authorities. After defining the fallback number, users can proceed to the **Activate Fallback Procedure** activity.

Depending on the country, the customs authority may communicate a fallback number that has to be printed on the fallback declaration. The fallback procedure can be switched on separately for all possible procedures, such as export, import, transit, EMCS, entry into a special customs procedure, and so on.

In some countries, customs authorities require subsequent e-filing after the fallback procedure is terminated. This process is called buffering and is also supported by SAP GTS.

All documents created when the fallback procedure is enabled are assigned to a fallback activity. This special fallback activity is designed to replace the e-filing activities with a printed form.

Supplementary Customs Declaration

Aside from the option of communicating a declaration to the customs authorities on a transactional basis (via e-filing, paper-based, or customs broker methods), there is an alternative approach for companies to consolidate and declare all relevant documents periodically, such as weekly or monthly. This type of declaration is called the *supplementary customs declaration* (SCD).

SAP GTS provides the functionality to facilitate this process. When users activate this functionality in the Customizing for a specific process, the system automatically groups together all the relevant shipments or customs declarations into an SCD. Within this global declaration, the system provides users with the flexibility to select and deselect shipments for inclusion in the declaration. Users can e-file the SCD with the customs authorities or generate a printed form, depending on the local authority's requirements.

Note that prior to setting up an SCD, users must obtain a specific authorization from the local customs authorities (e.g., authorization for simplified procedures, periodic declaration).

7.2 Import and Export Processes

Now that we've examined the customs documents related to Customs Management in SAP GTS, we'll look at import and export processes. This section delves into the business and system processes related to import and export management, the associated SAP GTS customization, and the essential master data necessary to execute the processes.

7.2.1 Import Processes

An import procedure involves a range of critical tasks, including the submission of detailed customs documentation. These documents typically include invoices, packing lists, certificates of origin, and transport documentation. It's important to note that the specific requirements mentioned for document submission may vary depending on the country and its customs regulations. It's essential for importers and trade participants to be aware of and adhere to the specific guidelines set forth by the customs authorities of the relevant country. Adapting to the unique requirements of each jurisdiction is crucial to ensure seamless and compliant import processing. Customs authorities meticulously scrutinize these records to verify the accuracy of the declared information, such as the nature, value, and origin of the goods.

Upon document approval, import duties and taxes are assessed based on the goods' classification under the Harmonized System (HS) codes, which categorize products for tariff purposes, as discussed in Chapter 5. Determining the correct classification is vital, as it directly impacts the applicable duty rates and trade policies.

The entity accountable for covering import duties is determined by the chosen *Incoterm*. Sometimes the applicable duty rate is 0%, which implies the entity has no obligation to remit duties to customs authorities. In addition to these import duties, often termed third-country duties, there might be supplementary charges such as antidumping duties.

Note

Incoterms are standardized trade terms established by the International Chamber of Commerce (ICC) that define the responsibilities of buyers and sellers in international transactions. Incoterms clarify aspects of transportation, delivery, insurance, customs, and payment to prevent misunderstandings and disputes.

After the duty assessment, customs clearance is granted, which allows the goods to physically enter a country. Customs officers may conduct physical inspections or use nonintrusive inspection technologies to ensure that the goods match the declared information and comply with safety and security standards.

In the following sections, we'll look at the business and system processes.

Business Process

Alpha Corporation procures the finished product GB-1003 from an esteemed vendor nestled in India. The chosen mode of transportation is maritime, entailing an approximate voyage of 30 days for the goods to reach their destination at the port of Antwerp. Figure 7.12 provides insight into the importing business process for Alpha.

Determining the most appropriate timing to start the import declaration process requires considering both a country's laws and a company's needs. When goods arrive at the customs office in the EU, for example, they are stored temporarily under customs supervision for up to 90 days. During this time, customs checks and monitors the goods. The goods are then assigned one of several customs procedures, which match their specific situation. This helps manage how the goods enter the EU, or they may be reexported to their origin. There are two primary choices for starting the import declaration process. The goods can be imported either at the customs office of entry (import declaration before goods receipt) or at the warehouse location within the customs union (import declaration after goods receipt).

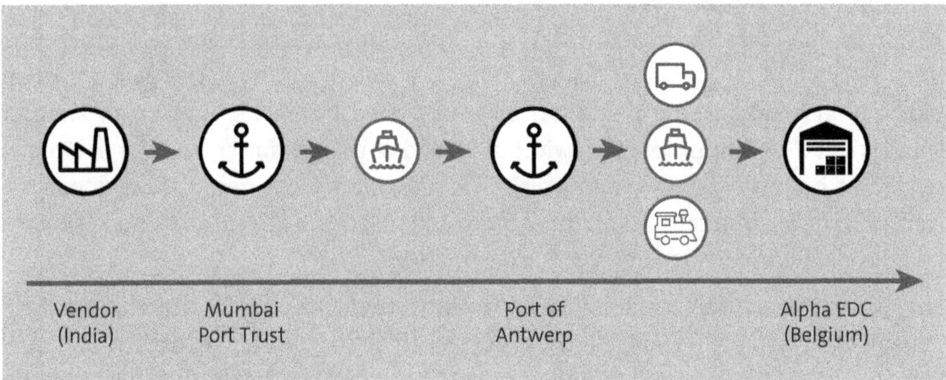

Figure 7.12 Import Business Process

Import Declaration before Goods Receipt

When companies based in the US or Asia prepare an import declaration, it typically must be done before the goods arrive at the warehouse. This means the declaration is made before the goods are physically examined and stored in a warehouse. The declaration is created at the port of entry.

In this situation, companies need to recognize that variations in quantity might arise between the formal declaration and the actual receipt of goods in the warehouse. If such discrepancies occur, companies may be required to report them depending on the import regulations of the country. These disparities can arise, for instance, when the supplying party sends more or fewer goods than were initially ordered.

Upon acceptance of the import declaration by the customs authority of the relevant country, the goods are then granted the freedom of movement within the customs union, in addition to being eligible for sale within the relevant market.

Import Declaration after Goods Receipt

Alpha is located in Belgium and has chosen to create an import declaration after the goods are received in its warehouse. This means that Alpha can receive the goods duty-suspended at their premises (inland) and do the customs clearance from its own location. In this scenario, the customs broker usually initiates an outbound *transit* procedure to ship the duty-suspended goods from the port of entry to Alpha's warehouse location.

Customs transit is a procedure that enables the movement of goods between two points within a customs territory via another territory or between different customs territories. This temporarily suspends duties, taxes, and commercial policies applicable upon import, allowing clearance procedures to occur at the destination rather than the entry point. Customs transit is useful in cases where a single customs territory has multiple fiscal areas by streamlining the movement of goods from entry to clearance while addressing customs and fiscal obligations.

Within the EU, there are two distinct transit procedures in place. First, the T1 procedure, often referred to as external transit, pertains to goods the entity hasn't paid duties on (non-Union goods). These items can be transported from one location to another within the EU's customs territory without incurring import duties, additional charges related to imports (e.g., internal taxes), or trade policy measures. If these goods are transported to a different EU Member State, the responsibility for customs clearance transitions to the customs office at the destination.

The second procedure, T2, is designed for the movement of goods for which duties have already been paid (Union goods). These goods can be transported within the EU's customs territory from one point to another without any alteration to their customs status. This includes the transportation of goods through another territory that falls outside the EU's customs territory.

Once the goods reach Alpha's premises, the transit procedure is discharged by notifying the relevant authorities. This notification is carried out either through e-filing or by submitting a paper form.

After the container is unloaded at Alpha's premises, the company starts recording the received items in its records. Once all the materials and quantities are logged, Alpha creates an import declaration and sends it to the customs authorities for approval. During this approval process, the customs authorities provide Alpha with a reference number for the shipment. In the EU, this number is called the *movement reference number (MRN)*, and in other countries, it's often just called the registration number.

Import Duty Calculations

Alpha's involvement in importing goods from India to the EU comes with the requirement to fulfill import duties and any associated taxes, as the goods are brought into free circulation within the European market without undergoing special customs procedures (Section 7.3). This principle of meeting duty and tax obligations is a common thread across various countries when goods are imported. The precise duty rates applicable depend on the specific destination country where the goods are being introduced.

When calculating import duties (or third-country duties) for a specific shipment, there are three crucial factors to consider: the customs value, the applicable customs tariff, and the origin of the goods. The calculation typically starts with the transactional value (value of the goods), and requires entities to add or deduct certain costs, such as insurance and freight, depending on the chosen Incoterm. This calculation results in the *customs value*, which is the base for determining the amount of duty owed to the customs authorities.

Note

The customs value comprises three distinct elements: (1) the transaction value of the goods, (2) the value of services linked to delivering the goods to the exporting country's border (if not already included in the transaction value), and (3) the value of services associated with transporting the goods from the exporting country's border to the importing country's border (also if not already included in the transaction value).

The adjustments to the transaction value of imported goods to calculate the customs value depend on the Incoterm being used. The Incoterm (also referred to as delivery terms) constitute an agreement between the seller and the buyer on who is responsible for the cost and risk of delivering the goods to the agreed place.

Delivery terms used in international commerce include those of the free on board (FOB) and cost insurance and freight (CIF) types, as defined by the ICC. To determine the customs value, entities only need to include the costs up to the place of introduction of the imported goods into the customs border.

When entities need to pay some sort of value-added tax (VAT) or sales tax, the base for applying this rate consists of the customs value and the duties relevant for that shipment. Therefore, VAT and sales tax also need to be paid on duties.

Table 7.1 presents the duty structure as perceived by Alpha while determining the customs value for the import of finished product GB-1003, originating from an Indian vendor.

	Duty Type	Amount	Percentage
	Value of the Goods	350.00 €	
+/–	Transportation Costs	3.50 €	
+/–	Insurance Costs	1.25 €	
=	Customs Value	354.75 €	
	Third-Country Duty	11.35 €	3.2%
+	Other Duties	0	0%
=	Total Duties	11.35 €	
	Customs Value	354.75 €	
+	Total Duties	11.35 €	

Table 7.1 Duty Framework for Import

	Duty Type	Amount	Percentage
=	Calculation Base for VAT or Sales Tax	366.10 €	
	Calculation Base for VAT or Sales Tax	366.10 €	
	VAT or Sales Tax	76.88 €	21%

Table 7.1 Duty Framework for Import (Cont.)

System Process

The Customs Management functionality in SAP GTS streamlines the process of submitting necessary import declarations by seamlessly integrating customs and logistics operations. Depending on the legal requirements, users can generate customs import declarations within SAP GTS either before or after posting the goods receipt. In some cases, such as when goods are to be delivered to a location (with duty suspension), a transit declaration might be needed. Import declarations are typically created before goods receipt in regions such as the US, Asia-Pacific (APAC), and other non-EU countries. However, within the EU, businesses can choose to create import declarations either before or after the goods receipt, based on their specific needs.

In the import process, products entering the customs territory of a country are assigned a customs procedure by submitting a customs declaration with the competent customs authorities (e.g., bringing goods into free circulation, placing goods under a BWH, placing goods under inward processing, etc.). Completion of the customs declaration differs depending on the specific customs procedure under which the products are placed. The customs procedure code is a data element that is included in the declaration.

Within the import procedure, as products cross into a nation's customs jurisdiction, they undergo a customs procedure allocation. This is achieved by filing a customs declaration with the appropriate regulatory bodies. The range of procedures encompass actions such as bringing goods into free circulation, securing goods within a BWH, or initiating inward processing. The way the customs declaration is finalized depends on the customs procedure governing the products' status.

Although entities can technically set up the creation of import declarations before and after goods receipt for the same foreign trade organization (FTO), they are typically not advised to do so. In the subsequent sections, the trigger points of the system and its behavior will be examined more comprehensively. But first, we'll explain general navigation to the apps that we'll discuss in the SAP Fiori launchpad.

Navigating Import Processes

In the SAP Fiori launchpad, users can conveniently access all tiles associated with import processes within the dedicated **Customs Import** space, as demonstrated in Figure 7.13.

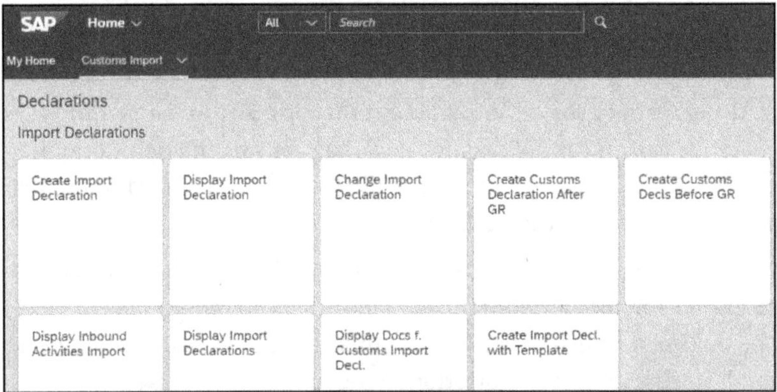

Figure 7.13 SAP Fiori Launchpad: Customs Import Space

The **Customs Import** space in the SAP Fiori launchpad serves as a comprehensive hub catering to the needs of users engaged in import operations. Depending on the user's needs, they can access pages for creating and displaying import declarations (**Declarations**) as well as for managing customs invoices and customs waybills (**Related Documents**, see Figure 7.14). Users can also explore pages related to **Customs Brokers**, **Master Data**, and **Monitoring**, which will be discussed in subsequent sections.

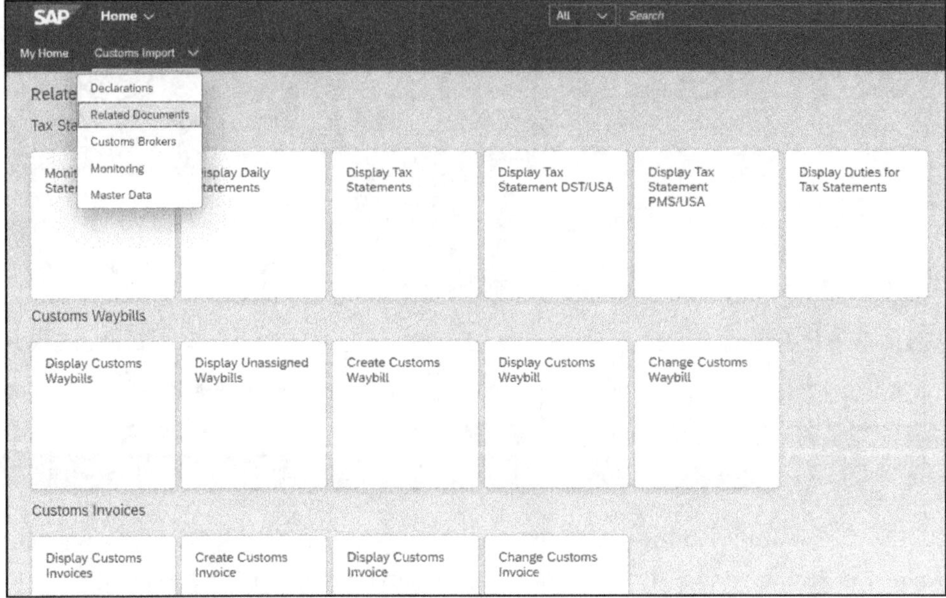

Figure 7.14 Customs Import: Related Documents Page

Import Declaration after Goods Receipt

The inbound process starts when Alpha creates a purchase order with its Indian supplier. When the supplier sends information about the shipment of the goods (shipping notification), an *inbound delivery* is created with reference to the purchase order. Creating an inbound delivery is a prerequisite for the customs declaration and can be done any time between the creation of the purchase order and the arrival of the goods at the receiving company's premises.

Within the EU, goods under duty suspension can be transported from the customs border to an inland warehouse by means of a transit declaration using the New Computerized Transit System (NCTS) procedure. This process allows the goods to maintain their duty suspension status throughout the transportation.

Figure 7.15 presents an overview of the primary business and system processes involved in generating an import declaration following the goods receipt. When the goods arrive at the border, a transit procedure is initiated. This transit document accompanies the goods during their transportation to the inland warehouse.

To create a transit declaration in SAP GTS, users need to define a preliminary document type and number for customs status and reference (e.g., T1 status and MRN) within the **Foreign Trade/Customs** tab at the header details level of the inbound delivery in SAP S/4HANA. Users should update this information before the entity receives the goods. The preliminary document type (T1 or T2) is determined by configuration per country.

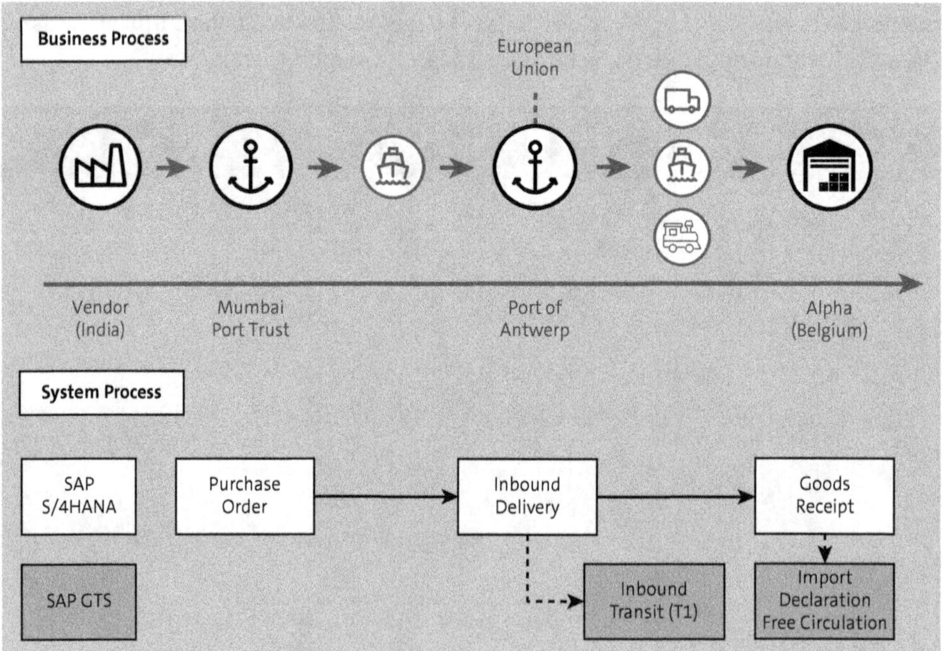

Figure 7.15 Business and System Processes: Import Declaration after Goods Receipt

After the goods arrive at the warehouse, an accompanying customs shipment document containing a unique number from the authorities is presented. Then, the transit declaration in SAP GTS created with a reference to the inbound delivery closes the transit procedure, and users can assign a customs destination to the concerned goods.

Receipt of products at the consignee's premises triggers the import declaration process. At that moment in time, two options are available:

- The posting of the goods receipt (material document) in the feeder system automatically triggers the creation of an import declaration in SAP GTS.
- The posting of the goods receipt document (material document) is transferred to a worklist in SAP GTS, and the import declaration is created manually, which allows users to group several items or documents.

The transit and import declarations created in SAP GTS follow the standard layout of any other customs declaration in the system.

The import declaration should have three types of data: data coming from the feeder system from the goods receipt material document, data that could be defaulted in SAP GTS (if determination rules are defined), and data that must be completed manually. An incompleteness check confirms that all the data is entered before users send information to the customs authorities.

Local customs regulations determine the declaration requirements and treatment of customs and other applicable taxes. Both paper-based and electronic declarations can be supported in the system, as well as generating broker instruction sheets that initiate a broker to file the actual customs declarations.

In Belgium, Alpha has chosen to streamline its import process by seeking approval for filing import declarations monthly. This simplification significantly reduces processing times. It eliminates the need for transactional communication, whether paper-based or electronic, replacing it with the creation of a periodic declaration or SCD. However, note that entities are still required to provide this information to the customs authorities.

Import Declaration before Goods Receipt

If the company opts to generate the import declaration before receiving the goods, the procedure still involves the creation of a purchase order and optionally an inbound delivery. The inbound delivery can be generated at any point before the goods arrive. This system process is illustrated in Figure 7.16.

Data from both the purchase order and the inbound delivery is transferred to SAP GTS. Import declarations can be generated using either of these two documents, or they can be based on a customs waybill or an invoice. Once the user makes a selection, the system compiles a list of all pertinent feeder system documents.

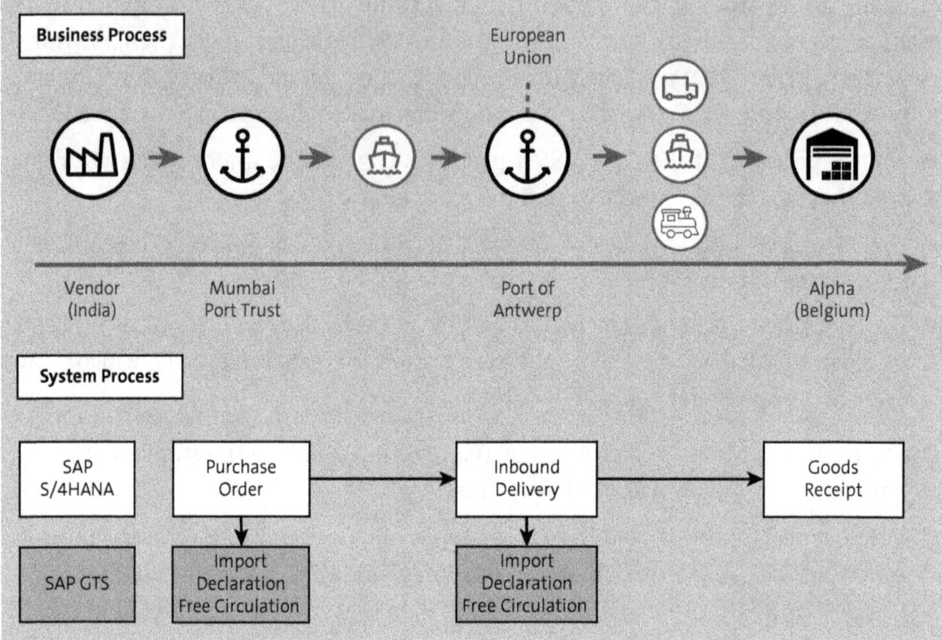

Figure 7.16 Business and System Process: Import Declaration before Goods Receipt

Because the inbound delivery contains crucial shipping and logistics information for completing the import declaration (e.g., more accurate quantity shipped by the supplier), we use the inbound delivery as a reference document in the Alpha example for the import declaration.

If the waybill number is maintained in the standard SAP field of the inbound delivery, this information can also be displayed in the worklist. This allows users to easily select all inbound deliveries with the same waybill number and group them into one import declaration.

In SAP GTS, the import declaration before goods receipt can be created based on the following different documents transferred from the feeder system:

- Purchase order
- Purchase order and inbound delivery
- Purchase order, inbound delivery, and waybill
- Purchase order, customs invoice, and waybill

These documents provide a large set of data to create an import declaration in SAP GTS prior to goods receipt.

The **Create Customs Declaration Before Goods Receipt** app allows you to select the documents used as the basis for the creation of the import declaration, depending on the selection criteria.

Figure 7.17 shows an example of the selection screen of the worklist for the creation of an import declaration based on a customs invoice. In the following section, we discuss the customs invoice in more detail.

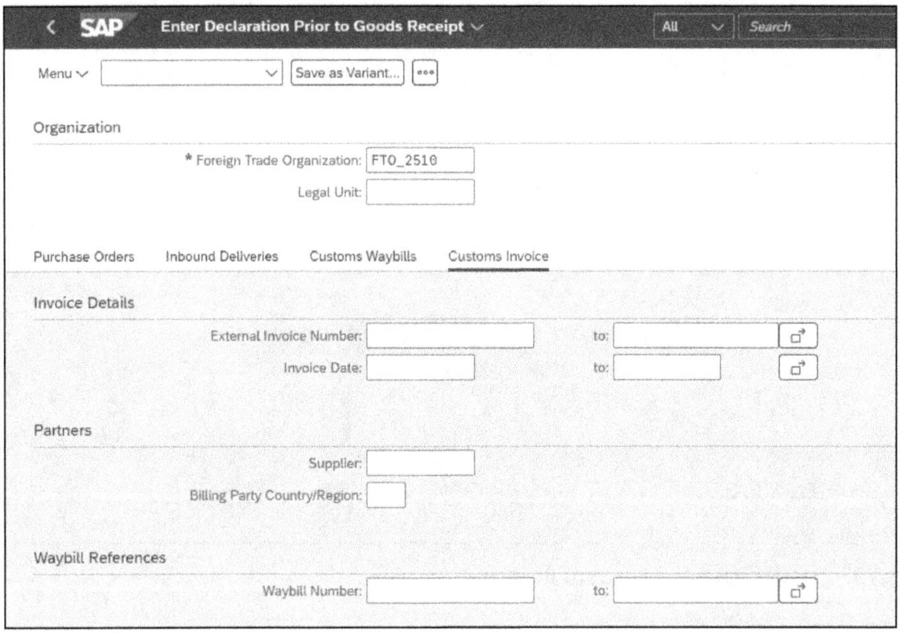

Figure 7.17 Selection Screen of the Worklist to Create an Import Declaration Prior to Goods Receipt: Customs Invoice

Customs Invoice

The commercial invoice used by the exporting party for the export declaration, which also accompanies the goods, can serve as the basis for creating the import declaration by the importing party. The functionality within SAP GTS transforms the commercial invoice into a customs invoice object in the worklist, even before the receipt of goods. This object encompasses essential components of the commercial invoice, such as the external invoice number and amounts.

The customs invoice can be generated either manually or through integration with a feeder system. Integration options include connecting with the company's feeder system for intercompany transactions or with the supplier's feeder system.

Leveraging integration with a feeder system offers significant advantages for customs invoice integration in SAP GTS by streamlining processes, reducing manual efforts, and enhancing data accuracy.

The customs invoice (accessible via the **Display Customs Invoices** tile) is divided into two main sections: header and item data. Figure 7.18 shows the custom invoice's header data under the **General Data** tab.

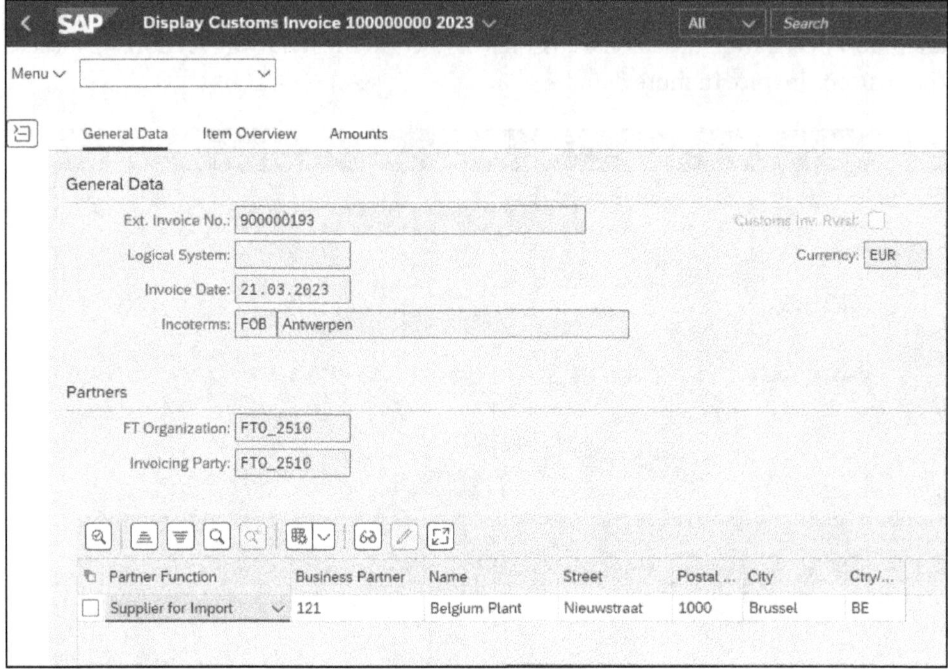

Figure 7.18 Header Data of a Customs Invoice

The header level of the customs invoice encompasses the following essential key fields:

- **Ext. Invoice No.**
  Enter the external invoice number, which is the reference number from the feeder system.
- **Invoicing Party**
  Enter the business partner who creates the invoice.

Figure 7.19 displays the customs invoice's item data. The item level of the customs invoice contains the following key fields:

- **Product Number**
- **Quantity/Open Quantity**
- **Net Weight/Gross Weight**
- **Ctry/Reg. of Origin**
- **Net Value**

In addition to the net value, some other costs used in the duty framework can be transferred on the header or item level of the customs invoice (e.g., transportation or insurance costs).

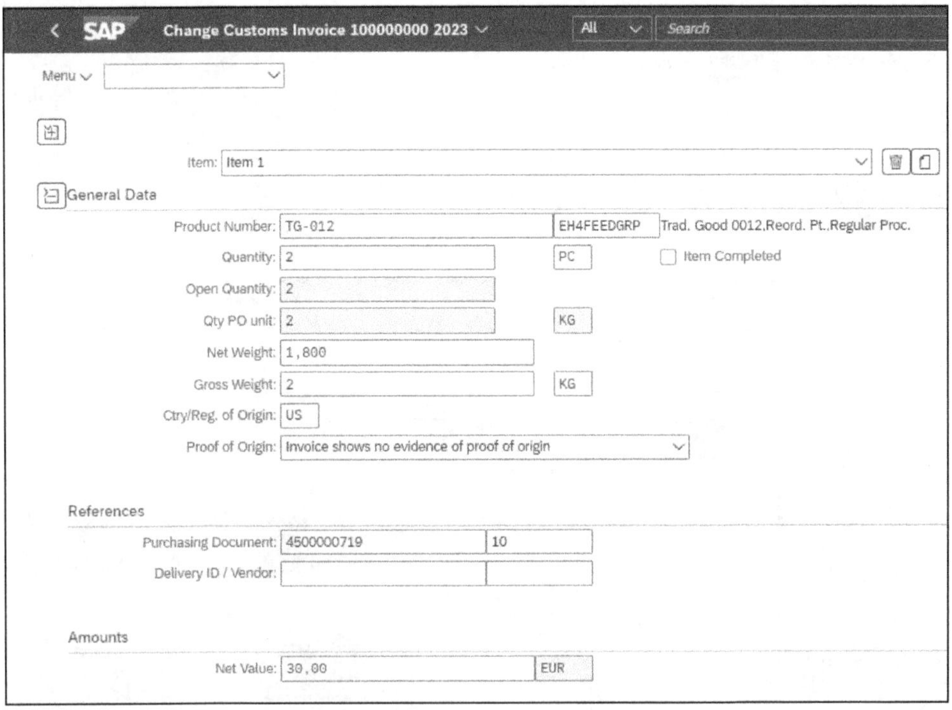

Figure 7.19 Item Data of a Customs Invoice

Customs Invoice with Alpha and Beta

Figure 7.20 illustrates the business and system processes of the integration of the customs invoice from the feeder system.

The Beta company, situated in the US, procures the finished product GB-1003 from the Alpha company, based in Belgium. Both entities are part of the same organization, making this transaction an intercompany operation. This section covers Beta's importation process. When a US-based company like Beta needs to initiate an import declaration, it must do so prior to receiving the goods. This means that the import declaration is generated before the goods are physically inspected and stored in the warehouse.

As a result, Beta initiates an import declaration when the goods arrive at the port in New York. A transport order is created between the two companies. Once transportation details become available, Alpha generates an outbound delivery with reference to the stock transport order. Following the picking, packing, and loading of the goods, they are prepared for export. Alpha subsequently generates an intercompany invoice, which is then transferred to SAP GTS and used for the creation of the export declaration.

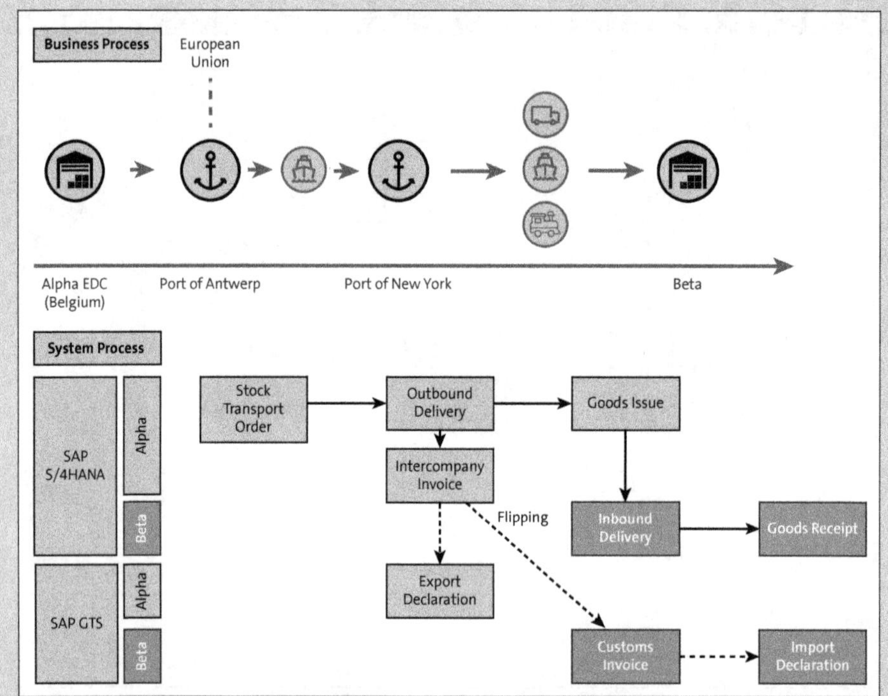

Figure 7.20 Business and System Processes: Customs Invoice Integration from the Feeder System

The intercompany invoice stemming from the outbound flow is converted into a customs invoice. This customs invoice is then used in the inbound process for Beta. In parallel, Beta initiates the inbound delivery.

Based on the data derived from the stock transport order and the customs invoice, Beta can create the import declaration before the actual receipt of goods in SAP GTS. Subsequently, the goods receipt is generated once the physical arrival of the goods at the warehouse takes place.

Once the purchase order and invoice are transmitted from the feeder system, they become accessible in the worklist and serve as the foundation for initiating the import declaration.

Figure 7.21 displays the worklist before goods receipt.

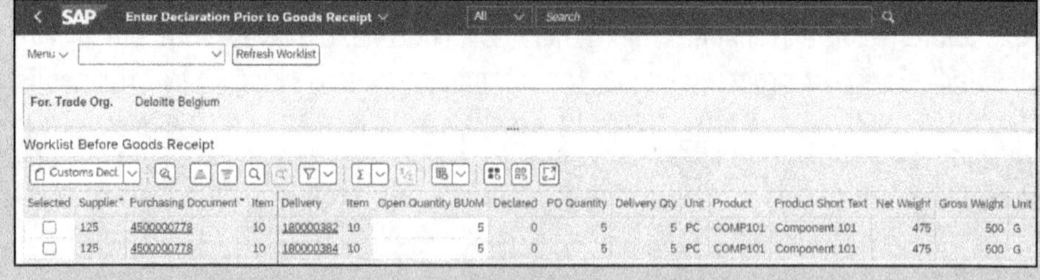

Selected	Supplier*	Purchasing Document*	Item	Delivery	Item	Open Quantity BUoM	Declared	PO Quantity	Delivery Qty	Unit	Product	Product Short Text	Net Weight	Gross Weight	Unit
☐	125	4500000778	10	180000382	10	5	0	5	5	PC	COMP101	Component 101	475	500	G
☐	125	4500000778	10	180000384	10	5	0	5	5	PC	COMP101	Component 101	475	500	G

Figure 7.21 Worklist before Goods Receipt: Customs Invoice

Customs Waybill

The customs waybill is generated within SAP GTS to incorporate waybill-related data into the import declaration. The specific fields available in the customs waybill can vary depending on the mode of transport used (e.g., sea, air, or road). For instance, in the case of sea shipments, the **Vessel Name** is a data element included in the customs waybill, while for road transportation, the **License Plate** field becomes accessible and relevant.

The customs waybill can be generated through manual input or integrated with a transportation feeder system. Within the waybill, references to logistics documents, such as those originating from a purchasing document, an inbound delivery, and/or the customs invoice, are included as part of the information.

Just like the customs invoice, Figure 7.22 illustrates the selection screen of the worklist before goods receipt, but this time focusing on the **Customs Waybills** tab.

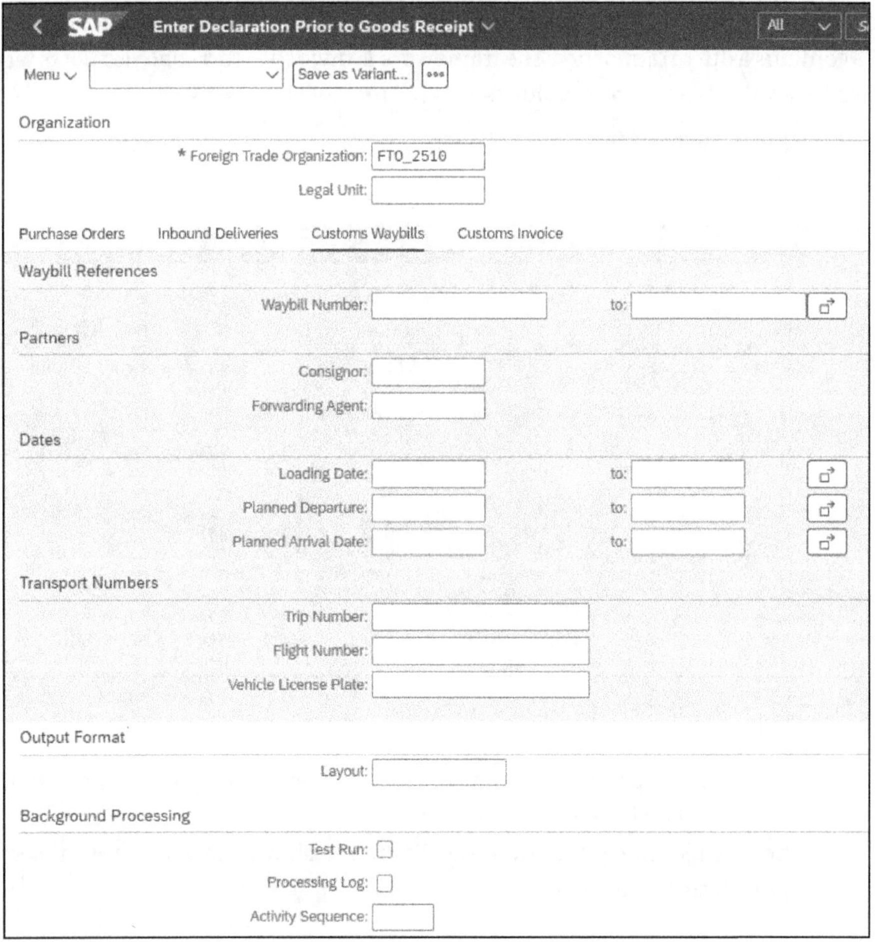

Figure 7.22 Selection Screen of Worklist to Create Import Declaration Prior to Goods Receipt: Customs Waybills

The selection can be performed based on at least one of the selection fields:

- **Waybill References**
  The master waybill or house waybill reference can be used.
- **Partners**
  The selection can be performed based on the consignor or the forwarding agent.
- **Dates**
  The planned arrival date can be used as a selection criterion to prioritize the incoming shipments.
- **Transport Numbers**
  References to transportation ID numbers can be used depending on the mode of transport (e.g., license plate is used for road transport, flight number is used for air shipments).

Import Duty Framework

To calculate the import duties and other applicable taxes, the customs declaration in SAP GTS contains a duty framework. The framework is divided into a **Logistics Costs** tab at the header level and a **Customs Value** tab at the item level.

Figure 7.23 displays the **Logistics Costs** tab, which is available at the header level of an import declaration for Alpha.

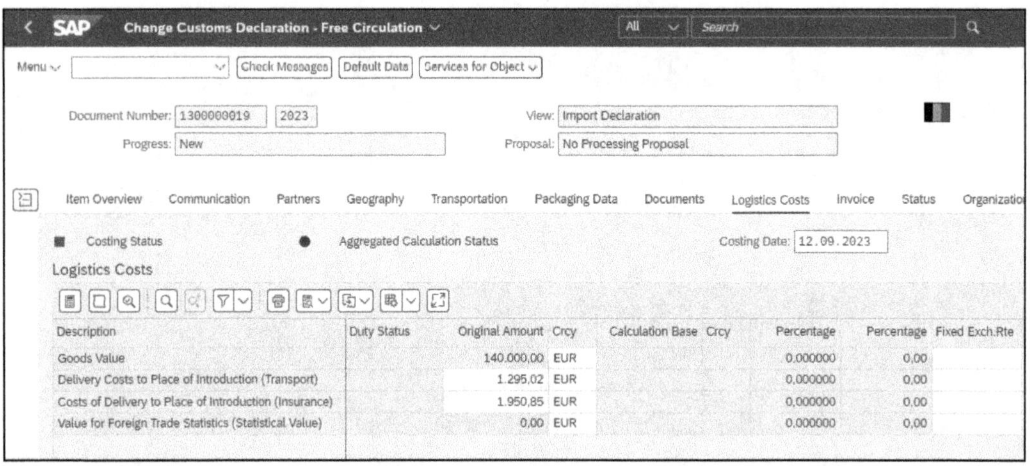

Description	Duty Status	Original Amount	Crcy	Calculation Base	Crcy	Percentage	Percentage	Fixed Exch.Rte
Goods Value		140.000,00	EUR			0,000000	0,00	
Delivery Costs to Place of Introduction (Transport)		1.295,02	EUR			0,000000	0,00	
Costs of Delivery to Place of Introduction (Insurance)		1.950,85	EUR			0,000000	0,00	
Value for Foreign Trade Statistics (Statistical Value)		0,00	EUR			0,000000	0,00	

Figure 7.23 Import Duty Framework: Header Level

Figure 7.24 and Figure 7.25 show the **Customs Value** tab, which displays the relevant costs on the item level of an import declaration for Alpha. Figure 7.24 shows all relevant values for the first item of the declaration, and Figure 7.25 shows the details for the second and last item of the declaration.

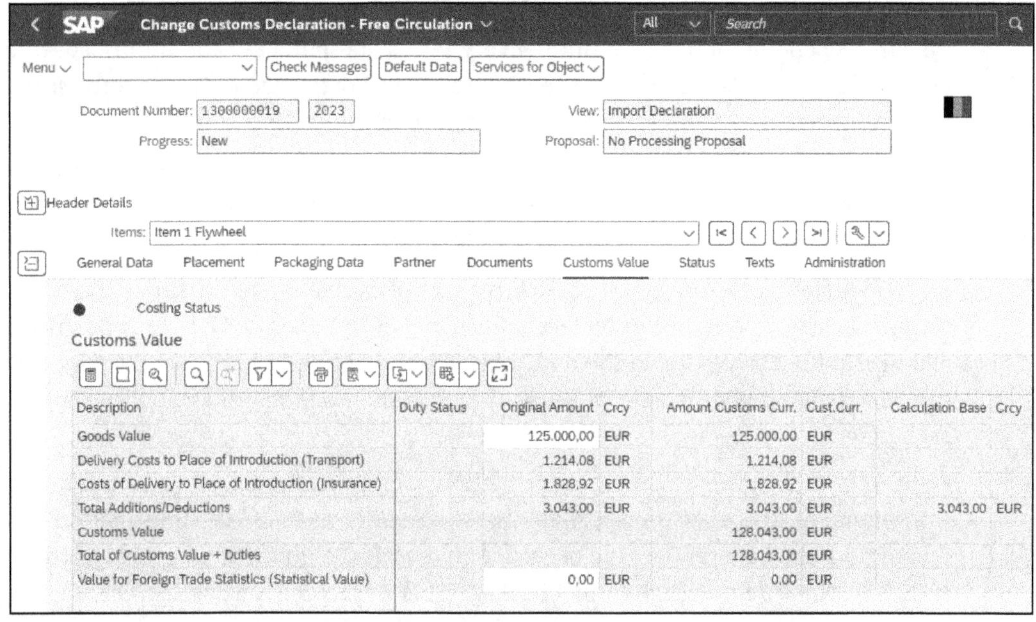

Figure 7.24 Import Duty Framework: Item Level for Item 1

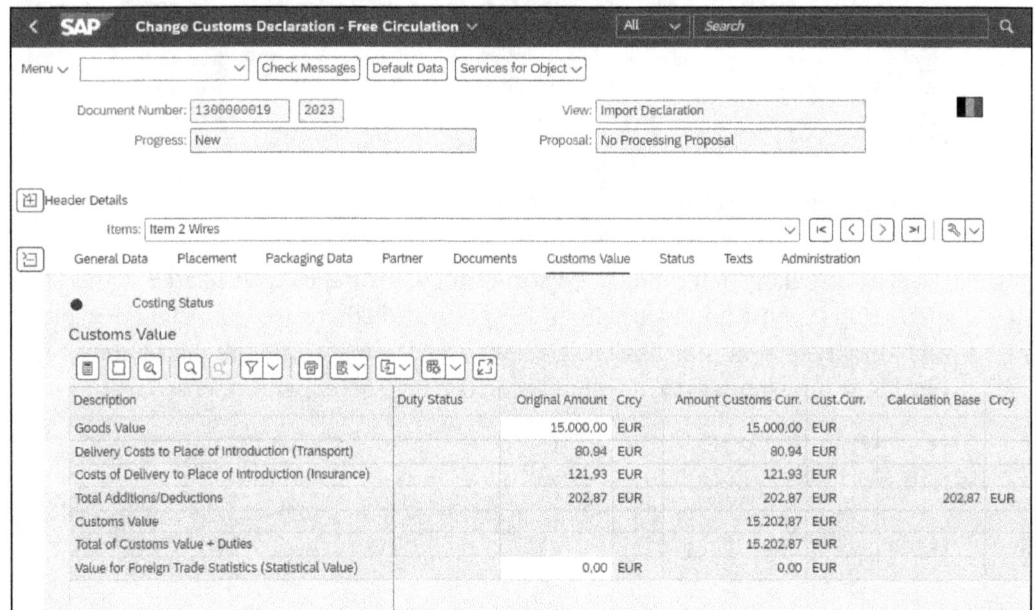

Figure 7.25 Import Duty Framework: Item Level for Item 2

The **Customs Value** tab provides a comprehensive overview of the duty calculation's essential components. Customs value calculations rely on the net price specified in the purchase order (representing the value of the goods), which is obtained from the feeder system. Depending on the relevant delivery terms and purchase conditions, specific elements can be added to or subtracted from the value of the goods. For instance, some varying elements include costs associated with shipping and insurance.

Certain data elements, such as the value of the goods and transportation costs, can be directly sourced from the feeder system. In contrast, other data such as duty rates, is typically managed and updated within SAP GTS.

SAP GTS offers the flexibility to conduct this type of calculation at two levels. Users can assign a specific cost at the header level, which means it applies to the entire document, and distribute it across the individual items. If users decide to distribute the cost, this can be based on the proportional value of each item or on the proportional weight of each item.

Figure 7.24 and Figure 7.25 present calculations demonstrating cost distribution. For this shipment, the total transportation cost amounts to €1,295.02. Information obtained from the **General Data** tab of each item reveals that item 1 has a gross weight of 15,000 kg, while item 2 weighs only 1,000 kg. Table 7.2 provides an overview of how the transportation costs are distributed based on weight.

	Weight (in kg)	Share (in %)	Distributed Transportation Cost (in Euros)
Item 1	15,000	93.75	1,214.08
Item 2	1,000	6.25	80.94
Total	16,000	100	1,295.02

Table 7.2 Distribution of Transportation Costs

Users can similarly determine the distribution of insurance costs related to this shipment. Users choose how to distribute these costs, whether based on weight or value, at each duty type level. This flexibility allows users to apply different distribution types within the duty framework, depending on specific requirements or criteria.

Note

For more information on the settings of this import duty framework, refer to Section 7.2.3.

7.2.2 Export Processes

Every time a company sells products to a customer located abroad or, in the case of the EU, outside of the customs union, an *export process* takes place.

The administrative requirements differ based on the local rules and regulations. Generally, a company that wants to export goods to a third country needs to file an export declaration with the customs authorities in the country they are exporting from. The options for getting customs authorities' approval described in Section 7.1 also apply here. An export declaration must accompany the goods during transportation to the customs office of exit, at the border of the customs union. Depending on the relevant country's legal requirements, commercial invoices, packing lists, and other documents might also be required.

In the following sections, we'll discuss the export business and system processes.

Business Process

Figure 7.26 shows Alpha's outbound process when it sells its finished products to a customer in Serbia.

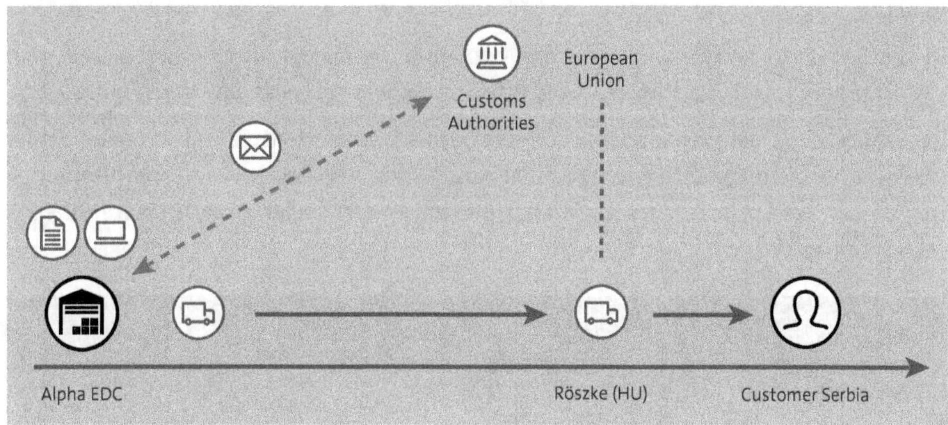

Figure 7.26 Export Business Process

The outbound process starts when Alpha receives an enquiry from its Serbian customer in which the customer expresses their willingness to purchase the goods offered by Alpha.

After the order has been placed and the goods are ready to be sent to the Serbian customer, Alpha needs to create an export declaration, which requires approval from the local customs authorities. Because the Belgian authorities allow e-filing, Alpha has decided to implement SAP GTS, enabling electronic communication with the customs authorities' platform.

As soon as Alpha receives approval from the authorities, the goods are allowed to leave the EU. Because the goods leave by truck, they take about two days to arrive in Serbia. In the example described in Figure 7.26, the goods cross the border to Serbia in Rözske, Hungary.

When the goods cross the border, the customs office of exit in Rözske sends a confirmation message to the Belgian customs authorities. The Belgian customs authorities send a similar confirmation message to Alpha, which has now received final confirmation that the goods have left the customs union. This closes the export procedure for the European customs authorities.

System Process

From a system perspective, the export process is initiated when Alpha creates a sales order in the feeder system. This sales order contains relevant information on the business partners, products, and quantities being sold. However, the integration with SAP GTS starts when the export relevant pro forma invoice is created.

As we did for import processes, we'll now explain navigation for export processes in the SAP Fiori launchpad.

Navigating Export Processes

In the SAP Fiori launchpad, users can conveniently access all tiles associated with export processes within the dedicated **Customs Export** space, as shown in Figure 7.27.

In essence, the **Customs Export** space serves as a comprehensive hub catering to the needs of users engaged in export operations. Depending on the users' requirements, they can access a page for creating and displaying export declarations (**Declarations**), as shown in Figure 7.27.

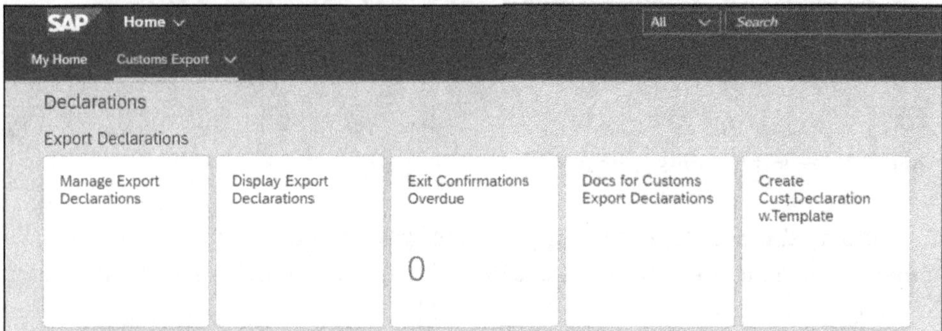

Figure 7.27 SAP Fiori Launchpad: Customs Export Space

Additionally, tiles are available for managing transit declarations and periodic declarations. There are dedicated pages under the **Customs Export** space for **Customs Brokers** and **Master Data**, as displayed in Figure 7.28.

Within the **Master Data** page, tiles are available to manage and monitor business authorizations and securities, as well as the cross-area tiles to manage business partners and products.

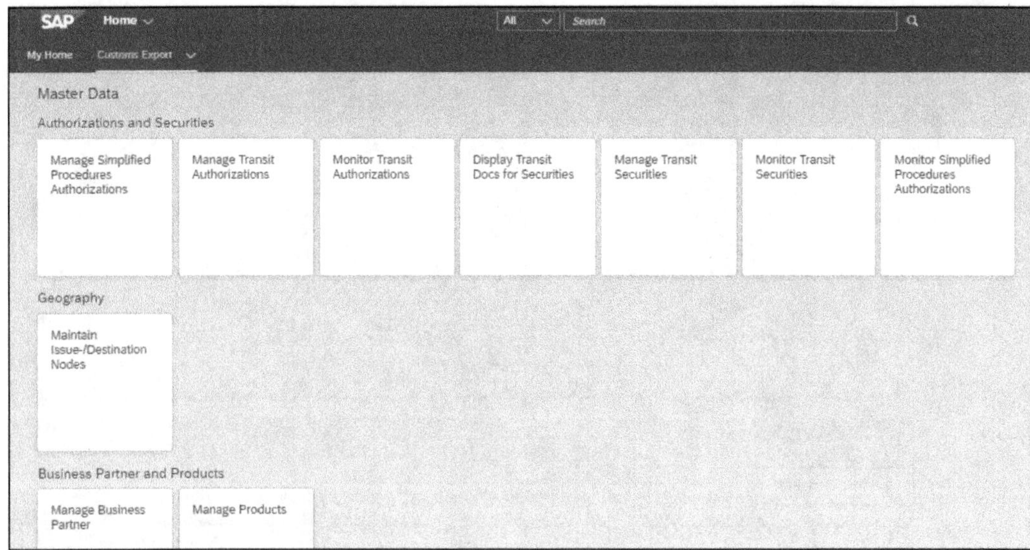

Figure 7.28 Customs Export: SAP Fiori Page for Master Data

Creation of the Export Declaration after Pro Forma Invoice

Users can create the outbound delivery with reference to the sales order. After the goods are picked, packed, and loaded, they are ready to be declared for export. Users can generate an export declaration by creating a billing document for this specific delivery.

In general, a *pro forma invoice*, which functions as the trigger to transfer information from the feeder system to SAP GTS, is created. An export declaration is created for relevant billing documents. This export declaration contains the relevant data that has been transferred from the feeder system but often needs to be completed in SAP GTS. To reduce the manual effort involved, defaulting data rules can be used.

Figure 7.29 illustrates the system steps that need to be performed to create an export declaration in SAP GTS.

After creating an export declaration, the system allows users to check whether all legally required fields are populated in the customs declaration. This functionality is called the *incompleteness check*. SAP GTS has defined a number of standard incompleteness checks that can be adapted or enhanced to suit the company's requirements. For countries where SAP GTS e-filing is localized, the incompleteness check covers all legally required data elements according to the relevant country's requirements.

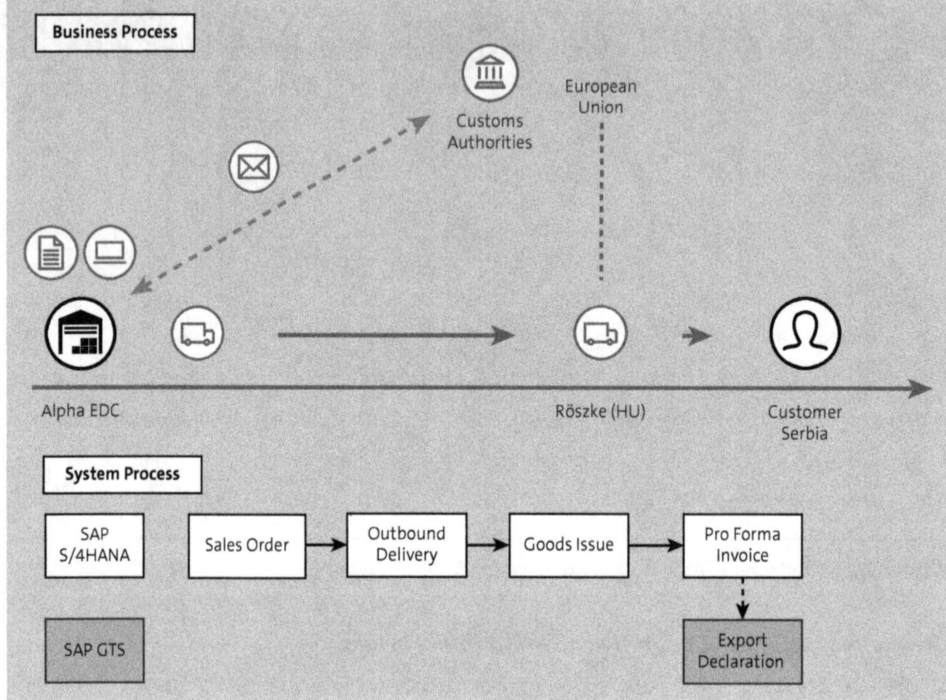

Figure 7.29 Business and System Processes: Export Declaration

Submission of the Export Declaration to the Customs Authorities

Once the export declaration is finalized, the company must submit this information to the customs authorities. Alpha interacts with the customs authorities using the e-filing method. The customs declaration is transmitted to a middleware partner in the form of an IDoc. This partner translates the SAP GTS IDoc into the correct customs authority XML format. The authorities confirm the receipt of this message by sending back an approval message, updating the export declaration with an MRN. Subsequently, the EAD is generated and printed.

These printed forms accompany the goods during transportation. After the goods have left the country (or the EU customs union, in this case), users need to enter an exit confirmation in the system to discharge the export process. For tax purposes, it's important to prove that the goods have left the EU. In addition, users can follow up on these export confirmations in SAP GTS. Once the export declaration is closed, it will no longer appear in the **Manage Export Declarations** tile, but it can be found via the **Display Existing Declarations** tile.

Transit Declarations

In the case of transporting non-Union goods from one point to another inside a customs union, customs authorities may require a transit declaration. SAP GTS allows the creation of this transit document either at the same time the export declaration is created or when the export declaration is fully approved by the customs authorities. If relevant, an *outbound transit* or *open transit procedure* document is created. The structure of the transit document is similar to the export document.

Like customs declarations, the transit declaration requires submission to and validation by the customs authorities. SAP GTS can send and receive all the necessary information through e-filing or by printing the required forms. Both **Customs Export** and **Customs Import** spaces contain several tiles to manage and display inbound and outbound transit declarations.

7.2.3 Customizing

Here we'll look at the most important Customizing steps for Customs Management. Note that prior to performing these steps, it's crucial to set up the correct legal regulations and document types. For a detailed explanation regarding these settings, refer to Chapter 3.

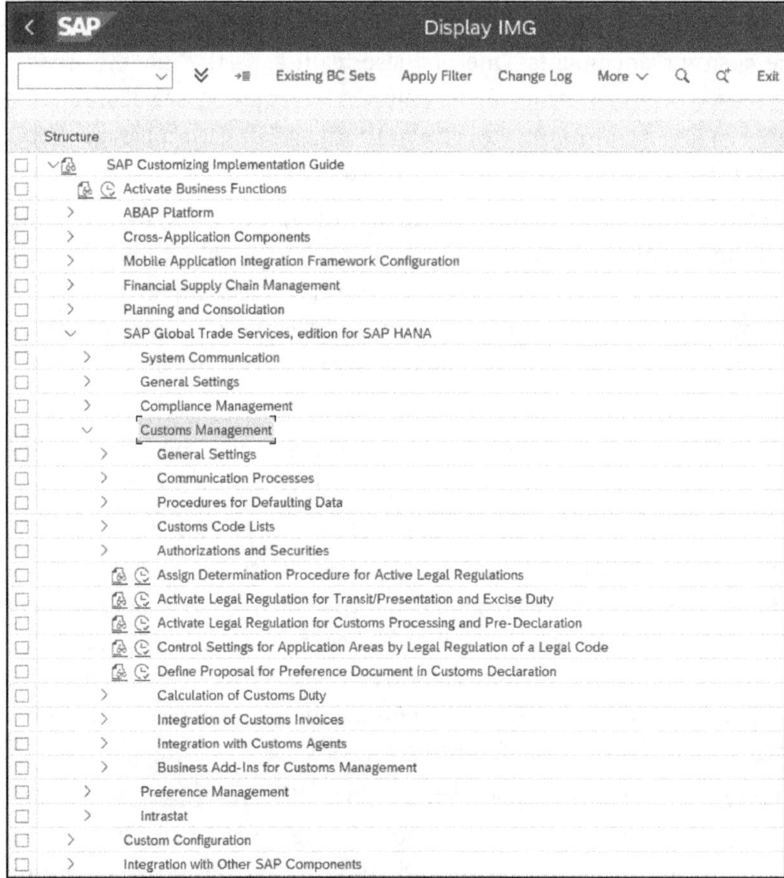

Figure 7.30 Customs Management Customizing

All the following steps must be completed prior to initiating import and export flows in the system. Please note that the sequence for performing the Customizing steps is important: all the steps should be done in the sequence in which they are displayed in Transaction SPRO and in the coming sections.

As shown in Figure 7.30, the Customizing for Customs Management includes the following sections: **General Settings**, **Communication Processes**, **Procedures for Defaulting Data**, **Customs Code Lists**, **Authorizations and Securities**, and **Calculation of Customs Duty**. These are the most important activities when setting up the Customs Management functionality. The main use of each of these sections is described next.

Clicking on the IMG activity documentation (icon on the left of each activity) provides additional details on each Customizing activity.

General Settings

The first section under Customs Management is **General Settings**. From there, many activities can be performed. However, this section will focus on one of the most important functionalities: the incompleteness check.

SAP GTS offers several functionalities to reduce manual effort as well as to check completeness of the customs documents. One of these features is the incompleteness check, which is used to confirm whether all the required information is available in the document before sending it out to the customs authorities. For each message that is available in the customs document, a separate check can be defined.

When e-filing, which was discussed in more detail in Section 7.1.4, the system requires users to remediate an incompleteness check before initiating the communication process with the customs authorities. This allows the system to check whether all data is available to confirm that users are sending a complete declaration to the authorities. This can result in significant time savings related to avoided error messages from the authorities.

Because most of the following settings to set up the incompleteness check are interdependent, it's important to perform the Customizing activities in the correct sequence.

In the Customizing activity **Control Incompleteness Checks for General Documents in Customs Management**, displayed in Figure 7.31, it's possible to set up everything relating to the incompleteness check. Users define the incompleteness procedure that groups a set of fields that need to be checked. At a later stage, users can assign these incompleteness procedures to specific messages. This means that for every message that is set up, it's possible to define and assign a new incompleteness procedure (more details on messages can be found in the following section).

This Customizing activity needs to be performed on two different levels. First, define an incompleteness procedure (**IncompProc**), as shown in Figure 7.32. Users link a **Function Module** to this incompleteness procedure when more complex verifications are necessary. However, the standard incompleteness procedure is sufficient for simple checks. Afterwards, users assign this incompleteness procedure to one or more messages in **Define Messages for Communication Process** (more details coming in the next section).

After creating the procedure, the next step is to define the fields that must be verified during every incompleteness check (in case the user opted for basic checks). SAP GTS provides a list of all the fields that can be added to the incompleteness check.

As shown in Figure 7.33, after navigating to the **Check Procedure – Field Assignment** pane, users can improve the quality of the check by adding validations such as the field length, type, or category to allow only a limited set of values to be maintained.

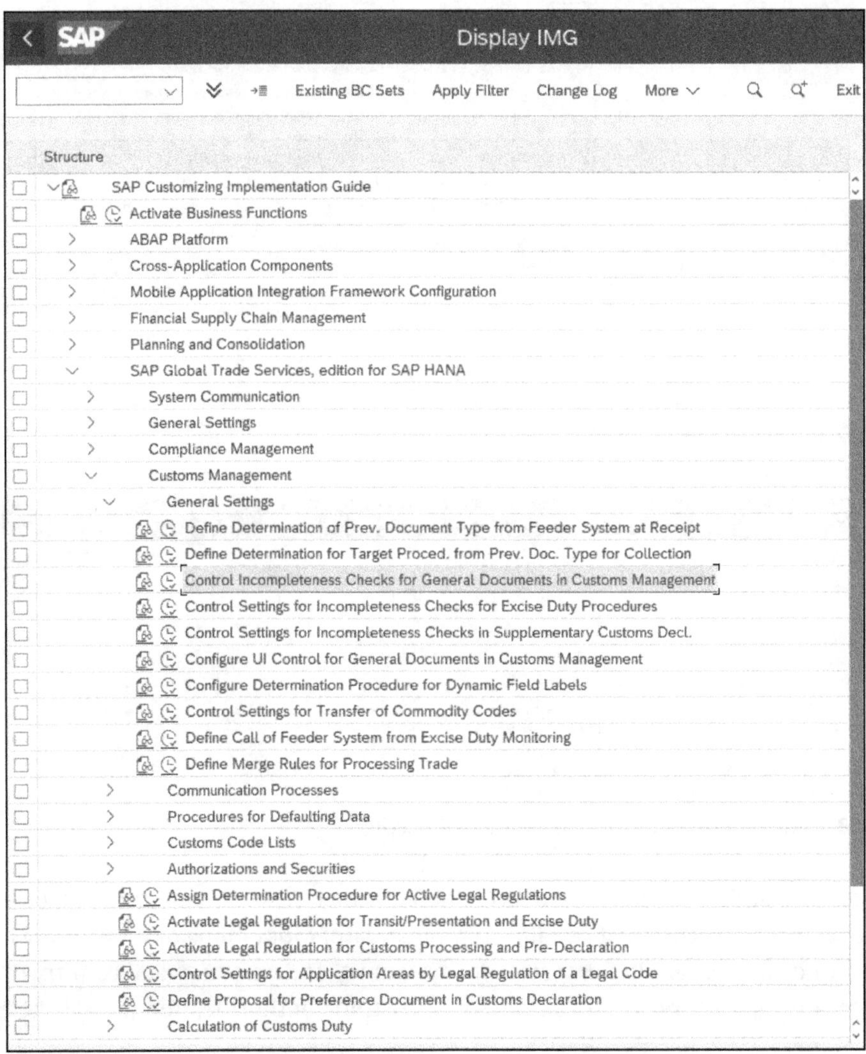

Figure 7.31 Control Incompleteness Checks for General Documents in Customs Management

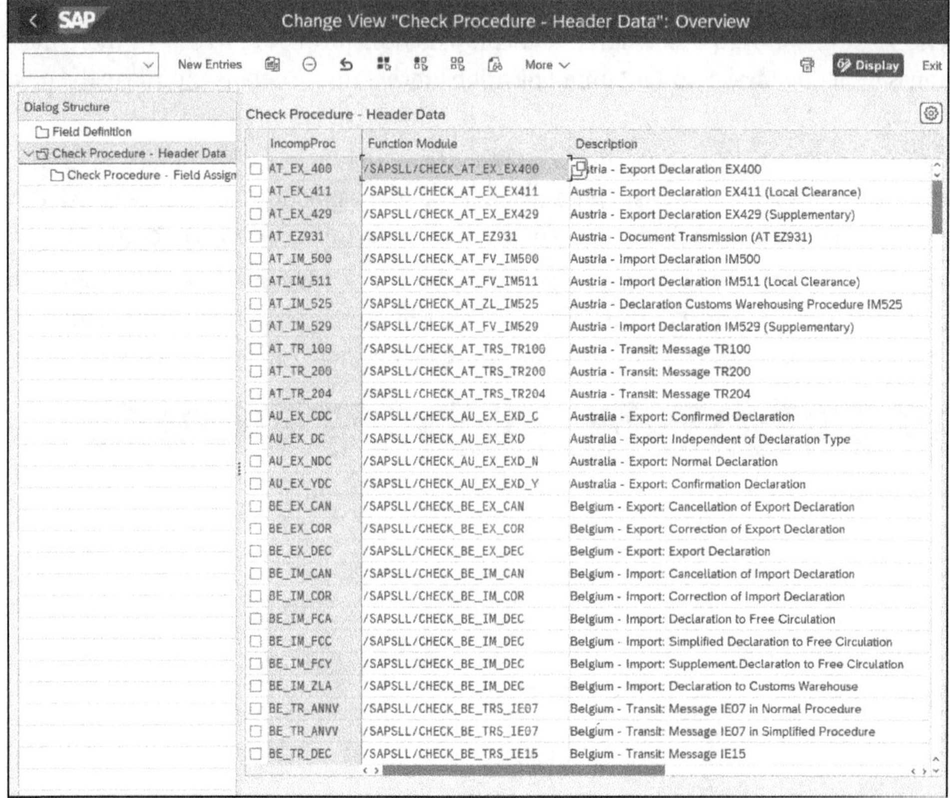

Figure 7.32 Incompleteness Procedures

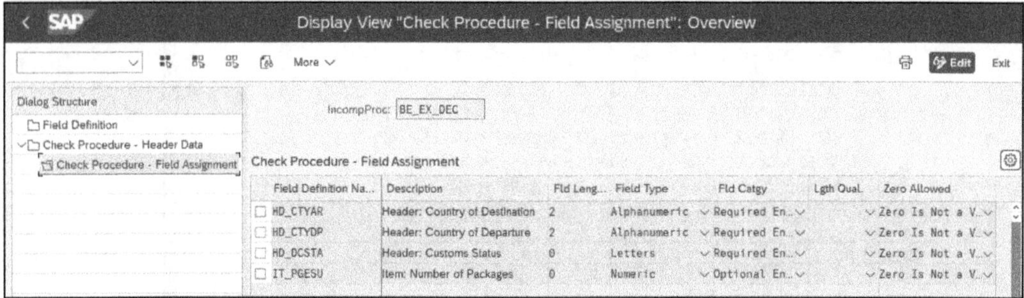

Figure 7.33 Field Assignment to Incompleteness Procedure

In the **Field Definition Name** column, users define which fields to add to the incompleteness check. This is useful if certain data is mandatory for e-filing or for printing a form. In our example, certain fields such as the country of destination and country of departure are mandatory to be submitted to customs authorities for the selected incompleteness procedure **BE_EX_DEC**.

Fld Length allows users to specify the number of characters needed when completing a specific field, while **Field Type** indicates whether the field needs to be alphanumeric, numeric, or letters only. For example, the **HD_CTYAR** (country of destination) field will only allow alphanumeric characters.

If a user indicates that a field is "required" in the **Fld Catgy** column, and it's not populated in the document, then the system won't execute the message in the **Communication** tab of the declaration. The incomplete field will result in a *hard block* when executing the incompleteness check, which is represented by a red traffic light in the customs declaration. The field needs to be populated before the declaration can be submitted to the authorities.

Fld Catgy can also be set as **Optional**. If an optional field isn't populated, the system displays a warning message when performing the incompleteness check, represented by an orange traffic light. This warning message doesn't prevent users from executing the message and proceeding with the business operations. This is called a *soft block*, or warning message, which is useful when a particular field is required only in very specific scenarios.

Note that SAP GTS offers a predefined set of incompleteness checks for the countries it supports based on the local legal requirements. However, this can be insufficient to meet a company's requirements, therefore additional checks may have to be added. As mentioned previously, this can be done either via configuration or via a customization of a function module.

A similar setup can also be done for SCDs in a separate Transaction SPRO activity (accessed via **SAP Global Trade Services, edition for SAP HANA • Customs Management • General Settings • Control Settings for Incompleteness Checks in Supplementary Customs Declaration**).

Communication Processes

The group of Customizing activities related to the communication processes and to the setup of the activity sequences and process activities is crucial in the implementation of Customs Management. Given the interdependency between these settings, it's important to perform the Customizing activities in the correct sequence, as we'll discuss in the next sections.

As with the standard Customizing of the incompleteness check, SAP GTS comes with predefined settings for the supported countries. A list of these countries can be found in Section 7.7.

Figure 7.34 provides an overview of all existing activities within the **Communication Processes**.

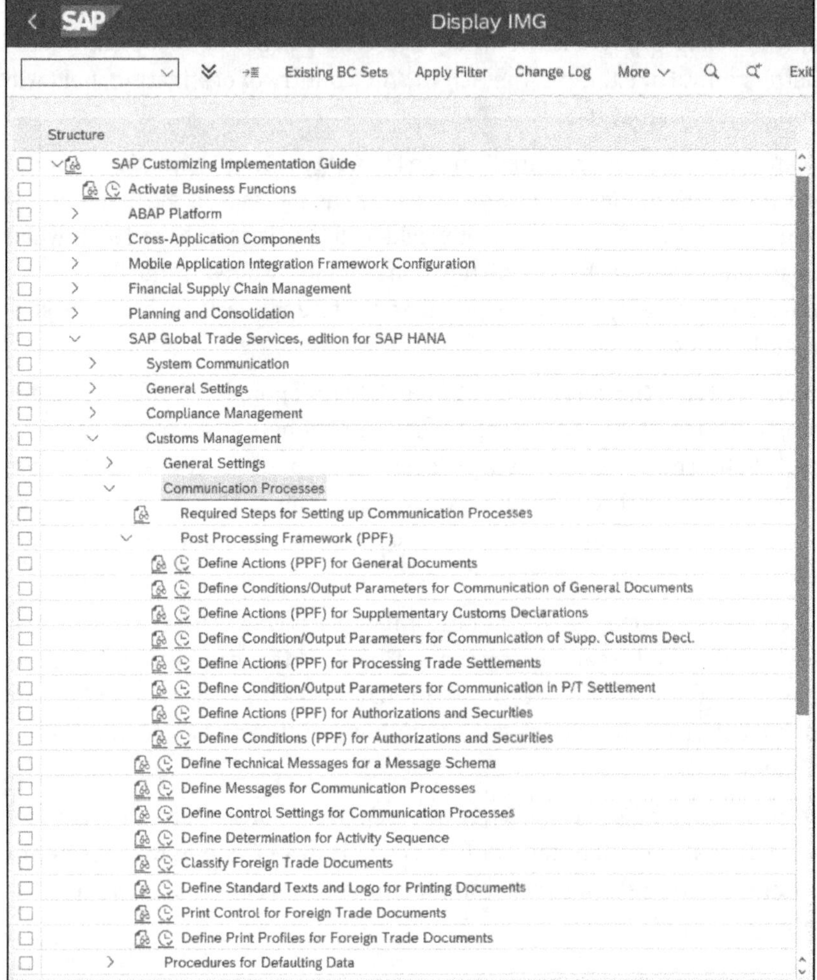

Figure 7.34 Communication Processes

Define Actions (PPF) for General Documents

The first step for setting up communication processes is to define the actions for general documents. This activity is found under the **Post Processing Framework (PPF)** settings. In this Customizing activity, messages such as printouts or EDI documents will be defined as action definitions in an action profile. This Customizing step is set up at multiple levels.

First, as indicated in Figure 7.35, an **Action Profile** needs to be defined. An **Action Profile** is a group of actions that can be executed for a customs declaration.

SAP GTS distinguishes between imports and exports and, where relevant, between transits and/or EMCS. The standard SAP GTS configuration contains **Action Profiles** with their respective **Action Definitions** and predefined **Processing Types** for several countries. When setting up the Customs Management functionality, if a country isn't yet available in the standard, the best practice is to extend SAP's naming convention and logic.

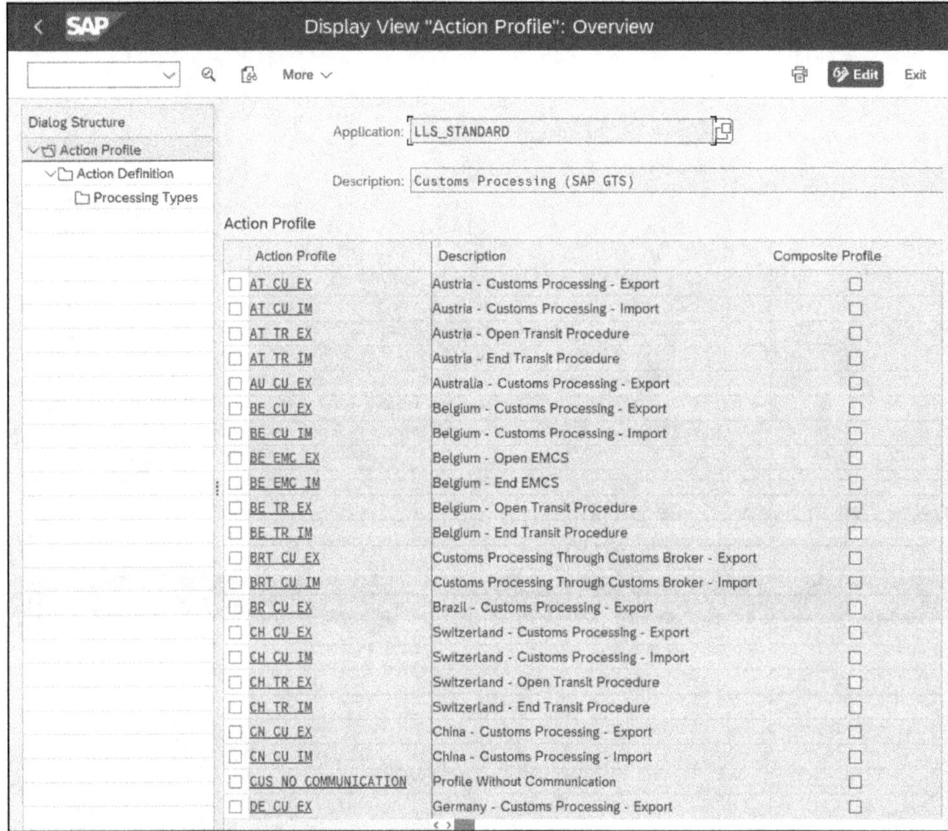

Action Profile	Description	Composite Profile
AT_CU_EX	Austria - Customs Processing - Export	☐
AT_CU_IM	Austria - Customs Processing - Import	☐
AT_TR_EX	Austria - Open Transit Procedure	☐
AT_TR_IM	Austria - End Transit Procedure	☐
AU_CU_EX	Australia - Customs Processing - Export	☐
BE_CU_EX	Belgium - Customs Processing - Export	☐
BE_CU_IM	Belgium - Customs Processing - Import	☐
BE_EMC_EX	Belgium - Open EMCS	☐
BE_EMC_IM	Belgium - End EMCS	☐
BE_TR_EX	Belgium - Open Transit Procedure	☐
BE_TR_IM	Belgium - End Transit Procedure	☐
BRT_CU_EX	Customs Processing Through Customs Broker - Export	☐
BRT_CU_IM	Customs Processing Through Customs Broker - Import	☐
BR_CU_EX	Brazil - Customs Processing - Export	☐
CH_CU_EX	Switzerland - Customs Processing - Export	☐
CH_CU_IM	Switzerland - Customs Processing - Import	☐
CH_TR_EX	Switzerland - Open Transit Procedure	☐
CH_TR_IM	Switzerland - End Transit Procedure	☐
CN_CU_EX	China - Customs Processing - Export	☐
CN_CU_IM	China - Customs Processing - Import	☐
CUS_NO_COMMUNICATION	Profile Without Communication	☐
DE_CU_EX	Germany - Customs Processing - Export	☐

Figure 7.35 Define Action Profiles

As shown in Figure 7.36, the next step is to define at least one **Action Definition** per **Action Profile**. This will need to be assigned to an executable message used in the customs declaration. For more information on messages, see the upcoming section "Define Messages for Communication Processes".

If, for example, a new printed form is defined as an additional message, users need to define a new action definition, which will then be linked to this message. Users may name this action definition **ZGTS_xx** (*xx* for the company's standard naming convention). Users can then execute the message correctly from the **Communication** tab of the customs declaration.

The final step of this Customizing activity is to define **Processing Types** for each action definition, as displayed in Figure 7.37. The commonly used processing types are **External Communication** (for printed forms) and **Method Call** (for e-filing declarations in the form of IDocs).

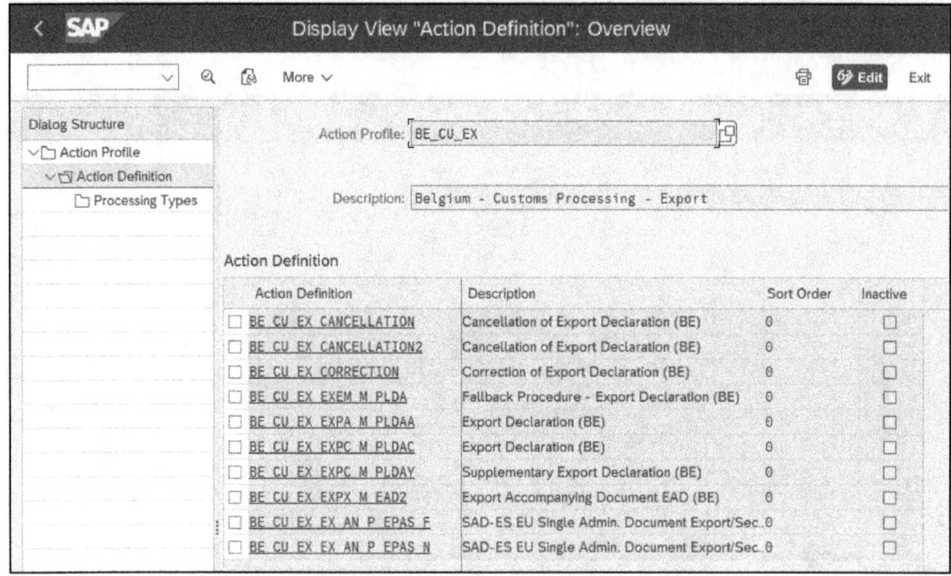

Figure 7.36 Define Action Definitions

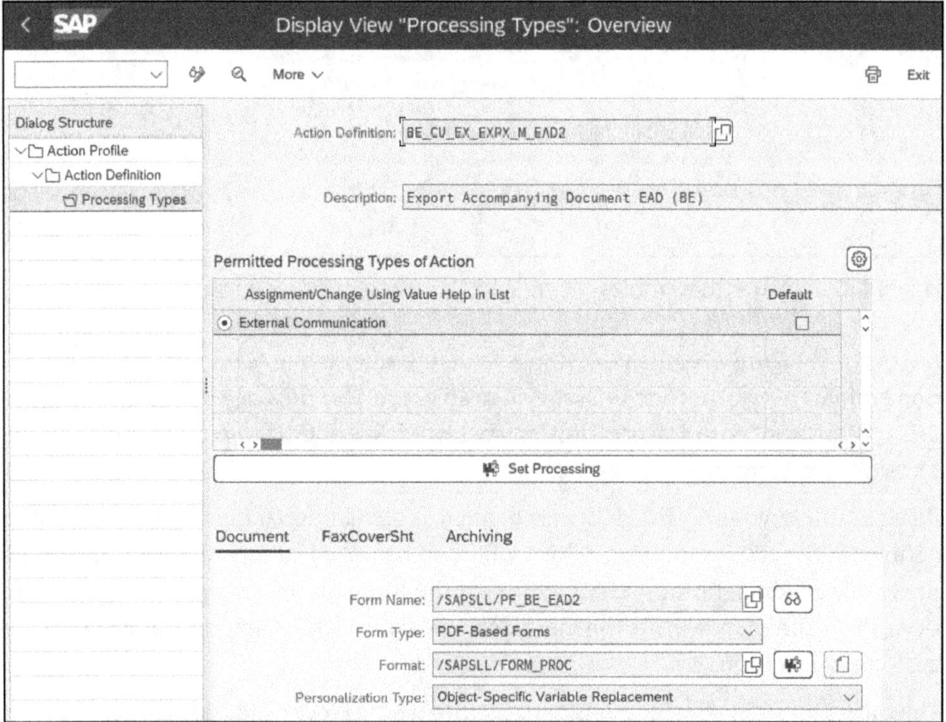

Figure 7.37 Define a Processing Type for Each Action Definition

In the lower section of the screen, users link the form that has been created to the action definition. Then, in the **Form Type** field, indicate whether the form is created as a **PDF-Based Form** or as a **Smart Form**.

Define Conditions/Output Parameters for Communication of General Documents

In this Customizing activity, users process the action templates created in the previous Customizing activity. This Customizing step is used to specify the **Start Condition** and **Schedule Condition** or printer specifications, at the **Action Definition** level.

As shown in Figure 7.38, the screen for Customizing of **Conditions for Actions** consists 7
of three parts. The upper-left part displays every **Action Profile** defined in the system. Only the action profiles defined in the previous activity will be displayed here. The upper-right part displays the **Action Definition(s)** of the selected **Action Profile**, while the lower part displays the details of the selected **Action Definition**. This again highlights the importance of performing the Customizing steps in the right sequence.

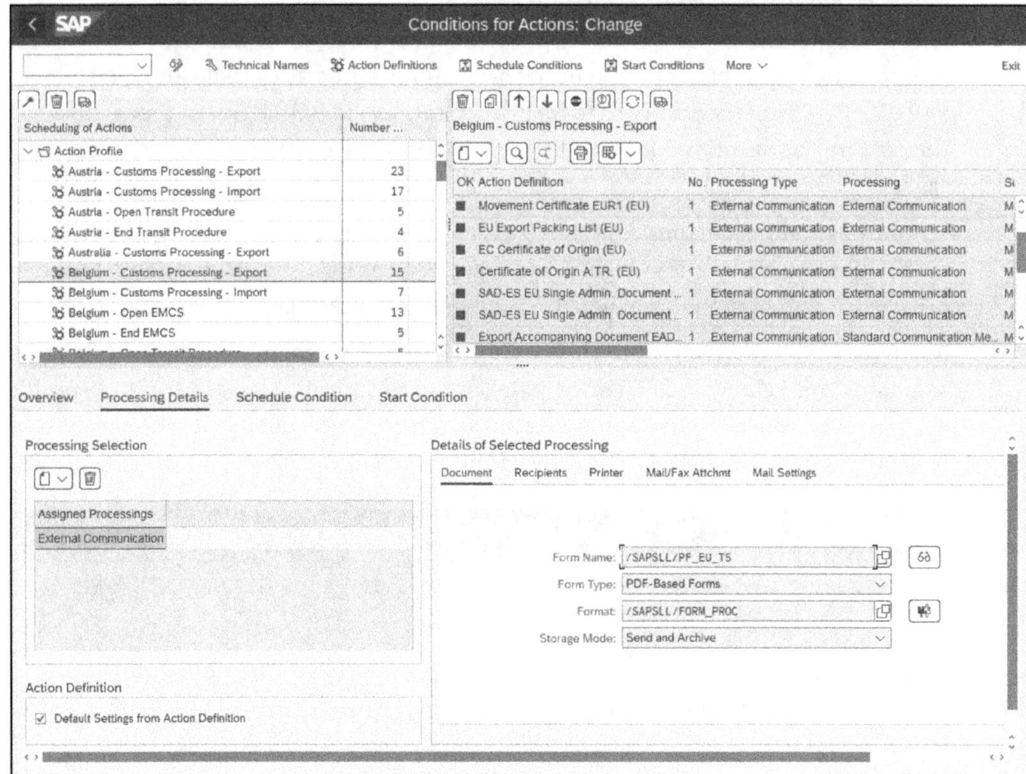

Figure 7.38 Customizing of Conditions for Actions

It can be useful to select the **Technical Names** option at the top of the screen to display the technical names of the action profiles and action definitions, as created in the previous Customizing step.

Most of the information in this Customizing activity is defined in the previous step. However, it's a system requirement to list all active action definitions here as well. Therefore, if a new action definition is created in the previous step, users must also add it in this Customizing activity.

Furthermore, in the lower part, users can select a few more details for the action definitions. For instance, in the **Processing Details** tab under **Printer**, users can specify which printer needs to be used to print a specific form or the number of copies it needs to print. These settings can also be set up via master data (Section 7.2.4).

If the customs authorities require an SCD or a periodic declaration (Section 7.1), the following activities need to be performed as well: **Define Actions (PPF) for Supplementary Customs Declarations** and **Define Condition/Output Parameters for Communication of Supp. Customs Decl.** These activities are similar to the ones described previously and need to be completed following the same logic.

Define Technical Messages for a Message Schema

SAP GTS generates IDocs for its communication with customs authorities. There are multiple messages that will be sent to the customs authorities that can be created in SAP GTS. However, the names that are given internally in SAP GTS aren't recognized by the customs authorities. Therefore, each country uses its own set of technical messages.

The purpose of this Customizing activity is to list the messages known by the customs authorities and group them logically (i.e., per country). This way, they can be used in other Customizing activities where they are linked to messages known by SAP GTS.

As Figure 7.39 shows, users first need to define a **Message Scheme** if it's not available by standard. This scheme will then contain all the technical messages for any specific country. When creating a **Message Scheme** for a company located in a country not supported by SAP, users should follow the naming convention used by standard SAP GTS.

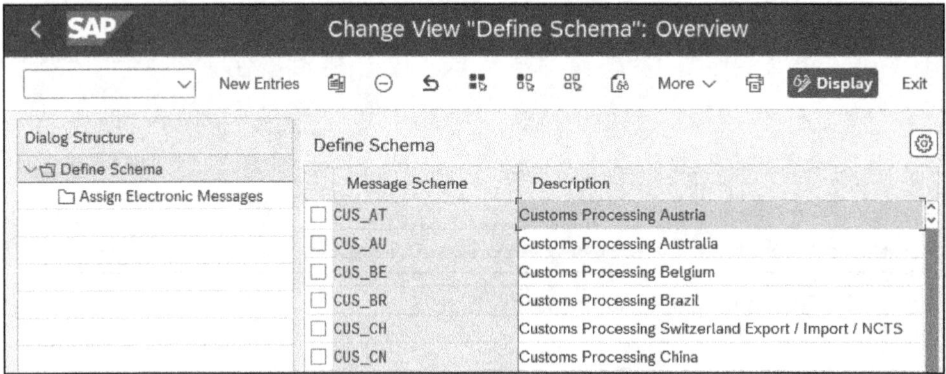

Figure 7.39 Define Schema

As displayed in Figure 7.40, once a **Message Scheme** is selected, users can then choose from a list of all the existing technical messages.

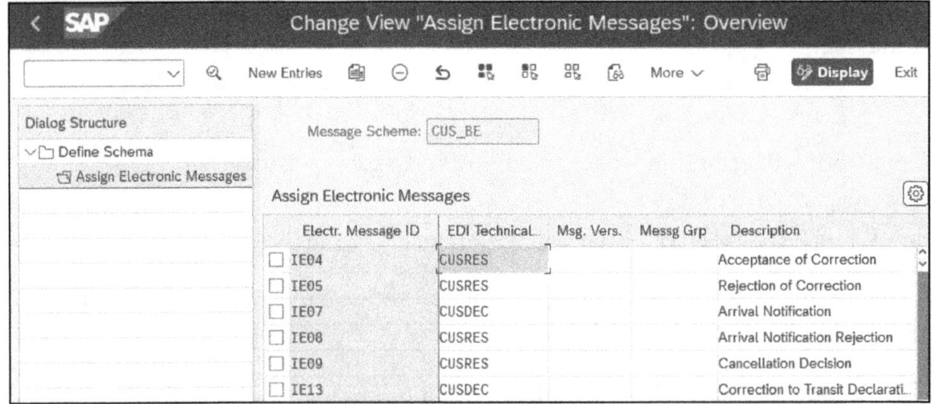

Figure 7.40 Assign Electronic Messages

The **Electr. Message ID** column refers to the technical message name known by the customs authorities. The **Description** column is related to this electronic message ID. The **EDI Technical** column refers to the type of IDoc recognized by SAP GTS. When defining a new SAP GTS message, users need to link this EDI technical name (also called message ID) to the newly created message, as we'll discuss in the next section.

Define Messages for Communication Processes

While the previous activity was used to maintain all messages provided by the customs authorities, the purpose of this activity is to maintain all internal SAP GTS messages. Here, the message will be linked to both a technical message, which is readable for the customs authorities, and to an incompleteness check.

As indicated in Figure 7.41, for every defined message, users can add a **Description**, **Message Type**, **Scheme**, **Message ID**, and **IncompProc** (incompleteness procedure).

Change View "Messages for Activity": Overview

Messages for Activity

Message	Description	Message Type	Scheme	Message ID	Cha..	Doct..	Clos..	IncompProc
I1610	Release for Transit (BE)	/SAPSLL/TRS_INBOUND	CUS_BE	IE29	☐	☑	☐	
I1620	Rejection of Release Request (BE)	/SAPSLL/TRS_INBOUND	CUS_BE	IE62	☐	☐	☐	
I1630	Decision About Cancellation (BE IE09)	/SAPSLL/TRS_INBOUND	CUS_BE	IE09	☐	☑	☐	
I1640	Acceptance of Correction (BE)	/SAPSLL/TRS_INBOUND	CUS_BE	IE04	☐	☐	☐	
I1650	Rejection of Correction (BE)	/SAPSLL/TRS_INBOUND	CUS_BE	IE05	☐	☐	☐	
I1660	No Permission for Transit (BE)	/SAPSLL/TRS_INBOUND	CUS_BE	IE51	☐	☐	☐	
I1670	Inspection Required (BE)	/SAPSLL/TRS_INBOUND	CUS_BE	IE60	☐	☑	☐	
I1680	Invalidity of Security (BE)	/SAPSLL/TRS_INBOUND	CUS_BE	IE55	☐	☐	☐	
I1690	Close Transit Procedure (BE)	/SAPSLL/TRS_INBOUND	CUS_BE	IE45	☐	☑	☐	

Figure 7.41 Define Messages for Communication Processes

In the **Message** column, the messages defined by standard SAP follow this logic:

- **M-message**
  For outgoing messages to the customs authorities in an e-filing setup.
- **I-message**
  For incoming messages from the customs authorities in an e-filing setup.
- **P-message**
  For all printed forms, used for both e-filing and a paper-based setup.
- **E-message**
  For all emails to be send from a declaration in an e-filing setup.

Aside from the standard messages, users can also define Z-messages, which are for custom messages.

Define Control Settings for Communication Processes

The **Define Control Settings for Communication Processes** activity is an important Customizing step in which users establish processes, activities, and activity sequences and then connect each activity to previously defined messages. Therefore, before performing this step, it's important to confirm that all previous Customizing steps are complete. This step sets up the company's communication processes with customs authorities in the system.

As displayed in Figure 7.42, this Customizing step occurs at multiple levels. The highest level is the **Process Schema**. It groups all the activities related to Customs Management for a specific country. For example, for Belgium, the process schema **CUSBE** is defined in our example.

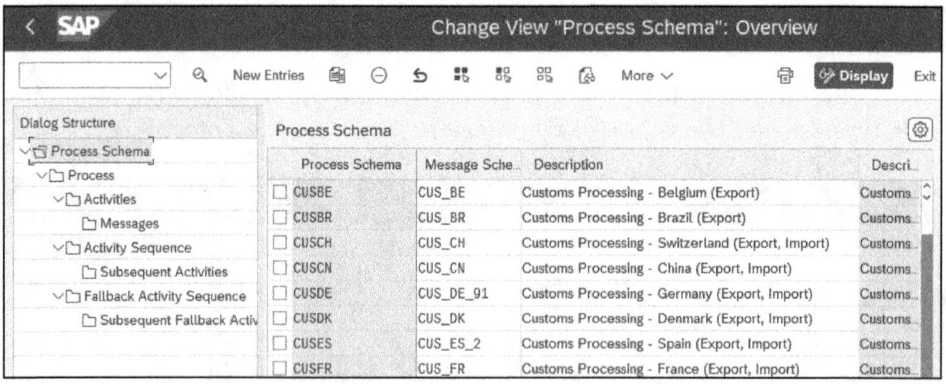

Process Schema	Message Sche...	Description	Descri...
CUSBE	CUS_BE	Customs Processing - Belgium (Export)	Customs...
CUSBR	CUS_BR	Customs Processing - Brazil (Export)	Customs...
CUSCH	CUS_CH	Customs Processing - Switzerland (Export, Import)	Customs...
CUSCN	CUS_CN	Customs Processing - China (Export, Import)	Customs...
CUSDE	CUS_DE_91	Customs Processing - Germany (Export, Import)	Customs...
CUSDK	CUS_DK	Customs Processing - Denmark (Export, Import)	Customs...
CUSES	CUS_ES_2	Customs Processing - Spain (Export, Import)	Customs...
CUSFR	CUS_FR	Customs Processing - France (Export, Import)	Customs...

Figure 7.42 Define Process Schema

For each **Process Schema**, several processes are defined, as displayed in the **Process** pane in Figure 7.43.

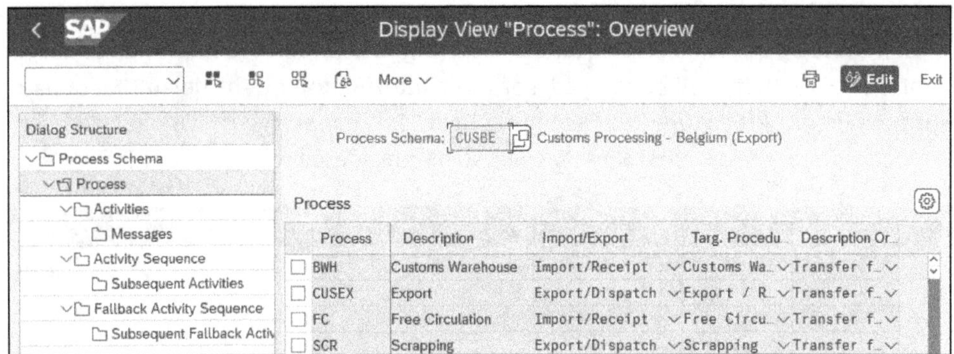

Figure 7.43 Assign Processes to a Process Schema

In this step, users define the processes and indicate whether they are related to either **Import/Receipt** or **Export/Dispatch**. The indication can be set in the **Import/Export** column so that the system knows how the process needs to be handled and, subsequently, in which corresponding menu (import or export) to show it.

Furthermore, users need to indicate the *target procedure* in the **Targ. Procedure** column. This determines which customs procedure the goods will be exported or imported with. For example, when an import declaration needs to be created into **Free Circulation** for a specific shipment, a process needs to be defined for creating this kind of import declaration. This process needs to be described as **Import/Receipt** with the **Targ. Procedure** set to **Free Circulation**. In the case of Alpha, which is a company operating with BWHs, it also needs to have an import process defined with **Customs Warehouse** as the **Targ. Procedure**.

At the next level (**Activities**), the activities, or *process activities*, are defined. Activities are the different steps that occur in a specific process. Each activity represents a specific status that an export, import, or transit declaration can have.

Each declaration needs to follow a predefined number of activities before the process is considered as complete. Those activities are the ones defined in this Customizing step.

As shown in Figure 7.44, for each of the **Activities**, users can indicate whether the activity takes place before presentation to the customs authorities (**Prem.Entry** [premature entry] checkbox), is part of a simplified procedure (**Smp. Procedure** [simplified procedure] checkbox, which is only allowed after approval of the customs authorities), or is optional (**Optional** checkbox). Users can also indicate whether the selected activity requires a license and, if so, which license in particular in the **License** column.

Optional activities are only executed in specific scenarios. Typical examples are **Cancellation** or **Correction** activities. These scenarios will only happen exceptionally and should therefore be set as optional. The logic behind these scenarios is determined by the determination data logic of the message. If the logic of the defaulting data determines that the message needs to be displayed in the customs declaration, then the process needs to be completed before users can continue with the next activity. However, if the logic of the defaulting data doesn't trigger the message in the customs declaration, this activity will be skipped. For more information on the defaulting of data, see the section "Procedures for Defaulting Data". If the message is manually added to the declaration, the optional activity will also be considered relevant for this flow.

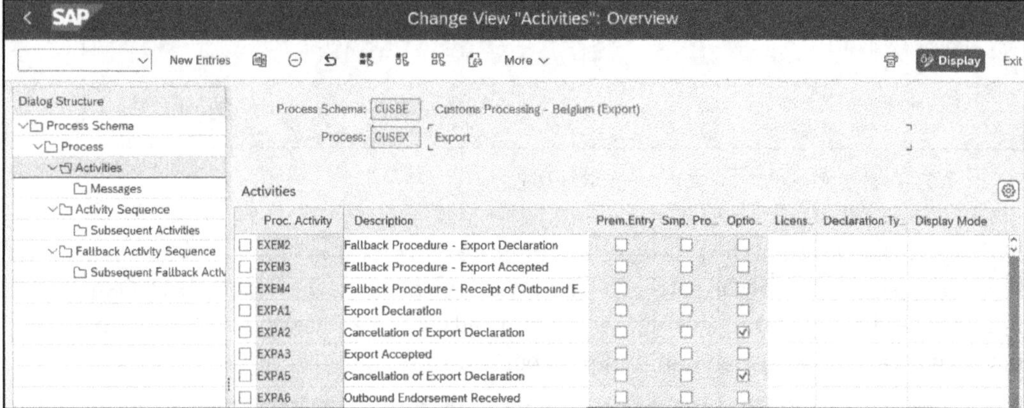

Figure 7.44 Assign Activities to a Process

At the next level of this Customizing activity (**Messages**), users can list all the relevant messages (defined in previous Customizing steps) per process activity, as displayed in Figure 7.45.

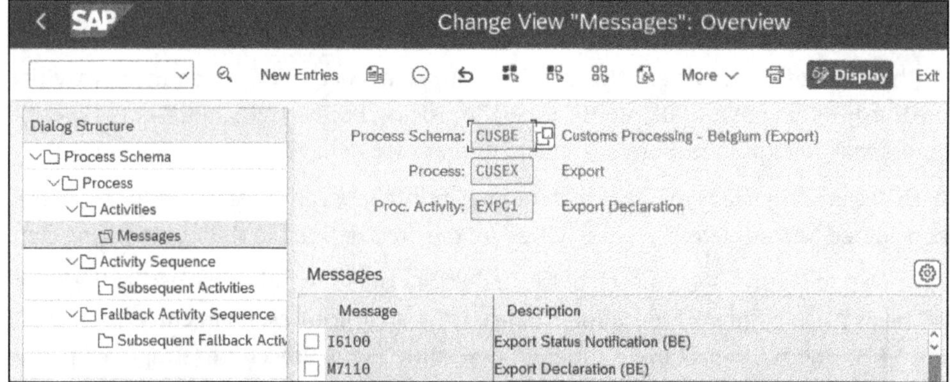

Figure 7.45 Assign Messages to an Activity

It's important to include all the incoming (I-messages), outgoing (M-messages), print (P-messages), as well as custom (Z-messages) messages that are expected in this activity. When the declaration is in a specific activity, executing or receiving messages that belong to a previous or later activity isn't possible. After all activities have been defined and assigned to one or more messages, users need to define the sequence in which these activities must be performed. Figure 7.46 shows that this is done via the definition of activity sequences. Each **Activity Sequence** will contain its own set of activities with a specific execution order.

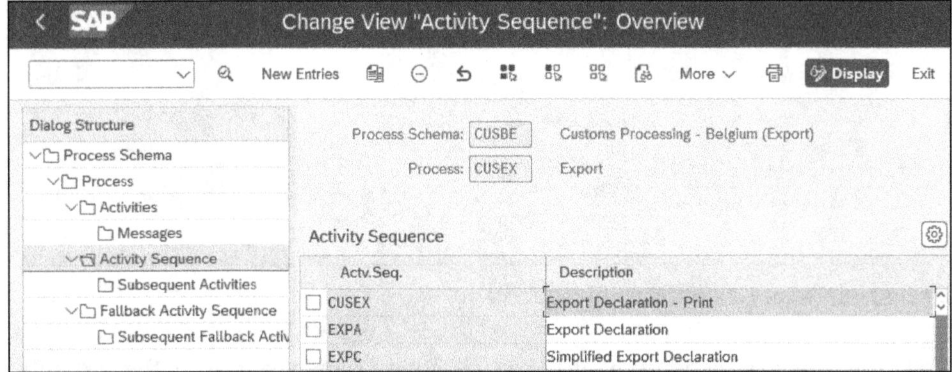

Figure 7.46 Define Activity Sequences

In the **Subsequent Activities** pane, users can group the required activities that were created in a previous step and assign them a number in the **Activ. Counter** column (see Figure 7.47). The system will then execute the activities in an ascending order based on the values assigned in the activity counter.

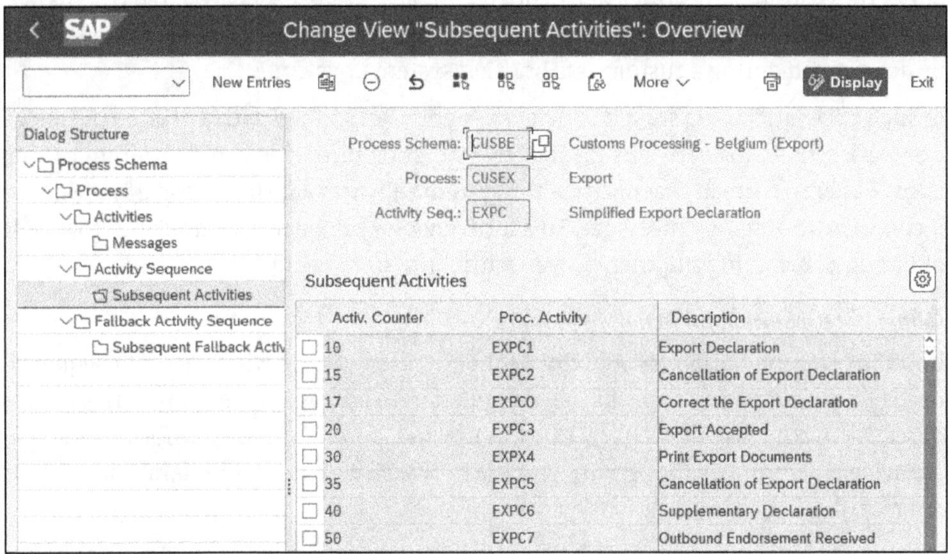

Figure 7.47 Assign Subsequent Activities

Next to the activity sequences used for e-filing, it's possible to maintain an activity sequence for the fallback procedure (**Fallback Activity Sequence**), as shown in the lower-left part of Figure 7.48.

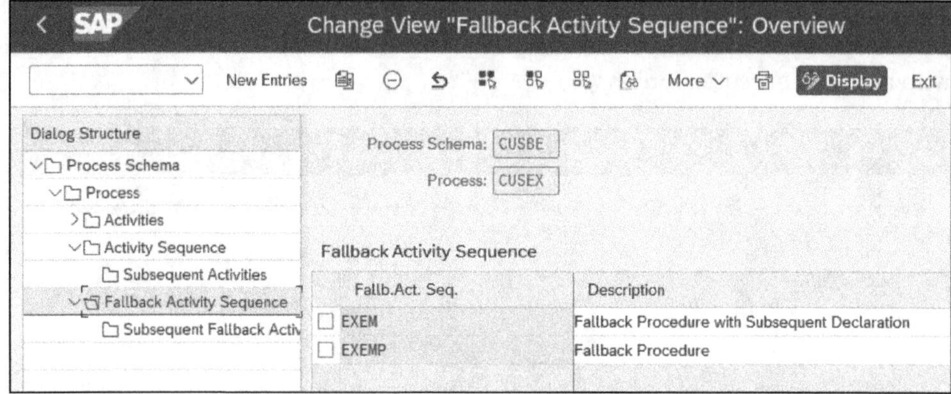

Figure 7.48 Define Fallback Activity Sequence

Assigning a fallback activity sequence to an activity sequence enables the system to determine which actions it should perform if the system is down and goes into the fallback procedure. After defining the fallback activity sequence and it's completed with its fallback activities, it's necessary to indicate the applicable fallback activity sequence within the normal activity sequence. This can be done by double-clicking on the specific activity sequence in the **Activity Sequence** pane. The fallback activity sequence is especially important to set up for activities containing electronic communication to either the customs authorities or to a broker. In these cases, users need to have this fallback procedure set up because the sending and receiving of messages will be temporarily deactivated during a customs authorities' system downtime.

In such a scenario, users should consider assigning an activity with a printed output as a fallback activity instead of using an activity that requires electronic communication. If this has been correctly set up, then the system will print a form instead of waiting for incoming and outgoing messages. This form enables the customs authorities to accept outgoing or incoming shipments, even with the system being down.

Define Determination for Activity Sequence

In **Define Determination for Activity Sequence**, users define which activity sequence needs to be assigned to a specific legal regulation in all possible relevant target procedures. By making use of the target procedure and, potentially, an authorization type, the system determines the activity sequence for which a customs declaration needs to be created.

Figure 7.49 shows one way to add a license type to these settings. In such an event, the system confirms whether the required authorizations are available in the system and uses this information to determine the **Activity Seq.**. In the Belgian example, for legal regulation **CUSBE**, if the goods are to be released into **Free Circulation** and no specific license type is assigned, then the **Import Declaration** will go into the **IFCA** activity sequence.

Change View "Determination of Activity Sequence": Overview

Determination of Activity Sequence

L.Reg.	Targ. Procedure	Lic. Ty...	Speci...	Activity Seq.	Descript.
CUSBE	Export / Reexport			EXPA	Export Declaration
CUSBE	Export / Reexport		S1	CUSEX	Export Declaration - Print
CUSBE	Export / Reexport	VVEX		EXPC	Simplified Export Declaration
CUSBE	Free Circulation			IFCA	Import Declaration
CUSBE	Free Circulation	VVIM		IFCZ	Local Clearance Notification - Import
CUSBE	Customs Warehouse	VVZL		IZLZ	Local Clearance Notification - Customs Warehouse

Figure 7.49 Determination of Activity Sequence

For these settings to work as described, it's crucial that you assign a **Process Schema** to the **Legal Regulation**, as shown in Figure 7.50. The required settings are already in place for countries supported by standard SAP GTS (Section 7.7). For countries that aren't supported by standard SAP GTS, this mapping needs to be performed. This is done in the Customizing activity **Control Settings for Application Areas by Legal Regulations of a Legal Code**, which comes later in the sequence of Customizing steps. In this Customizing activity, the assignment of **Exchange Rate Type** to **Legal Regulation** is also performed.

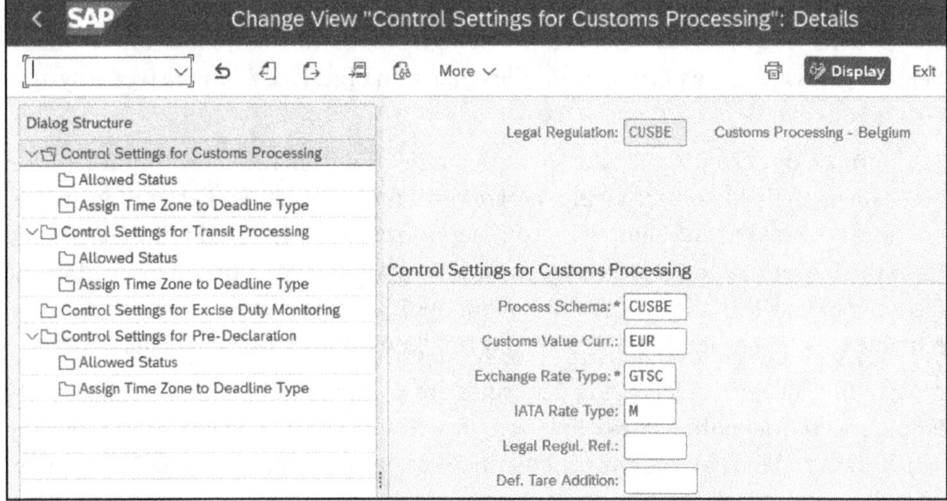

Figure 7.50 Control Settings for Customs Processing

Classify Foreign Trade Documents

When a printed form is created in the system, it must be classified to indicate the type of customs document it belongs to. Figure 7.51 shows the **Foreign Trade Documents: Classification** area, where users can add an official document type to the newly created print message. The list of form types is predefined by the system and is based on the country. Doing so enables the system to perform a plausibility check at the document type level. If a **Message** (P-message) can't be assigned to any of the official document types, the **Type** called **Customer Document** can be used.

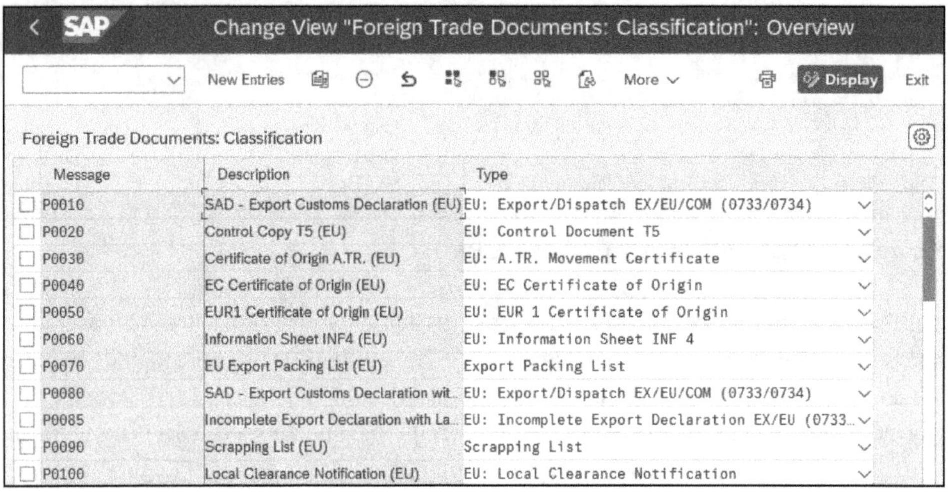

< SAP Change View "Foreign Trade Documents: Classification": Overview

New Entries More Display Exit

Foreign Trade Documents: Classification

Message	Description	Type
P0010	SAD - Export Customs Declaration (EU)	EU: Export/Dispatch EX/EU/COM (0733/0734)
P0020	Control Copy T5 (EU)	EU: Control Document T5
P0030	Certificate of Origin A.TR. (EU)	EU: A.TR. Movement Certificate
P0040	EC Certificate of Origin (EU)	EU: EC Certificate of Origin
P0050	EUR1 Certificate of Origin (EU)	EU: EUR 1 Certificate of Origin
P0060	Information Sheet INF4 (EU)	EU: Information Sheet INF 4
P0070	EU Export Packing List (EU)	Export Packing List
P0080	SAD - Export Customs Declaration wit...	EU: Export/Dispatch EX/EU/COM (0733/0734)
P0085	Incomplete Export Declaration with La...	EU: Incomplete Export Declaration EX/EU (0733...
P0090	Scrapping List (EU)	Scrapping List
P0100	Local Clearance Notification (EU)	EU: Local Clearance Notification

Figure 7.51 Classify Foreign Trade Documents

Procedures for Defaulting Data

Defaulting data is an important feature of the Customs Management functionality in SAP GTS that enables considerable time savings. This feature automatically populates fields or messages in customs declarations that aren't received from the feeder system document.

The purpose of defaulting data is to populate fields automatically with information based on predefined rules, reducing the workload for each customs declaration. In general, most data elements required for completing a customs declaration are transferred from the feeder system. However, sometimes certain information isn't stored in the feeder system, which means that the corresponding fields in the customs declaration remain blank and have to be manually filled in if no default rules are predefined.

All activities related to defaulting data rules are grouped in the Customizing activity **Procedures for Defaulting Data.** However, once this has been correctly configured, the actual data to be defaulted must be defined via master data in the SAP GTS menu (described in Section 7.2.4).

This Customizing activity is composed of different subactivities, as shown in Figure 7.52.

As already mentioned earlier, the setup of defaulting data is performed in two steps: Users first define the logic or rules in the Customizing section. After these rules have been defined, users have to maintain the data that should be defaulted in the customs declarations in the master data.

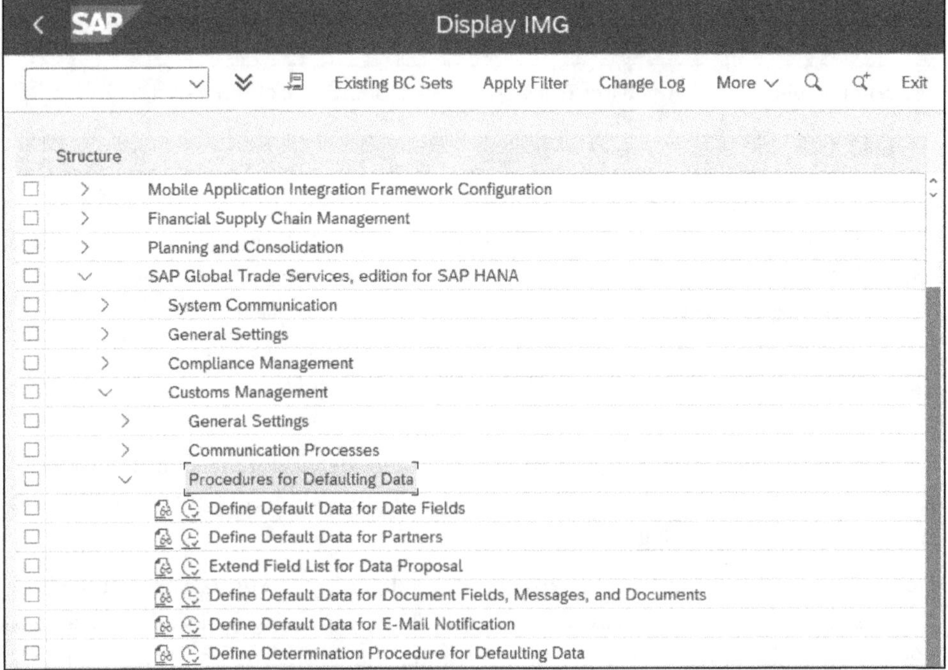

Figure 7.52 Procedures for Defaulting Data

Defaulting Alpha's Data (Part I)

For Alpha, the customs office of exit is one of the mandatory data elements of its export customs declaration. This represents the customs office from which the goods leave the country or customs union. Because this information isn't available in the feeder system and will therefore not be transferred to SAP GTS, defaulting data rules could be beneficial.

Alpha knows that road transport shipments from its EDC in Belgium to its customer in Serbia always exit through the same customs office in Hungary. Therefore, the following rule can be set up: If the **Country of Destination** is "Serbia" and if the **Mode of Transport** is "Truck" (expressed in a specific code), then enter "Customs Office Hungary" in the **Customs Office of Exit** field of the export declaration.

In SAP GTS Customizing, users need to define that the **Customs Office of Exit** field can be populated based on the combination of **Country of Destination** and **Mode of Transport**. If this hasn't been defined via Customizing, then the users can't set up this combination in master data.

In the master data of SAP GTS, users can populate the rule as shown in Table 7.3.

Country of Destination	Mode of Transport	Customs Office of Exit
RS (for Serbia)	03 (for Truck)	Customs Office of Hungary

Table 7.3 Default Data Example

This rule results in the automatic population of the **Customs Office of Exit** field based on the information available in the **Country of Destination** and **Mode of Transport** fields. More specifically, the **Customs Office of Hungary** is defaulted in every export declaration to Serbia where goods are moved via truck. Departure country Belgium doesn't need to be specified in the rule, as the rules are defined per country. In this case, the rule was defined for the Belgian defaulting procedure.

Furthermore, the system distinguishes between date fields, partners, messages, email notifications, and normal data fields. For each field, there is a separate SAP GTS Customizing step, as shown in Figure 7.52.

Each of these types of defaulting data requires a default procedure. These procedures are assigned to a country-process combination. Thus, for each country, there is at least one import and one export defaulting procedure. The defaulting procedure groups all the default rules for that country-process combination.

If there is no predefined default procedure because the relevant country isn't supported by standard SAP, users need to define these default procedures first in the Customizing step **Define Determination Procedure for Defaulting Data**. Here, based on the legal regulation, the document type, and the process, users can identify a default data procedure for dates, partners, general data fields, messages, and even email notifications.

Note

The defaulting possibilities for date fields and partners are less flexible than the defaulting for general fields and messages.

Users can access the defaulting data of the general data fields and messages via menu option **Define Default Data for Document Fields, Messages, and Documents**. This Customizing activity is built into several levels. At the highest level on the left, as shown in Figure 7.53, users can define all **Field Combinations** needed when setting up default data.

All sorts of field combinations can be defined. An example is the field combination **FC_0014**, which is based on **Country of Departure** and **Customs Status**.

The next level (**Default Procedure**) shows the overview of all the defaulting data procedures, grouped per procedure (e.g., export or import) and per country (see Figure 7.54).

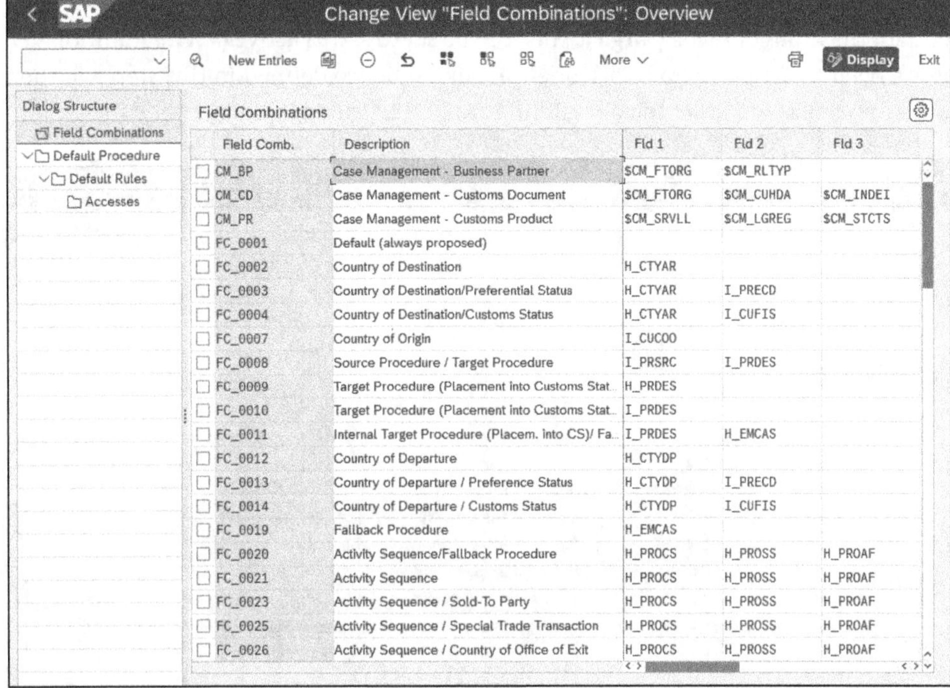

Change View "Field Combinations": Overview

Dialog Structure: Field Combinations / Default Procedure / Default Rules / Accesses

Field Combinations

Field Comb.	Description	Fld 1	Fld 2	Fld 3
CM_BP	Case Management - Business Partner	$CM_FTORG	$CM_RLTYP	
CM_CD	Case Management - Customs Document	$CM_FTORG	$CM_CUHDA	$CM_INDEI
CM_PR	Case Management - Customs Product	$CM_SRVLL	$CM_LGREG	$CM_STCTS
FC_0001	Default (always proposed)			
FC_0002	Country of Destination	H_CTYAR		
FC_0003	Country of Destination/Preferential Status	H_CTYAR	I_PRECD	
FC_0004	Country of Destination/Customs Status	H_CTYAR	I_CUFIS	
FC_0007	Country of Origin	I_CUCOO		
FC_0008	Source Procedure / Target Procedure	I_PRSRC	I_PRDES	
FC_0009	Target Procedure (Placement into Customs Stat...	H_PRDES		
FC_0010	Target Procedure (Placement into Customs Stat...	I_PRDES		
FC_0011	Internal Target Procedure (Placem. into CS)/ Fa...	I_PRDES	H_EMCAS	
FC_0012	Country of Departure	H_CTYDP		
FC_0013	Country of Departure / Preference Status	H_CTYDP	I_PRECD	
FC_0014	Country of Departure / Customs Status	H_CTYDP	I_CUFIS	
FC_0019	Fallback Procedure	H_EMCAS		
FC_0020	Activity Sequence/Fallback Procedure	H_PROCS	H_PROSS	H_PROAF
FC_0021	Activity Sequence	H_PROCS	H_PROSS	H_PROAF
FC_0023	Activity Sequence / Sold-To Party	H_PROCS	H_PROSS	H_PROAF
FC_0025	Activity Sequence / Special Trade Transaction	H_PROCS	H_PROSS	H_PROAF
FC_0026	Activity Sequence / Country of Office of Exit	H_PROCS	H_PROSS	H_PROAF

Figure 7.53 Field Combinations

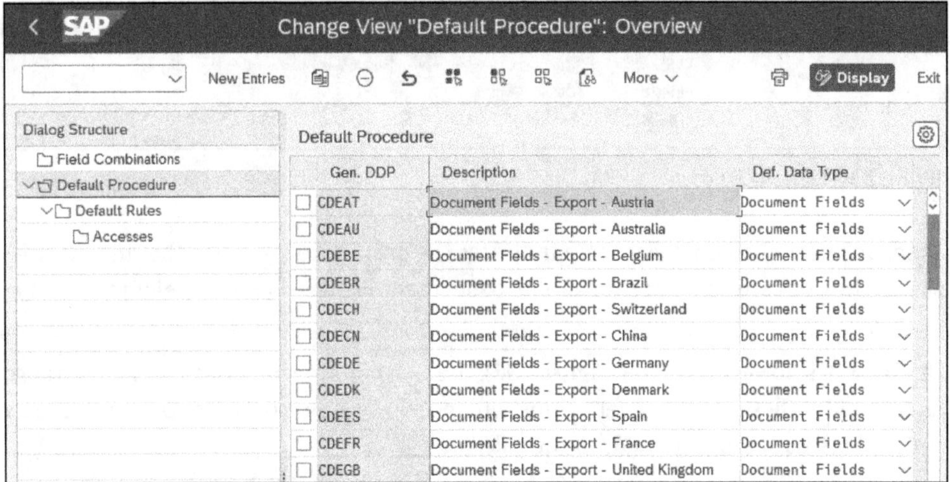

Change View "Default Procedure": Overview

Dialog Structure: Field Combinations / Default Procedure / Default Rules / Accesses

Default Procedure

Gen. DDP	Description	Def. Data Type
CDEAT	Document Fields - Export - Austria	Document Fields
CDEAU	Document Fields - Export - Australia	Document Fields
CDEBE	Document Fields - Export - Belgium	Document Fields
CDEBR	Document Fields - Export - Brazil	Document Fields
CDECH	Document Fields - Export - Switzerland	Document Fields
CDECN	Document Fields - Export - China	Document Fields
CDEDE	Document Fields - Export - Germany	Document Fields
CDEDK	Document Fields - Export - Denmark	Document Fields
CDEES	Document Fields - Export - Spain	Document Fields
CDEFR	Document Fields - Export - France	Document Fields
CDEGB	Document Fields - Export - United Kingdom	Document Fields

Figure 7.54 Default Procedure

As was mentioned earlier, for each country-process combination, one procedure is in place. For example, as shown in Figure 7.54, Belgian exports use default procedure **CDEBE**. Furthermore, in this step, the distinction between **Document Fields** (general data fields) and messages is made.

For each default procedure, a set of *target fields* or *rules* is defined. As Figure 7.55 shows, each of these target fields (**Targ.Fld**) need to be set to status active (**Actv.**). Additionally, users need to assign them to a sequence. The sequence defined in this step will then determine in which order the default rules will be executed in the master data.

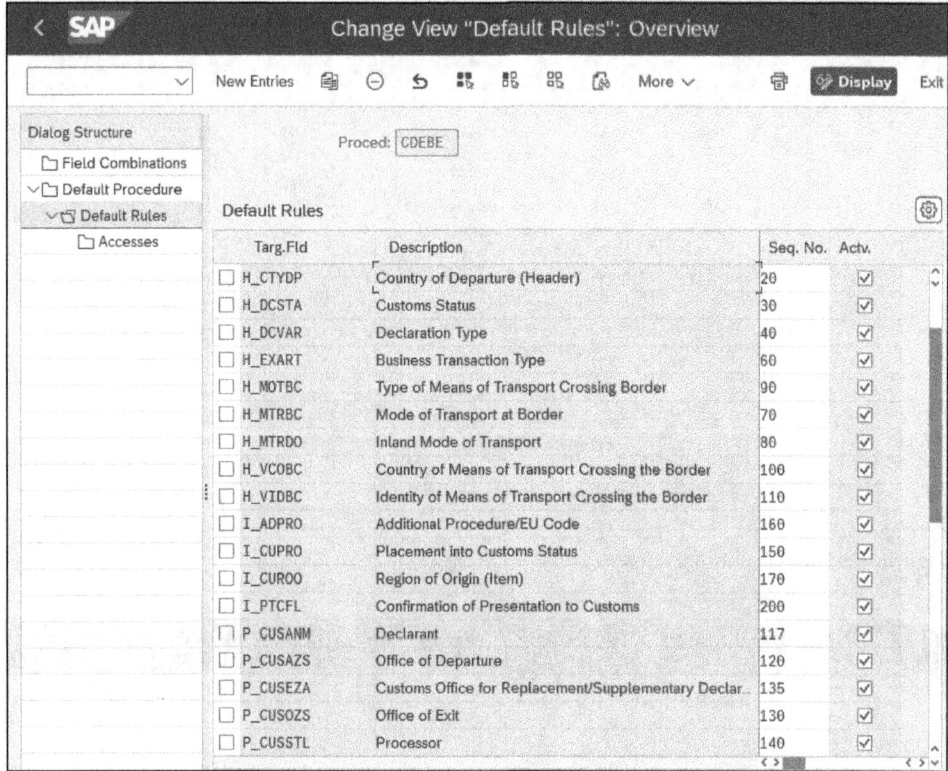

Figure 7.55 Default Rules

At the top of the screen, the system indicates the procedure for which the rules are displayed. In this case, the rules for **CDEBE** (**Document Fields—Export—Belgium**) are displayed.

Finally, at the lowest level (see Figure 7.56), users can define **Accesses** for each of these target fields. These accesses determine which field or field combinations (previously defined) are used as input fields in the rule.

By assigning a sequence number, users determine the sequence in which the defaulting data logic will run through the accesses. The field combination (**Field Comb.**) with the lowest number will be considered first. If no determination can be performed for the first access (i.e., first level), the system will consider the subsequent accesses (i.e., levels) and use them to determine the default value. The logic goes sequentially through each level until it finds a match and defaults a value into the field. If there is no match, the field remains unpopulated.

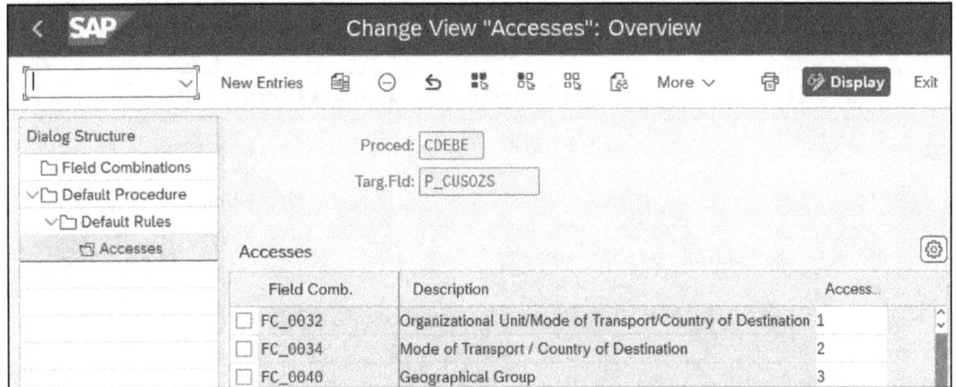

Figure 7.56 Accesses

Figure 7.56 shows the access for **Targ.Fld** (target field) **P_CUSOZS** (office of exit). In this example, the **Organizational Unit/Mode of Transport/Country of Destination** combination shown in the **Description** column is accessed first. The system only continues its determination by considering the following accesses if no value can be populated in the customs declaration based on the access for the target field. For example, in the case of Alpha (see Table 7.3), if it's an export to Serbia but via air instead of truck, then there will be no match in this access, and the office of exit won't be defaulted.

When the company knows how to identify the customs office of exit based on the mode of transport and country of destination, these fields will be the preferred choice for input.

SAP GTS comes with a variety of predefined field combinations. An important one is **FC_0001 Default (Always Proposed)**, which indicates that a certain target field should always be populated in the same way without making use of any combination of input fields.

Defaulting Alpha's Data (Part II)

As an addition to the first, simple example in which goods from Belgium were shipped by truck through Hungary to Serbia, it's more likely that the defaulting rules will be a combination of several simple rules.

Assume the same company also frequently ships goods to Turkey, and that if these goods move by truck, they always leave the EU via a customs office in Bulgaria. The same defaulting rule as in the previous example can be applied, and in the Customizing activity of SAP GTS, nothing changes. In the **Master Data** section of the defaulting data, a line needs to be added, stating that if the **Country of Destination** is "Turkey" and the **Mode of Transport** is "Truck," the system should default to **Customs Office Bulgaria**, as shown in Table 7.4.

Rule Based on Country of Destination/Mode of Transport		
Country of Destination	**Mode of Transport**	**Customs Office of Exit**
RS (for Serbia)	03 (for Truck)	Customs Office Hungary
TR (for Turkey)	03 (for Truck)	Customs Office Bulgaria

Table 7.4 Default Data Extended Example

Besides these shipments by truck to Serbia and Turkey, all other shipments of the company are done by sea. All these sea shipments leave the territory via the port of Antwerp.

Users can add the code for sea shipments as the **Mode of Transport** in the first rule and add a new line in the **Master Data** for each country to which the company ships. Users can also select the **Default (Always Proposed)** rule instead. This rule would default **Customs Office of Antwerp** in every other case, as shown in Table 7.5. Because there would be no hit in the first access (level), the logic would simply continue to the next one where it would default to **Customs Office of Antwerp**.

Default (Always Proposed)
Customs Office of Antwerp

Table 7.5 Default Data: Default (Always Proposed)

Note that the sequence of the default data accesses determines which field combination applies first. If the **Default (Always Proposed)** rule was applied first, then the truck shipments to Serbia or Turkey would also have the customs office of Antwerp as their customs office of exit, which is incorrect.

As mentioned, standard SAP GTS already has a predefined number of rules or field combinations. However, if a needed field combination isn't yet defined, users can create the combination. In the same Customizing activity, in the **Field Comb.** section on the left (refer to Figure 7.56), users can create a name and description for the field combination and add up to eight different fields to it.

Sometimes, a specific source or target field isn't standardly defined by SAP GTS. In this case, users can create these fields in the Customizing activity **Extend Field List for Data Proposal**. If the defaulting logic is too complex and can't be obtained via the standard SAP GTS Customizing, there is a specific business add-in (BAdI) available as well.

Customs Code Lists

SAP GTS always uses the same document types for the same processes (e.g., CULOEX for export declarations), independently of the country. However, every country has specific data that is accepted and required by the customs authorities. For certain fields, SAP GTS provides a list of allowed input values. This means that these fields only accept the values defined in this Customizing activity. These *customs code lists* are specific to legal regulations. The following are examples of common fields that use customs code lists:

- **Customs Status**
- **Placement into Customs Status**
- **Packaging Material**
- **Types of Seals**
- **Mode of Transport**
- **Type of Previous Document**

Using customs codes avoids entering incorrect values in a declaration and, therefore, avoids longer processing times. Because these code lists are aligned with the values used by the customs authorities, SAP GTS enables users to communicate "in the same language" as the final recipient.

All activities related to the customs code lists can be found in the **Customs Code Lists** group in the Customizing area of SAP GTS, as shown in Figure 7.57.

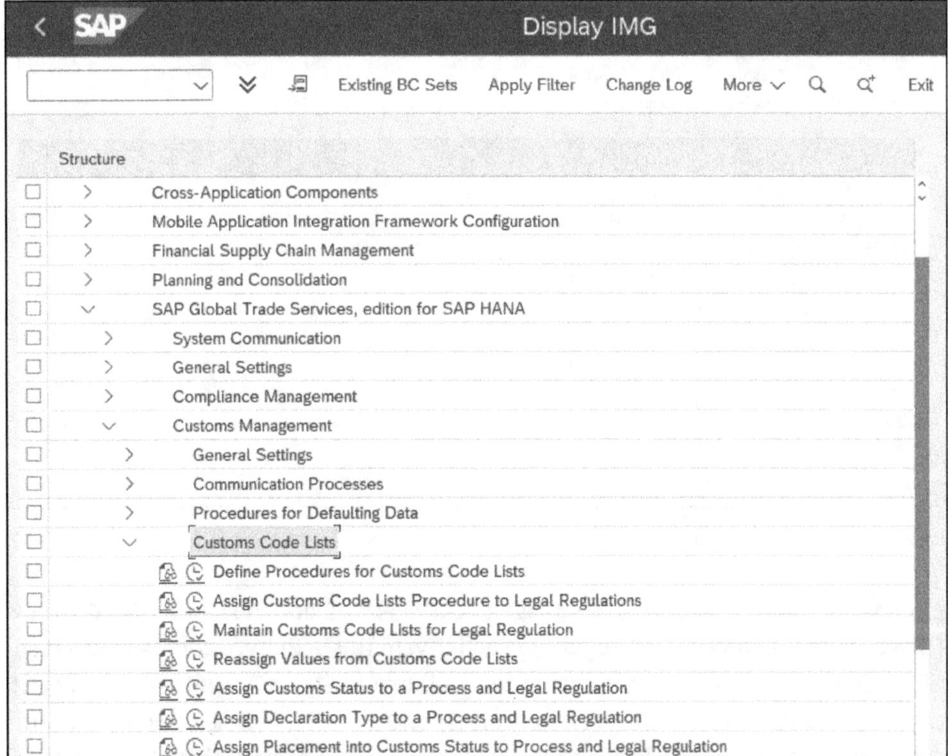

Figure 7.57 Customs Code Lists

Before defining the code lists, users need to define a country-specific procedure. The Customizing activity **Define Procedures for Customs Code Lists** (see Figure 7.58) allows users to create these procedures for customs code lists.

Separate procedures are created per country, but in special cases, it's useful to have additional customs code lists for different activities in a country. For example, users could have a standard procedure triggered if there is an import or export declaration, and an additional one for transit or for EMCS activities. This last customs code list is then valid only for transit declarations.

Figure 7.58 displays the overview of the different customs code list procedures available in standard SAP GTS.

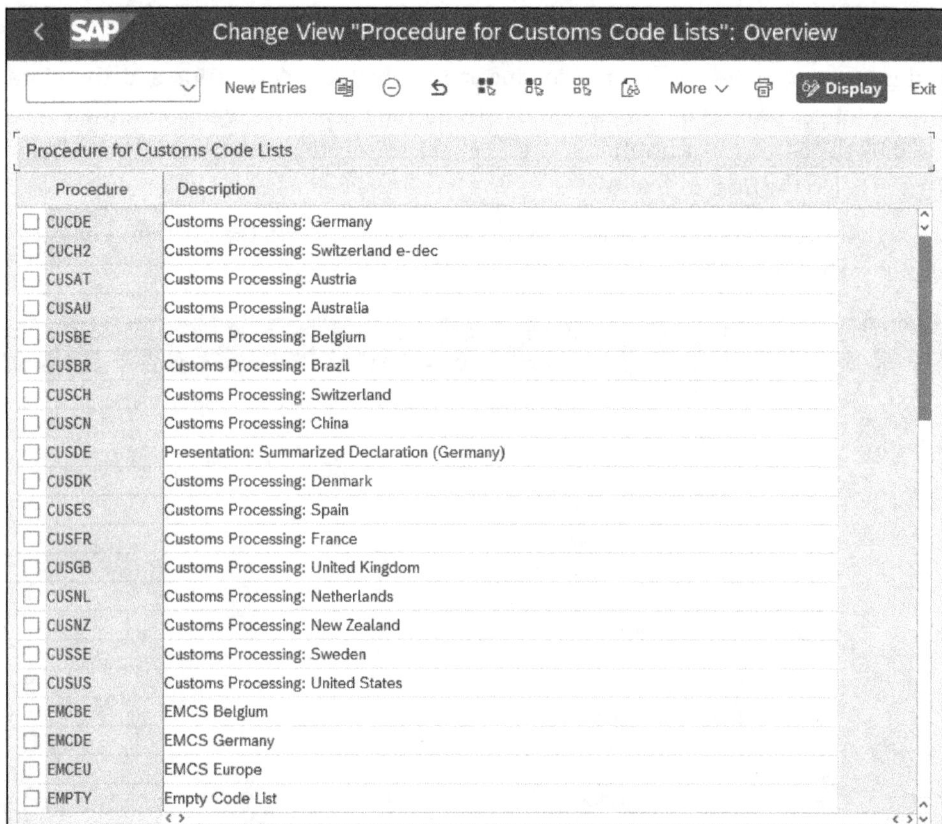

Figure 7.58 Define Procedure for Customs Code Lists

After defining the procedure, users assign it to a legal regulation via the **Assign Customs Code Lists Procedure to Legal Regulation** activity. In this activity, a company can identify which code list procedures it will use, as shown in Figure 7.59. Standard SAP GTS comes with a range of data objects that correspond to individual fields of the customs declaration. To maintain allowed values of a data object, users list that object first. For example, for **CUSBE**, if users want the **Mode of Transport at Border** field to only have a set of allowed values in the declaration, users first need to add the field to this list, before assigning the exact values in the next Customizing step.

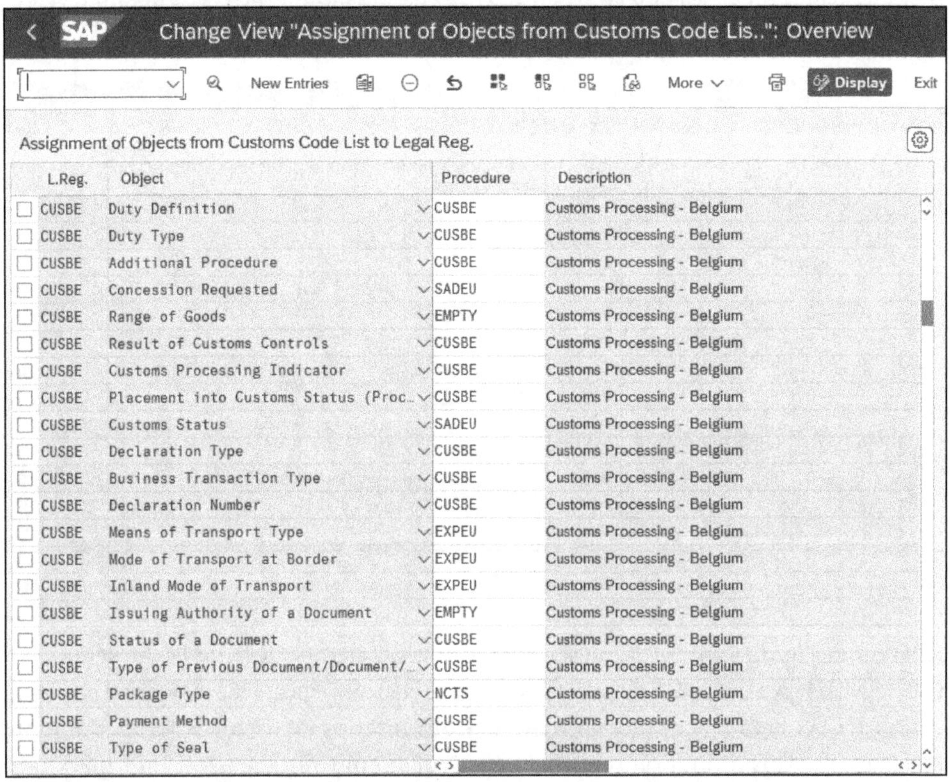

Change View "Assignment of Objects from Customs Code Lis..": Overview

New Entries More Display Exit

Assignment of Objects from Customs Code List to Legal Reg.

L.Reg.	Object	Procedure	Description
CUSBE	Duty Definition	CUSBE	Customs Processing - Belgium
CUSBE	Duty Type	CUSBE	Customs Processing - Belgium
CUSBE	Additional Procedure	CUSBE	Customs Processing - Belgium
CUSBE	Concession Requested	SADEU	Customs Processing - Belgium
CUSBE	Range of Goods	EMPTY	Customs Processing - Belgium
CUSBE	Result of Customs Controls	CUSBE	Customs Processing - Belgium
CUSBE	Customs Processing Indicator	CUSBE	Customs Processing - Belgium
CUSBE	Placement into Customs Status (Proc...	CUSBE	Customs Processing - Belgium
CUSBE	Customs Status	SADEU	Customs Processing - Belgium
CUSBE	Declaration Type	CUSBE	Customs Processing - Belgium
CUSBE	Business Transaction Type	CUSBE	Customs Processing - Belgium
CUSBE	Declaration Number	CUSBE	Customs Processing - Belgium
CUSBE	Means of Transport Type	EXPEU	Customs Processing - Belgium
CUSBE	Mode of Transport at Border	EXPEU	Customs Processing - Belgium
CUSBE	Inland Mode of Transport	EXPEU	Customs Processing - Belgium
CUSBE	Issuing Authority of a Document	EMPTY	Customs Processing - Belgium
CUSBE	Status of a Document	CUSBE	Customs Processing - Belgium
CUSBE	Type of Previous Document/Document/...	CUSBE	Customs Processing - Belgium
CUSBE	Package Type	NCTS	Customs Processing - Belgium
CUSBE	Payment Method	CUSBE	Customs Processing - Belgium
CUSBE	Type of Seal	CUSBE	Customs Processing - Belgium

Figure 7.59 Assign a Customs Code Lists Procedure to Legal Regulation

Finally, in the **Maintain Customs Code Lists for Legal Regulation** activity, users create the possible values that will be available in the customs declaration for that field. Figure 7.60 displays the entries that will be available in the **Mode of Transport** field for customs declarations created for a Belgian company (legal regulation CUSBE). The system will only allow the values maintained here to be entered in the customs declaration. In the declaration, a dropdown will be available for the relevant field, presenting only the possible values. If the user enters another value manually in the field, an error message will be displayed. Figure 7.60 shows a case in which only values from 1 to 9 will be allowed in the field; otherwise, the user will receive an error message.

There are a few more Customizing activities related to the customs code lists. These additional steps are related to the customs status and customs placement. These fields, respectively on the header and item levels of customs declarations, inform the customs authorities of the origin and destination of the goods. In this case, the origin and destination of the goods aren't considered from a geographical point of view, but rather from a customs procedure point of view. For instance, the customs status indicates that the goods are exported out of the country, while the customs status placement indicates that the goods were stored in a customs warehouse. The possible values for both of these fields are set up in separate Customizing steps.

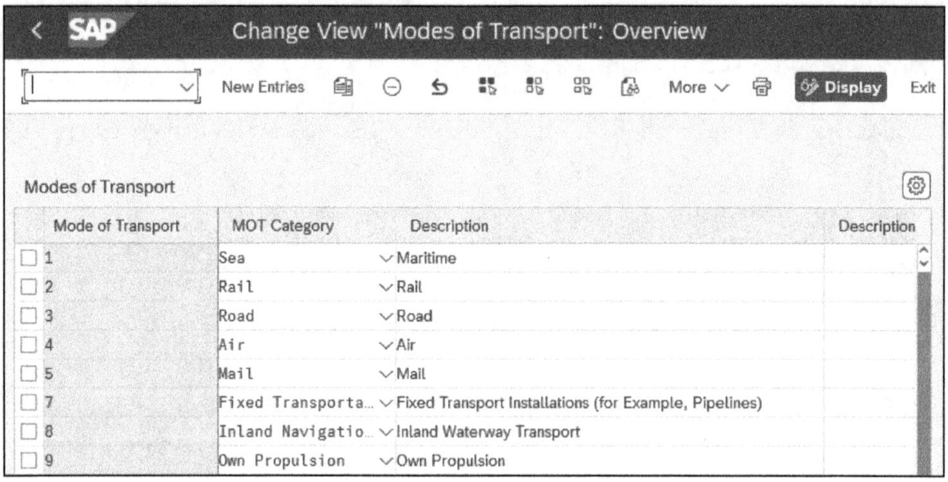

Change View "Modes of Transport": Overview

New Entries More Display Exit

Modes of Transport

Mode of Transport	MOT Category	Description	Description
1	Sea	Maritime	
2	Rail	Rail	
3	Road	Road	
4	Air	Air	
5	Mail	Mail	
7	Fixed Transporta...	Fixed Transport Installations (for Example, Pipelines)	
8	Inland Navigatio...	Inland Waterway Transport	
9	Own Propulsion	Own Propulsion	

Figure 7.60 Maintain Customs Code Lists: Mode of Transport

Looking at Figure 7.61, in the customs declaration created for the **Legal Regulation CUSBE** with **Process CUSEX** (export process), the customs status (**Cust. Status**) can have values **CO**, **EU**, or **EX**. This prevents the user from entering **IM** for an export declaration, for example. This limitation reduces the risk of submitting incorrect information to the customs authorities.

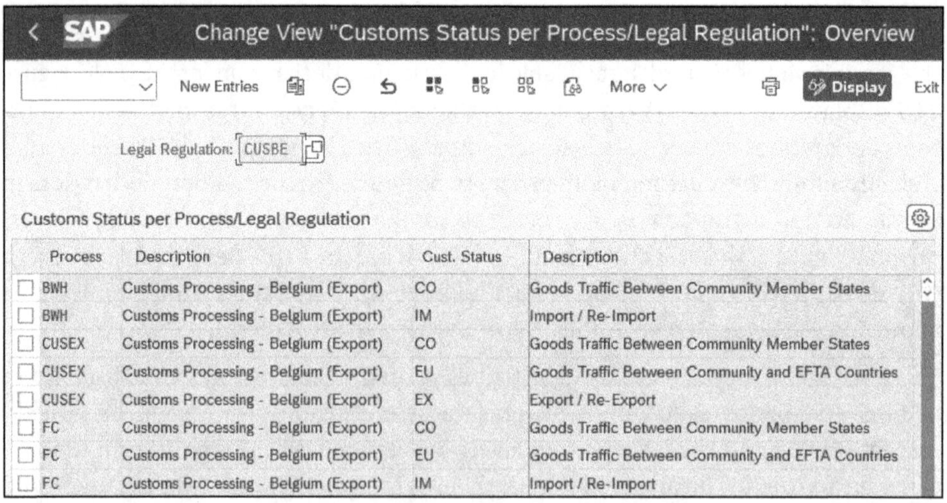

Change View "Customs Status per Process/Legal Regulation": Overview

New Entries More Display Exit

Legal Regulation: CUSBE

Customs Status per Process/Legal Regulation

Process	Description	Cust. Status	Description
BWH	Customs Processing - Belgium (Export)	CO	Goods Traffic Between Community Member States
BWH	Customs Processing - Belgium (Export)	IM	Import / Re-Import
CUSEX	Customs Processing - Belgium (Export)	CO	Goods Traffic Between Community Member States
CUSEX	Customs Processing - Belgium (Export)	EU	Goods Traffic Between Community and EFTA Countries
CUSEX	Customs Processing - Belgium (Export)	EX	Export / Re-Export
FC	Customs Processing - Belgium (Export)	CO	Goods Traffic Between Community Member States
FC	Customs Processing - Belgium (Export)	EU	Goods Traffic Between Community and EFTA Countries
FC	Customs Processing - Belgium (Export)	IM	Import / Re-Import

Figure 7.61 Assign Customs Status to a Process

Authorizations and Securities

Customs authorities require various types of authorizations to perform certain activities. This is the case, for example, for activities such as simplified procedures for import/export/transit, periodic declaration processes, and economic customs procedures. Users should consider maintaining these authorizations in SAP GTS as they will depend on the activities that are performed (e.g., BWH or IPR).

In addition, an economic operator might need to include securities when performing specific activities (e.g., simplified procedures, economic customs procedures, deferred duty payment schemes, etc.). Securities can also be maintained in SAP GTS.

SAP GTS has identified several standard types of authorizations and securities. Users can define additional types via Customizing if these predefined types don't meet their company's requirements.

The **Authorizations and Securities** group of SAP GTS Customizing consists of multiple steps to specify the details of a specific authorization or security. As shown in Figure 7.62, the steps are separated by process, including, for example, transit procedure, import processing, or inward processing.

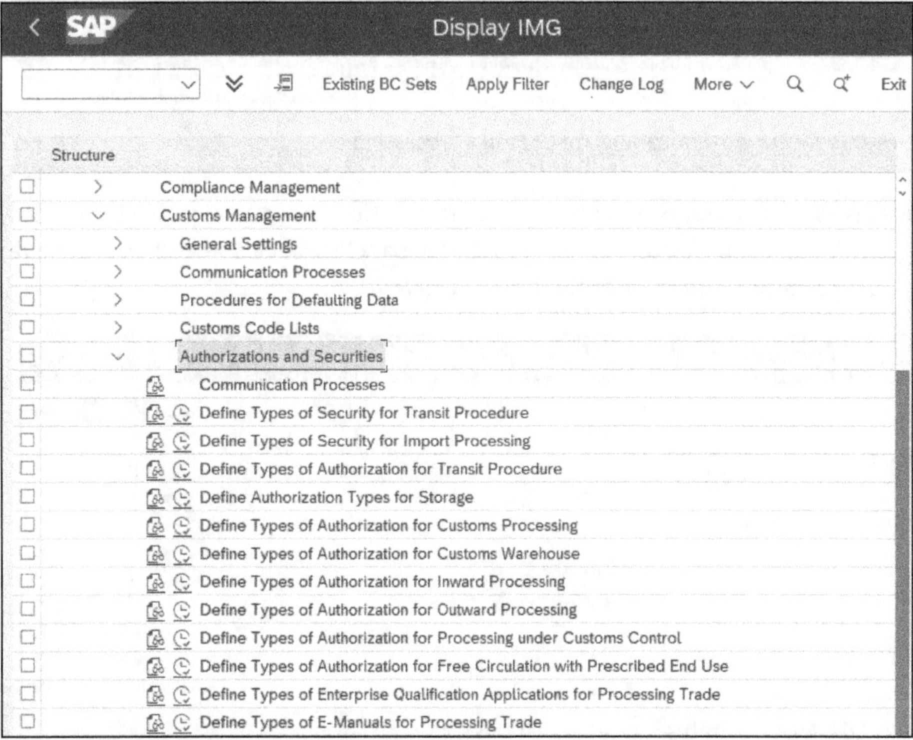

Figure 7.62 Authorizations and Securities

In each of these activities, users can indicate whether the selected authorization or security applies in a certain country, based on the legal regulation.

As indicated in Figure 7.63, users need to choose the applicable **Authorization Type** in addition to specifying the relevant legal regulation (**Legal Reg.**).

Change View "Type of Authorization for Customs Processing": Overview

Type of Authorization for Customs Processing

Legal Reg.	Authorization Type	Description
ATLAS	AZ	Approved Exporter (Export)
ATLAS	VV	Simplified Customs Declaration Procedure (Import)
CUSAT	VVEX	Simplified Customs Declaration Procedure - Export
CUSAT	VVIM	Simplified Customs Declaration Procedure - Import
CUSAU	CES	Status as Authorized Exporter
CUSBE	VVEX	Simplified Customs Declaration Procedure - Export
CUSBE	VVIM	Simplified Customs Declaration Procedure - Import
CUSCH	VVEX	Authorized Consignor - Export
CUSCH	VVIM	Presentation at Domicile - Import
CUSES	VVEX	Simplified Customs Declaration Procedure - Export
CUSES	VVIM	Simplified Customs Declaration Procedure - Import
CUSFR	VVEX	Simplified Customs Declaration Procedure (PDDE) - Export
CUSFR	VVIM	Simplified Customs Declaration Procedure (PDDI) - Import
CUSGB	EIR	Declaration as LCN in Accounting by Declarant
CUSGB	SDE	Simplified Customs Declaration
CUSNL	DPEXS	Simplified Procedure for Export Declaration
CUSNL	DPFCR	Simplified Procedure for Declaration to Free Circulation
CUSNZ	SES	Secure Export Scheme Partner (SES)
CUSSE	VVEX	Simplified Customs Declaration Procedure - Export

Figure 7.63 Define Authorizations for Customs Processing

Users also specify a status for each of the authorizations or securities (see Figure 7.64) by going into the **Allowed Status** folder in the left pane. These can be **Authorization Created**, **Authorization Active**, or **Authorization Expired.**

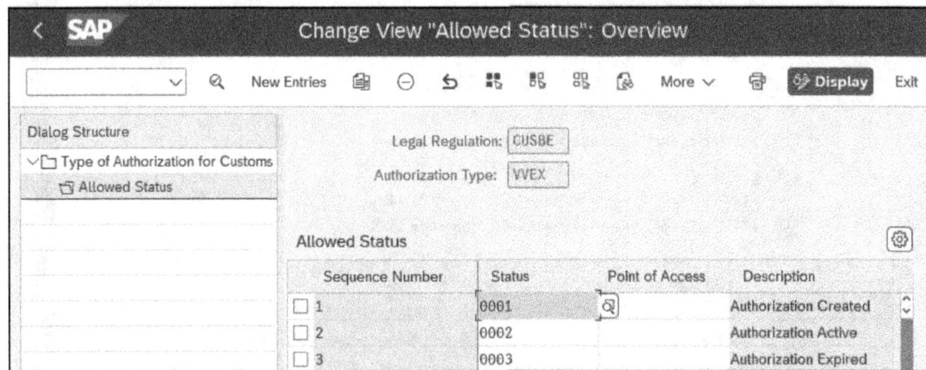

Figure 7.64 Allowed Status for Authorizations

After selecting these settings in the Customizing section of SAP GTS, users have to create these specific authorizations or securities in the master data section of Customs Management (Section 7.2.4).

Calculation of Customs Duty (Duty Framework)

The duty framework functionality in SAP GTS groups all the values and calculations that the customs authorities require a company to report. Common required values include the value of the goods, shipment-specific insurance and freight costs, and import duties and taxes. Users create a *duty framework* by combining several of these values and creating subtotals.

Because the legal requirements for the declaration of certain values are country-specific, the settings related to the duty framework are performed at the legal regulation level.

Recall from Section 7.2.1 earlier that each customs declaration comes with a tab on the header level (**Logistics Costs**) and on the item level (**Customs Value**). These tabs are available in both import and export declarations. A framework for capturing values and calculating the duties can be created. SAP GTS allows users to fully design this framework, however, users should consider how to comply with the company's legal and business needs.

All Customizing settings for this duty framework can be found back in the **Calculation of Customs Duty** step. This group of Customizing activities contains several important steps (see Figure 7.65).

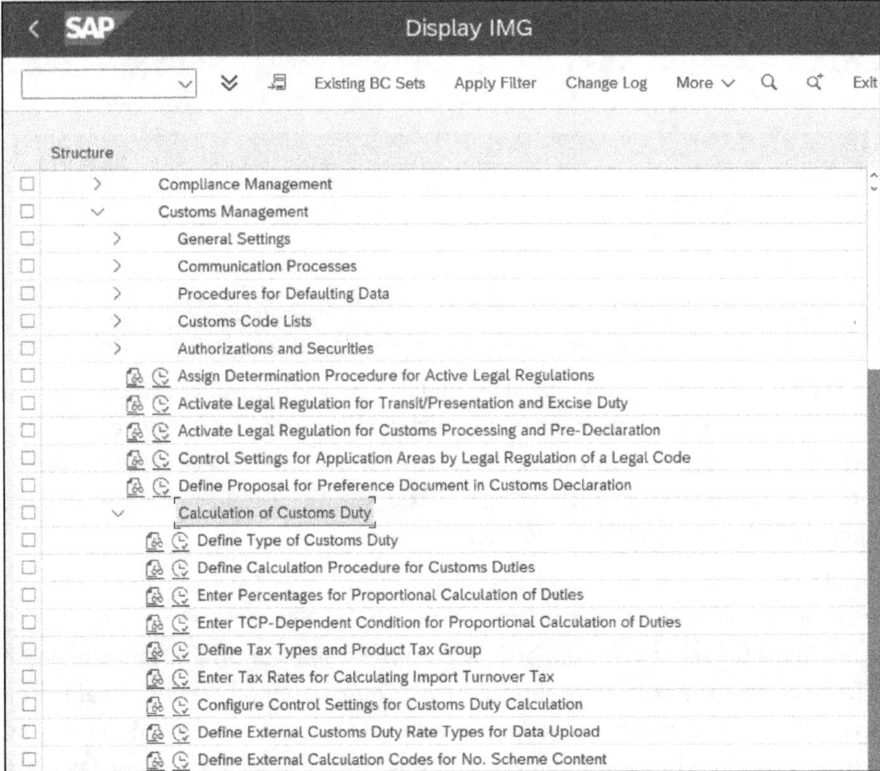

Figure 7.65 Calculation of Customs Duty

The duty framework is composed of duty types that can be configured into a framework. Therefore, the first step is to define these duty types in the Customizing activity **Define Type of Customs Duty**. SAP GTS comes with an extensive list of predefined duty types, but users can set up additional duty types depending on the company's requirements. Figure 7.66 shows an overview of the existing duty types, such as **A001 Goods Value, A002 Transportation Costs,** or **A003 Insurance Costs.**

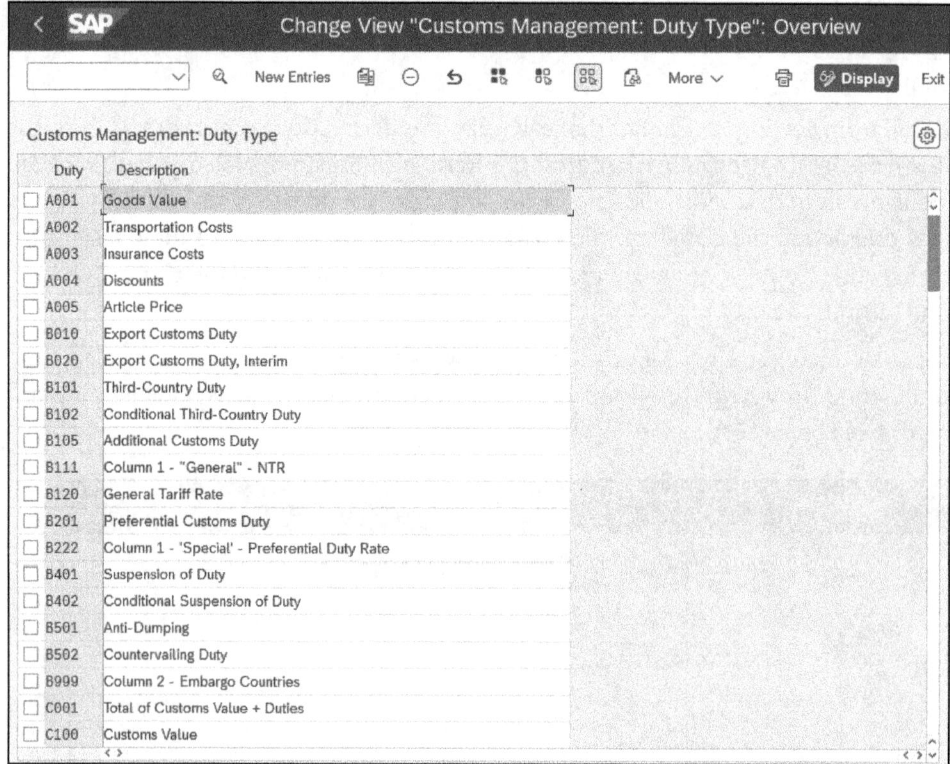

Figure 7.66 Define Duty Types

During the definition of the duty type, users need to indicate the duty category, which can be selected from a predefined list of categories. This is done by double-clicking on a duty type to open its details, as shown in Figure 7.67. The **Duty Category** helps the system identify the nature of the duty types. For example, these duty categories include **Anti-Dumping** or **Preferential Customs Duty**.

Users need to define other characteristics in addition to **Duty Category.** In the **Distrib. Type** field, for example, users select how the value on the header level gets distributed among the items (whether it should be based on the weight or on the value, or there should be no distribution). If users choose to distribute based on weight or value, the system calculates the proportional weight or value of each item compared to the total weight or value of the entire shipment. This way, the costs at the header level are allocated to each of the items as a percentage of the total weight or value. In the **Maint. Level** field, users select whether the duty type needs to be included in the duty framework on the header level, on the item level, or on both.

Change View "Customs Management: Duty Type": Details

New Entries More Display Exit

Duty Type: A001

Customs Management: Duty Type

Description:	Goods Value
Duty Category: *	Transaction Amount
Distrib. Type:	No Distribution
Maint. Level:	Specify Duty Type in Item, and Cumulate in Header
Access w/ Ctry:	SAP Standard Access
Tax Type:	
Pro-Rata:	Duty not Pro-Rata
Explanation:	

Figure 7.67 Duty Type Details

After creating separate duty types, users need to combine them into a framework in line with the company's legal and business requirements via the Customizing activity **Define Calculation Procedure for Customs Duties.** After selecting the correct legal regulation, users can create separate calculation procedures for import, export, and transit.

In this activity, users need to list all duty types (**Duty**) that should be included in the framework and assign them a sequence number (**Seq.No.**), as shown in Figure 7.68.

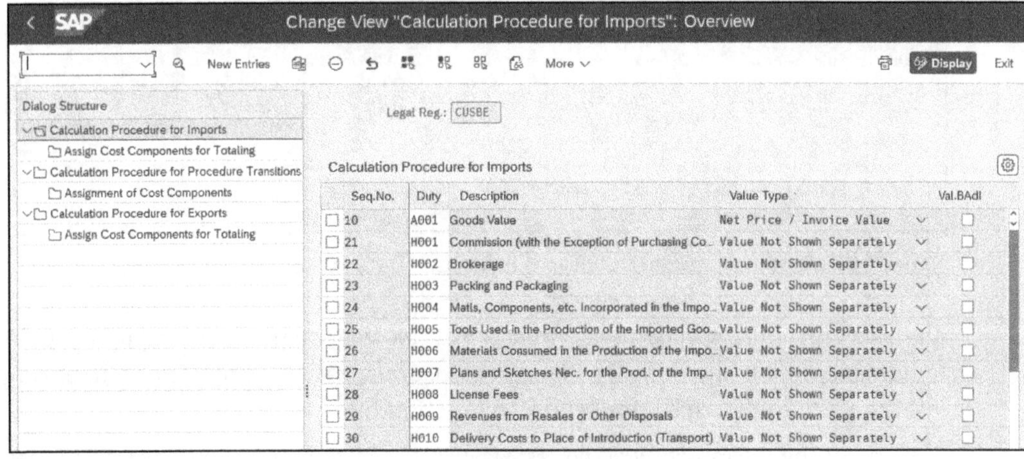

Change View "Calculation Procedure for Imports": Overview

New Entries More Display Exit

Dialog Structure
- Calculation Procedure for Imports
  - Assign Cost Components for Totaling
- Calculation Procedure for Procedure Transitions
  - Assignment of Cost Components
- Calculation Procedure for Exports
  - Assign Cost Components for Totaling

Legal Reg.: CUSBE

Calculation Procedure for Imports

Seq.No.	Duty	Description	Value Type	Val.BAdI
10	A001	Goods Value	Net Price / Invoice Value	
21	H001	Commission (with the Exception of Purchasing Co...	Value Not Shown Separately	
22	H002	Brokerage	Value Not Shown Separately	
23	H003	Packing and Packaging	Value Not Shown Separately	
24	H004	Matls, Components, etc. Incorporated in the Impo...	Value Not Shown Separately	
25	H005	Tools Used in the Production of the Imported Goo...	Value Not Shown Separately	
26	H006	Materials Consumed in the Production of the Impo...	Value Not Shown Separately	
27	H007	Plans and Sketches Nec. for the Prod. of the Imp...	Value Not Shown Separately	
28	H008	License Fees	Value Not Shown Separately	
29	H009	Revenues from Resales or Other Disposals	Value Not Shown Separately	
30	H010	Delivery Costs to Place of Introduction (Transport)	Value Not Shown Separately	

Figure 7.68 Define Calculation Procedure for Customs Duties

As usual, the duty type with the lowest sequence number will appear first in the framework. In our example, the **A001 Goods Value** will be the first duty type displayed in a Belgian import declaration. For each of the duty types, users need to assign a **Value Type**. Standard SAP GTS provides a range of types for the system to identify the duty type.

Besides adding duty types, the standard SAP GTS duty framework functionality allows the creation of subtotals of previous duty types. However, users can make use of a BAdI to perform more complex calculations. In this example, the user would not select the **Val.BAdI** checkbox column.

Users can create a subtotal in the **Assign Cost Components for Totaling** activity by selecting the duty type from the **Calculation Procedure for Imports** and drilling down to **Assign Cost Components for Totaling**. Here, you need to add the elements that are to be summed up. Note that certain standard duty types, such as **C100 Customs Value** and **D001 Total Duties**, are set up as subtotaling duty categories. Therefore, in such cases, using the totaling functionality isn't required.

Another important activity when setting up a duty framework is **Define External Customs Duty Rate Types for Data Upload**. When a company purchases duty rates from a content provider, linking the duty types used in SAP GTS to the ones used by the content provider enables the system to function accurately. Matching these duty types is done in this Customizing activity, as displayed in Figure 7.69.

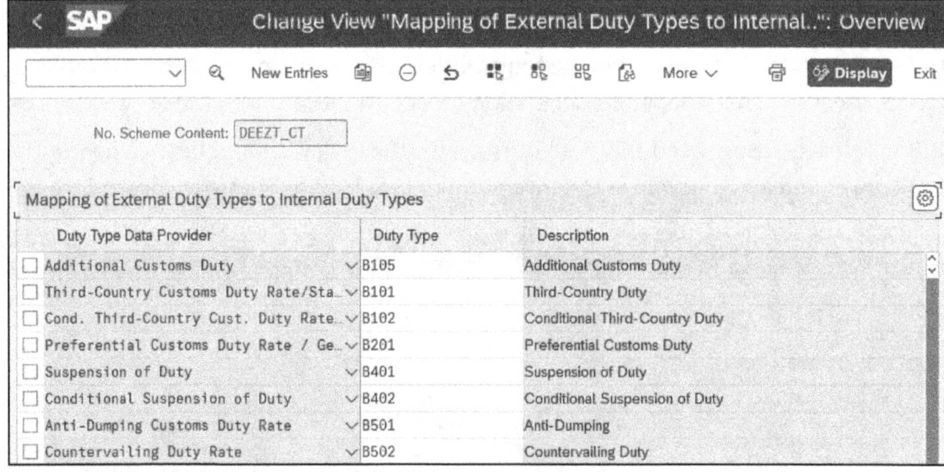

Figure 7.69 Mapping External Duty Types to Internal Duty Types

All these Customizing activities are the most important steps for setting up the Customs Management functionality. Remember that they need to be performed in the correct sequence, as described throughout this section.

In many of the Customizing activities, we've referred to corresponding settings that need to be performed in master data as well. Therefore, in the following section, we'll highlight the master data elements that are relevant for general Customs Management processes.

7.2.4 Master Data

The previous sections of this chapter discussed import and export transactions in detail. To support these transactions, users need to maintain master data in advance. The following data elements are captured in the **Master Data** page of the **Customs Export** and **Customs Import** spaces:

- **Authorizations/Securities**
- **Geography**
- **Business Partners and Products**
- **Product Catalog**

More information about master data elements can be found in Chapter 4. Here, we'll focus on some of the data relevant for supporting import and export transactions.

Partners

Besides the standard customers and vendors transferred into SAP GTS from a feeder system, Customs Management allows users to create, change, and display other types of business partners, for example, customs offices.

A customs office can have several roles, including customs office of departure, destination, exit, and transit. It's usually mandatory to declare one of these types of customs offices to the customs authorities, meaning that companies generally specify the customs office type in the customs declaration.

Customs offices can be created either manually or uploaded into the system via XML. Specifically for the EU, the EU customs authorities provide an XML file, including an overview of EU customs offices. As the format of this file differs from the one that users need for the system, SAP GTS has a function to convert the file format.

Manually creating customs offices is done through the **Manage Business Partner** app in SAP Fiori, just like how other business partners are created in SAP GTS. Note that, when creating a customs office, users need to select the correct business partner role (for the customs office partner, SLLCOF [customs office]) and assign a customs office number to it.

Authorizations and Securities

The master data of the authorizations and securities is a key prerequisite to use certain procedures and simplifications. The master data for authorizations and securities includes the following:

- Maintenance of authorizations
- Maintenance of securities for import processes
- Maintenance of securities for transit procedures

Authorizations and securities can be managed and monitored via the SAP Fiori launchpad. Users can access these tiles through the **Master Data** page from the appropriate customs space (e.g., **Customs Export**, **Customs Import**).

The layout and logic are similar predefined authorizations and securities in SAP GTS. As shown in Figure 7.70, the screen for license/authorization creation is divided into three sections. The upper box of the screen provides general information about things such as the legal regulation and the authorization/security type. Besides an internal SAP GTS number, the official number of the authorization/security provided by the customs authorities needs to be maintained in the **Authorization Number** field.

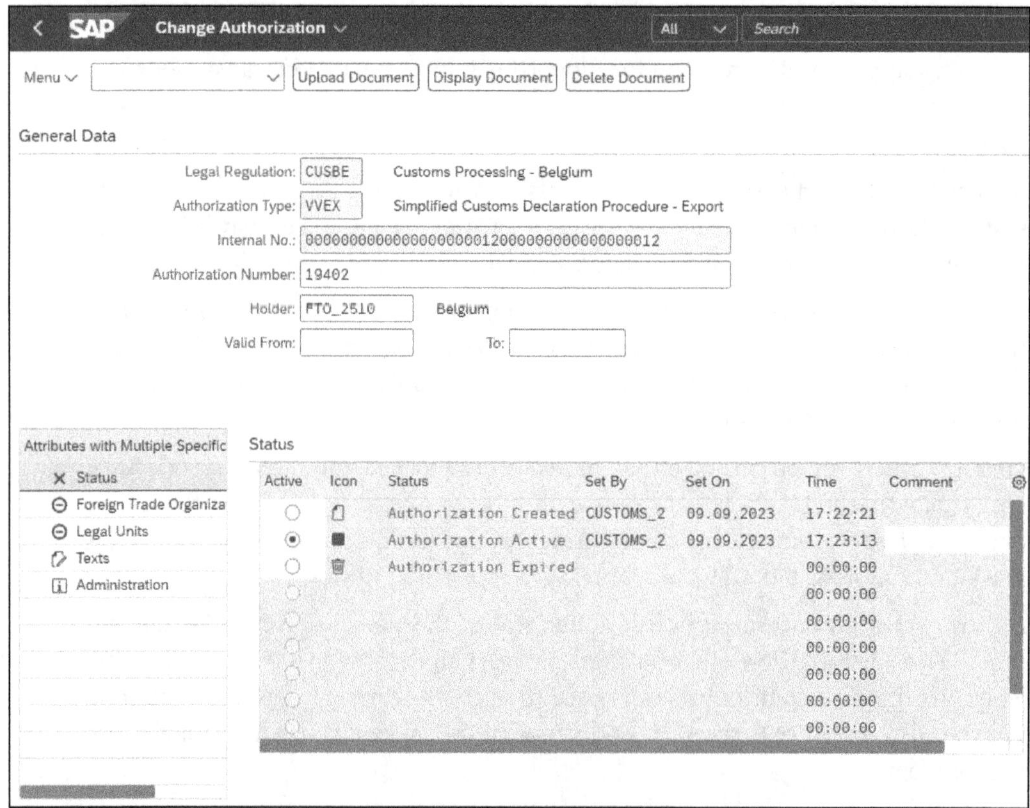

Figure 7.70 Authorization for a Simplified Procedure

The attributes and their details are listed in the lower-left section of the screen. Both authorizations and securities consist of multiple attributes that allow you to specify to which organizational unit they apply to and the status of this authorization. Although the setup is similar for authorizations and securities, the required attributes can differ.

Every authorization or security contains the **Status** attribute. Note that creating an authorization/security isn't sufficient. To use an authorization/security, users need to set the **Status** to **Active** and include a current validity period. The **Foreign Trade Organization** and **Legal Units** attributes contain the applicable organizational units.

Figure 7.70 displays the type of authorization that was specified in the Customizing. The legal regulation and authorization type were assigned in the SAP GTS Customizing, and the statuses have been created. Here, users can see the **Authorization Created**, **Authorization Active**, and **Authorization Expired** statuses, as specified in Section 7.2.3.

Besides the information copied over from the feeder system, SAP GTS offers a ruleset to automatically complete certain information on the customs document, for instance, if a sea/air shipment always leaves a country via the same harbor/airport, a specific customs office is defined for sea transport, and another is defined for air transport. SAP GTS enables users to default data for document fields, message determination toward authorities, and email notifications.

These defaulting data rules are created in Customizing, but the application of these rules with specific inputs is performed in master data.

Defaulting Alpha's Master Data

Looking back at the Alpha company, we can summarize the following business logic:

- If the goods are transported by truck (**Mode of Transport** = **3**) to Serbia, then the customs office of exit is Customs Office Hungary (**CU_DEMO_PL**).
- If the goods are transported by truck (**Mode of Transport** = **3**) to Turkey, then the customs office of exit is Customs Office Bulgaria (**CU_DEMO_BG**).
- In all other cases, the goods leave the customs union via the Customs Office Antwerp (**BE101000**).

Figure 7.71 and Figure 7.72 display how these rules need to be set up in the master data section of SAP GTS.

To access this activity, users need to navigate via the administrator role directly in the SAP GUI to **SAP Global Trade Services • Settings • Default Data • Define Data Defaulting Rules for Document Fields**, as there are no standard tiles available for this function in SAP Fiori. Prior to maintaining any value, users need to select for which default data procedure and legal regulation they wish to define the default data.

The first two bullet points (goods trucked to Serbia and Turkey) are based on **Mode of Transport / Country of Destination** access (see Figure 7.71).

The last bullet point can be obtained by using the **Default (always proposed)** access (see Figure 7.72).

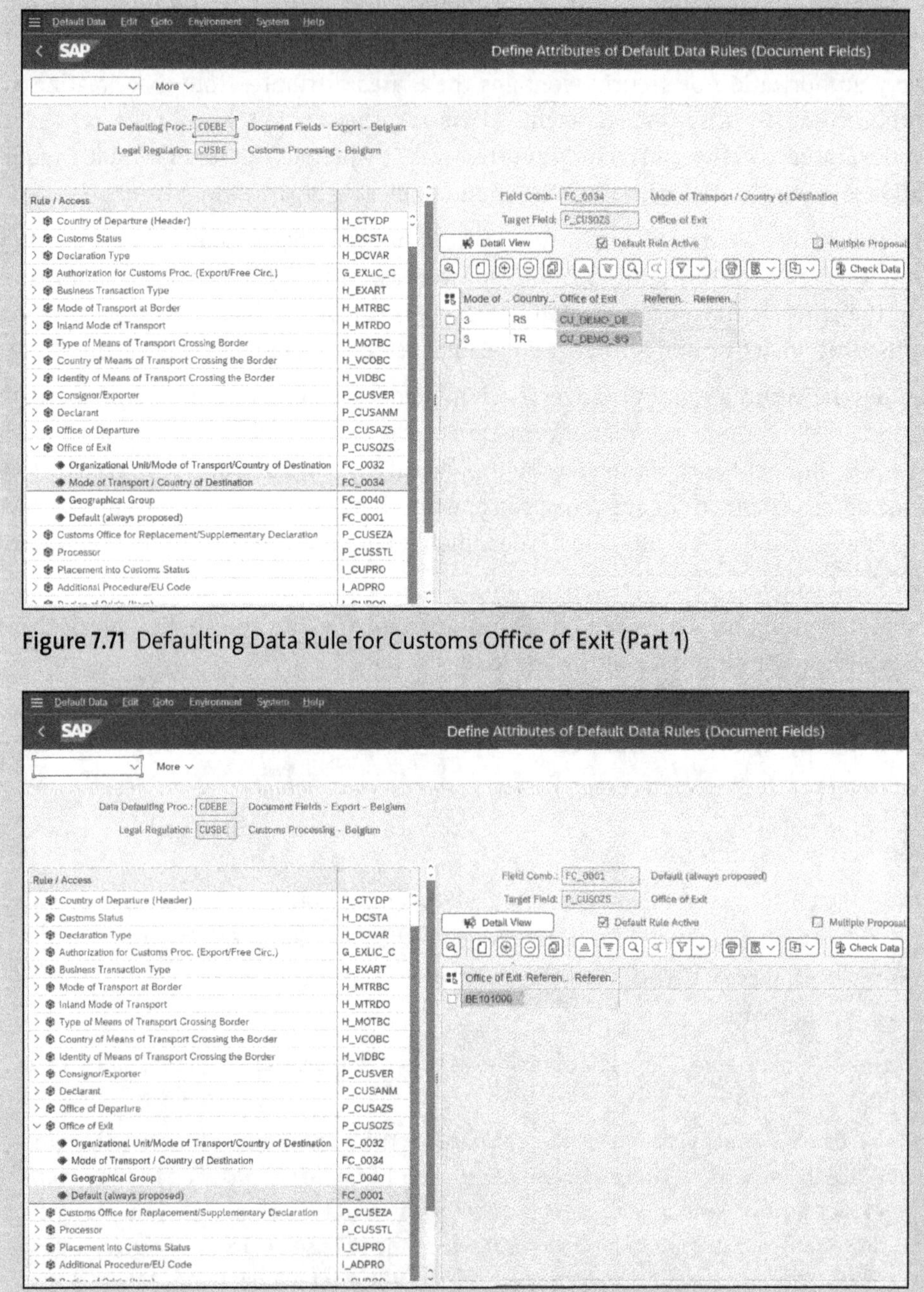

Figure 7.71 Defaulting Data Rule for Customs Office of Exit (Part 1)

Figure 7.72 Defaulting Data Rule for Customs Office of Exit (Part 2)

The data defaulting procedure and the legal regulation created during Customizing are shown at the top of the screen. By selecting these input parameters, users can influence the defaulting data of all export declarations for Belgium. The box on the left gives an overview of all the fields that can be defaulted (based on which fields are selected in the SAP GTS Customizing). By drilling down on a certain field (or rule), users reach the accesses. In the box on the right, users can add or modify the default data logic applicable to the field that will be defaulted into the customs declaration.

Remember from Section 7.2.3 that users need to default the correct messages into the customs declarations. To populate the rules for communication messages, navigate directly in the SAP GUI to **SAP Global Trade Services • Settings • Default Data • Define Data Defaulting Rules for Messages** via the administrator role. Figure 7.73 displays the outcome of this transaction.

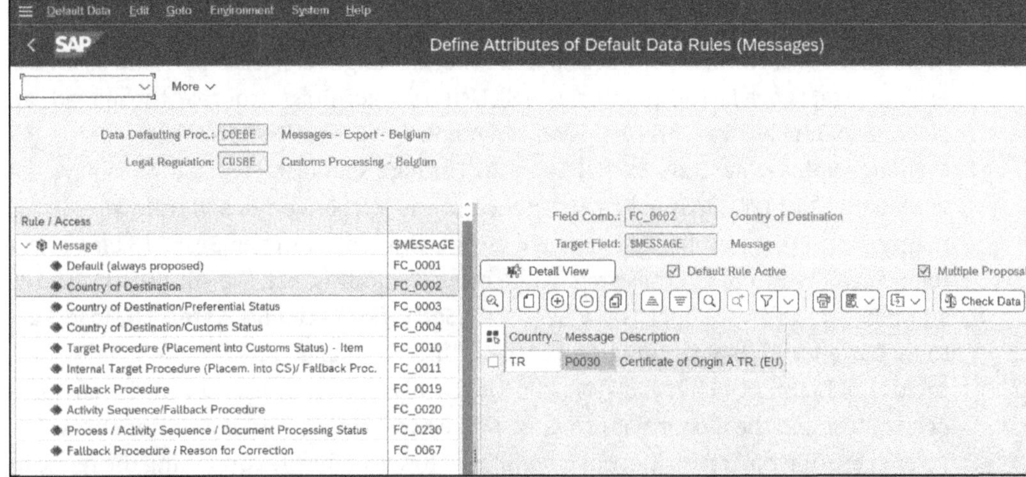

Figure 7.73 Default Data Rule for Messages

Again, the legal regulation and data defaulting procedure are shown at the top of the screen. The box on the left shows the only applicable rule in this activity (more specifically, **Message**, and all the different accesses below it). As with the defaulting of data fields, **Default (always proposed)** can be a useful access. In the box on the right-hand side, the actual values are populated into the access.

Figure 7.73 shows that, if an export declaration is created that captures the information of a shipment between Belgium and Turkey (the country of destination), then a specific printed form (**Certificate of Origin A.TR (EU)**) is automatically defaulted in the **Communication** tab of the customs declaration to print.

By designing the defaulting functionality of the messages correctly, users can limit the risk of user error (or, in this case, printing unnecessary forms).

Exchange Rates

All declared values in a customs declaration must be converted into the local customs currency. Individual companies aren't allowed to decide which exchange rate they want to apply to their transactions; the customs authorities publish an exchange rate. This exchange rate must be maintained and updated in SAP GTS on a monthly basis, so the correct exchange rate is applied.

In Customizing, there are some prerequisites that need to be executed first. Users need to configure a worklist with a maintenance interval and tolerance percentage for each user, as well as source and target currencies with an exchange rate quotation for each exchange rate. This is done through the following Customizing activities:

- **ABAP Platform • General Settings • Currencies • Define Worklist for Exchange Rate Entry**
- **ABAP Platform • General Settings • Currencies • Assign Exchange Rate to Worklist**

Because most customs authorities update the exchange rates on a periodic basis, the exchange rates need to be updated as well. Using the administrator role, the upload can be done manually, in a worklist, via admin menu **SAP Global Trade Services • Settings • Exchange Rates • Maintain Exchange Rates Through Worklist.** Figure 7.74 displays the worklist in SAP GTS. Manual maintenance of the exchange rates via the worklist is manageable if the number of exchange rates is limited. However, SAP GTS also offers the possibility to upload these exchange rates. To do this, use the administrator role and navigate to **SAP Global Trade Services • Settings • Exchange Rates • Upload Customs Currency Exchange Rates.** Select the desired exchange rate type for the upload, the desired currency that the system should convert the values into in the customs declaration, and the file format provided by customs authorities or a content provider. After executing the transaction, users will need to select the file containing the relevant authorities' exchange rates for the system to upload the exchange rates. As a result, the current exchange rate will be applied in the customs calculations.

Figure 7.74 displays the customs exchange rates in SAP GTS.

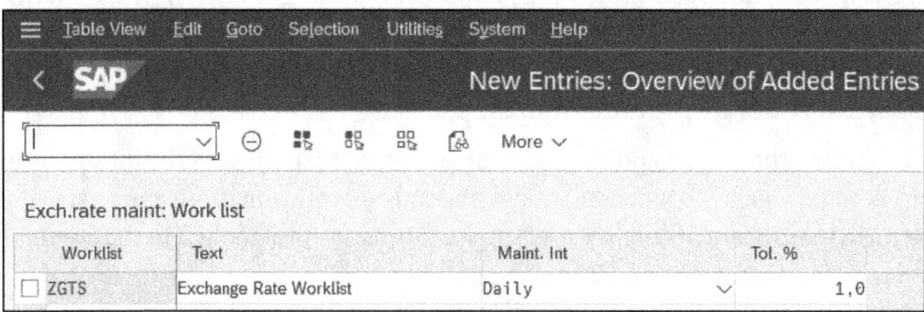

Figure 7.74 Display Customs Exchange Rates

If the system is using compensatory interest rates, these can also be maintained by using the administrator role and navigating to **SAP Global Trade Services • Settings • IP: Maintain Interest Rates for Compensatory Interest.**

7.3 Special Customs Procedures

In addition to the standard import and export processes applicable to all international companies, some organizations also operate under *special customs procedures*. These procedures suspend and/or reduce the import duty and applicable tax exposure.

Before a company can operate a special customs procedure, a customs authorization needs to be obtained from the customs authorities. One of the requirements the customs authorities impose prior to approving an authorization is an accurate bookkeeping of all goods that enter and leave that customs procedure.

Once granted, this authorization states what a company can and can't do under the procedure. SAP GTS can support a company's ability to obtain and manage an authorization.

This section provides an overview of some terms you've already seen: customs warehouses (CWHs) or bonded warehouses (BWHs), inward processing and outward processing, FTZ, and processing trade regimes (PTR). This section also highlights the master data elements related to these special customs procedures and the corresponding Customizing process.

7.3.1 Bonded Warehouse

A customs warehouse, or *bonded warehouse* (BWH), is where companies can store goods under suspension of customs duties and other taxes. A BWH is different from other special customs procedures because of the limitations on the types of activities that can take place under this procedure. In a BWH, manufacturing activities can't be performed, but minor logistics activities, such as repacking and regrouping materials and keeping materials in good condition, are allowed. The result of this arrangement is that a BWH is useful for a trading company or a company's distribution center, but not for performing manufacturing activities.

This section introduces the concept of duty-paid and duty-unpaid stock as a crucial element of the BWH functionality before explaining the concept of a BWH in more depth.

Stock in SAP GTS

From a customs point of view, in the case of special customs procedures, companies can identify two types of stock: duty-paid and duty-unpaid goods.

Companies have the option to not pay import duties (suspend duties) by storing specific goods in a CWH. These goods are referred to as *duty-unpaid goods*. A company can increase this type of stock by buying products from suppliers located in a third country or if goods are transferred from another BWH or other special customs procedure. In the latter case, users will create an inbound transit declaration—also known as a T1 declaration.

When looking at *duty-paid stock*, companies can increase their stock of *duty-paid goods* either by paying import duties and taxes at the moment of importation of bonded or duty-unpaid stock or purchasing goods from a local supplier in the same country or customs union. Because the stock types are generally not differentiated in the feeder system's stock keeping, it's important for users to maintain this information in SAP GTS. For accuracy, the sum of duty-paid and duty-unpaid stock in SAP GTS should align with the total stock level in the feeder system.

Depending on the type of BWH, the relevant customs authorities may legally require the company to physically separate the duty-paid and duty-unpaid stock. This isn't required in most cases in Europe.

Note

Here, we'll cover a brief overview of the most common flows possible in a BWH. In practice, there is a wide range of flows, depending on the location of the warehouse and the country to which the goods are shipped.

Inbound Movements

Using our previous example, whenever Alpha receives goods from its vendor in China, these goods are added to the BWH as duty-unpaid goods.

Figure 7.75 shows both the business and system processes of the import process into the BWH.

The inbound process into the BWH starts when Alpha creates a purchase order in its feeder system with the Chinese vendor. The system creates an inbound delivery before the goods arrive at the plant.

When the goods arrive in Alpha warehouse from the port of Antwerp, Alpha asks a customs broker to create an inbound transit declaration (T1 document) to bring the goods to Alpha's premises under suspension of duties. This inbound transit is also represented by a declaration in SAP GTS. As soon as the goods arrive at Alpha's premises, Alpha must notify the local customs authorities by clearing the transit declaration via a transit discharge declaration. The customs authorities, in turn, inform Alpha whether and when the company is allowed to open the container. Because the shipment involves duty-unpaid goods, the customs authorities may perform a physical check to confirm the shipment was cleared correctly.

When posting the goods receipt, users create a material document in the feeder system that will trigger an entry into the BWH. The entry will increase the duty-unpaid stock in the system. Within the standard BWH functionality, SAP GTS only supports creating import declarations after goods receipt.

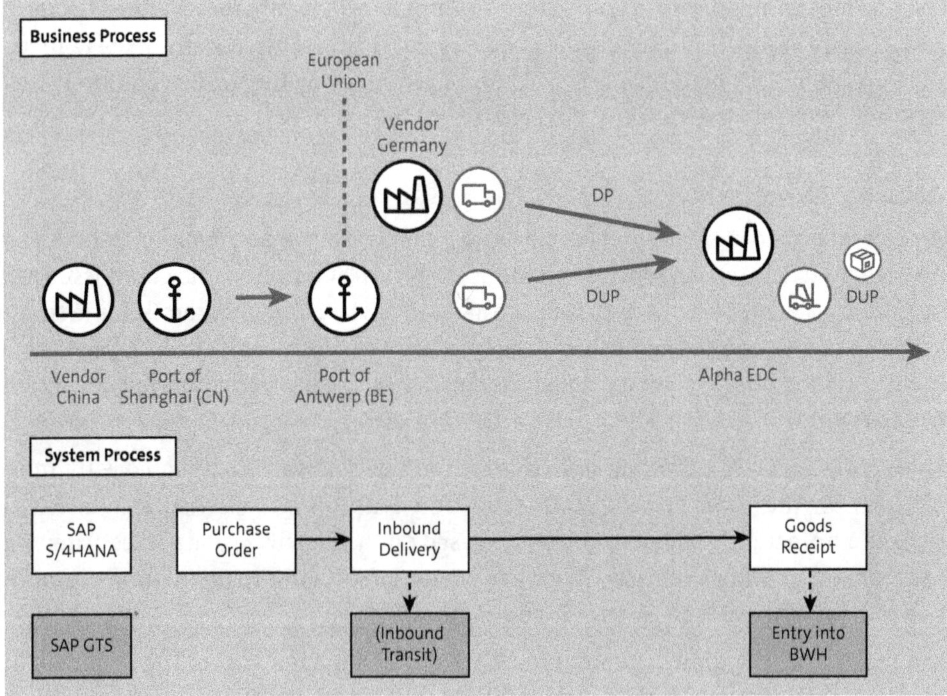

Figure 7.75 Bonded Warehouse: Inbound Process

The BWH entry document is a variant of an import declaration. To create this document, users can select the correct shipment in the **Enter Declaration after Goods Receipt** worklist in SAP GTS. This worklist can be accessed by navigating to the Create Customs Declaration after GR app under **Customs Import • Declarations • Import Declarations.** Every material document that is relevant for the BWH (meaning that the material is flagged as BWH-relevant, and the storage location is bonded) appears in this worklist. Refer to Chapter 3, Section 3.1.4, for more information on whether a storage location is bonded.

In this BWH entry declaration, Alpha declares the exact quantity of the goods and the related value. Calculation of the customs value and corresponding duties is an important step in the declaration and can be performed when the goods are released from the warehouse.

For every BWH entry declaration created in SAP GTS, the e-filing functionality can be used to file the declaration. Companies have the option to request a specific simplification from the customs authorities that allows them not to declare entries into the BWH on a transactional basis. In this kind of arrangement, companies can create an SCD (Section 7.1.4).

If Alpha purchases duty-paid goods from its German vendor, there is no need to create a BWH entry declaration. Posting the goods receipt in the feeder system creates a material document. This material document won't generate any import declaration in SAP GTS, but it will increase the duty-paid stock in SAP GTS.

Outbound Movements

Alpha's outbound reexport process starts with the creation of a sales order, followed by an outbound delivery. As soon as the goods are picked and packed, the goods issue can be posted. However, before the goods are allowed to leave Alpha's premises, the transaction needs to be filed to customs authorities. Alpha can then create a pro forma invoice in the feeder system to trigger the creation of a reexport declaration document in SAP GTS.

As with the standard export process, this reexport declaration is populated with information from the feeder system, master data from SAP GTS, manually entered information, or automatically added information based on the defaulting data rules. After the reexport declaration is complete, Alpha can communicate this to the customs authorities with printed forms or electronic communication.

In a reexport declaration document in SAP GTS, a new tab is added to the item details. The **CWH Closing Portions** tab contains information about the closing portion.

From a financial point of view, it's more beneficial for Alpha to send duty-unpaid goods to Serbia, so the system automatically selects duty-unpaid goods when a user initiates a shipment to Serbia. This concept is called *smart picking*.

Figure 7.76 shows both the system and business view of the outbound reexport process. SAP GTS supports creating outbound transit documents, which can be relevant when sending goods to a specific country with an EU-specific process. Users can simultaneously create this transit document with the reexport document or can postpone creating the transit document until the customs authorities approved the reexport declaration.

When Alpha sells its finished goods to a customer in France (which is considered a domestic sale as both countries are EU Member States belonging to the same customs union), there is no requirement to create a reexport declaration. In this case, the material document created when posting the goods issue decreases the stock. There is no need to create a pro forma invoice for this process. Because the customer in France doesn't want to be burdened with paying the import duty, Alpha prefers to send duty-paid goods. The smart picking functionality will automatically determine the preferred type of stock, depending on the country of destination.

To summarize, the company prefers to sell duty-unpaid stock if the customer is located outside its own country/customs union but prefers duty-paid stock for reexport domestic sales.

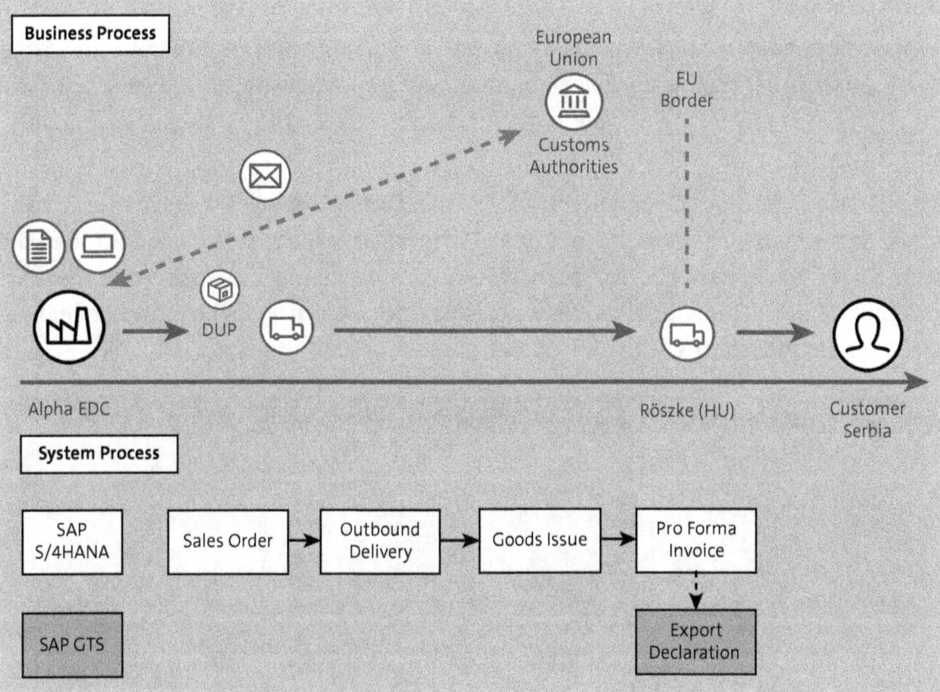

Figure 7.76 Bonded Warehouse Reexport Process

The following two scenarios can occur with these preferences:

- A customer in the same country or customs union wants a specific product, but only duty-unpaid stock is currently available in the BWH. In this case, the system automatically creates a *free-circulation declaration* (a type of import declaration), and the required duties and applicable taxes are calculated and paid. This implies that the duty-unpaid stock decreases, and, with this, the duty-paid stock increases. However, as the transaction is related to a sales flow, the duty-paid stock immediately decreases again.
- A customer outside the customs union wants to purchase a product, but at that moment, only duty-paid stock is available. The company can choose to wait to send out the goods until it receives a shipment with duty-unpaid goods. However, this may not always be possible due to the promised delivery dates.

  The alternative is to use the duty-paid stock, but in this event, the potential of the BWH won't be used—that is, duties are paid in a scenario when it wasn't legally required. Therefore, the option a company chooses will depend on its specific business strategy and preferences.

Stock Overview

When a bonded warehouse is activated in SAP GTS, the stock is kept in both the feeder system and SAP GTS. The stock keeping in the feeder system always indicates the total stock levels; in SAP GTS, the system distinguishes between duty-paid goods and duty-unpaid goods.

The Display CWH Stock Overview app shown in Figure 7.77 enables users to monitor the stock movements for a specific material. Furthermore, the system also indicates the number of goods reserved for both stock types. Selecting the **Receipts/Issues List** option displays all individual movements. This stock overview can be found under **Special Customs Procedures • Customs Warehouse • Stock Monitoring**.

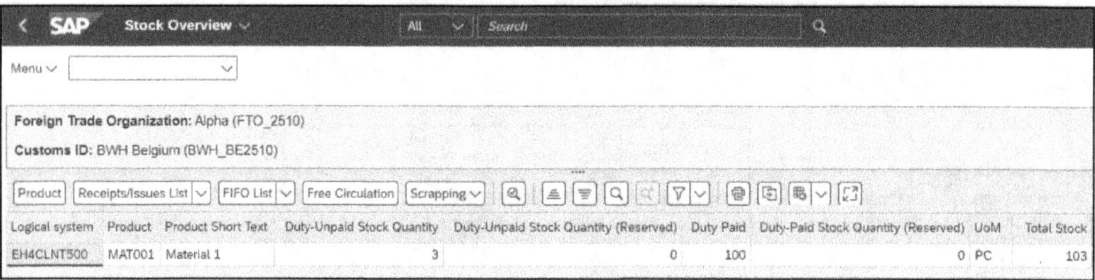

Figure 7.77 Stock Overview BWH

> **Note**
> See Chapter 9 on reporting for an overview of all other relevant stock transactions and reports.

Besides the stock overview per item, standard SAP GTS provides a stock report that includes all materials relevant for BWH. This report is called Determine Duty – Unpaid Individual Stock Item for Customs Warehouse and is accessible via the administrator menu under **SAP Global Trade Services, edition for SAP HANA • Background Processing • Customs Warehousing**.

Closing Portion

The purpose of setting up a BWH is to suspend duties and taxes until further notice. When the goods arrive in the BWH, the company may not know where the goods will be destined yet. Once the company determines the destination of the goods, and the destination calls for the payment of duties and taxes, the system retrieves the value of goods used at the entry of the goods to perform the duty calculation.

In Alpha's case, the system retrieves the value of goods calculated in the BWH entry document and reuses them in the declaration for bringing the goods into free circulation to calculate the duties. Each outbound document, therefore, is linked to one or more inbound documents. This concept is called *closing portion*.

The closing portion links the quantities of each material in an outbound document with the quantities of that same material in an inbound document using first-in, first-out (FIFO) logic. In this way, there is a clear link between the goods that enter and exit the BWH. The open balance of all these inbound documents represents the stock available in SAP GTS.

Closing Portion: Example

To see the full extent of this closing portion concept, we'll return to Alpha. In the past few weeks, Alpha has received incoming shipments with only duty-unpaid goods, as shown in Table 7.6.

Entry into Bonded Warehouse Declarations				
Date	**SAP GTS Document Number**	**Product**	**Quantity**	**Unit of Measure**
03.01.2024	1300000004	Material A	40	PC
		Material B	10	PC
03.01.2024	1300000005	Material A	5	PC
		Material C	8	PC
03.01.2024	1300000006	Material B	90	PC

Table 7.6 Inbound Shipments

Now, a customer has placed an order with Alpha, and after the creation of the pro forma invoice, a reexport declaration is created in SAP GTS. The information displayed in Table 7.7 is deducted from the reexport declaration.

Export Declaration				
Date	**SAP GTS Document Number**	**Product**	**Quantity**	**Unit of Measure**
03.01.2024	1400000004	Material A	42	PC
		Material B	17	PC
		Material C	6	PC

Table 7.7 Outbound Shipments

An additional tab on the item level is created in the reexport declaration stating the details in Table 7.8 for material A.

Document Number	**Year**	**Item**	**Release Date**	**Quantity**	**Unit of Measure**
1300000004	2024	1	03.01.2024	40	PC
1300000005	2024	1	03.01.2024	2	PC

Table 7.8 Reference on Additional Tab

After the determination of the closing portion, Table 7.9 displays the remaining open quantity per the BWH entry declaration.

Closing Portion		
Material	**Document**	**Remaining Quantity of Document**
Material A	1300000004	0
	1300000005	3
Material B	1300000004	0
	1300000006	83
Material C	1300000005	2

Table 7.9 Remaining Open Quantities

Companies can time the closing portion determination based on their business requirements. Companies could consider delaying the determination of the closing portion as long as possible to allow for last-minute changes. Doing so helps to avoid last-minute stock cancellations or corrections. However, when needed, SAP GTS enables users to cancel the BWH outbound document after the closing portions have been determined. The system automatically performs the necessary stock updates.

Companies can also see the closing portion determination as effectively writing off the entry declaration. The stock is removed from the BWH and, therefore, from SAP GTS.

For the domestic issue process, no reexport declaration is created, but an import (free circulation) declaration is created. In this case, the import declaration determined by the closing portion indicates the applicable duty rate for the free circulation document.

Exceptional Procedures

Sometimes, stock levels aren't accurately synchronized between systems because stock movements are tracked in both the system and the warehouse. When this happens, users can address it with *stock-correcting procedures* in SAP GTS.

Scrapping is a process in which materials, such as damaged goods, are physically destroyed because they have lost their commercial value. As the goods in a BWH are usually duty-unpaid goods, companies need to inform the customs authorities of scrapping activities in advance for approval.

From a technical point of view, the scrapping procedure starts within the stock overview in SAP GTS where users indicate the material and quantity to be scrapped. This creates a *scrapping document*. Then, it's necessary to perform the same scrapping activity in the feeder system by posting a confirmation of the scrapping movement. The feeder system transfers the movement to SAP GTS, which permanently removes the stock.

Stock correction – inventory count is a process that will take place throughout the year. Periodically, users need to compare the stock in the feeder system and the physical stock by doing a stock count. Because stock is kept in both the feeder system and SAP GTS, it's important to capture any stock differences in the stock levels of SAP GTS. Therefore, both positive and negative stock differences should be registered.

The inventory count process starts in the feeder system, but the posting with inventory differences is transferred to SAP GTS. The material document created in the feeder system adjusts the stock levels in SAP GTS. Depending on the type of stock difference, the system creates a corresponding inbound or outbound declaration to justify the unaccounted stock changes in the warehouse. If the duty-unpaid stock decreases, the company is required to pay the import duties and taxes for these items.

In addition to the stock correcting functionality and the scrapping process, a third important procedure is manually bringing the goods into free circulation without a reference to the feeder system. Users can initiate this process from the stock overview. A free circulation document is generated for the quantity of goods selected. As a result, the company is required to pay duties.

7.3.2 Inward Processing

From May 1, 2016, the new Union Customs Code (UCC) has replaced the former Community Customs Code (CC), introducing various modifications, including a noteworthy change affecting the special customs procedure of inward processing. Processing under customs control (PUCC) has been consolidated with IPR into a singular concept of *inward processing*.

Nevertheless, from a system perspective, the differentiation between PUCC and IPR still persists. Therefore, both concepts will be explained. Companies can base the decision on implementing the PUCC or IPR functionality by considering their main business activities. If a company is focused on importing components for production and sales to the local market (domestic intra-EU sale), the PUCC functionality will be suitable. If a company is mostly focused on importing components for production and reexport to non-EU customers, the IPR functionality may be better.

Note

This is only a brief overview of the most common flows possible within inward processing trade. In practice, there is a wide range of specific flows, depending on authorization specifics and the country to which the goods are shipped.

The following section provides an overview of inward processing and all relevant functionalities within SAP GTS.

Inward Processing Relief

The IPR functionality in SAP GTS allows companies to avoid paying import duties for components consumed in the production of finished goods that are subsequently reexported. In the context of our business scenario, displayed in Figure 7.78, the Alpha company can consider implementing the IPR functionality for the production facility in Belgium.

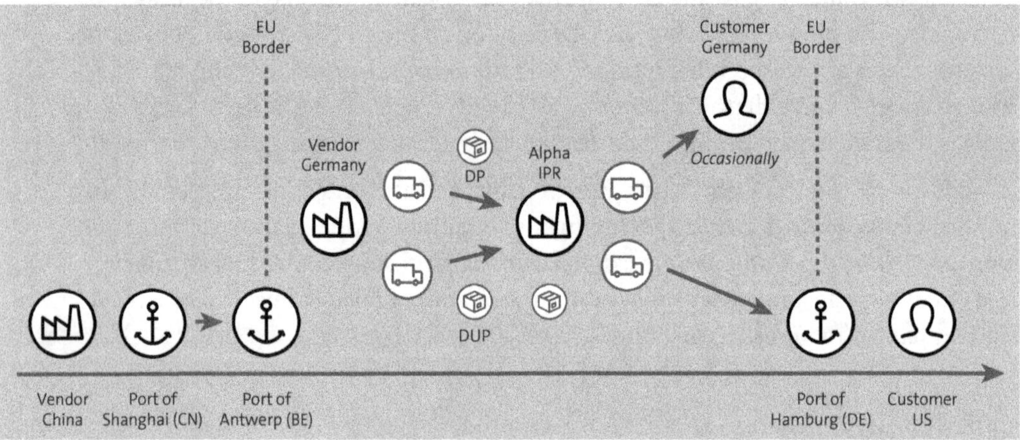

Figure 7.78 Business Process for Inward Processing

Alpha has to apply for an IPR authorization from the customs authorities (in this case, in Belgium). One of the requirements is to present a list of materials (commodity codes) and determine the maximum volume of materials that can be received under the IPR procedure. Furthermore, Alpha will have to identify precisely the bills of materials (BOMs) in which these components are used and list all the finished goods that contain IPR components.

Once the authorization is granted, Alpha has to reflect the authorization in SAP GTS. For this to be completed, different master data steps have to take place. The authorized products will need to be flagged as relevant for the IPR procedure (both components and finished products). For each IPR-relevant finished product, a BOM needs to be maintained in SAP GTS. The maintenance of the BOM can be done in two ways:

- **Static/theoretical approach: static BOM maintained in SAP GTS**
  A static/theoretical BOM is maintained on the master data level in SAP GTS. For each finished product, a BOM is created, including all IPR-relevant components and their theoretical consumption. The stock will be written off based on the quantities maintained in this static/theoretical BOM.
- **Dynamic approach based on real consumption: external BOM from feeder system**
  In SAP GTS, in the product master of the finished product, select the **External BOP**

checkbox. This will trigger the stock to be written off based on actual production consumption, which is delivered from the production orders created in the feeder system.

Inbound Movements

The process that allows companies to receive components under the IPR suspension procedure is similar to the process described for BWH in Section 7.3.1. Like the BWH process, SAP GTS doesn't support the entry into IPR until the feeder system receives the goods. The goods receipt document triggers the option to create an IPR entry from the worklist after goods receipt. This worklist can be accessed under **Customs Import • Declarations • Import Declarations • Create Import Declaration**. Once the IPR entry is created in SAP GTS, it will contain a reference to Alpha's IPR authorization setup on the master data level.

The placement of components into the IPR procedure will trigger a stock impact in SAP GTS. Note that unlike the BWH functionality, the IPR stock that is tracked within SAP GTS is limited to the duty-unpaid component stock. The creation of the IPR entry will trigger an increase in the IPR component stock.

In an IPR procedure, you also keep track of the time of entry as the company has a defined processing period to reexport the finished goods (included in the IPR authorization). The goods need to be discharged within the agreed time frame to avoid payment of import duties and taxes, and a potential compensatory interest. Figure 7.79 shows the IPR stock overview of material **COMP101** after an entry into IPR. The **Stock Quantity** of **COMP101** is increased with **57** pieces. This overview can be accessed under **Special Customs Procedures • Inward Processing • Stock Monitoring • Display IP Stock Overview.**

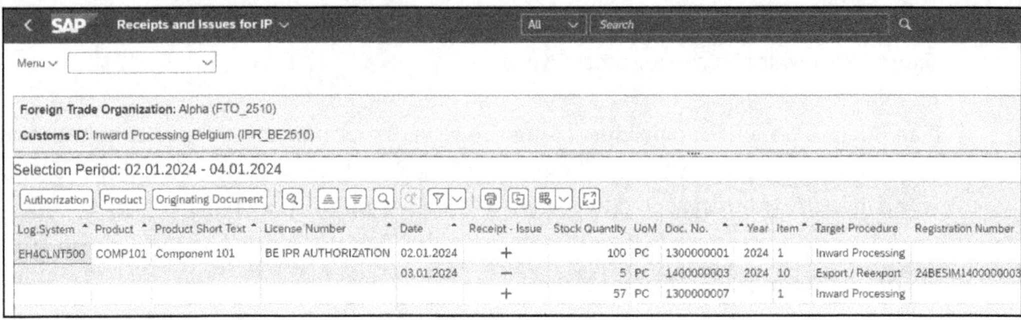

Figure 7.79 Inward Processing Stock Overview

Production Process

During the production of IPR finished goods, customs authorities sometimes allow a certain percentage of waste, which is also known as the rate of yield. These quantities should be identified in the feeder system based on the production orders. To avoid paying import duties for the waste quantities, users must indicate the correct commodity code.

Outbound Movements

After the production process, Alpha will reexport the finished products to a customer outside of the customs union. Following the standard sales flow, the creation of the pro forma billing document in the feeder system will transfer this information to SAP GTS, which creates the reexport declaration. This newly created reexport declaration will contain the reference to the required authorization.

After the closing portion is determined for the IPR component in the export declaration, the stock posting in SAP GTS is done, the IPR inventory is reduced and can't be reversed, and the document can't be canceled. Figure 7.80 shows the IPR stock overview of material **COMP101** after a reexport. The **Stock Quantity** of **COMP101** is decreased with the quantity determined by the closing portions. This overview can be accessed under **Special Customs Procedures • Inward Processing • Stock Monitoring • Display IP Stock Overview**.

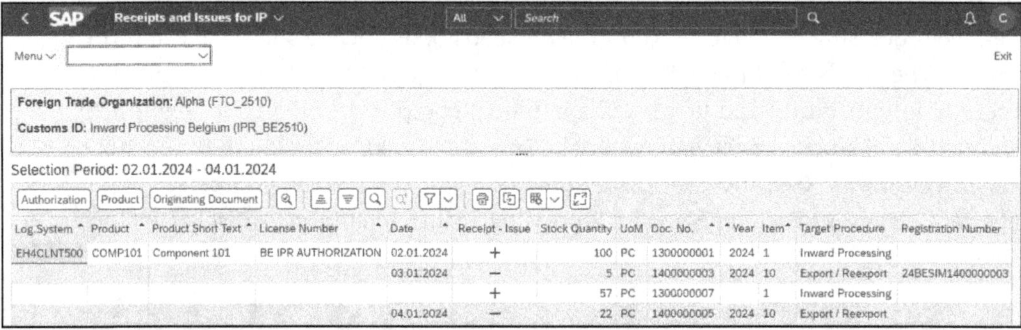

Log.System	Product	Product Short Text	License Number	Date	Receipt - Issue	Stock Quantity	UoM	Doc. No.	Year	Item	Target Procedure	Registration Number
EH4CLNT500	COMP101	Component 101	BE IPR AUTHORIZATION	02.01.2024	+	100	PC	1300000001	2024	1	Inward Processing	
				03.01.2024	–	5	PC	1400000003	2024	10	Export / Reexport	24BESIM1400000003
					+	57	PC	1300000007		1	Inward Processing	
				04.01.2024	–	22	PC	1400000005	2024	10	Export / Reexport	

Figure 7.80 Inward Processing Stock Overview

If an IPR component or a product containing IPR components is shipped to a domestic customer, the import duties and taxes have to be paid based on the components, and compensatory interests could be levied by the customs authorities.

Bill of Discharge and Settlement

The holder of an inward processing authorization is required to present a bill of discharge to their supervising customs office, including detailed information on various customs data such as the quantity of each type of goods which were placed under the special procedure in respect of which discharge is claimed.

SAP GTS provides a standard bill of discharge report with the details on all inbound placements of IPR components and outbound consumption flows for a specific period of time. This report can be used as a basis to prepare the reporting according to the data and formatting requirements of the local countries/customs authorities. The bill of discharge report can be accessed under **Special Customs Procedures • Inward Processing • Stock Monitoring • Display IP Individual Stock.**

As seen earlier within this section, the production of the IPR finished goods is linked to a predefined processing period (available in the customs authorization). To fulfill this requirement, the *IPR settlement* report has to be executed on a regular basis. The report can identify the IPR components that have expired—that is, aren't yet reexported—after the completion of the processing period. For these components (received by the IPR entry declaration), the duties have to be paid, and the declaration has to be made to the customs authorities. The IPR settlement report can be accessed via the administrator menu under **SAP Global Trade Services, edition for SAP HANA • Background Processing • Inward Processing/Processing under Customs Control • Create Settlement for Special Customs Procedures (IP/PUCC).** To display the settlement run, the user can navigate in SAP Fiori to **Special Customs Procedures • Inward Processing • Document Monitoring**, and select the **Display IP Settlement Run** tile.

7

Equivalent Goods and Prior Export

Equivalent goods are duty-paid goods, which are stored, used, or processed instead of the non-duty-paid goods placed under a special procedure. To be determined as equivalent goods, the products must be identical in description, quality, and technical characteristics.

Prior export is an arrangement that enables the trader to manufacture using equivalent goods and reexport them before the replacement goods have been imported. It's commonly referred to as export before import (EX/IM). This process is supported within SAP GTS: if there is no more IPR component stock available at the moment of reexport, the system automatically identifies this scenario as an EX/IM or prior export. There will be no closing portions added in the SAP GTS reexport declaration, but anticipated stock will be increased in the stock report. Once the replenishment has been done, the anticipated stock will decrease, and the closing portion will be added to the prior export document in SAP GTS (referring to the replenishment IPR entry document).

Processing under Customs Control

The PUCC functionality allows companies to keep track of products produced under suspension of duties and eventually bring the finished products into free circulation and pay duties on the latter. This could be beneficial if the duty rates for a finished product are lower than the duty rates for the components from which the finished product is made; duties will then be paid on the duty rate for the finished product and not the one for the components. PUCC has the same transactions as IPR and is therefore found under **Special Customs Procedures • Inward Processing**.

7.3.3 Outward Processing

Outward processing relief (OPR) allows a company to obtain a partial relief from import duties and/or VAT when reimporting finished goods that contain previously exported components within a certain period. This procedure is used by companies that outsource the manufacturing or assembly of the finished products in a foreign country before reimporting the finished products. For instance, a company located in Italy manufactures high-quality fabrics but outsources its sewing activities to China. The company will export the fabrics manufactured in Italy to a toller located in China. As soon as the fabrics are turned into wearable clothes, the company will reimport these clothes in Europe and sell them on the European market. Upon reimporting these garments in Europe, the company will obtain partial relief from import duties and/or VAT as the clothes are made of fabric manufactured in the customs union (Italy).

SAP GTS supports OPR for both outbound (export of the component) and inbound (reimport of the finished product) processing. SAP GTS tracks the OPR stock increases (through outbound processes) and stock decreases (through inbound processes). To determine the partial relief from import duties, and/or VAT, SAP GTS uses the added value method, whereby the value of the exported components is subtracted from the product value of the finished product for the calculation of the customs value upon reimportation. This will in turn determine the customs duties and VAT to be paid upon reimport.

In our present business scenario, if Alpha exports components that are used to manufacture a finished product to a toller located located in China (Manufacturer B) and then reimports the finished product in the customs union, the company can apply the OPR procedure, as displayed in Figure 7.81.

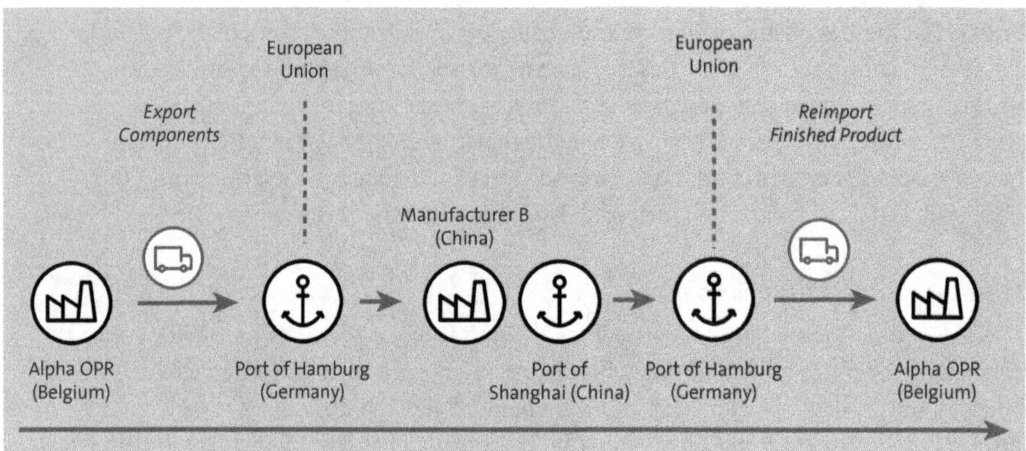

Figure 7.81 Business Process for Outward Processing

From an outbound process perspective, the components are exported to China under OPR. From an inbound process perspective, the finished product is reimported in the customs union as an OPR import. Upon reimport, the customs value of the finished product is reduced by the value of the initially exported components, which reduces the amount of import duties owed.

From a system perspective, several key prerequisites need to be fulfilled to leverage OPR:

- The OPR customs ID needs to be set up in Customizing.
- The OPR license needs to be set up in master data. The following needs to be included in the OPR authorization: relevant FTOs, legal units (LUs), product numbers of both OPR components and finished products, commodity codes of the finished products, and OPR business partners (the vendors that the company will export their components to and reimport the finished product from).
- Both the OPR components and OPR finished products need to be flagged as such in master data.
- The BOM of the OPR finished product needs to be available in SAP GTS master data.

In the previous section, we briefly explained the OPR business case, and in the following two sections, we'll explain the OPR outbound and inbound movements, respectively, from a system perspective.

Outbound Movements

The outbound movements for OPR components follow the typical export flow in the feeder system. Upon export, users can decide whether to export these goods under the OPR procedure. Note that the customs procedure code (CPC) for an OPR export is different from the CPC for export without a special customs procedure.

If users choose to export OPR components, the stock for this OPR component is increased. The OPR stock can be accessed via **Special Customs Procedures • Outward Processing • Stock Monitoring • Display OP Stock Overview**. Essentially, these components are sent out of the customs union to a production facility and can be used to reduce the duties when a finished product containing them is reimported. Furthermore, the OPR authorization and reimport period are calculated in the OPR declaration. The reimport period defined in the authorization indicates when the finished product that contains the OPR components must be reimported.

Inbound Movements

The process of reimporting the finished goods is similar to the standard import process. To create an import declaration, users must use a specific OPR worklist. A popup with the list of components and default quantities appears, according to the predefined BOMs, which are consumed from the SAP GTS inventory. Users can adjust the quantities and validate the screen. Similar to the outbound process, the CPC indicates if a company is reimporting OPR finished products. If a company chooses to reimport an

OPR finished product, the customs value of the finished product will be reduced by the value of the initially exported components (based on the BOM), which decreases the customs duties the company owes. Additionally, the stock of the OPR components contained in the OPR finished product will be decreased accordingly.

7.3.4 Foreign Trade Zone

The standard FTZ functionality supported by SAP GTS currently covers US-specific FTZ requirements. An FTZ is a zone in the US authorized by the Foreign Trade Zones Board and supervised by US Customs and Border Protection (CBP), where the goods are considered legally outside of the US customs territory.

Note

A free trade zone is a dedicated area where import, storage, manufacturing, and export activities can take place without being subject to certain specific regulations and taxes. In the US, the correct terminology to be used is foreign trade zone. This terminology is also used in SAP GTS because the functionality is specific to supporting US localization.

Using a fictitious company, Epsilon, this section walks through an example of how SAP GTS supports FTZ business processes.

Epsilon's Global Trade Requirements

Epsilon is an independent subsidiary of an international company operating in the life sciences industry. It's headquartered in Michigan, where the company has an FTZ. In Epsilon's FTZ, production activities take place, which means that Epsilon manufactures and sells final goods made out of components purchased from domestic and foreign vendors. The business case captures the following inbound and outbound transactions:

- Epsilon purchases some components (e.g., COMP-3011 and COMP-3022) from domestic vendors (located in the US). Other components (e.g., COMP-3033 and COMP-3044) are purchased from foreign vendors located in Germany.
- Manufacturing activities are executed in the FTZ, consuming the components to produce the final product (e.g., FP-3111).
- Final products are sold to domestic and foreign customers.

Figure 7.82 visually represents this business case.

By using an FTZ, Epsilon can defer (in case of a later entry for consumption in the US) or suspend (in case of a later reexport) the import duties and fees payment (e.g., reduced merchandise processing fee [MPF]). As in a BWH (refer to Section 7.3.1), the FTZ inventory consists of two types of stock: the duty-paid stock and the duty-unpaid stock. However, the FTZ adds an additional layer by allowing different categories of duty-unpaid goods, such as *privileged foreign* (PF) goods and *nonprivileged foreign* (NPF) goods. For PF goods, the zone user must pay the duties using the tariff rate of foreign goods based on the condition as admitted to the zone. For NPF goods, on the other hand, the user must pay the duties using the duty rate of the manufactured final product. Note that the assignment of the PF or NPF foreign zone status of the goods is performed at the entry of the goods into the FTZ. This can't be changed unless specific conditions, provided for by the FTZ rules, are met.

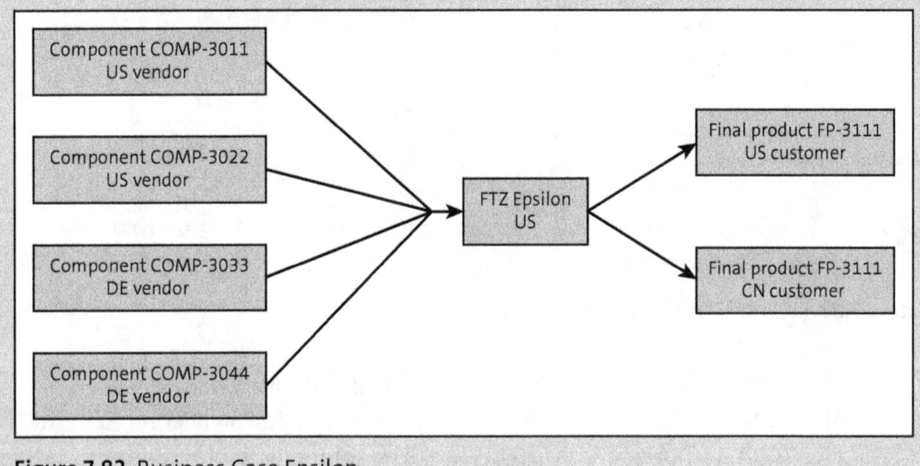

Figure 7.82 Business Case Epsilon

An FTZ procedure differs from a BWH procedure in that it has the following characteristics:

- A wide range of manufacturing activities are allowed.
- There is no time limit for the goods to remain in the FTZ.
- The domestic goods (or duty-paid goods) and the foreign goods (or duty-unpaid goods) can commingle.

An FTZ provides other major benefits to Epsilon in the following ways:

- **Enablement of the company to use the inverted tariff duty relief mechanism**
  Depending on the zone status (PF or NPF) of the component, this mechanism can generate savings. For example, this mechanism generates savings when the tariff rate of the NPF components is higher than the tariff rate of the final product. Figure 7.83 shows three components (A, B, and C) entering the FTZ as NPF with duty rates of 8%, 12%, and 10%, respectively. These components are consumed in the manufacturing of a final product Z, whose import duty rate (5%) is lower than the duty rates of its components. At the time of import, because the components have an NPF status, the duty calculation is performed based on the duty rate of the final product, resulting in a lower duty amount.

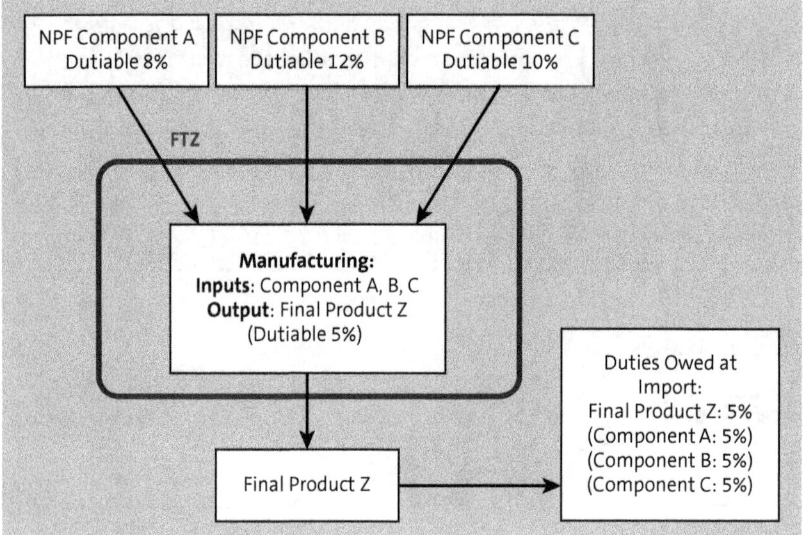

Figure 7.83 Inverted Tariff Duty Relief Mechanism

- **Reducing the cost due to merchandise processing fees (MPFs)**
  The MPF is a fee charged by the CBP for all merchandise formally entered in the US. When an FTZ is authorized for weekly entry, the company is allowed to file a single consumption entry for goods withdrawn from the FTZ (corresponding to duty payment on goods that enter the US market). By grouping these entries, savings are realized by the company. More precisely, as the maximum MPF per entry is $614.35, the cost can drop from $614.35 per individual entry to $614.35 per weekly entry by using groups (regardless of the number of individual entries contained in the weekly entry). The savings can be considerable for companies performing thousands of entries per year. Note that the MPF minimum and maximum change annually.
- **Reducing or eliminating duties on scrap and waste material generated during production processes**
  Where scrap or waste, whether recoverable or unrecoverable, is created during the manufacturing or manipulation of foreign status merchandise in the FTZ, there are several methods available for companies to use to reduce or eliminate the duties on those waste or scrapped items. The process of disposal can be physical, chemical, or any other method approved by CBP. Regardless of the method used, it's recommended that companies consult with the FTZ administrator and CBP for guidance specific to their business structures to confirm compliance.
- **Deferring the duty payment for production equipment**
  The duty payment for the production equipment can be deferred until actual production use of the equipment.

We'll continue with a discussion of the business and system processes in the next sections.

Business Process

Let's go through the business process of Epsilon, as shown in Figure 7.84.

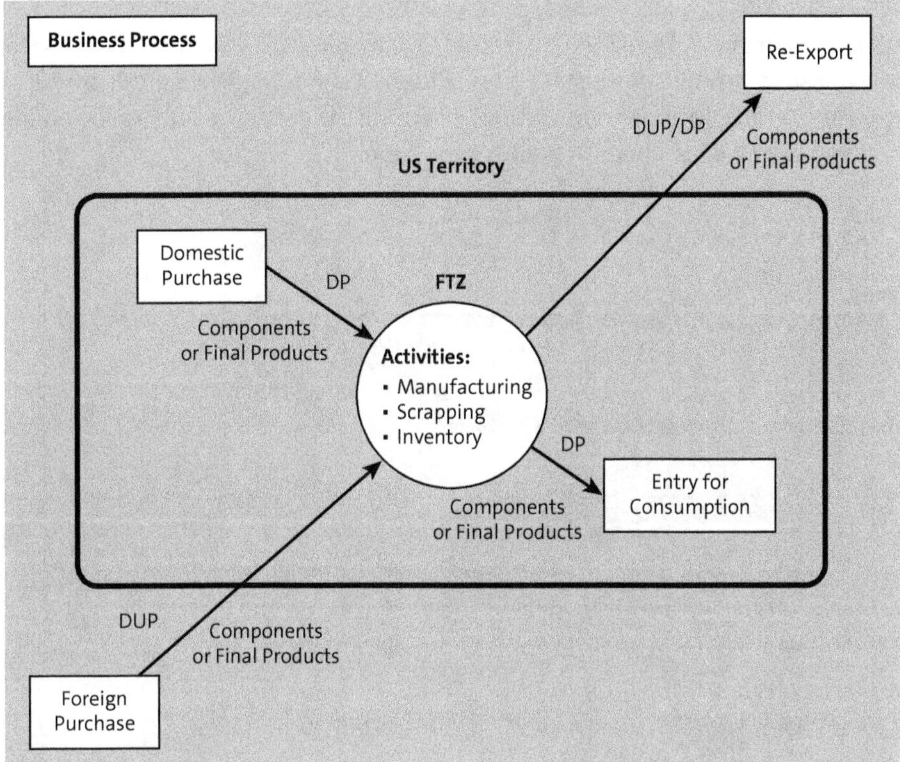

Figure 7.84 Business Process for FTZ

For Epsilon, the goods can be admitted in the FTZ through two types of flows. If Epsilon receives goods from a domestic vendor, the duty-paid stock increases, whereas a foreign admission increases the duty-unpaid stock. These goods can be final products ready for resale or components intended for a production process in the FTZ. On the outbound side, the company can withdraw FTZ stock through the following processes: a reexport from the US (requires a reexport declaration) or an entry for consumption (requires an import declaration to pay the duties).

Note that CBP requires an annual customs form (CF216) to be approved for activities in the zone. This application specifically requests the permission to manipulate, manufacture, exhibit, or destroy merchandise in a zone. The operator is responsible for filing an application on form CF216 with the port director.

Additionally, companies must create and store an annual reconciliation report. It's the operator's responsibility to prepare a reconciliation report within 90 days after the end of a zone year. The operator must maintain the annual reconciliation report in case any customs audits take place. The operator must also notify the port director that the task has been completed.

Inbound Movements

The domestic purchase of goods is illustrated in Figure 7.85. This inbound process is initiated by Epsilon through a purchase from a domestic vendor. During this business process, no customs documents are required because the goods are duty-paid and the transportation isn't done under customs supervision.

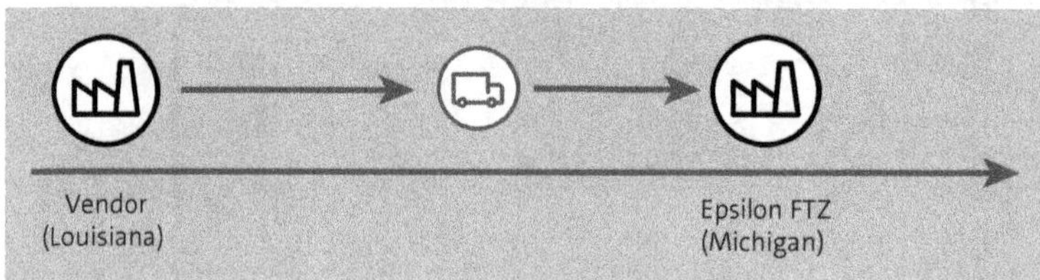

Figure 7.85 Business Process: Domestic Purchase

> **Note**
>
> It's also possible to domesticate foreign goods prior to bringing them into the FTZ. For example, in the point-of-entry customs clearance process, you can release duty-unpaid goods from overseas into free circulation and transfer them duty-paid to an FTZ.

The foreign goods can enter the FTZ in the following three ways:

- **Foreign admission: Nondirect delivery**
  In this scenario, the company must file a customs admission document (CBP form 214) with the customs authorities while the goods are still at the US port of unloading, for example, at the Port of Detroit (see Figure 7.86).

  ❶ After the admission is approved by the customs authorities, a transit procedure (either permit to transfer [PTT] or in-bond [IB]) can be opened in Detroit to deliver the goods to Epsilon.

  ❷ A PTT is a document that authorizes the movement of goods into, out of, or between FTZs within the same customs port of entry in the US. Cargo movements across different customs ports require an in-bond procedure (CBP form 7512). In our example, to deliver the goods to Epsilon, a PTT is required because Epsilon's FTZ is located in the same customs port where the goods were offloaded—the port of Detroit.

  ❸ After all goods have arrived at the FTZ, the PTT procedure is closed, and the admission process can be completed by transmitting the "concurrence" message confirming the quantity of the goods received. Going forward, we'll refer to the generic "transit procedure" covering both IB and PTT, unless we're outlining details for one of the specific processes.

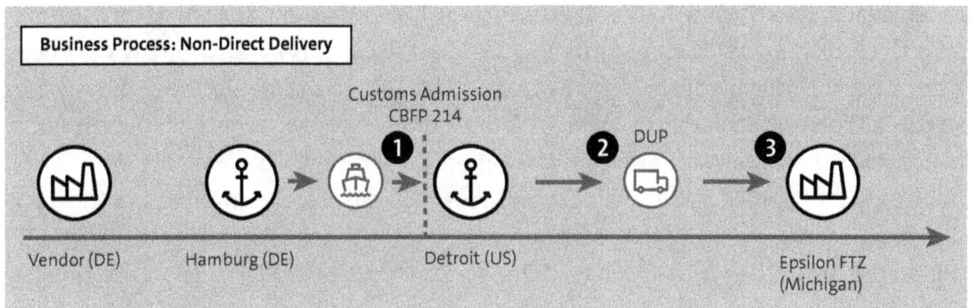

Figure 7.86 Business Process: Nondirect Delivery

- **Foreign admission: Direct delivery**
  The direct delivery process covers a flow that can be used after the company completes an approval process with the customs authorities. Applying for a direct delivery is subject to several criteria (e.g., the goods are well known in advance and don't require customs examination). The benefit is that for a direct delivery, the PTT can be immediately created (without prior CBFP 214 admission approval), and the goods can be delivered to Epsilon. When the goods arrive on Epsilon's premises and are put into storage, Epsilon creates a local clearance admission, as shown in Figure 7.87. At the end of the day, all local clearances are merged into a daily admission that is filed with the customs authorities. Consolidating the deliveries into a daily admission reduces the administrative work and, consequently, the administrative cost.

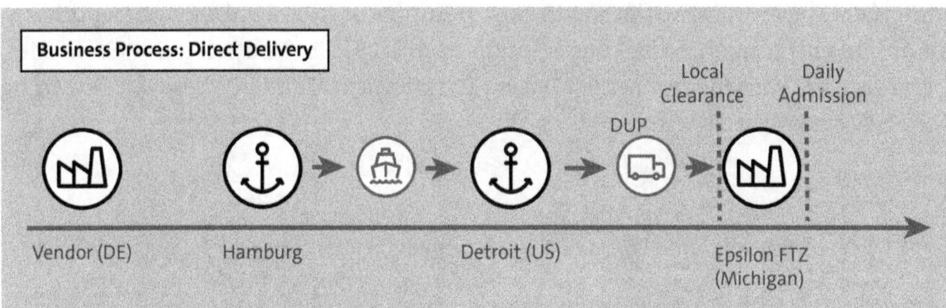

Figure 7.87 Business Process: Direct Delivery

- **Inbound zone-to-zone transfer**
  Goods shipped from one FTZ to another must file an entry for immediate transportation (IB) followed by an admission into the receiving FTZ. The receiving FTZ should also receive documents from the originating FTZ operator (e.g., the original 214) that detail the history of the products entering the FTZ.

Outbound Movements

If the goods are shipped outside of the US territory (to China, e.g., as shown in Figure 7.88), the company needs to file an export declaration. Depending on the status of the goods available according to the FIFO rule (there is no smart picking based on the duty-paid stock/duty-unpaid stock status and destination country), the company may need a transit procedure. If domestic goods are the first ones in the queue, the company files a standard export declaration, whereas if duty-unpaid goods are picked, the company needs an IB document followed by a reexport declaration. The reexport declaration must be accompanied by a transit procedure, which enables the shipment of the duty-unpaid goods from the FTZ to the US port of departure. The responsible party (depending on the Incoterms) must open an IB for goods to leave the FTZ when it's leaving the US territory.

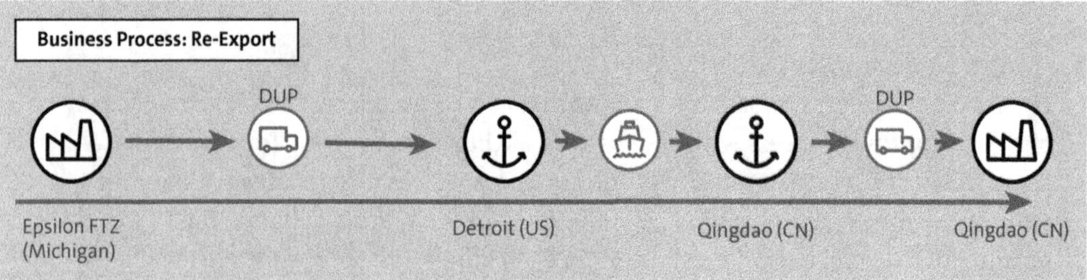

Figure 7.88 Entry for Consumption into the US

If the goods are shipped to a US customer, for example, in Chicago (see Figure 7.89), goods must be cleared with customs. If duty-unpaid goods are picked, they must go through a procedure called *entry for consumption*, which confirms the company pays import duties according to the goods' zone status (PF or NPF). Similar to an import consumption entry, an entry for consumption from an FTZ is filed using CBP Form 3461, CBP Form 7501, and/or another equivalent form to clear with CBP those goods leaving the FTZ.

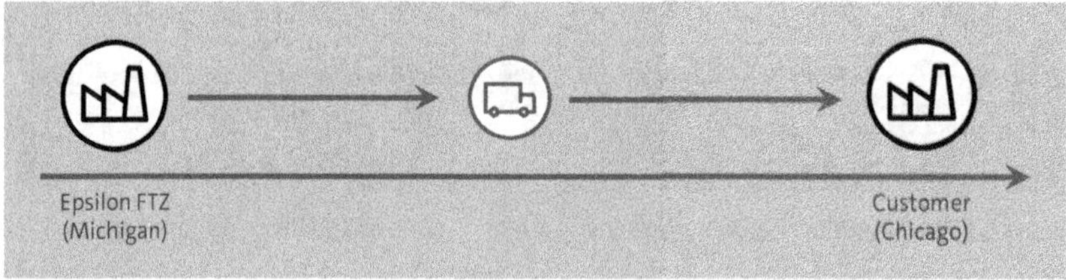

Figure 7.89 Business Process: Entry for Consumption into the US

An alternative procedure for entry for consumption is using a weekly entry. Prior to the removal of goods from the FTZ, the company files the goods estimated to be leaving the FTZ during the following calendar week on CBP Form 3461. Once approved by CBP, transfers of the goods from the FTZ may proceed. (If the goods transferred from the FTZ are larger than the forecast filed and approved on CBP Form 3461, an additional entry will be needed to cover the extra merchandise.)

Goods shipped from one FTZ to another require an entry to be filed for immediate transportation (IB). The business process for a zone-to-zone transfer is shown in Figure 7.90. SAP GTS can also be used to process zone-to-zone transfers.

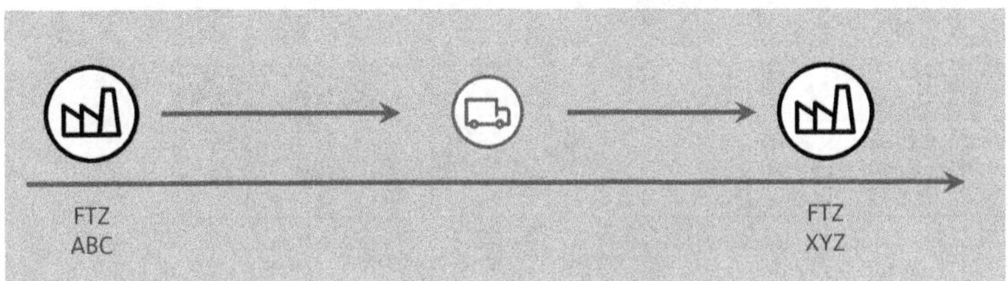

Figure 7.90 Business Process: Zone-to-Zone Transfer

System Process

The SAP GTS FTZ functionality is typically linked to a process in the feeder system. The admission into the FTZ is linked to the purchasing process, the zone inventory management is linked to the production process, and issuing the goods out of the FTZ is linked to the sales process.

To keep track of stock levels for components and final products, Epsilon must define the BOM of each final product that consumes components during production. In the following sections, we'll illustrate a scenario where Epsilon purchases components, manufactures a medical setscrew, and sells some to a foreign customer and some to a domestic customer. The final product, FP-3111, is a setscrew that consists of four components, as shown in Figure 7.91.

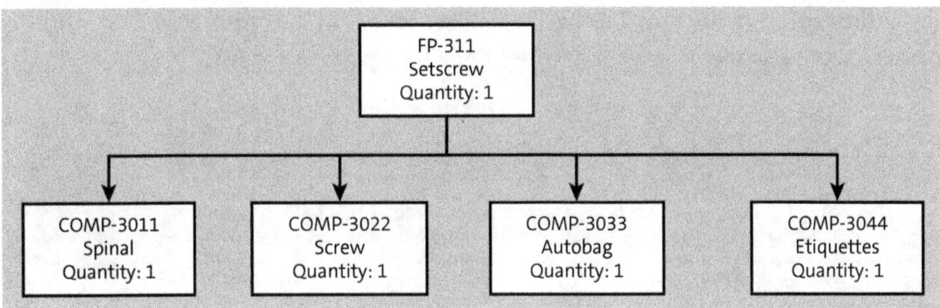

Figure 7.91 BOM for Medical Setscrew

Inbound Movements

Figure 7.92 represents Epsilon's domestic purchase flow from a business and system point of view. The process starts when Epsilon creates a purchase order in the feeder system to purchase goods from a domestic vendor located in Louisiana. After the purchase order, the inbound delivery is created followed by the goods receipt in the FTZ. This process increases the duty-paid stock.

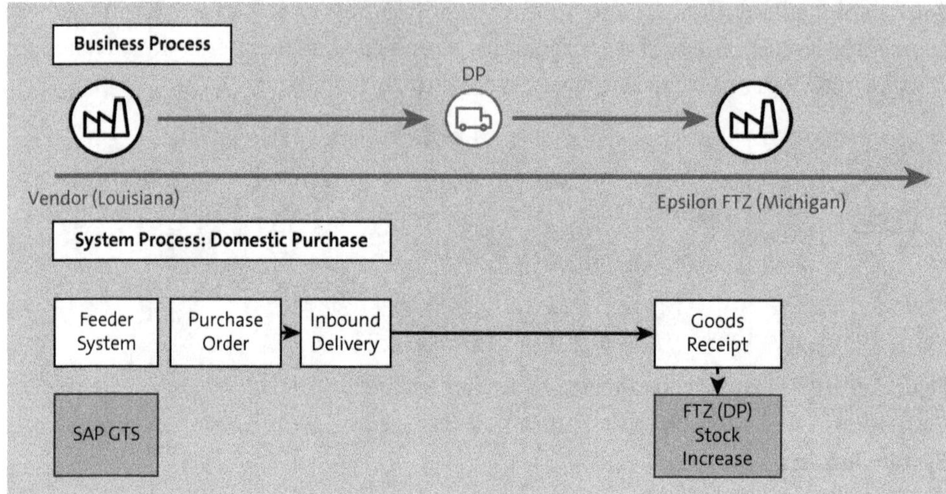

Figure 7.92 Business and System Process: Domestic Purchase

The goods receipts posting in the feeder system is integrated with SAP GTS by means of mapping movement types from the feeder system to SAP GTS. However, the stock update in SAP GTS will only occur after the run of two system monitoring programs, which are typically scheduled via a background job: **Process Stock Receipts Into Foreign-Trade Zone** and **Post Stock To FTZ**. These programs can be found under **System Monitoring • Background Processing • Foreign-Trade Zone**.

Next, we'll discuss the business and system process for a direct delivery, which is represented in Figure 7.93. When goods are received via a direct delivery, no admission filing is required prior to the transit filing. In Epsilon's case, a PTT is required, which is immediately created to allow the goods to be shipped to Epsilon's facility.

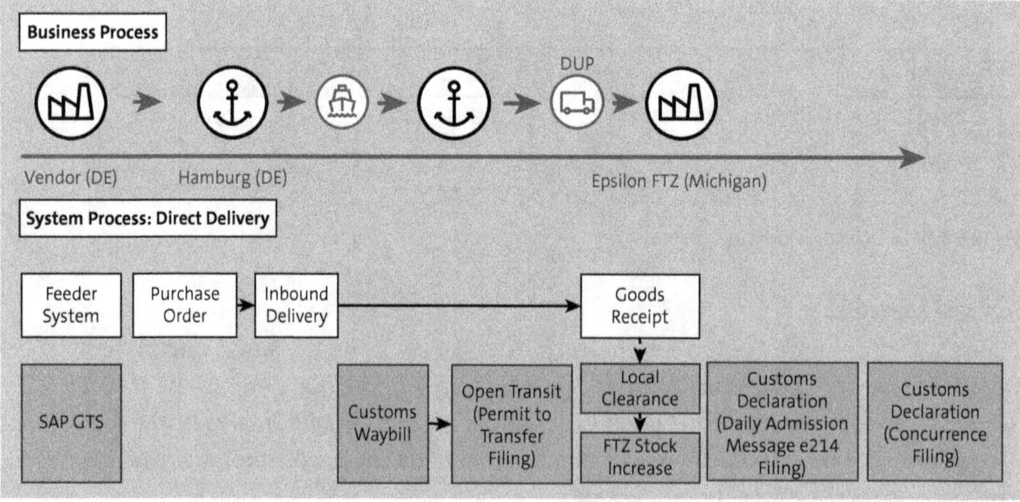

Figure 7.93 Business and System Process: Direct Delivery

Upon goods receipt, Epsilon creates a local clearance admission document. All local clearance admission documents are consolidated into one daily admission declaration, which is e-filed, as well as the concurrence message at the end of the day, to the customs authorities.

Next, we demonstrate a nondirect delivery by purchasing the COMP-3033 and COMP-3044 components to later produce our final product. This process starts when Epsilon creates the purchase order and the inbound delivery in the feeder system. Because this is a nondirect delivery scenario, Epsilon is only allowed to receive the goods into the zone after submitting and receiving approval of the CBFP 214 (FTZ local clearance admission message) by the customs authorities (see Figure 7.94). After receiving approval to bring the goods into the FTZ, a transit declaration can be created to request approval to bring the goods to the FTZ.

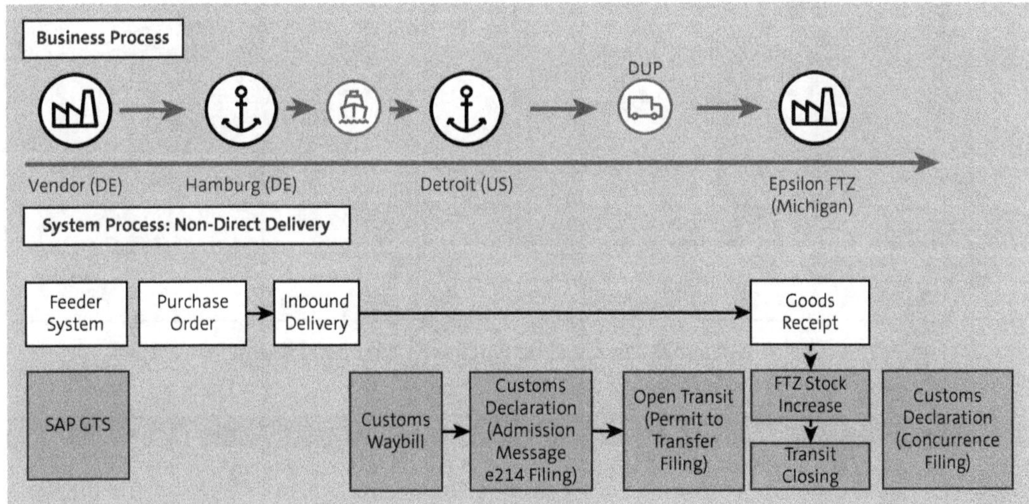

Figure 7.94 Business and System Process: Nondirect Delivery

A customs waybill is required to allow for FTZ activities in SAP GTS (see Figure 7.95). The customs waybill requires **Freight Data** such as packaging information. The **Forwarding Agent** must be entered on this screen, and the **Reference Number** for the **Purchase Order** document and **Inbound Deliv.** document must be added. This can be accessed in SAP Fiori via **Customs Import • Related Documents • Customs Waybill • Display Customs Waybill.**

An admission document can be created via the **Create Customs Declaration before GR** app by entering the waybill number in the corresponding field. This can be done in the SAP Fiori launchpad by navigating to **Customs Import • Declarations • Import Declarations.** After clicking the **Execute** button, the system displays the available waybills for which a local clearance admission declaration can be created. Figure 7.96 displays the **Worklist Before Goods Receipt** area.

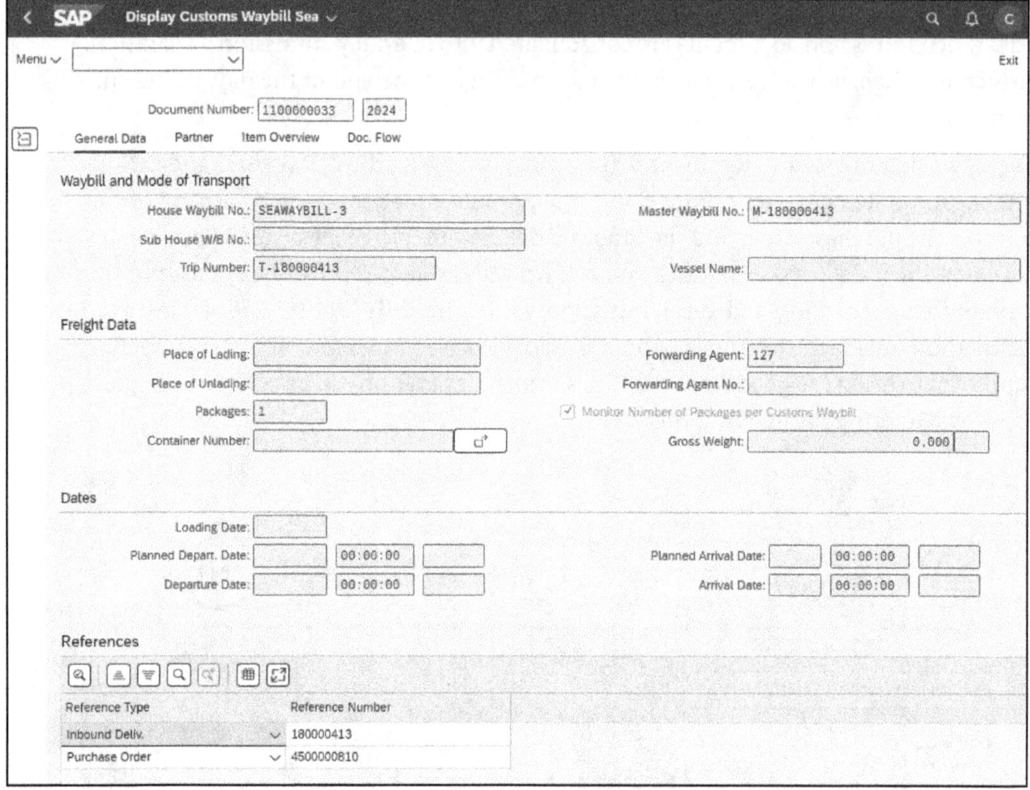

Figure 7.95 Customs Waybill Based on the Purchase Order and Inbound Delivery

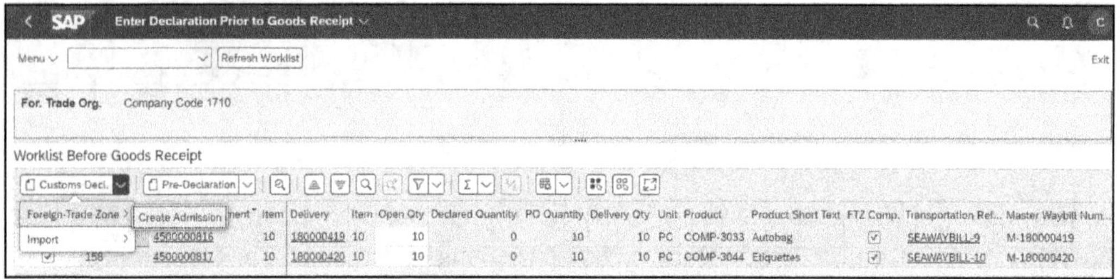

Figure 7.96 Admission before Goods Receipt

After the local clearance admission declaration is created, Epsilon can initiate the electronic communication with customs authorities to submit and receive the approval of form CBFP 214. After the customs authorities approve the admission, the goods are transferred from the unloading port to the FTZ under a transit procedure. In Epsilon's case, this transit procedure is opened by filing a PTT with customs authorities. The PTT can be created from the local clearance admission declaration (see Figure 7.97).

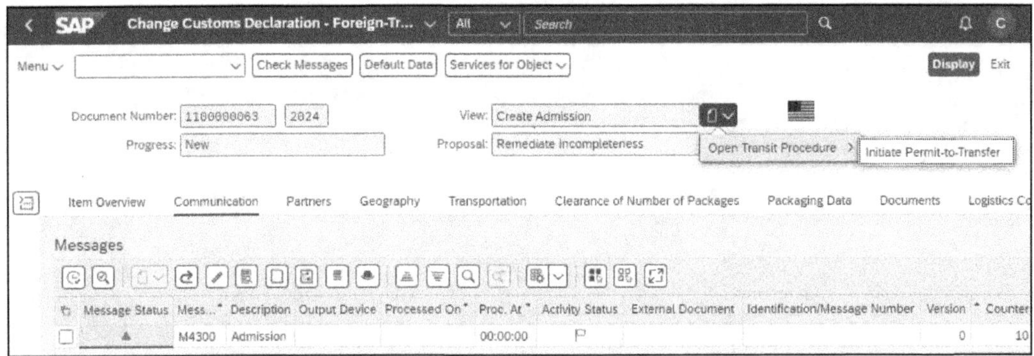

Figure 7.97 Permit to Transfer

After the PTT has been created from the local clearance admission, it can be found back in the **Manage Transit Declarations** app (see Figure 7.98) under **Customs Export • Declarations • Transit Declarations** in the SAP Fiori launchpad.

This PTT transit procedure is closed after all goods that are part of the PTT have arrived at Epsilon's FTZ facility and the transit discharge messages are e-filed. The goods receipt needs to be processed and background jobs need to run for the stock to be posted in the system.

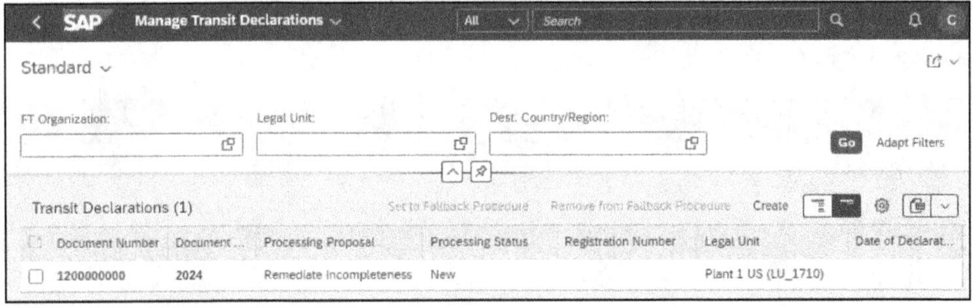

Figure 7.98 Manage Transit Declarations

After running these transactions, users can display the stock update via the **Display FTZ Receipts/Issues List** app in the SAP Fiori launchpad under **Customs Special Procedures • Foreign-Trade Zone • Stock Monitoring.** Figure 7.99 shows the stock receipts and issues list with the entry of our four components in the FTZ. For component **COMP-3011** and **COMP-3022**, Epsilon purchased 100 pieces per component of duty-paid goods via domestic purchase. For **COMP-3033** and **COMP-3044**, 100 pieces per component of duty-unpaid goods were entered in the FTZ via an admission. The zone status (FP or NFP) is automatically assigned to the stock based on the information maintained at the master data level (Section 7.3.7).

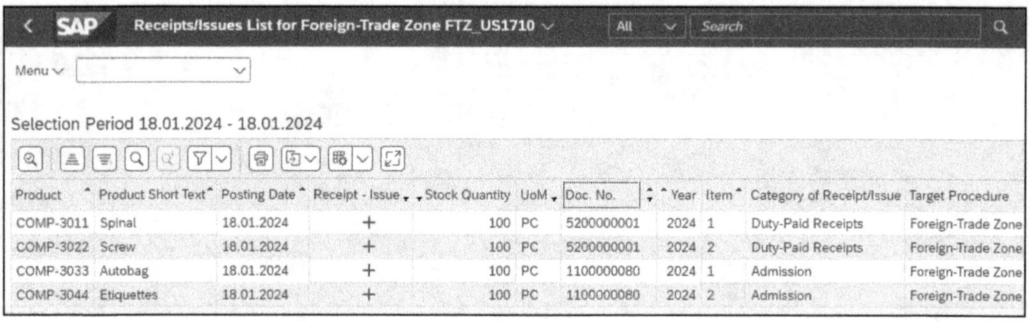

Receipts/Issues List for Foreign-Trade Zone FTZ_US1710

Selection Period 18.01.2024 - 18.01.2024

Product	Product Short Text	Posting Date	Receipt - Issue	Stock Quantity	UoM	Doc. No.	Year	Item	Category of Receipt/Issue	Target Procedure
COMP-3011	Spinal	18.01.2024	+	100	PC	5200000001	2024	1	Duty-Paid Receipts	Foreign-Trade Zone
COMP-3022	Screw	18.01.2024	+	100	PC	5200000001	2024	2	Duty-Paid Receipts	Foreign-Trade Zone
COMP-3033	Autobag	18.01.2024	+	100	PC	1100000080	2024	1	Admission	Foreign-Trade Zone
COMP-3044	Etiquettes	18.01.2024	+	100	PC	1100000080	2024	2	Admission	Foreign-Trade Zone

Figure 7.99 Receipts and Issues List for FTZ

Production Process

The FTZ procedure keeps track of the goods' time of entry. After the production order is released in the feeder system, the goods issue of the components posted against the production order will trigger the switch from the available stock (PF or NPF) to work in process (WIP) stock in SAP GTS. During manufacturing, the goods must be consumed according to the FIFO rule regardless of the type of stock (duty-paid stock or duty-unpaid stock). Therefore, no smart picking is in place for the FTZ functionality. Table 7.10 illustrates the FIFO functionality in which duty-paid goods are picked because the duty-paid stock has been entered first in the FTZ inventory. As mentioned, because strict FIFO rules must be followed, the system won't use smart picking because it's not allowed for FTZ.

Date	Receipt/Issue	Stock	Quantity	Unit of Measurement
2/01/2024	+	Duty-paid stock	20	EA
3/01/2024	+	Admission	10	EA
5/01/2024	- (export)	Duty-paid stock	10	EA

Table 7.10 FIFO Rule in FTZ

Figure 7.100 illustrates the WIP stock creation in SAP GTS after the production of 60 pieces of the final product FP-3111 in the feeder system. **Work In Process** shows what is currently being consumed to produce the final product. As a result, the total stock available is the sum of all available duty-paid and duty-unpaid stock, excluding the WIP quantity as these goods are unavailable because they are being used in the production process. This overview can be found in the SAP Fiori launchpad by navigating to **Special Customs Procedures • Foreign Trade Zone • Stock Monitoring • Display FTZ Stock Overview**.

Stock Overview for Foreign-Trade Zone FTZ_US1710

Menu

Individual Stock | Receipts/Issues List | WIP Stock List

Product	Product Short Text	UoM	Total Stock	Work in Process	Privileged Foreign	Non-Privileged Foreign	Duty Paid
COMP-3011	Spinal	PC	100	60	0	0	100
COMP-3022	Screw	PC	100	60	0	0	100
COMP-3033	Autobag	PC	100	60	100	0	0
COMP-3044	Etiquettes	PC	100	60	100	0	0

Figure 7.100 Stock Overview for FTZ Products

Outbound Movements

In our scenario, Epsilon plans to sell finished products to a customer in China. As this is a foreign customer, the reexport process is triggered. Figure 7.101 shows the full business and system process. The flow starts with the creation of a sales order. This is followed by an outbound delivery and a pro forma invoice. Because the pro forma invoice is integrated with SAP GTS, the system automatically creates a reexport customs declaration. The goods issue is also integrated with SAP GTS. This integration updates the correct stock posting in SAP GTS. Along with the reexport declaration, in our example, Epsilon is responsible for filing the open transit movement. The transit procedure can be opened directly from the reexport declaration.

To update the stock in SAP GTS, the goods issue needs to be processed by running the **Process Stock Issues From Foreign-Trade Zone** transaction under the administrator menu **SAP Global Trade Services, edition for SAP HANA • Background Processing • Foreign Trade Zone.**

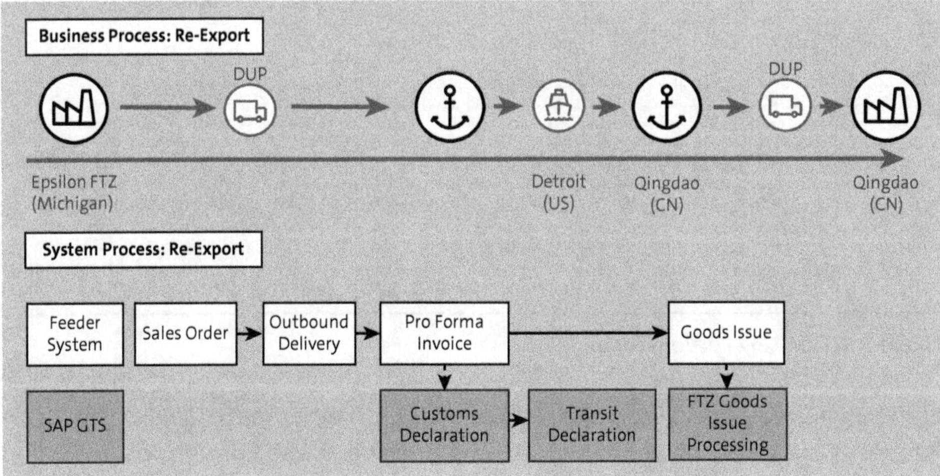

Figure 7.101 Business and System Process: Reexport

This transaction updates the stock that can be displayed via the **Display FTZ Receipts/ Issues List** app. Figure 7.102 shows that the components and WIP stock are consumed, and this consumption is categorized as an **Export**. This app can be accessed in the SAP Fiori launchpad by navigating to **Special Customs Procedures • Foreign Trade Zone • Stock Monitoring**.

Receipts/Issues List for Foreign-Trade Zone FTZ_US1710

Selection Period 18.01.2024 - 18.01.2024

Product	Product Short Text	Posting Date	Receipt - Issue	Stock Quantity	UoM	Doc. No.	Year	Item	Category of Receipt/Issue	Target Procedure
COMP-3011	Spinal	18.01.2024	+	100	PC	5200000001	2024	1	Duty-Paid Receipts	Foreign-Trade Zone
			–	10	PC	1400000034		10	Export	Export / Reexport
COMP-3022	Screw	18.01.2024	+	100	PC	5200000001	2024	2	Duty-Paid Receipts	Foreign-Trade Zone
			–	10	PC	1400000034		10	Export	Export / Reexport
COMP-3033	Autobag	18.01.2024	+	100	PC	1100000080	2024	1	Admission	Foreign-Trade Zone
			–	10	PC	1400000034		10	Export	Export / Reexport
COMP-3044	Etiquettes	18.01.2024	+	100	PC	1100000080	2024	2	Admission	Foreign-Trade Zone
			–	10	PC	1400000034		10	Export	Export / Reexport

Figure 7.102 Receipts and Issues of Components in FTZ

After SAP GTS has processed the FTZ goods issue records from the feeder system, the reexport customs declaration item level displays the **FTZ Closing Portions** tab, which shows the link between the admission document of the consumed components and the exported final products (see Figure 7.103).

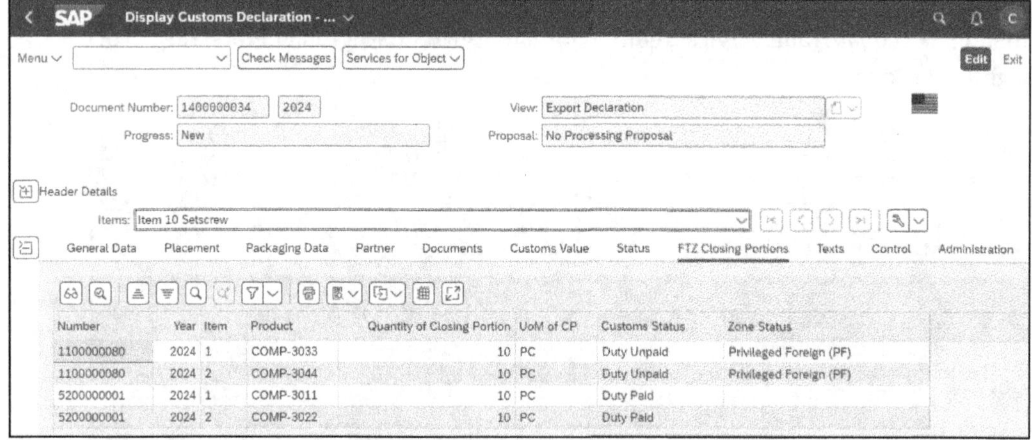

Number	Year	Item	Product	Quantity of Closing Portion	UoM of CP	Customs Status	Zone Status
1100000080	2024	1	COMP-3033	10	PC	Duty Unpaid	Privileged Foreign (PF)
1100000080	2024	2	COMP-3044	10	PC	Duty Unpaid	Privileged Foreign (PF)
5200000001	2024	1	COMP-3011	10	PC	Duty Paid	
5200000001	2024	2	COMP-3022	10	PC	Duty Paid	

Figure 7.103 Closing Portions for an FTZ

Epsilon will also ship 10 pieces of duty-unpaid finished products (FP-3111) to a customer in Chicago. To process this outbound movement and entry for consumption into the US, an import declaration will need to be created in SAP GTS. As stated in the business process overview, this can be done on a shipment by shipment basis, that is, transactionally via an import declaration into the US. The more cost beneficial way is to process a weekly entry via a weekly entry summary declaration. This is done by the creation of a domestic sales order, followed by an outbound delivery and eventually a goods issue posting that triggers the creation of an entry summary declaration. The business and system processes are displayed in Figure 7.104.

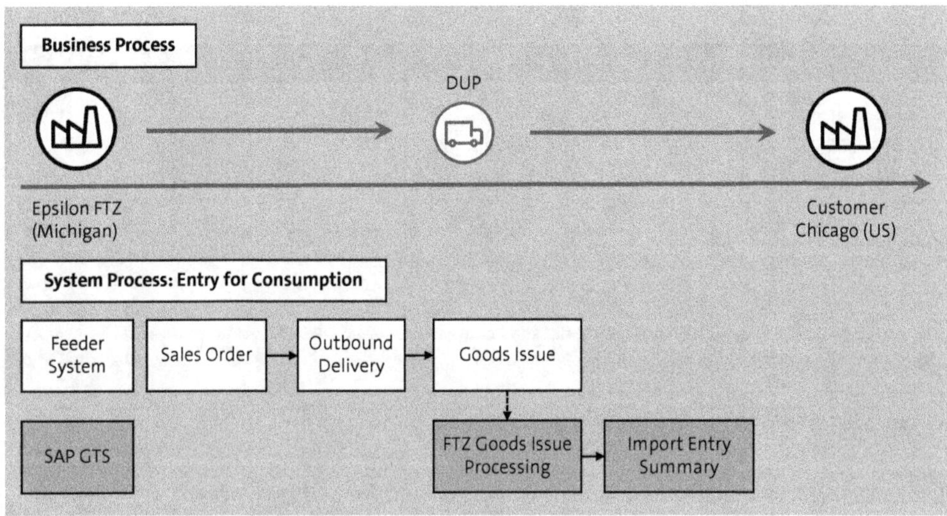

Figure 7.104 Business and System Process: Entry for Consumption into the US

Before filing the weekly entry summary, a FTZ forecast can be created in SAP GTS and filed with customs authorities to communicate and get approval for the estimated quantity and/or value that will get imported into the US during a certain period. Products declared in an entry for consumption for the next period, for example, next week, need to be approved during the current period. Figure 7.105 shows an FTZ forecast in SAP GTS, which can be accessed via the SAP Fiori launchpad by choosing **Special Customs Procedures • Foreign Trade Zone • Worklist Processing**, and then selecting the **Manage FTZ Forecast** tile.

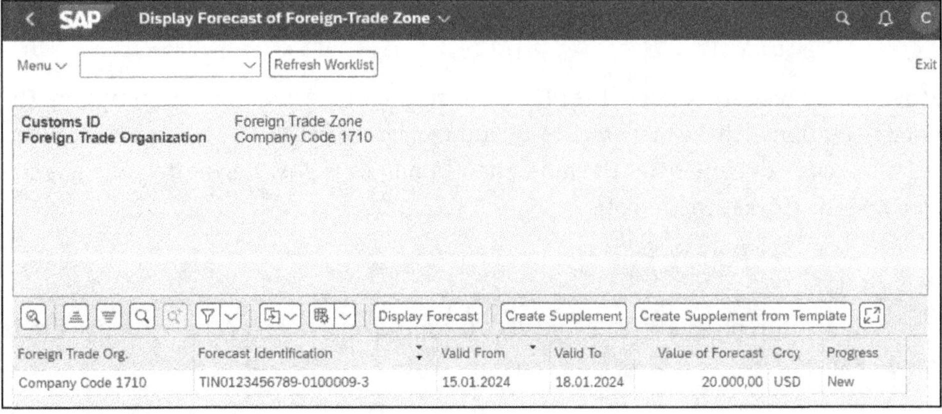

Figure 7.105 FTZ Forecast

Once the goods issue is posted in the feeder system for the domestic consumption, the stock will be removed, and an entry summary declaration will be created in SAP GTS (see Figure 7.106). The entry summary can be accessed in the SAP Fiori launchpad through **Customs Import • Declarations • Import Declarations • Display Import Declarations.**

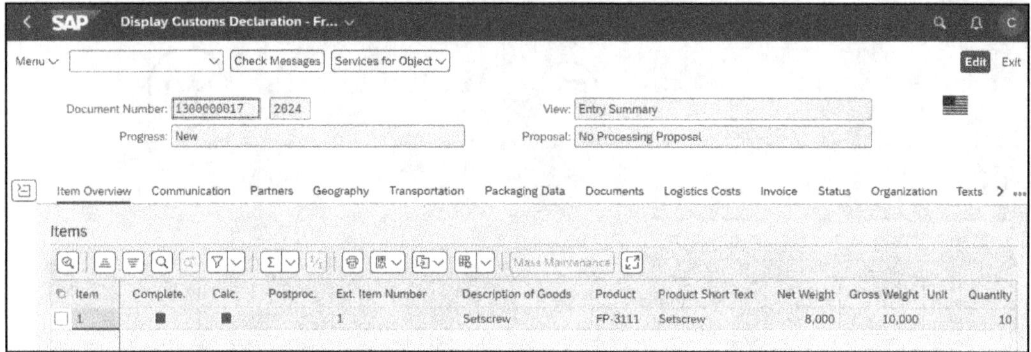

Figure 7.106 Entry Summary Declaration

Looking at the Display FTZ Receipts/Issues List app for Epsilon's FP-3111, we can see the production process of 60 pieces, the reexport of 10 pieces, and the entry for consumption for 10 pieces in Figure 7.107.

Menu Exit

Selection Period 18.01.2024 - 19.01.2024

Product	Prod.Short Text	Posted On	Rcpt/Issue	Stock Quantity	UoM	Doc. No.	Year	Item	Category of Receipt/Issue	Target Procedure
FP-3111	Setscrew	18.01.2024	+	1.890	PC	9000000005	2024	1	Finished Product from Production	Foreign-Trade Zone
			+	60	PC	9000000006		1	Finished Product from Production	Foreign-Trade Zone
			–	10	PC				Export	
			–	10	PC				Release to Free Circulation	

Figure 7.107 Receipts and Issues of the Finished Product in FTZ

Moving duty-paid stock out of the FTZ isn't included in the entry for consumption. The documentation of those movements for audit purposes in SAP GTS is done via an internal transaction document. At the moment of goods issue posting, the document is created and the stock is written off.

Reporting

Three standard reports are available via **Customs Special Procedures • Foreign-Trade Zone • Reports:**

- **Annual Reconciliation report**
  The Annual Reconciliation report can be generated in SAP GTS and contains a description of merchandise for each zone, lot or unique identifier, zone status, quantity on hand at the beginning of the year, cumulative receipts and transfers (by unit), quantity on hand at the end of the year, and cumulative positive and negative adjustments (by unit) made during the year. A report status can be maintained to track if the report is still in draft mode, already submitted to the authorities, or already accepted by the authorities.
- **FTZ Annual Board report**
  The Annual Board report can be generated in SAP GTS and is typically used for key performance indicator (KPI) measurement. The report shows total value of initial stock, receipts from non-US vendors, domestic receipts from other zones, issues to non-US customers, domestic issues, and the final stock. A report status can be maintained to track if the report is still in draft mode, already submitted to the authorities, or already accepted by the authorities.
- **Manage Harbor Maintenance Fee (HMF) Quarterly Summary report**
  For goods that have arrived on a vessel and are subsequently admitted into an FTZ, companies are obligated to pay the HMF on a quarterly basis using CBP Form 349.

With the FTZ process and the reporting explained in the previous section, we've completed the US-specific customs procedure.

7.3.5 Processing Trade Relief: China Localization

This section explains the China-specific customs procedure called processing trade regimes (PTR). PTR allows companies located in China to suspend the payment of import duties and/or VAT when importing components from overseas into Chinese processing trade warehouses and reexporting the finished products to other overseas parties.

There are two types of PTRs: contract manufacturing (the Chinese manufacturing entity purchases and owns the raw material provided by an overseas company) and tolling manufacturing (the overseas company owns the raw materials provided to the Chinese manufacturing entity). Using a fictitious company, Zeta, this section describes and explains the PTR business processes supported by SAP GTS. A brief overview will be given of the most common flows possible within PTR; in practice, there is a wide range of specific flows, depending on the location of the manufacturer and the country to which the goods are shipped.

Zeta's Global Trade Requirements

Zeta is an independent subsidiary of an international company operating in the manufacturing industry headquartered in Shenzhen, China, where the company also has a manufacturing site under PTR. This means that Zeta uses components purchased from domestic and foreign vendors to manufacture finished products before selling them. The business case in this section captures the following inbound and outbound transactions:

- Zeta will purchase components to build up its production stock. These components can be locally procured (COMP-215) or purchased from a foreign vendor located in Croatia (COMP-211, COMP-212, COMP-213, COMP-214).
- Manufacturing activities are carried out in the processing trade manufacturing site. The end product of this production process is final product FP-211.
- Finally, the components or final products are sold to domestic or foreign customers. An overview of the BOM used in the production of welder FP-211 is given in Figure 7.108.

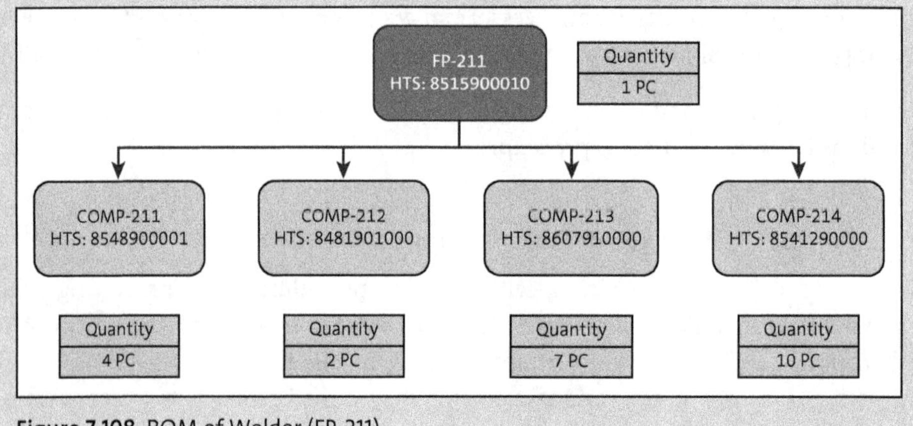

Figure 7.108 BOM of Welder (FP-211)

Business Process

This section details Zeta's PTR business process, which consists of both inbound and outbound processes.

A prerequisite of manufacturing under PTR is for the company to obtain the necessary authorizations. To register for a PTR authorization, the processing company must have a contract with a client. This contract must include a list of all components required under processing trade, the finished goods to be reexported, and the timeline of the contract. Each contract must be registered individually with the customs authorities. A processing company can have multiple contracts at the same time. When a processing company has multiple contracts, each contract will be tracked by a separate authorization.

When all authorizations are verified and approved by China's customs authorities, the company can start its production under processing trade.

When the contract timeline ends, or production quantity stated in the contract is completed, final settlement with customs authority for this contract will take place. During the settlement process, component consumption and remaining stock are reported to authorities. Based on the settlement result, customs will take the necessary actions (e.g., assessing duty payment for components not consumed).

Inbound Process

As a first step in its business process, Zeta, located in Shenzhen, China, purchases components (COMP-211, COMP-212, COMP-213, COMP-214) from a vendor located in Zagreb, Croatia. The inbound process is graphically shown in Figure 7.109. When the components cross the Chinese border, a PTR entry declaration is created where a reference to the manufacturing contract (customs authorization) is required for customs authorities to verify that these components have been registered for PTR.

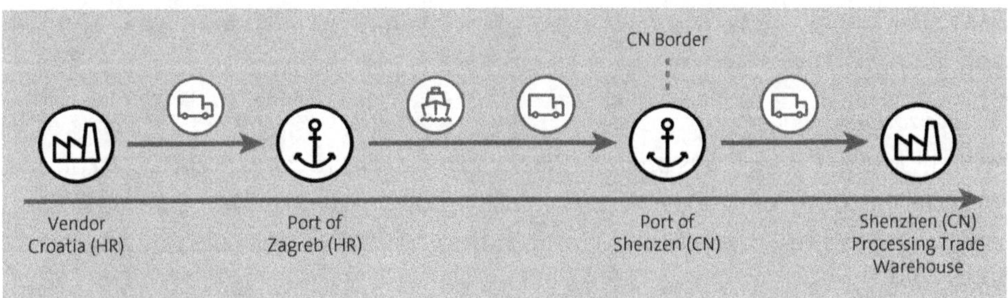

Figure 7.109 PTR Inbound Process

Outbound Process

After completing the inbound process, Zeta will start the production process for the finished product. After Zeta carries out these activities, the company can sell the finished product to either a foreign or a domestic customer. This outbound process is graphically shown in Figure 7.110.

If the goods are being shipped out of China, for example, to a German customer, a reexport declaration will need to be created to release the goods for reexport and consume the initial imported PTR components.

If the goods are being shipped within China, for example, to a Shanghai customer, the goods must go through a procedure called *entry for domestic consumption,* in which the company files a request with the customs authorities to release the goods for domestic sales. In most cases, this approval can be obtained post-processing and requires the company to pay the earlier exempted import duties.

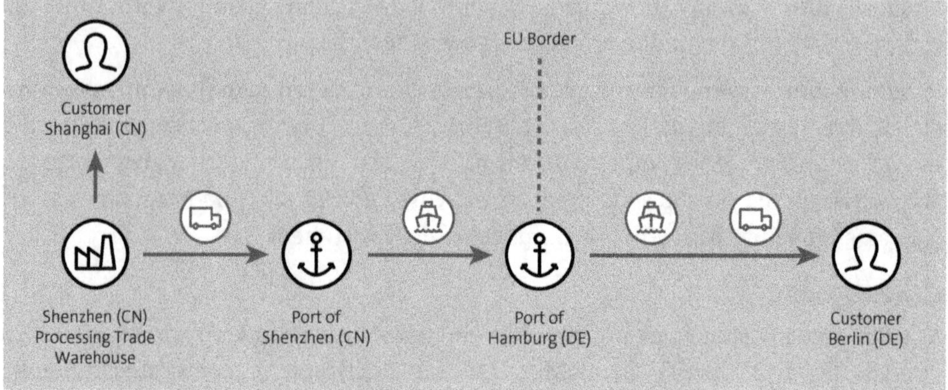

Figure 7.110 PTR Outbound Business Process

System Process

Each of the steps described earlier in the business process can be linked to a transaction in the PTR functionality. This section discusses the links between the business process and the system process.

Authorizations

As indicated in the business process, the company needs to register each PTR contract with the customs authorities prior to starting production activities under PTR. Users can handle this registration through the authorization maintenance functionality in SAP GTS.

In SAP GTS, two kinds of authorizations are dedicated to the PTR functionality: Enterprise Qualification Application (EQA) and the e-manual. The PTR functionality enables these authorizations to be created and maintained in the system. Moreover, it enables the approval process to be filed using electronic communication. The following list describes these authorizations in greater detail:

- **Enterprise Qualification Application (EQA)**
  The EQA is the legal requirement in China for listing all permitted goods using the processing trade customs procedure and for submitting customs declarations. The EQA is registered per company and records which components will be imported under PTR and which finished products will be reexported. In the EQA, companies can aggregate similar products by making use of a merge rule. Chinese custom authorities provide standard merging rules (e.g., HS codes; value of the goods) that allow companies to declare the products on a merged basis. After the company submits the EQA to the customs authorities for review, the authorities will review the material information and decide whether to grant approval.

  Figure 7.111 provides an example of the EQA as maintained in the system. The EQA can be found in the **Enterprise Qualification Application** app under **Customs Special Procedures • Master Data • Authorizations** in the SAP Fiori launchpad.

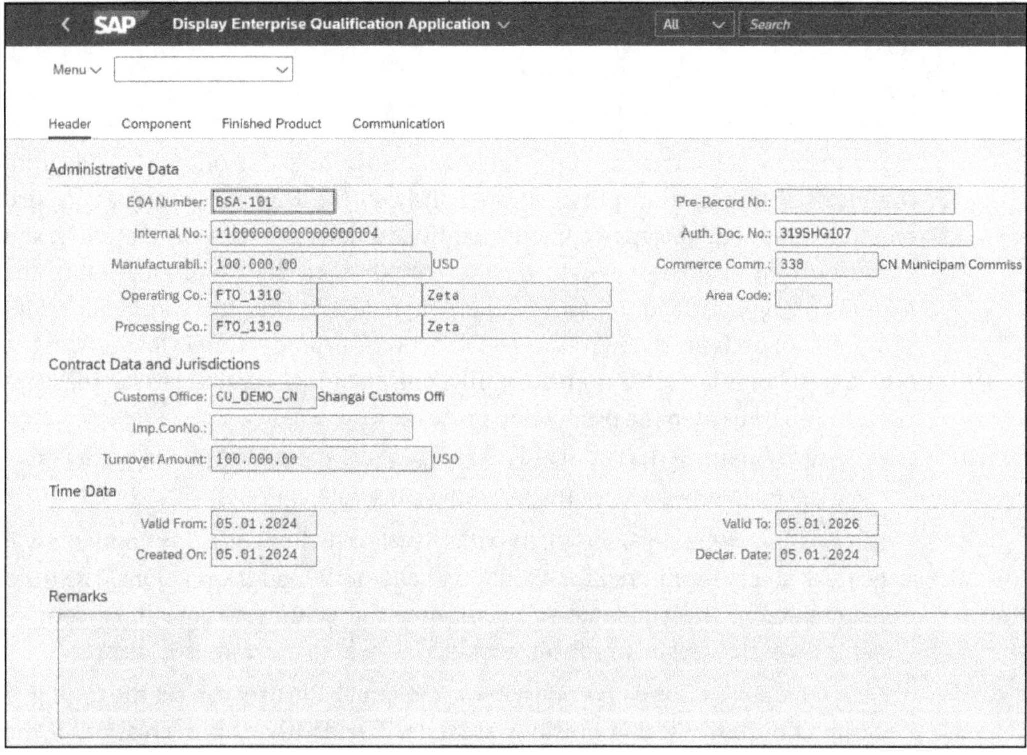

Figure 7.111 Enterprise Qualification Application: Header Tab

The EQA consists of several tabs, each containing specific content related to the authorization:

- **Header**: Contains a general overview of the EQA characteristics. This tab includes information about the manufacturing company, the customs office filing the declaration, and the validity date of the EQA.
- **Component**: Contains detailed information about the components and the merged components. **Sequence Number** and **Description** of the merged components can also be found in this tab.
- **Finished Product**: Contains detailed information about the finished product. **Sequence Number** and **Description** of the merged finished product can also be found in this tab.
- **Communication**: Contains electronic communication between SAP GTS and the customs authorities. This tab includes the status of the EQA and the return message from the customs authorities.

After the EQA is created, users can communicate the EQA information maintained in SAP GTS to customs authorities via the **Communication** tab. This tab displays all inbound and outbound messages. When the customs authorities approve the EQA, the system receives an approval message. This message triggers the status of the authorization to be set to **EQA Approval**, allowing users to create e-manuals linked to this EQA.

- **E-manual**

  The e-manual is based on the data of the EQA. These are registered per manufacturing contract, which means multiple e-manuals can be linked to one EQA. Aside from the material information provided previously in the EQA, the e-manual is also required to provide the approximate quantity, value of the relevant materials, and validity period of the e-manual (which corresponds to the validity period of the manufacturing contract). Moreover, the e-manual specifies the component consumption for each finished product and a theoretical/production BOM. As with the EQA, the customs authorities must approve the e-manual. After approval, the components can be used in the production process.

  The e-manual is one of the key elements in the setup of the PTR functionality in SAP GTS. It's the main reference for the following stock calculations:

  - *Stock maintenance*: In SAP GTS, the stock levels of PTR-relevant components will be tracked and monitored for all inbound and outbound transactions. The BOM maintained in the e-manual will determine the volume of consumed components for each finished product reexported or sold to the domestic market.
  - *Determination of closing portions*: In the e-manual, the user can set the timing of closing portions. Closing portions are determined either upon creation of the export document or after triggering a worklist for customs declarations items.

  Users can set up electronic communication with customs authorities regarding the e-manual. After the e-manual is maintained, users can send this information to customs. After approval, a message is received that confirms the approval of the e-manual and an e-manual reference number, changing the status of the e-manual to **Approved**. Only then is further processing allowed both from a business and system point of view.

  The e-manual is accessed through the **Manage E-Manual** app found under **Customs Special Procedures • Master Data • Authorizations**. The e-manual consists of several tabs, each containing specific content related to the authorization (see Figure 7.112):

  - **Header**: Contains a general overview of the e-manual characteristics. In this tab, users specify which plant the contract relates to as well as the value and validity of the contract.
  - **Component**: Contains detailed information about the components and the merged components. **Sequence Number** and **Description** of the merged components can also be found in this tab.
  - **Finished Product**: Contains detailed information about the finished product. **Sequence Number** and **Description** of the merged finished product can also be found in this tab.
  - **Bill of Material**: Contains the BOM for each finished product.

- **BOM of Merged Products**: Contains the BOM for each merged finished product. In this BOM, the components are also merged products.
- **Communication**: Contains the electronic communication between SAP GTS and the customs authorities, as well as the status of the e-manual.

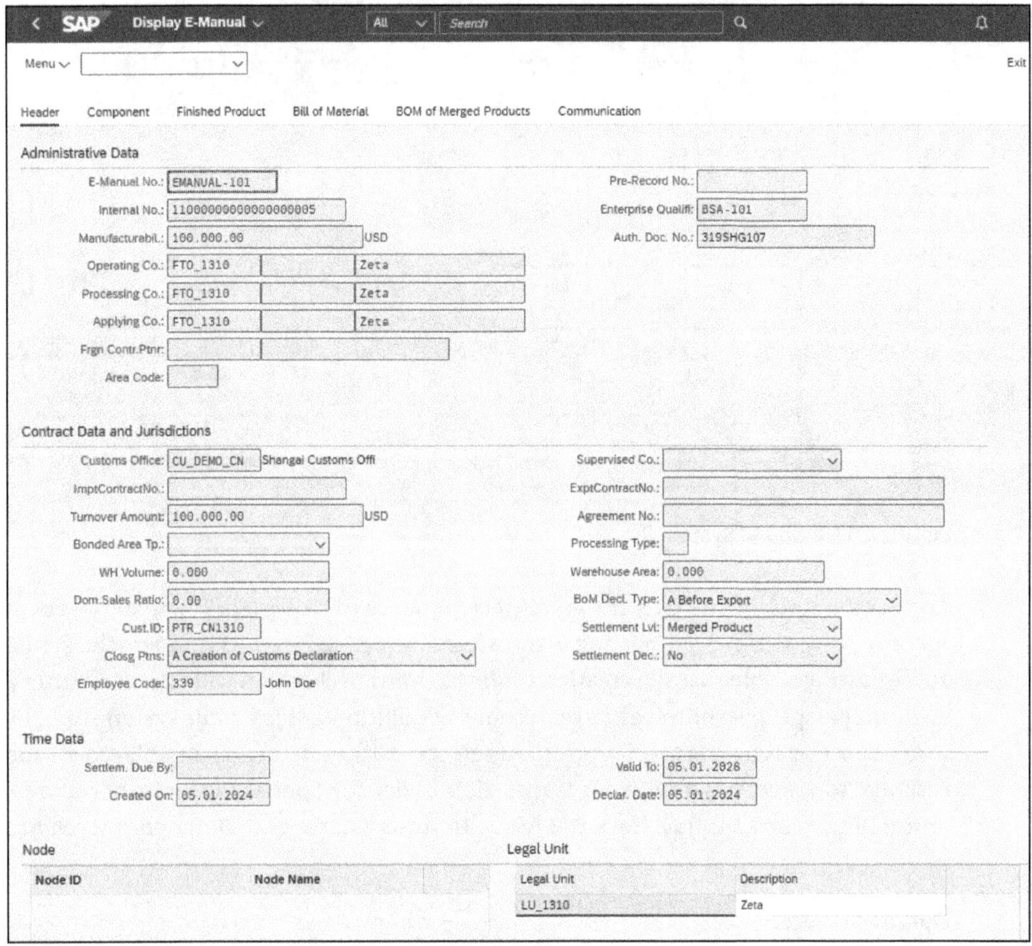

Figure 7.112 E-Manual: Header (Maintained for FP-211)

Inbound Process

After the EQA and e-manual are created and approved in SAP GTS, the company can start the inbound process.

As shown in Figure 7.113, the inbound process into the processing trade warehouse starts when the company creates a purchase order with the previously specified vendor in Croatia in its feeder system. An inbound delivery is created before the goods arrive at the Shenzhen plant in China. When the goods arrive at the Chinese border, the PTR entry declaration is created in SAP GTS based on the purchase order or the inbound delivery and filed with the Chinese customs authorities under suspension of import duties. The customs authorities will release the declaration.

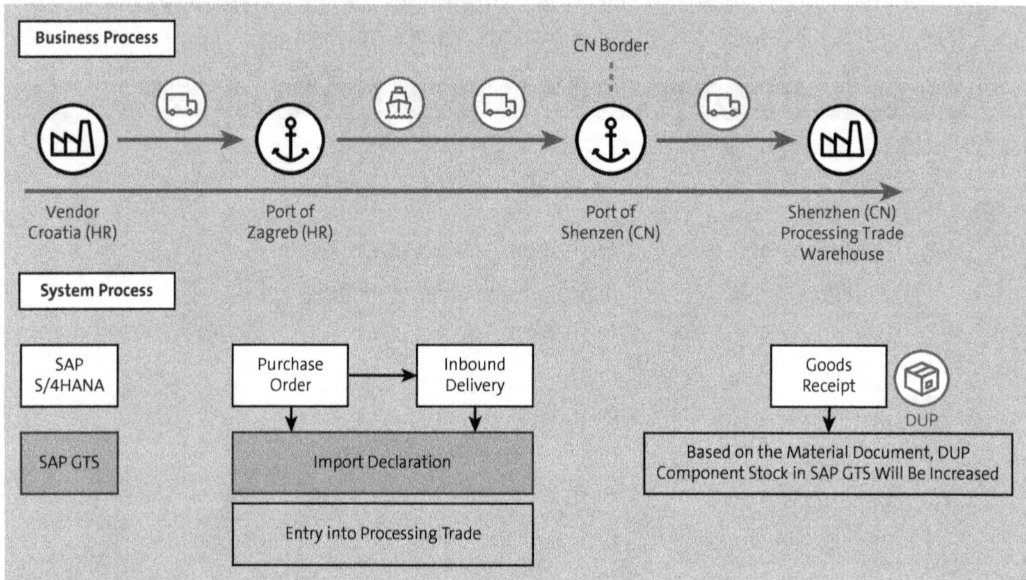

Figure 7.113 Inbound System Process

To post the stock in SAP GTS, the goods receipt needs to be posted. The material document of the goods receipt posting is transferred to SAP GTS, which triggers the actual stock postings. Note that the creation of the customs declaration will add stock from a customs perspective; however, to represent this additional stock in the system, a goods receipt posting is necessary. After this is done, the stock update can be displayed via the **Display Stock Receipts and Issues** transaction under the **Special Customs Procedure • Processing Trade • Display Stock Overview**. In this scenario, the component stock has increased.

Outbound Process

After completing the inbound process, Zeta can start manufacturing with the imported PTR-relevant components. After the production is done at the processing trade warehouse, the finished products are ready for outbound shipment. As shown in Figure 7.114, the outbound process starts with the creation of a sales order, followed by an outbound delivery.

The following two scenarios are involved in the outbound shipment of the goods:

- **Reexport**
  As stated, customs authorities require a reexport document when companies sell the finished product to a foreign customer, in this example, a customer in Berlin, Germany.

After picking and packing of the finished products, the goods issue can be posted. Before the goods are allowed to leave the processing trade warehouse, a pro forma invoice is created in the feeder system. This triggers the creation of a reexport document in SAP GTS. The structure of the reexport declaration is the same for any other export document created in SAP GTS (refer to Section 7.1). Depending on the setting maintained in the e-manual, the closing portions will be immediately determined upon creation of the reexport document or after triggering a worklist for customs items. For more information on the determination of closing portions, refer to Section 7.3.1.

After the determination of closing portions, the **P/T Closing Portions** tab on the item level of the reexport declaration will be populated with deducted stock postings. Note that these calculations are based on the finished product BOM maintained in the e-manual.

- **Domestic sales**
  In some cases, the finished products aren't reexported and are sold in the Chinese market. This process is referred to as a domestic issue in SAP GTS. The components were initially imported under PTR and thus with duty payment exemption. When these components are taken out of processing trade and enter the Chinese market, the company must pay import duties. In SAP GTS, a dedicated transaction processes domestic goods issues: **Special Customs Procedures • Processing Trade • Processing of Goods Movements.**

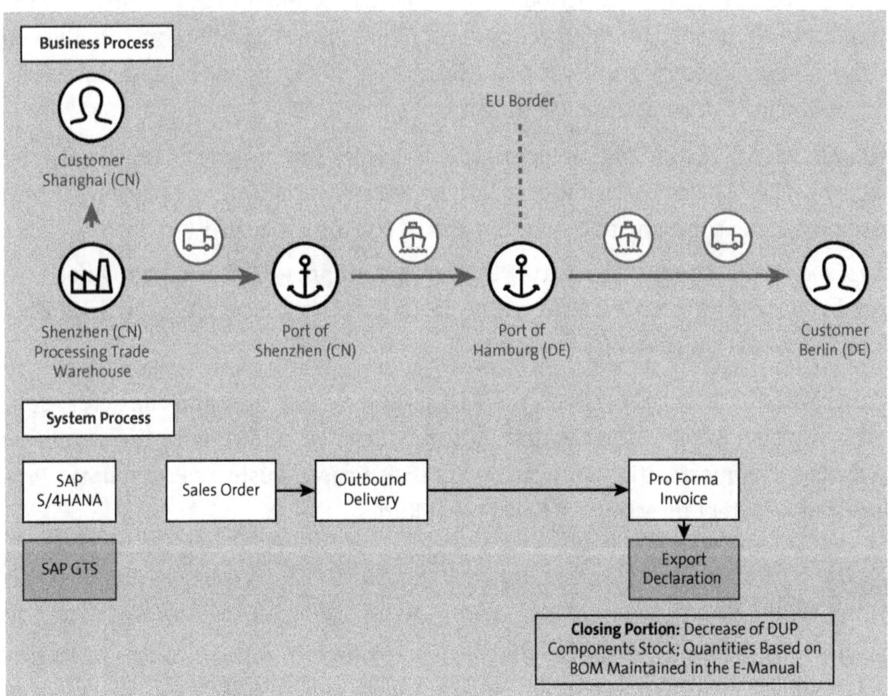

Figure 7.114 PTR Outbound System Process

Settlement Process

As described in the business process, a company can only manufacture under processing trade as long as they have an active PTR contract approved by customs authorities. After the contract timeline ends or the production quantity stated in the contract is completed, final settlement with customs authorities will take place.

The system represents the end of a business contract by the e-manual's validity period. After this validity period expires, a settlement process is conducted. In this settlement, companies summarize stock quantities of all imported and exported items during the validity period of the e-manual. In the SAP GTS system, a settlement run can be created via **System Monitoring • Background Processing • Create Settlement Run for Processing Trade**. The communication and validation of these quantities with customs authorities takes place in two phases:

1. **Presettlement (centralized declaration)**
   The presettlement document can be displayed via **Special Customs Procedures • Processing Trade • Display Settlement Run**. The company has to provide customs authorities with a list of all import and export declarations, as well as all domestic issues created over one predefined period, for validation and comparison.
2. **Settlement (reconciliation)**
   The settlement document can be displayed via **Special Customs Procedures • Processing Trade • Display Settlement Run**.

At the beginning of the validity period of the e-manual, companies have to indicate the components and finished products relevant for PTR. At the end of the contract, the settlement provides the opportunity for rectification of stock variances. Companies can provide following information to customs:

- *Actual BOM*: At creation of the e-manual, companies can provide customs with the theoretical BOM of finished products. If there are any variations, the settlement process enables the company to declare the actual consumed quantities.
- *Scrapping and lost quantities*: During the settlement process, the company needs to explain all stock differences between the declared and actual stock levels, including declaring the scrapped or lost quantities.

If there are stock variances between the PTR declared quantities and the actual consumed quantities, a correction declaration can be created, allowing for the necessary rectifications of import duties. This can be done via **Special Customs Procedures • Processing Trade • Enter Correction for Settlement Run**.

7.3.6 Customizing

In the previous sections, we've completed the explanation of the business processes related to special customs procedures, which are supported by SAP GTS. In this section, details on how these special customs procedures can be configured in SAP GTS will be elaborated on. There are a few Customizing steps specific to special customs procedures for both the feeder system and SAP GTS.

Define Customs ID

To indicate in SAP GTS that a special customs procedure is applicable, users must add an additional organizational element, the customs ID (**Cust. ID**), to the organizational structure. Without this customs ID, it's not possible to set up the special customs procedures.

A user can define customs IDs in **Define Customs ID for Customs Procedures with Economic Impact** via **SAP Global Trade Services, edition for SAP HANA • General Settings • Organizational Structure**.

A user can assign the newly created customs ID to both a logical system group and an FTO in the **Control Customs ID for Customs Procedures with Economic Impact** activity (**SAP Global Trade Services, edition for SAP HANA • General Settings • Organizational Structure**).

To enable the FTZ functionality, users have to define additional elements on top of assigning the customs ID to a logical system group and an FTO. First, the Facility Information and Resources Management System (FIRMS) code needs to be defined. This enables the communication with the US customs port of entry. The FIRMS code per FTZ can be retrieved from the CBP and represents the exact location of goods. Secondly, the FTZ ID needs to be maintained and represents a number assigned by the CBP to an FTZ.

Note

With regards to the FTZ functionality, most of the Customizing activities related to communication processes and structuring the operative cockpit come configured out of the box by SAP GTS.

Users can define the FTZ customs IDs additional elements in **Control Customs ID for a Foreign-Trade Zone** (**SAP Global Trade Services, edition for SAP HANA • General Settings • Organizational Structure • Control Customs ID for a Foreign-Trade Zone**).

Communication Processes

For each special customs procedure setup, a new process within the process template will need to be created to capture additional processes. For example, in the case of BWH, a process called **Customs Warehouse** needs to be added to capture the inbound process in the BWH. For an FTZ, users must add the **Foreign Trade Zone** process to capture the inbound process in the FTZ. For PTR, a process called **Processing Trade** needs to be set up to capture the inbound process in the PTR warehouse. Although the content of the settings are different, the logic to set up these new processes is identical to the

Customizing described in the general Customs Management Customizing. Consult Section 7.2.3 for more information on this Customizing activity.

Several additional Customizing steps can be performed to configure the solution to meet companies' needs. Two additional Customizing steps are given as examples.

For authorized direct delivery, SAP GTS allows users to automatically add to or create a local clearance admission when a goods receipt is posted to the FTZ. This local clearance admission contains all goods receipts of the same waybill for the current FTZ posting day. To activate the automatic addition/creation of the local clearance admission, navigate via the process template to the correct activity sequence as explained in the general Customs Management Customizing section (Section 7.2.3). This option can be activated for the configured activity sequence **FTLCA** (**Create Local Clearance Admission**), which is provided in standard SAP GTS via **SAP Global Trade Services, edition for SAP HANA • Customs Management • Communication Processes • Define Control Settings for Communication Processes.**

When a finished product is sold in the US, depending on the zone status of the goods receipt, an entry summary may be necessary. To enable this, a configuration step needs to be taken in the **USSUM** (**Entry Summary**) activity sequence in Customizing activity **SAP Global Trade Services, edition for SAP HANA • Customs Management • Communication Processes • Define Control Settings for Communication Processes.**

Define Merge Rule

As explained in Section 7.3.5, specific merge rules can be used in both the EQA and the e-manual to declare a similar set of products "as one" to customs authorities. Each merge rule can consist of different criteria, for example, HS code or weight. In Customizing, the different merge rules can be set up based on the customized criteria via **SAP Global Trade Services, edition for SAP HANA • Customs Management • General Settings • Define Merge Rules for Processing Trade.**

Authorizations and Securities

As explained in Section 7.2.3, in different areas of customs and global trade, a company requires various types of authorizations for performing certain activities, such as simplified procedures for import, export, and transit. For special customs procedures, it's also necessary to define and maintain the authorization type in Customizing, in the same way as described in Section 7.2.3. This activity can be done by accessing Customizing via the following path: **SAP Global Trade Services, edition for SAP HANA • Customs Management • Authorizations and Securities • Define Types of Authorization for [Special Customs Procedure].**

To maintain the two authorizations dedicated to PTR, as discussed in Section 7.3.5, the authorization types need to be set up in the dedicated Customizing transactions before the authorization can be maintained as master data. The following Customizing activities are relevant: **SAP Global Trade Services, edition for SAP HANA • Customs Management • Authorizations and Securities • Define Types of Business Scope Accounts for Processing Trade/Define Types of E-Manuals for Processing Trade.**

In the previous explained Customizing steps, users can configure settings that are inherent to the authorization (e.g., import/export). Moreover, the merge rules used in the authorizations can be defined in this Customizing step. After the detailed information of the authorization is maintained, a status needs to be set for each of the authorizations. These can be **Authorization Created**, **Authorization Active**, or **Authorization Expired**.

Customizing in SAP S/4HANA (Plug-In)

In the SAP S/4HANA plug-in, users must define and assign the same customs ID as defined in SAP GTS. This can be done in the **Customs: Customs Procedures with Economic Impact: Define Customs ID** activity.

The next Customizing step is to complete the **Customs: Assign Customs ID to Customs Procedures with Economic Impact** activity. Note that the system makes a distinction between BWH, PTR, FTZ, PUCC, and IPR on the one hand and OPR on the other.

In this second activity, the system connects the customs ID to a combination of a plant and storage location, which means the storage locations indicated here are special customs procedure-relevant storage locations.

> **Note**
>
> These Customizing activities are additional to the general Customizing required for customs processing. For an overview of the general Customizing activities, refer to Section 7.2.3.

7.3.7 Master Data

Besides the master data described in Section 7.2.4, we need to discuss a few special topics regarding the special customs procedures.

Customs Products

To indicate whether a material is relevant for one of the described special customs procedures, users need to flag the material. Flagging a material is very important for the correct handling of stock, so all activities need to be performed in the correct sequence. The product master in SAP GTS contains the **Special Customs Procedures** tab for this purpose, as shown in Figure 7.115.

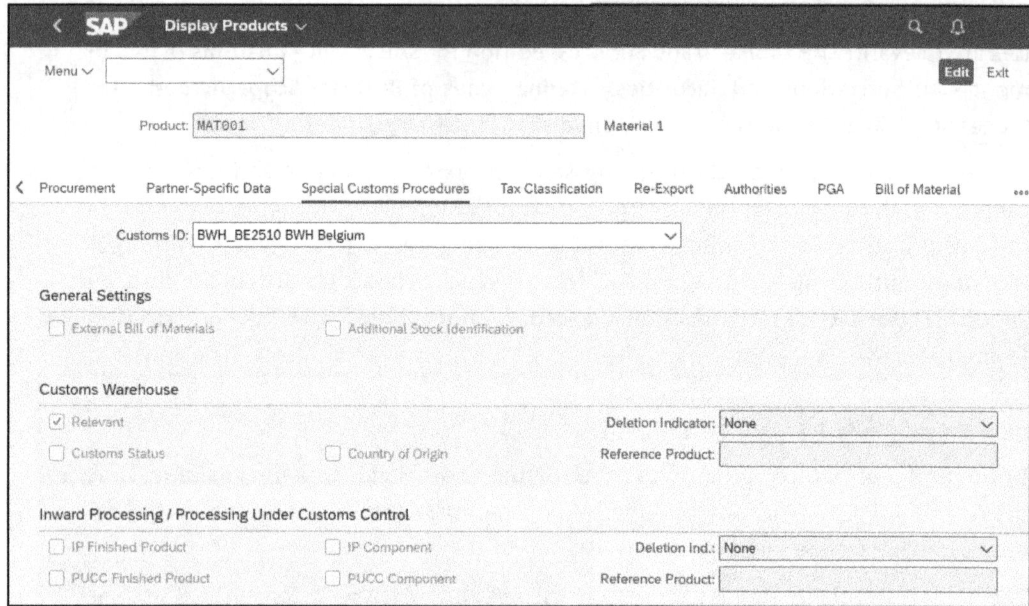

Figure 7.115 Material Master with the Special Customs Procedures Tab

In this tab, you can define for which special customs procedure the material is relevant. Furthermore, for OPR, IPR, PUCC, FTZ, and PTR, users must indicate whether the product is a component or a finished good. Users need to enter the **Customs ID**, check the box for the procedure users want to activate, and decide on the stock keeping level (e.g., country of origin or batch).

After all relevant materials are flagged for a special customs procedure, these flags need to be transferred back to the feeder system. This retransfer can be done either on an individual basis or for all materials by executing the **Redistribution of Products for Special Customs Procedure** transaction. This transaction can be accessed through the administrator menu: **SAP Global Trade Services • Synchronization • Master Data Synchronization**.

Now, in both systems, the materials are indicated as relevant for a special customs procedure, so there is one last step that still needs to be performed. Most likely, some stock is already physically available in the warehouse, and it's represented by the stock level in your feeder system. Because SAP GTS is also keeping stock, this initial stock needs to be transferred to SAP GTS as well. This transfer of initial duty-paid stock is performed in the plug-in (in SAP S/4HANA) via the **Initial Transfer of Duty-Paid Stock to GTS** activity. This can be accessed through **GTS: Global Trade Services Area Menu • Initial Transfer of Master Data**.

Bill of Material

For IPR, OPR, PUCC, FTZ, and PTR, users must create the BOM at the level of the finished goods in the product master in SAP GTS. Here, users link the IPR, OPR, PUCC, FTZ, and PTR components to the finished goods. In the BOM master data, a theoretical component usage is also entered. This component ratio is used when an inbound or outbound flow of the IPR, OPR, PUCC, FTZ, or PTR process is initiated because the consumption of the component must be registered in SAP GTS.

The discussed BOM has to be created manually in SAP GTS, and it's not identical to the BOM in the feeder system. This BOM in SAP GTS links the finished goods only to its IPR, OPR, PUCC, FTZ, and PTR components and doesn't contain the other components of the finished goods that aren't relevant for the procedure. This BOM contains only the primary components that have been purchased (from a supplier outside the customs union for IPR, PUCC, FTZ, PTR); it doesn't contain any semifinished or intermediate products that haven't been purchased initially. For OPR, it contains only the components exported to the warehouse of the external manufacturer. The components added in the manufacturer's premises aren't considered.

Business Partners

For the OPR functionality to work as designed, users have to flag the business partner to which the OPR components are intended to be exported. Similarly, users have to flag which OPR business partners you'll import OPR finished products from. These flags are set in the product master in SAP GTS.

Authorizations and Securities

Although the concept behind the different types of authorizations and securities is very similar and follows the same instructions provided in Section 7.2.4, there are minor differences between authorizations and securities for the different special customs procedures. Mainly, the differences concern the attributes of the applicable authorization or security (e.g., processing period, settlement period).

If a company decides to set up a BWH, it's legally required to apply for an authorization with the customs authorities. In SAP GTS, this authorization is called the **Authorization for Customs Warehousing Procedure**. In this authorization, users need to indicate the responsible customs office and type of BWH. This license is valid only for the LU specified in the attributes.

The authorizations for IPR, OPR, and PUCC have some specific requirements:

- The products (components and finished goods) have to be listed.
- The processing period (IPR, PUCC) and reimport period (OPR) have to be specified (this is the life span of OPR, IPR, and PUCC products).

Another important indicator for IPR is **Closing Portion Determination**, which regulates when the closing portion is created. There are two options: automatically at the creation of the export declaration or manually at a later stage.

For the OPR process, all the relevant business partners have to be entered in the authorization as well; these are the partners who receive components from the company and those who supply the finished goods back to the company.

Foreign Trade Zone

If a company decides to set up an FTZ, all products relevant for the FTZ are required to be made relevant for the zone on the master data level. The user can select the customs ID that is specific to the FTZ under the **Special Customs Procedures** tab of the material master in SAP GTS. As indicated in Figure 7.116, users have to specify whether the product is a finished product, semifinished product, or a component, as well as what the zone status is of that specific product. Users can indicate whether the product is PF, NPF, or no zone status at all.

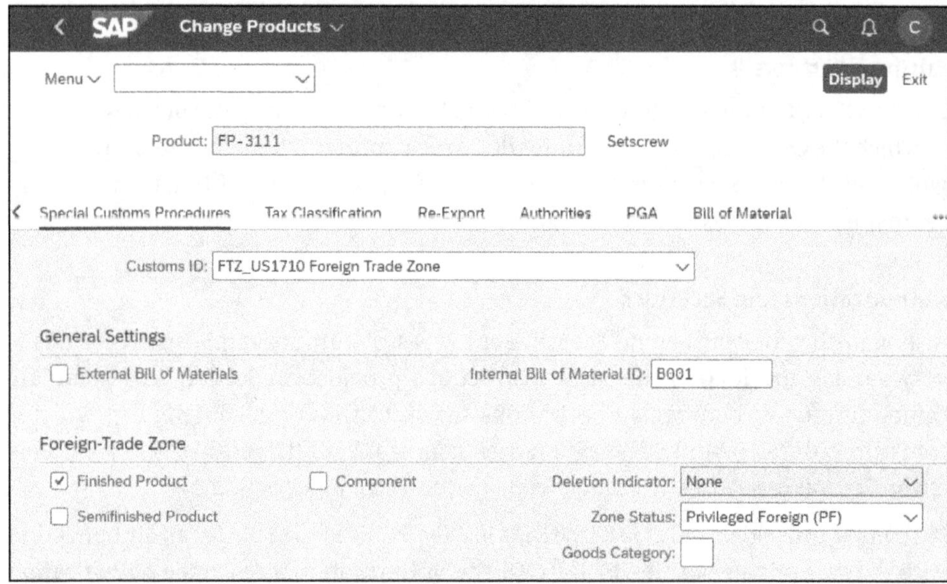

Figure 7.116 Special Customs Procedure Master Data: FTZ

To support the production scenario, an additional requirement is to assign the internal BOM to the FTZ-specific customs ID. This BOM can be set up in its dedicated tab (**Bill of Material**) in the product master in SAP GTS and should be configured for the finished product. The user has to maintain which components result in a finished product. Additionally, the BOM needs to be activated for Customs Management.

Processing Trade Regime

When a company decides to set up manufacturing under PTR, all components and finished products relevant for PTR have to be flagged on a master data level for PTR. In the product master, the user can select the PTR-specific customs ID under the **Special Customs Procedure** tab. Moreover, whether the products are components or finished products, trading under PTR must be indicated. Note that this action isn't sufficient to make the product available under PTR.

In addition to the flag set in the product master, the products have to be included in the designated authorizations (EQA and e-manual). For a detailed overview of these PTR-specific authorizations, refer to Section 7.3.5.

7.4 Excise Duty Monitoring

Some companies are also obliged to have a process in place to monitor the transport of excise goods under excise duty suspension. This process of *excise duty monitoring* is discussed in this section.

Customs authorities in the EU Member States have an Excise Movement Control System (EMCS) to monitor the movement of goods under excise suspension. This excise duty processing applies to alcohol, tobacco, and energy products, for example. EMCS is used only in the EU.

Let's continue with the business and system processes, as well as relevant Customizing and master data.

7.4.1 Business Process

Figure 7.117 displays the overall business process for EMCS, including all the parties involved and the sequence in which all transactions take place.

For example, when a manufacturer of tobacco in Belgium wants to send its products to a customer located in Poland, the following activities typically occur (as numbered in Figure 7.117):

❶ The consignor submits a notification to the customs office of departure in Antwerp, Belgium, which indicates that the goods are relevant for EMCS.

❷ The local customs office of departure replies with a confirmation message if the shipment of the goods is accepted. The customs office also provides the company with an Administrative Reference Code (ARC) number that has to be printed; this is a unique reference number of the declaration.

❸ After receiving this acceptance, the company can dispatch the excise goods. The goods are physically shipped to the location of the consignee.

❹ The customs authorities send a message to the customs office of destination in Poland. The purpose of this message is to provide the information regarding the EMCS-relevant shipment to the customs office of destination.

❺ The customs office of destination forwards this message to the consignee.

❻ The goods arrive at the place of destination.

❼ The consignee submits a report of receipt.

❽ The customs office of destination sends a receipt of the notification to the consignee.

❾ The customs office of destination forwards this report of receipt to the customs office of departure.

❿ The customs office of departure closes the loop by sending the information back to the consignor. These last two steps run simultaneously.

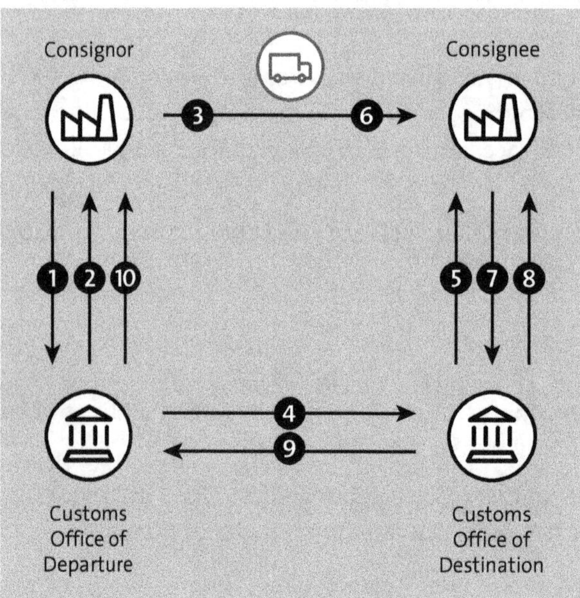

Figure 7.117 Business Process for EMCS

In this way, all concerned parties are informed about the departure and receipt of the excise goods under excise duty suspension.

The section of EMCS process flow relevant for the consignor is illustrated in Figure 7.118.

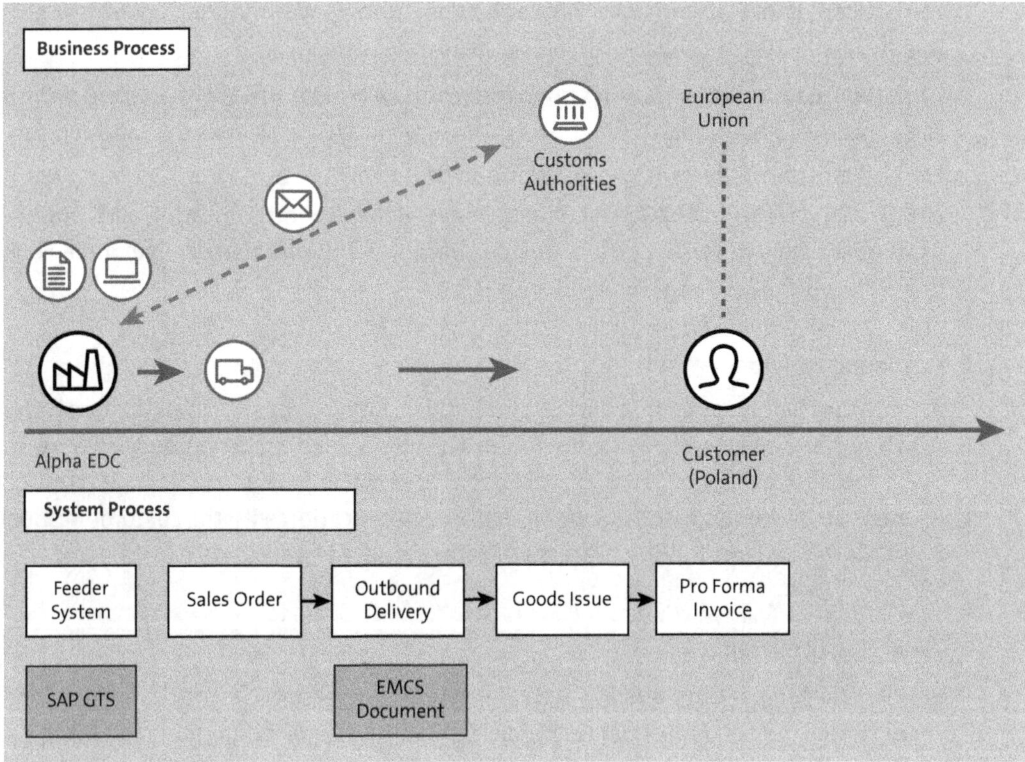

Figure 7.118 EMCS Process Flow

7.4.2 System Process

Like the previously described functionalities in the Customs Management functionality, every business process step can be accessed via the EMCS specialist role in the SAP Fiori launchpad. In this section, a link will be made between the EMCS business process and the system process.

Inbound Activities

When a supplier sends goods under duty suspension, an electronic incoming message from the local customs authority appears in SAP GTS. This implies that the process starts in SAP GTS without integration with the feeder system. In this way, your company is informed of the incoming shipment. The message contains the ARC number provided by the customs authorities.

After the goods arrive, the receiving company can compare the number of goods that are physically delivered to the expected number. The receiving company can accept the shipment if all goods are delivered in the correct quantities. But if there is a discrepancy in quantity of goods delivered, the consignee can accept the goods with restrictions. If the discrepancy isn't acceptable, the goods are rejected. This status update needs to be provided to the local customs authorities.

After the customs authorities approve the report of receipt filed by the consignee for the concerned shipment, the inbound process is closed. By closing the inbound process, the consignor is released from its responsibilities.

In SAP GTS, an inbound EMCS document from which the communication with the customs authorities can take place is created. From the inbound perspective, there is no link between the feeder system and SAP GTS.

Outbound Activities

A company that wants to ship excise goods under duty suspension to its customer has to ask permission from the customs authorities before sending these goods. In the outbound process, a transfer of information from the feeder system to SAP GTS is required to create an outbound EMCS document. The communication with the customs authorities follows the same e-filing process as for the standard export process.

7.4.3 Customizing

Besides the general Customizing activities required to set up Customs Management, there is no need for additional Customizing activities to facilitate the EMCS process. However, it's important to note that the interaction between the feeder system and SAP GTS doesn't occur via the standard SAP GTS plug-in. You have to set up a customized interface to enable the transfer of information and company-specific triggers to transfer information. SAP GTS provides a function module to enable this process (/SAPSLL/API_6800_EMC).

SAP GTS provides a standard interface between EMCS and customs exports to communicate the ARC number in the export document. As such, the export exit confirmation also determines the closing of the EMCS document for the outbound flow.

7.4.4 Master Data

To successfully perform e-filing, both the consignor and consignee of the process must have a tax warehouse number maintained in the system. There is no standard interface for this information, meaning that the tax warehouse number must be entered manually in SAP GTS. A customer-specific development is required to automate the transfer from the feeder system to SAP GTS if the data is maintained in the feeder system. To identify whether a product is EMCS relevant, you must classify the products with excise product codes.

7.5 US Reconciliation Process

Reconciliation is a program that allows an importer to file entry summaries with the applicable government authorities after the actual receipt of the goods. This can be the result of several factors, including the post-importation transfer pricing adjustments, valuation of royalties or assists in particular accounting periods, limited returned goods provisions, and/or certain FTA qualification.

The following sections will discuss the business and system processes and Customizing.

7.5.1 Business and System Processes

The US reconciliation procedure is relevant for the import process. Once the declaration is created in the SAP GTS system, the information can be found back at the header level in the **Transportation** tab (accessible via the **Display/Change Import Declaration** tile) where information such as the reason for the reconciliation and the due date must be provided. If the reconciliation is based on USMCA, this must be indicated in a separate field as well.

For every import declaration in SAP GTS, a user must enter specific reconciliation data. One approach is called *entry-by-entry flagging* because each individual declaration is flagged separately. Users can set a default value for the **Reason for Reconciliation Summary** via the **Manage Customs Declaration** app. If you use this option, one bill of lading is tied to one customs declaration.

Besides the entry-by-entry approach, another approach is called *blanket reconciliation flagging*, whereby an Automated Broker Interface (ABI) automatically flags all entries of an importer for a given period. In this approach, several bills of lading are tied to one customs declaration.

The reason for reconciliation could be based on four things:

- **Value**
  This flag is open to all value issues (assists, royalties, and computed value) and any other factors affecting CBP valuation, such as indirect payments.
- **Classification**
  An entry summary may be flagged for classification only when one of the following issues have been established: pending administration ruling (including pre-classification rulings), protest, or pending a court action.
- **9802**
  Subheadings 9802.00.60-9802.00.90 permit a reduced duty treatment for the value of components manufactured in the US and assembled abroad. The focus is on the value aspect of the 9802, from estimates to actual figures.

- **FTA**
  This indicates that a claim may be made at the time of reconciliation filing; the filing of a FTA claim is optional. Duties are paid at the time of filing entry summary on the flagged entry summaries, and if the goods qualify, a refund can be requested upon filing the reconciliation entry.

Flagging for reconciliation must be done at the time of filing the import declaration with the customs authorities. The **Reconciliation** flag alerts the customs authorities that a reconciliation entry will be filed at a later stage.

Then, after the goods are received, the reconciliation summary must be filed with the customs authorities in the predefined period of 12 months for USMCA reconciliation and 21 months for any other reconciliation. In SAP GTS, the reconciliation summary is managed using the SCD functionality via the **Create Suppl. Customs Declaration** app. There are two ways to manage the reconciliation summary:

- **Entry-by-entry principle**
  Each declaration is reconciled separately.
- **Aggregate reconciliation**
  Several declarations can be reconciled together, and only an absolute increase is taken into consideration. *Absolute increase* refers to changes or adjustments between line items on a given entry that result in an increase to the entry.

Note

USMCA reconciliations and non-USMCA reconciliations have to be grouped separately.

7.5.2 Customizing

As a prerequisite, the legal regulation for US imports (i.e., the US Automated Commercial Environment [ACE]) must be configured. There are a couple of additional steps that build on standard SAP GTS. Following are the Customizing steps relevant for the setup of the reconciliation process:

- **SAP Global Trade Services, edition for SAP HANA • General Settings • Document Structures • Define Document types for Application Areas**
  The document type for SCD must be set up. The document type USRECN is delivered in standard SAP GTS with the correct action provided (`US_PD_IM_RECON`).
- **SAP Global Trade Services, edition for SAP HANA • General Settings • Document Structures • Define Document types for Application Areas**
  The SCD type for the US reconciliation must be linked to the legal regulation for US import (ACE).

- **SAP Global Trade Services, edition for SAP HANA • Customs Management • Communication Processes • Define Control Settings for Communication Processes**
  The activity sequences for the customs import declarations should be periodic (**B** relevant for periodic **SCD—Confirmed**).
- **SAP Global Trade Services, edition for SAP HANA • Customs Management • Communication Processes • Define Control Settings for Communication Processes**
  The definition of the activity for reconciliation is attached to the import process with a relevant outbound and incoming message as a reply from the customs authorities.

These Customizing activities are additional to the general Customizing required for customs processing. For an overview of the general Customizing activities, refer to Section 7.2.3.

7.6 Electronic Communication

Over the course of this chapter, we've discussed electronic communication on several occasions. Besides electronic communication with the customs authorities, companies can also build an interface that allows you to transfer data to a customs broker. In this case, the customs broker lodges the official declaration with the customs authorities.

To transfer all the required information, SAP GTS uses IDocs. The system provides a standard IDoc with its specific structure; users can use this standard IDoc when a custom-built interface (e.g., one with a broker) is required.

The following sections describe both the IDoc interface and monitoring functionality as well as the standard broker enablement solution.

7.6.1 Standard Communication

An IDoc consists of two important records: the data record, which contains all the data elements captured from the customs declaration, and status records. Status records are predefined statuses that an IDoc has to go through before it's considered successfully transferred.

Navigate to the IDoc overview from the specific customs declaration document by clicking on **Display IDocs** (via the **Change/Display Customs Declaration** tile). Figure 7.119 shows the IDoc details. The **IDoc display** area on the left-hand side shows both the **Data records** and **Status records**. The **Content of Selected Segment** area in the lower-right part of the screen contains the SAP GTS field name and actual content that is entered in that specific field. The upper-right **Short Technical Information** area provides more technical information on the message types, basic IDoc type, partner number and type, and port.

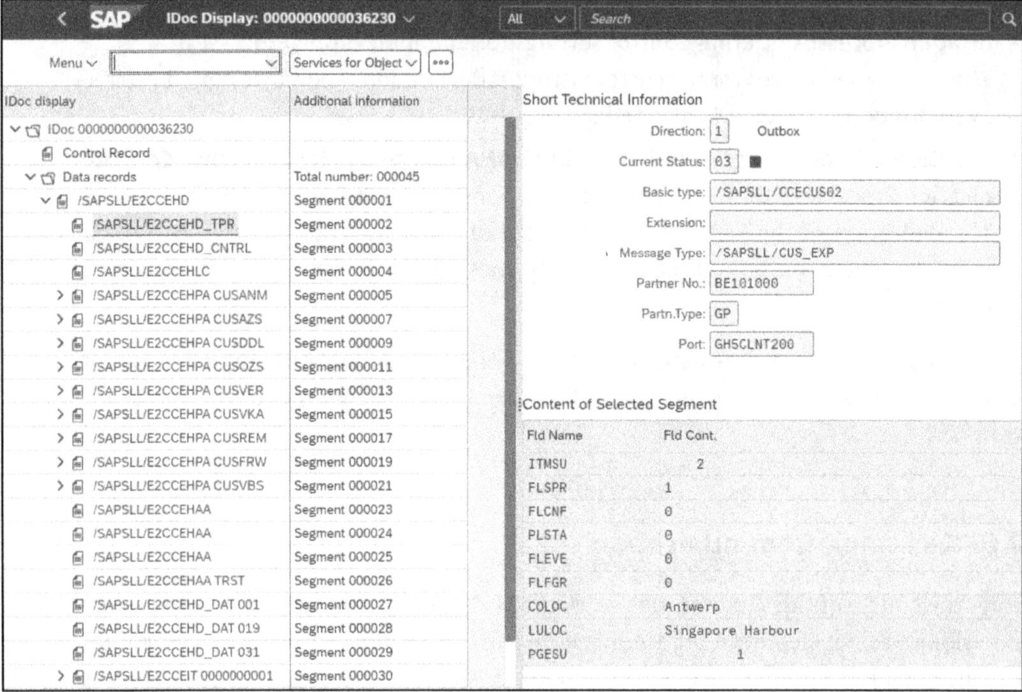

Figure 7.119 Standard IDocs

To understand which information an IDoc contains, it's useful to look into every data record element of an IDoc. Table 7.11 provides an overview of a few IDoc segments that make it easier to retrieve the information from the IDoc structure.

Data Records Element from IDoc	Description
/SAPSLL/E2CCEHD_TPR	Information from the **Transportation** tab at the header level in a customs declaration
/SAPSLL/E2CCEHD_CNTRL	Technical information on an IDoc
/SAPSLL/E2CCEHPA_...	Information on partner details that are used in customs declarations
/SAPSLL/E2CCEHAA	Information on values stored in the **Logistics Costs** tab at the header level in a customs declaration
/SAPSLL/E2CCEHD_DAT	Information related to document dates
/SAPSLL/E2CCEIT	Information on item details in which each item gets its own separate line with more information at a lower level

Table 7.11 IDoc Data Records Structure

7.6.2 Electronic Communication Transactions

There are a few transactions to set up electronic communication and monitor the status of incoming and outgoing IDocs as listed in Table 7.12 and Table 7.13.

When setting up the connection with an external party, the transaction codes shown in Table 7.12 are used.

Transaction Code	Description
WE21	EDI Port Definition
WE20	EDI Partner Profile
/SAPSLL/ECC_POLL_CUS	Assign RFC Connections for Mailbox Connect

Table 7.12 Electronic Communication: Connection Setup Transaction Codes

In addition, there are also a set of transactions that display the IDocs, which is useful for issue-solving activities, as shown in Table 7.13.

Transaction Code	Description
WE05	IDoc List
WE09	Find IDoc
WE07	IDoc Statistics
BD87	Status Monitor
/SAPSLL/ECC_EDI_MES	Assign EDI Message Files

Table 7.13 Electronic Communication: Displaying IDocs Transaction Codes

Transactions Related to Interface Setup

In **EDI Port Definition** (Transaction WE21), users can add the port that is used during the EDI. In **EDI Partner Profile** (Transaction WE20), users can select which message type (refer to Section 7.2.3) can be sent out and received back per profile.

To set up electronic communications with either the customs authorities or a customs broker, it's crucial to add all the required partner profiles here. Figure 7.120 displays the **Partner profiles** screen.

To set up a partner profile, first choose the partner type (**Partn.Type GP**) for the business partner. For example, a user can give a business partner the name of the customs office of departure. Users need to set up this same partner in the master data as a customs office as well. In the customs export declaration, it must state that this particular customs office is the customs office of departure. In a customs import declaration, this particular customs office should be stated as the customs office of destination.

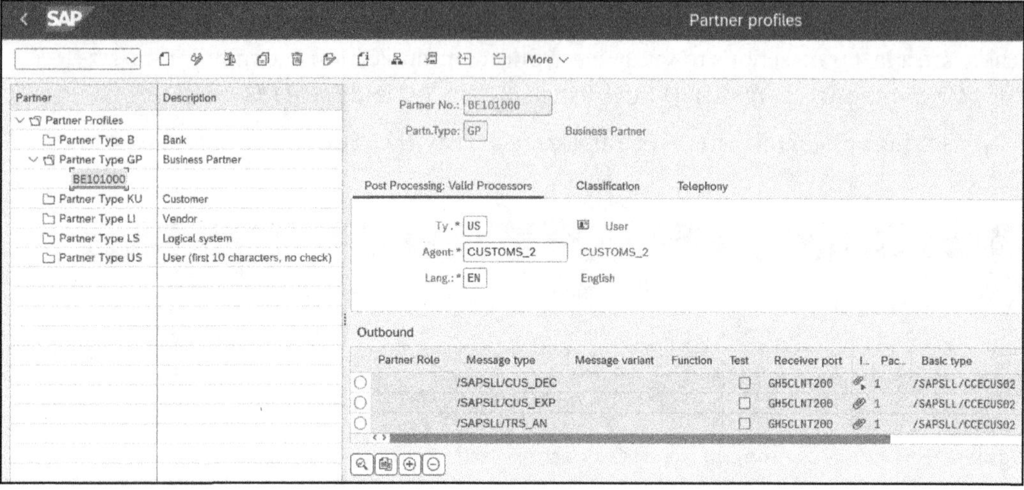

Figure 7.120 Partner Profiles

After the partner profile is created via the administrator menu **SAP Global Trade Services • System Connection • EDI – Communication • EDI – Partner Profiles**, add all the message types from Customizing to either the outbound parameters for outbound messages (M-messages) or the inbound parameters for incoming messages (I-messages), as shown in Figure 7.121.

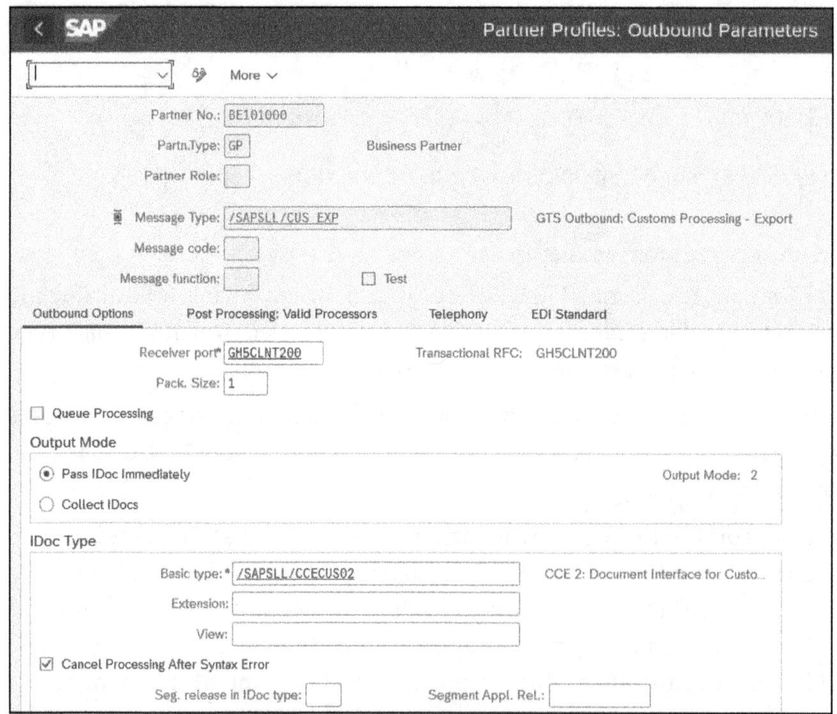

Figure 7.121 Partner Profiles: Outbound Parameters

When adding these message types, the user has two options to output the IDoc. When selecting the **Pass IDoc Immediately** option, all IDocs are immediately transferred to the EDI partner. When selecting **Collect IDocs**, the IDocs are transferred to the customs authorities only after a background job is performed.

Transactions Related to IDoc Monitoring

The **IDoc List** transaction (Transaction WE05) allows you to search for all IDocs created within a specific time slot. However, it's also possible to search for one particular IDoc using the IDoc number in the selection screen. The results of the query are grouped in outbound and inbound IDocs, as displayed in Figure 7.122 For multiple message types, the system also takes this parameter into account to further structure the IDocs. By selecting just a single IDoc, you can see the data records and status records, as indicated earlier in Figure 7.119.

While the **IDoc List** transaction gives a structured overview of the IDocs, **Find IDocs** (Transaction WE09) shows an unstructured, flat list of the selected IDocs. The layout of this list is similar to the monitoring option for customs declarations.

IDoc Statistics (Transaction WE07) gives a structured overview of all the IDocs with an error message. The output of the list is similar to the output displayed in Figure 7.122. By selecting the IDocs, you can see the actual structure and the data that is transferred.

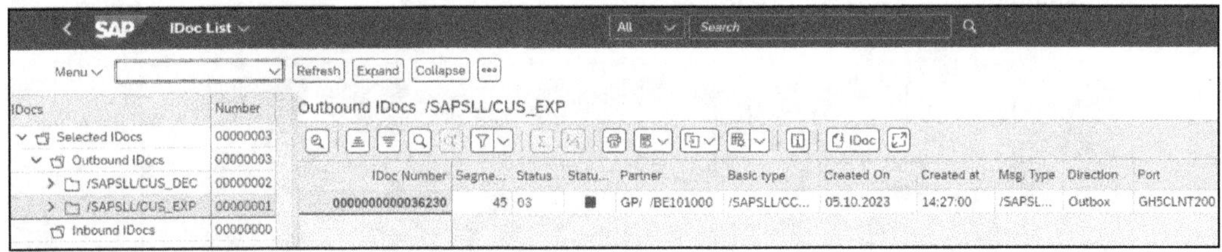

Figure 7.122 IDoc List

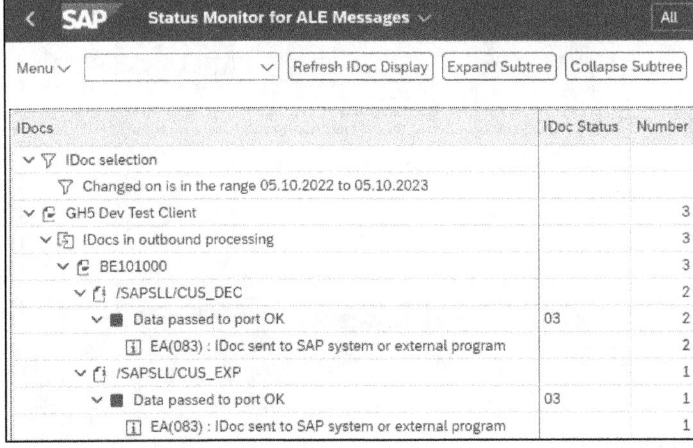

Figure 7.123 IDoc Status Monitoring

At last, in **Status Monitor** (Transaction BD87), you can get additional explanations about whether an IDoc is processed correctly and, in case of an error, the detailed reason of the error. Figure 7.123 shows the results of the query.

The system groups the IDocs based on their processing statuses. If there is an error message, users get the information on the cause of the issue.

7.6.3 Standard Broker Enablement

Broker filing has historically been a customer-specific development to enable the use of printed forms (e.g., broker instruction sheets) or electronic interfaces (e.g., by using IDocs and SAP Process Integration or other third-party middleware).

In SAP GTS, communication with customs brokers is supported for export and import into free circulation before goods receipt. The communication is based on web services, and companies can send or receive the declaration data using an XML format.

For import processing, the solution covers the following two scenarios:

- **Customs broker-centric scenario**
  As shown in Figure 7.124, the customs broker initiates the import declaration and sends a copy (replicate of the customs declaration) to the importer after customs clearance. The importer provides product master data such as classification and weights.

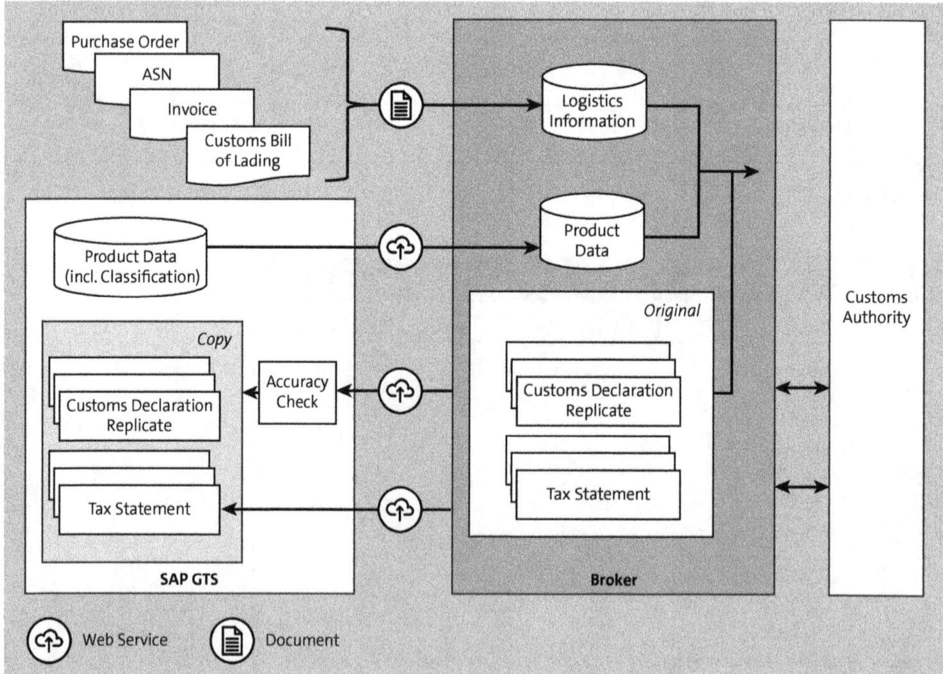

Figure 7.124 Customs Broker-Centric Scenario

- **Importer-centric scenario**
  As shown in Figure 7.125, the importer initiates the process and sends the original import declaration request to the customs broker, who receives a copy. The customs broker requests customs clearance and sends updates about processing status, customs registrations number, and so on back to the importer. The customs broker may not change the structure of the import declaration request; that is, adding/deleting line items isn't permitted.

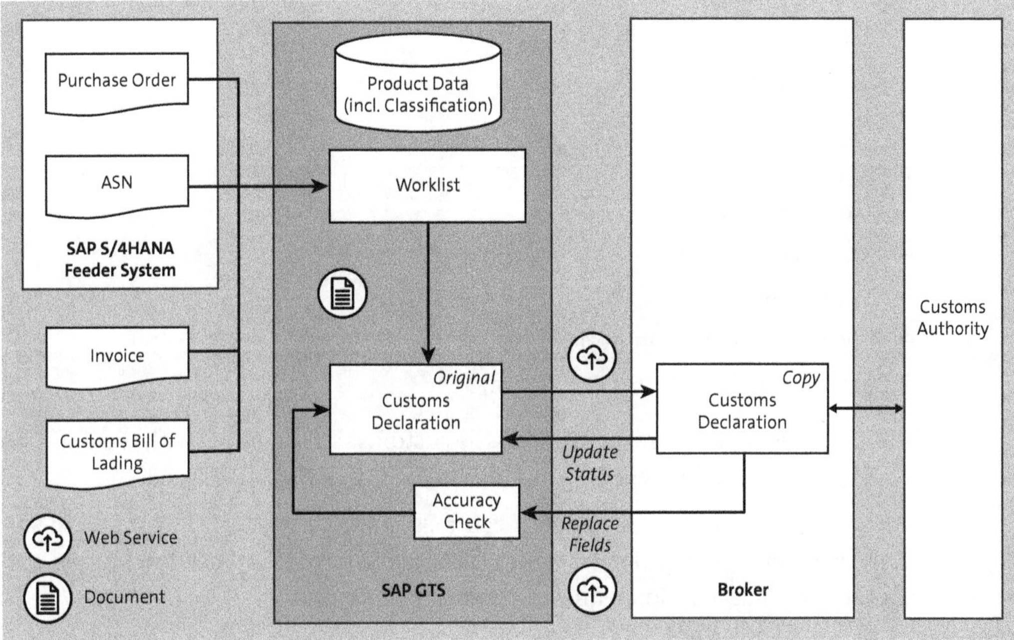

Figure 7.125 Importer-Centric Scenario

In an importer-centric scenario, if the customs broker has changed data in the customs declaration, the following data in the customs declaration is transferred from the customs broker's reply:

- Means of transport at the border, including the following:
  - Mode of transport
  - Type
  - ID
  - Country
  - Number
  - Description
  - Mode of transport

- Domestic means of transport, including the following:
  - Mode of transport
  - ID
  - Country
  - Starting point
  - Destination point
  - Place of loading
  - Place of unloading
  - Point of arrival
  - Customs offices
  - Incoterm
  - Date of declaration
  - Date of acceptance
  - Date of release
  - Processing status of authorities
  - Item number of customs declaration

In addition, the customs broker can send the customs tax statement with the duties to be paid in both scenarios, resulting in a tax statement on the SAP GTS side.

For the export processing, the exporter-centric scenario is supported by standard SAP GTS. In this scenario, depicted in Figure 7.126, the exporter initiates the process and sends the export declaration request to the customs broker.

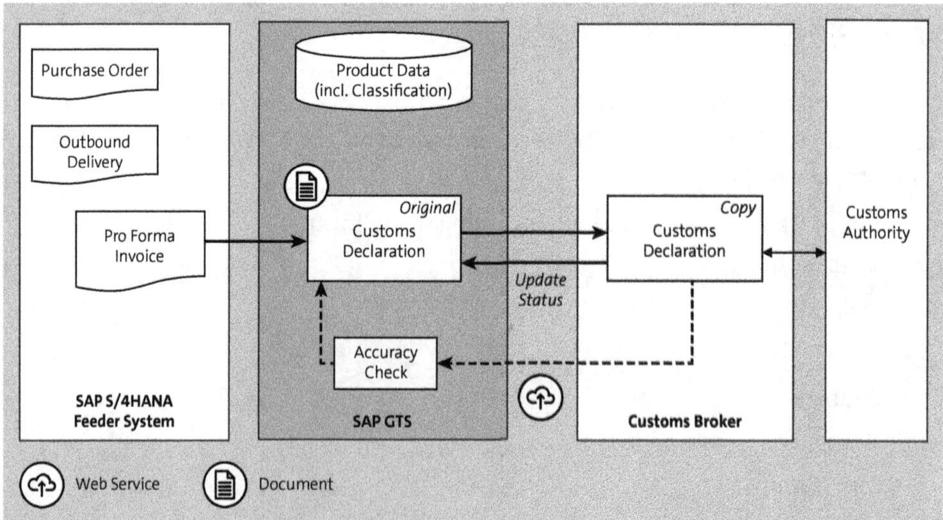

Figure 7.126 Exporter-Centric Scenario

The key feature of the broker integration solution is that it allows integrity checks to be run based on the responses from the customs broker, so the importer and exporter can validate and monitor the content of the responses. In addition, the importer and exporter can include performance ratings based on the response from the customs broker, which can be monitored and analyzed to also verify existing service-level agreements (SLAs). Moreover, the customs broker can send the customs tax statement with the duties to be paid in case of import scenarios.

The advantages of the broker enablement are that it enables broker harmonization and follows the SAP standard instead of modifying EDI. In addition, the implementation can be considered easier because you can rely on the following:

- Well-described interfaces
- Complete and documented data model
- Industry standard format (XML/IDoc)

Process Example

This example showcases the possibilities of the broker enablement setup and focuses on the following main steps of the broker filing process:

1. The standard broker message is sent.
2. The broker updates missing information.
3. The inbound message from the broker is processed, and the declaration is updated in SAP GTS.

In Figure 7.127, we see the **Communication** tab of the customs declaration. The **MB001** message has been processed and sent to the broker.

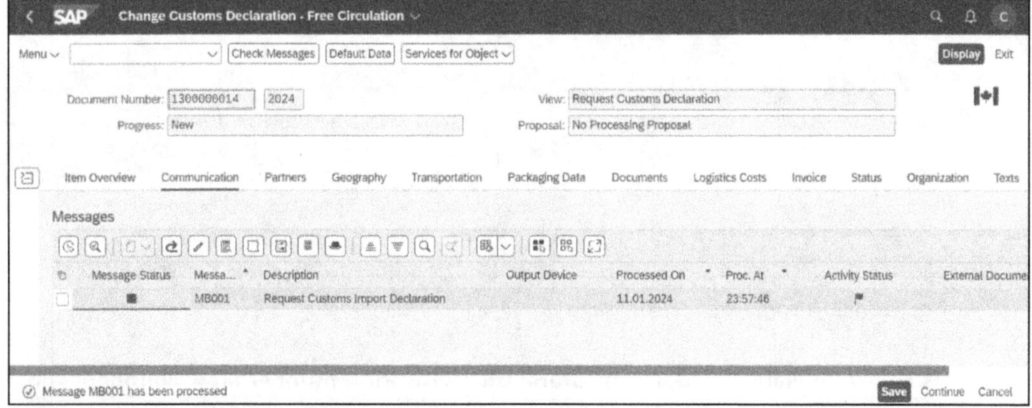

Figure 7.127 Initial Import Declaration in SAP GTS

Figure 7.128 shows that required information is missing from the declaration.

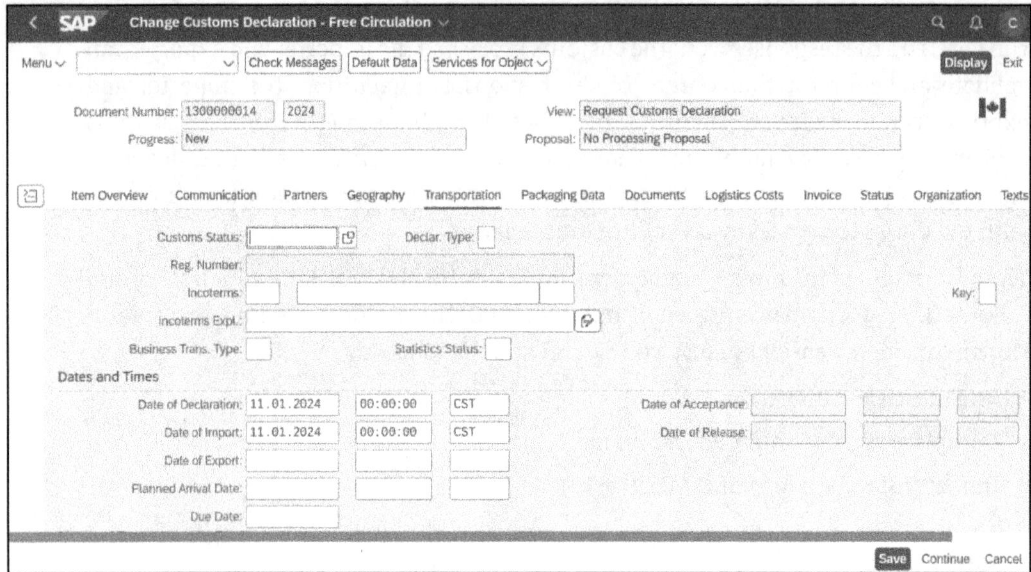

Figure 7.128 Example of Missing Incoterms in a Customs Declaration

The broker updates the declaration in their system. The changes are returned to SAP GTS as an inbound **IB001** message, as shown in Figure 7.129.

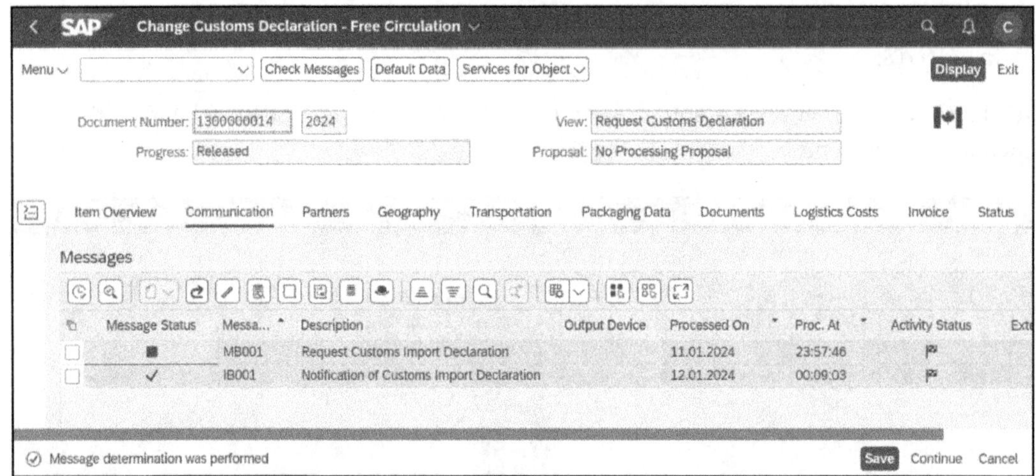

Figure 7.129 Incoming IB001 Message

As shown in Figure 7.130, the **Incoterms**, a registration number (**Reg. Number**), and a customs broker transaction number (**Cust. Broker TransNo**) were added to the declaration by the broker.

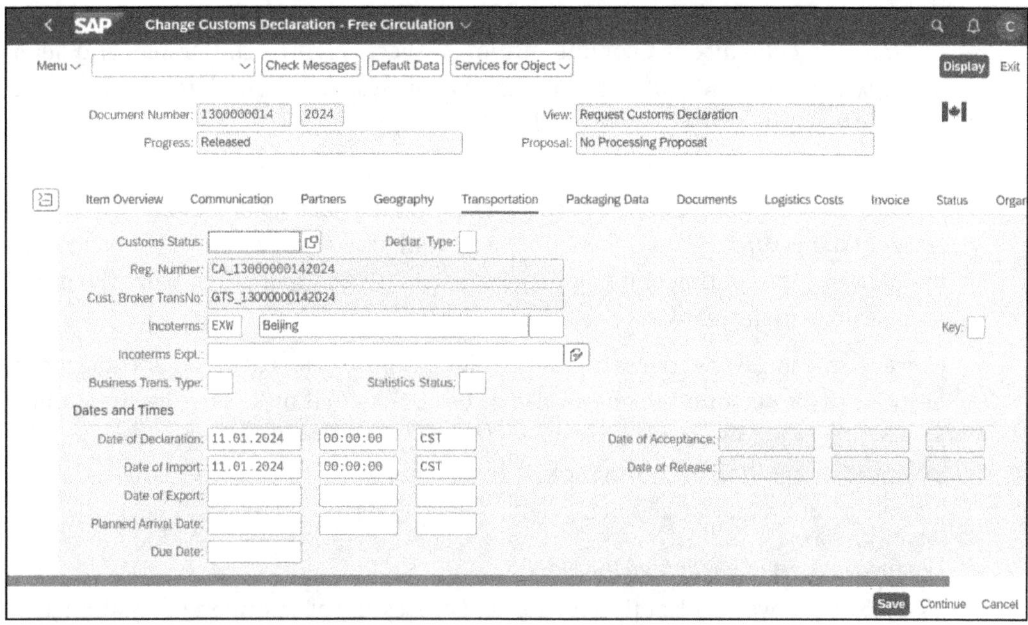

Figure 7.130 Updated Declaration with Incoterms, Registration Number, and Customs Broker Transaction Number

System Setup

As mentioned earlier, the interaction with the broker in a customs broker enablement (CBE) scenario is done via web services, which are predefined in the SAP GTS system's Enterprise Services Repository (ESR).

All broker scenarios feature bidirectional communication, as shown in Figure 7.131, meaning that data flows from SAP GTS to the broker and from the broker back to SAP GTS.

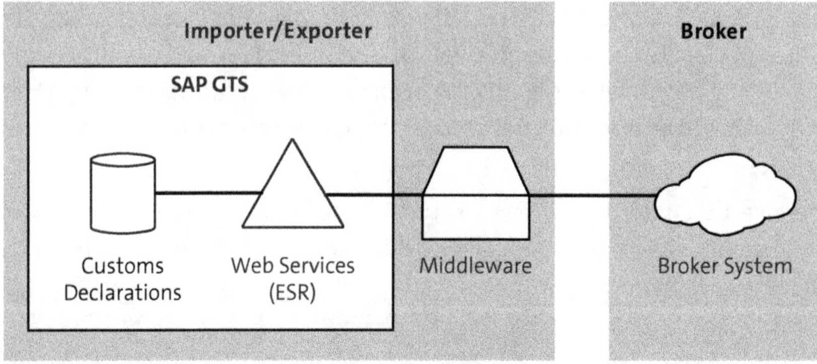

Figure 7.131 Broker Communication Data Flowing from Left to Right through Outbound Services and from Right to Left through Inbound Services

Because we consider all data flows from the perspective of the SAP GTS system, we call this *outbound* and *inbound* communication, respectively. The chosen scenario determines whether the SAP GTS system or the broker system is considered the main system that will initiate the customs declaration.

Because the outbound services have to be connected to our customs broker, all of those services have to be mapped to a particular destination. Instead of mapping all flows to target systems directly from the SAP GTS system, users can set up a connection to a middleware. The routing and mapping between SAP GTS and the broker system will then be done in the middleware platform.

However, routing web services through a middleware remains optional, and direct connections to the customs broker can also be used. Note that these services offer limited support for scenarios where multiple endpoints or brokers have to be addressed. For advanced integration or routing scenarios, such as publish-subscribe patterns, a middleware solution may be adopted.

Example Importer-Centric Scenario Flow

As an example, we'll look at the creation and processing of an import declaration in an importer-centric scenario. The communication with the broker is broken down into several steps, as you can see in Figure 7.132. From a system point of view, the broker's internal processes and communication with the customs authority is considered a black box: how the broker handles declarations internally is of less importance as long as the relevant information is sent and received.

When we create a customs declaration, the MB001 message created in SAP GTS will trigger an `ImportDeclarationRequest_Out` service call to the broker system. The declaration will then be processed by the broker, and there will be a call out to the `ImportDeclarationNotification_In` service, which SAP GTS will register as an IB001 message. That service can actually be called by the broker multiple times as they receive new information. The first call must provide certain key information, such as customs authority identifiers and the date of declaration. Additional calls could update the customs authority processing status or change transport information. The final call would usually set the date of release. Giving the broker the ability to perform partial notifications or updates is something that should be discussed during the integration setup because it can complicate the process.

A separate service called `ImportTaxAssessmentNotification_In` exists to receive tax assessment information as an IB002 message, which will fill in the tax statement in SAP GTS with duty information.

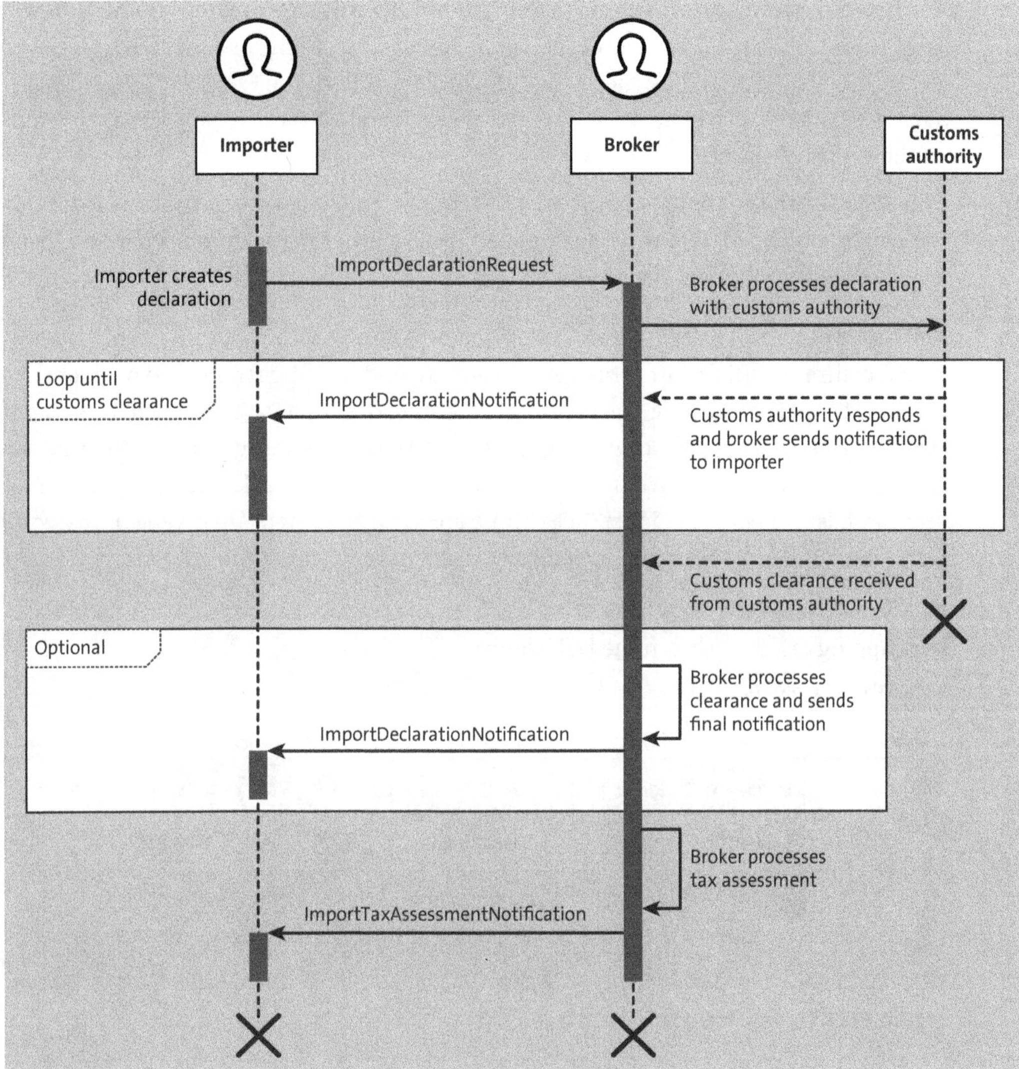

Figure 7.132 Sample UML Sequence Diagram Showing Import Declaration Creation in an Importer-Centric Scenario

Example Exporter-Centric Scenario Flow

In this example, we'll look at the creation and processing of an export declaration in an exporter-centric scenario. The communication with the broker is broken down into several steps. From a system point of view, the broker's internal processes and communication with the customs authority is considered a black box: how the broker handles declarations internally is of less importance as long as the relevant information is sent and received.

When we create a customs declaration, the MB010 message created in SAP GTS will trigger an `ExportDeclarationRequest_Out` service call to the broker system. Once the declaration is processed by the broker, the broker will call the `ExportDeclarationNotification_In` service, which SAP GTS will register as an IB010 message providing details of the export clearance.

Finally, when the goods are exported, the broker will call `ExportDeclarationExitNotification_In`, which SAP GTS will register as an IB015 and confirms the shipment has been exported and completed the sequence.

Enhancement

When communicating with a broker, a user may find that the standard data exchange is insufficient. For example, the broker may require data that isn't included in standard web services, or users set up custom fields in SAP GTS containing information that the broker needs to provide. We note the enhancement for the broker communication to support custom fields in Figure 7.133. There are additional steps involved compared to a standard SAP enhancement. Specifically, users have to adjust the following:

- Web service definition
- Mapping of data to/from the web service
- SAP GTS objects

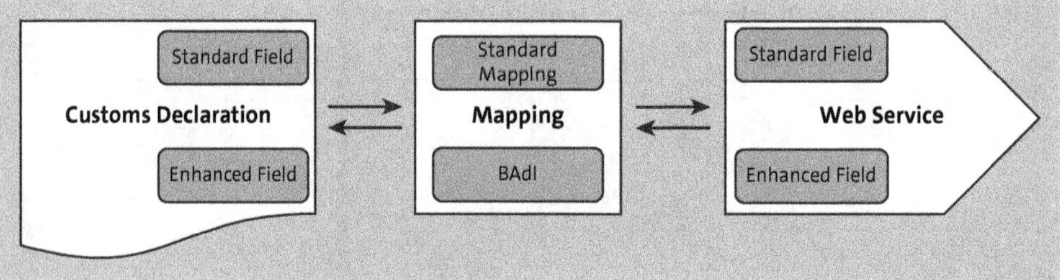

Figure 7.133 Custom Field Enhancement

The changes to standard objects are handled using the typical enhancement framework. Custom fields can be added to SAP GTS objects, such as CUHD, CUIT, or LEGCUS, using the existing append structures. If adding custom fields to outbound web services, users usually already have those defined in the SAP GTS objects. Nonstandard data from inbound services might be mapped to a standard location in SAP GTS or in an enhanced field.

The web service has to be enhanced accordingly to align with the expected data, typically referred to as data type enhancements. Once applied, the proxy in SAP GTS has to be regenerated.

Finally, the mapping between SAP GTS and the web service has to be adjusted to account for new fields, whether that's filling a field in an outbound service or processing a new field in an inbound service. The following BAdIs are available for this:

- /SAPSLL/CU_SE_IMPRTDECL_NOTIF
- /SAPSLL/CU_SE_IMPRTDECL_REQ
- /SAPSLL/CU_SE_EXPRTDECL_NOTIF
- /SAPSLL/CU_SE_EXPRTDECL_REQ
- /SAPSLL/CU_SE_EXPRTEXIT_NOTIF
- /SAPSLL/CU_SE_TAXASSMT_NOTIF
- /SAPSLL/MD_SE_BRKRPRODRPL_REQ (product replication)

IDoc processing for Customs Management is mainly used in SAP GTS supported countries. If SAP GTS doesn't support specific countries, the CBE solution can be implemented.

7.7 Countries Supported by SAP GTS, Edition for SAP HANA

The import and export processes of countries in Table 7.14 are supported by SAP GTS, edition for SAP HANA.

Country	Legal Regulations
Austria	CUSAT (e-zoll connector required [ProData, AT])
Australia	CUSAU (only supported for export)
Belgium	CUSBE
China	CUSCN (paper-based filing or declaration management)
Denmark	CUSDK
France	CUSFR
Germany	ATLAS
Netherlands	CUSNL
New Zealand	CUSNZ (only supported for export)
Spain	CUSES
Sweden	CUSSE
Switzerland	CUSCH
United Kingdom	CUSGB
United States	ACE (includes electronic filing for FTZ and importer security filing [ISF])

Table 7.14 Countries Supported by SAP GTS for Import and Export

In addition to the standard import and export processes, SAP GTS, edition for SAP HANA also supports the transit functionality for the following countries. Table 7.15 also includes their corresponding legal regulations.

Country	Legal Regulations
Austria	TRSAT
Belgium	TRSBE
Denmark	TRSDK
France	TRSFR
Germany	TRSDE
Netherlands	TRSNL
Slovenia	TRSSI (not under standard SAP maintenance)
Spain	TRSES
Sweden	TRSSE
Switzerland	TRSCH
United Kingdom	TRSGB
United States	TRSUS

Table 7.15 Countries Supported by SAP GTS for Transit

SAP GTS, edition for SAP HANA also supports the EMCS functionality for the following countries. Table 7.16 also includes their corresponding legal regulations.

Country	Legal Regulations
Belgium	EMCBE
Germany	EMCDE
Netherlands	EMCNL
United Kingdom	EMCGB

Table 7.16 Countries Supported by SAP GTS for EMCS

For each of these countries, SAP GTS comes with a predefined communication approach.

If the country for which you want to set up the SAP GTS Customs Management capabilities isn't in this list, you need to perform the configuration as explained in Section 7.2.3.

7.8 Summary

Throughout the chapter, we provided an overview of the main concepts and business and system processes of the Customs Management functionality in SAP GTS.

For customs documents (Section 7.1), we described all the basic information as well as the details of the different business processes. We focused on the available SAP Fiori apps and monitoring functionalities.

Section 7.2 provided an in-depth overview of the standard import and export processes. Section 7.2.3 provided all the frequently used and important Customizing settings for Customs Management.

Section 7.3 on special customs procedures provided a high-level overview of the functionalities of the following customs procedures: bonded warehouse (BWH; also known as customs warehouse), inward processing relief (IPR), outward processing relief (OPR), production under customs control (PUCC), foreign trade zones (FTZs), and processing trade (PTR).

We also described the excise duty monitoring in Section 7.4. In Section 7.5 we covered the US reconciliation process.

Section 7.6 explained how the electronic communication with the customs authorities works and how to monitor incoming and outgoing IDocs. We described what an IDoc looks like and discussed some transactions to monitor IDocs closely. This section also showed that with the standard broker integration solution, companies can set up communication with the broker using standard SAP GTS.

To conclude this Customs Management chapter, we listed all the countries that are currently supported by SAP GTS, edition for SAP HANA and its Customs Management functionality, along with their legal regulations.

In the next chapter, we'll discuss the final major SAP GTS module: Preference Management.

Chapter 8
Preference Management

To encourage global trade, countries establish free trade agreements (FTAs) to provide importers with the benefits of reduced or eliminated duties. The Preference Management functionality in SAP GTS supports companies in calculating and certifying goods under various FTAs.

The Preference Management functionality in SAP GTS is used to request supplier declarations (i.e., solicitation), calculate the preferential eligibility of products under an FTA (i.e., qualification analysis), and issue proof of preferential origin in the form of blanket or transactional statements (i.e., certification).

We'll explain the Preference Management functionality in this chapter, starting with the basics of how preferential trade agreements work (Section 8.1). We'll then walk through the business processes in Section 8.2, the system process flow in Section 8.3, and Customizing instructions in Section 8.4. Finally, we'll discuss management of preference data for product identifiers in Section 8.5 and cover relevant master data in Section 8.6.

8.1 Preferential Trade Agreements

As mentioned in Chapter 1, the most favored nation (MFN) principle established by the World Trade Organization (WTO) sets up a framework whereby equal treatment among WTO members is established, which, in practical terms, means that MFN countries apply consistent duties to products imported from other MFN countries. However, notable exceptions to MFN treatment are the usage of preferential trade agreements (commonly known as FTAs) whereby countries grant reduced or eliminated duties on certain products provided that requisite conditions are met.

Companies use different types of preferential trade agreements, which may vary based on the number of beneficiary countries. There are generally three types of agreements:

- **Multilateral agreements**
  Provides benefits across more than two countries that are party to the FTA. An example of a multilateral agreement is the United States-Mexico-Canada Agreement (USMCA) in North America.

- **Bilateral agreements**
  Provides benefits for a maximum of two countries that are party to the FTA. An example of a bilateral agreement is the United Kingdom's agreement with Japan (UK-FTA).
- **Unilateral agreements**
  Unilateral agreements typically focus on providing benefits for only a single party to the agreement. These are commonly referred to as a Generalized System of Preferences (GSP) program (for more information, refer to Chapter 1, Section 1.1.3) whereby a less developed economy/nation gains reduced duty treatment upon importation into a more developed economy/nation.

As shown in Figure 8.1, we can identify several multilateral agreements and trading blocs around the globe that have a range of countries involved. For instance, the European Union (EU) includes many countries and only negotiates as a bloc, as compared to other regional examples such as the Southern Common Market (MERCOSUR), which has fewer countries and can negotiate as individual countries or as a bloc.

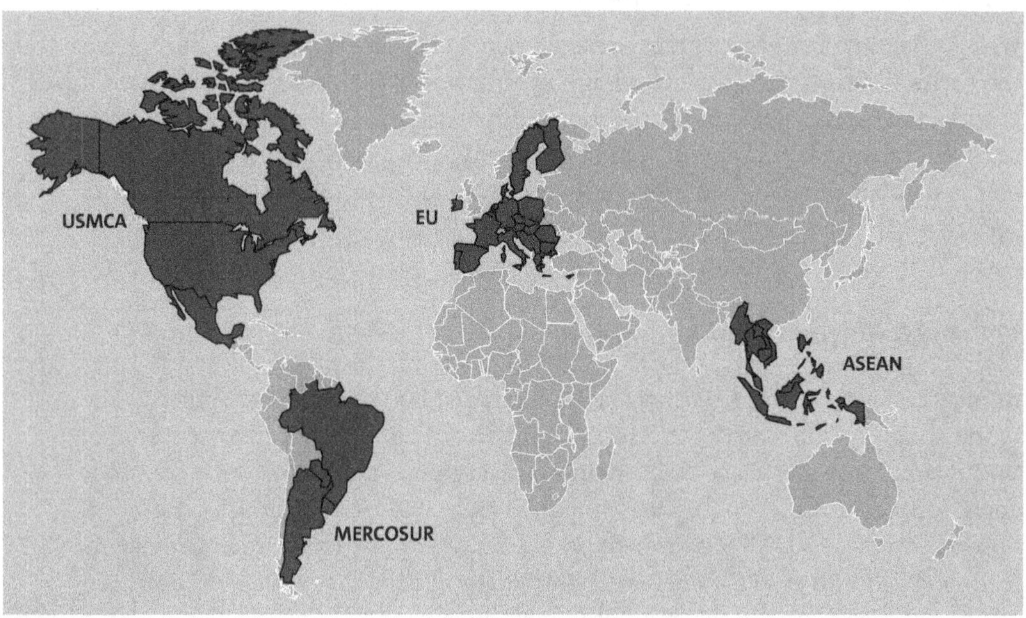

Figure 8.1 Examples of Multilateral Agreements and Trading Blocs

Table 8.1 displays the countries included in these trade blocs and multilateral agreements.

United States-Mexico-Canada Agreement (USMCA)	Association of Southeast Asian Nations (ASEAN)	Southern Common Market (MERCOSUR)	European Union (EU)	
▪ Canada ▪ Mexico ▪ United States	▪ Brunei ▪ Cambodia ▪ Indonesia ▪ Laos ▪ Malaysia ▪ Myanmar ▪ Philippines ▪ Singapore ▪ Thailand ▪ Vietnam	▪ Argentina (founding member) ▪ Brazil (founding member) ▪ Paraguay (founding member) ▪ Uruguay (founding member) ▪ Bolivia ▪ Chile ▪ Colombia ▪ Ecuador ▪ Guyana ▪ Peru ▪ Surinam ▪ Associated states	▪ Austria ▪ Belgium ▪ Bulgaria ▪ Cyprus ▪ Croatia ▪ Czech Republic ▪ Denmark ▪ Estonia ▪ Finland ▪ France ▪ Germany ▪ Greece ▪ Hungary ▪ Ireland	▪ Italy ▪ Lithuania ▪ Luxembourg ▪ Latvia ▪ Malta ▪ Netherlands ▪ Poland ▪ Portugal ▪ Romania ▪ Slovakia ▪ Slovenia ▪ Spain ▪ Sweden

Table 8.1 Member States of Major Regional Trade Blocs (September 2023)

Each country may initiate the process to establish preferential trade agreements with other countries. There are various examples of preferential trade agreements:

- **North America**
  The USMCA is in place between the US, Canada, and Mexico. Many of these countries also have bilateral FTAs with other partners, such as South Korea (e.g., CA-KR FTA, Korea-US FTA).
- **Latin America**
  Several trade agreements provide preferential duty treatment for intra-zone movement of goods, with the largest trading bloc being MERCOSUR, which includes, among others, Argentina, Brazil, Paraguay, and Uruguay. Other emerging trading blocs include the Andean Community, the Caribbean Community (CARICOM), and the Pacific Alliance.
- **Europe**
  The EU has various agreements in place both within Europe as well as with other regions and countries around the world. For example, the EU has an agreement with

Southern African Development Community (SADC) covering countries such as Botswana, Lesotho, Mozambique, Namibia, South Africa, and Eswatini. Like other countries, the EU has an extensive list of FTAs in place with third countries all over the world, alongside a comprehensive negotiation agenda for future agreements.

- **Asia Pacific**
  This region has various agreements that often overlap and provide varying degrees of duty relief to signatory countries. For example, the region signed the world's largest FTA known as the Regional Comprehensive Economic Partnership (RCEP) in 2020 that covers approximately 30% of the global gross domestic product (GDP), is made up of 15 countries in the region, and covers nations making up a third of the world's population.[1]
- **Cross-regional**
  The Comprehensive and Progressive Agreement for Trans-Pacific Partnership (CPTPP) is a FTA in force between countries in the Indo-Pacific region, including Australia, Canada, Brunei, Chile, Japan, Malaysia, Mexico, New Zealand, Peru, Singapore, Vietnam, and the United Kingdom (subject to ratification).

When companies ship products between countries under an FTA and these products fulfill certain conditions, these companies and their customers can take advantage of preferential treatment. Such treatment allows companies to pay a reduced or zero duty rate when importing eligible products, which is determined by the preferential status of the product in question as well as the preferential duty rate allowed for under the agreement. Preferential rules of origin typically consider the extent that manufacturing processes in the FTA countries change the product, the location of manufacturing, the sourcing of components, and the shipment routing/details from the country of export or production to the country of importation. It's important to note that the preferential country of origin differs from the economic or non-preferential country of origin because the preferential country of origin isn't always the country where the last substantial production step occurred. In addition, preferential country of origin is determined based on separate criteria that are established among the FTA beneficiary countries rather than the WTO Agreement on Rules of Origin (ROO Agreement), which established harmonization and general principles regarding non-preferential origin determination.

The Preference Management functionality in SAP GTS allows companies to manage the collection of supplier product information, perform qualification analysis, and certify goods under particular agreements with the necessary integrations and audit trails to meet the rules set forth under individual agreements. SAP GTS determines the preferential status of products shipped to customers (including parties both related and

1 "A New Centre of Gravity: The Regional Comprehensive Economic Partnership and Its Trade Effects." UNCTAD, December 15, 2021. *https://unctad.org/publication/new-centre-gravity-regional-comprehensive-economic-partnership-and-its-trade-effects.*

unrelated to the exporter or producer) located in a member country of a specific FTA and provides those customers with the required documentation to allow importation of eligible products at a reduced or zero import duty rate.

Customs authorities typically require a certificate or statement of origin as the legal documentation to qualify products for reduced or eliminated duties upon importation of goods with preferential origin. In most cases, this proof of origin can be issued only by specifically approved persons, institutions, or governmental organizations. Specific requirements around certification method and permitted certifying parties are established in the language of each agreement. Issuing the proof of origin requires extensive administrative efforts, which can be automated through Preference Management in SAP GTS.

Let's discuss a hypothetical business scenario that demonstrates the complex concepts, functionalities, and business and system processes of Preference Management. Later, we'll discuss how the functionality of SAP GTS supports the hypothetical business scenario via Customizing, master data, and product identifiers preference data.

Gamma's Global Trading Requirements

Gamma manufactures high-end leather shoes for men. In its US manufacturing facility in North Carolina, it converts the lamb leather into stylish designer shoes. To manufacture the finished product, the company reworks and assembles several materials procured from suppliers all over the world. As shown in Figure 8.2, the rubber soles are procured in Vietnam, and shoelaces come from suppliers in Peru and Mexico. The lamb leather is purchased from a local US supplier, and the wool is sourced from Canada.

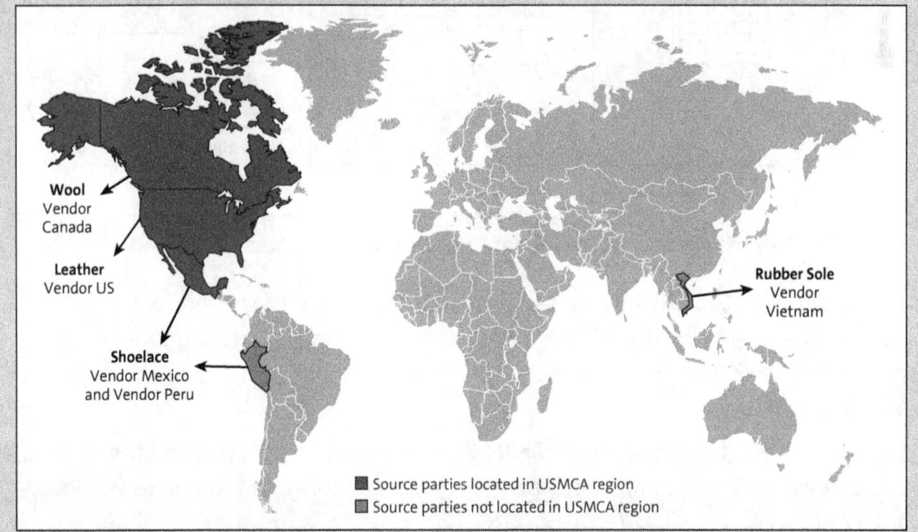

Figure 8.2 Gamma's Source Countries

Because Gamma is located in the US and some of the components are procured locally in the US or from Mexico or Canada, Gamma believes it may benefit from the USMCA agreement.

If Gamma could issue certificates of origin to its customers stating that their products are of preferential origin in the USMCA agreement, those customers could avoid paying import duties on the imported goods. This would significantly increase Gamma's attractiveness in the market as a supplier of high-end shoes as well as potentially allowing the company to increase its margins or decreasing the cost of its goods (i.e., as duties would no longer be part of Gamma's cost or embedded within its customers purchase price).

Gamma's main consumer market is Canada, a USMCA member country, but a minor share of its products is destined for the European market, more specifically for Germany, as shown in Figure 8.3.

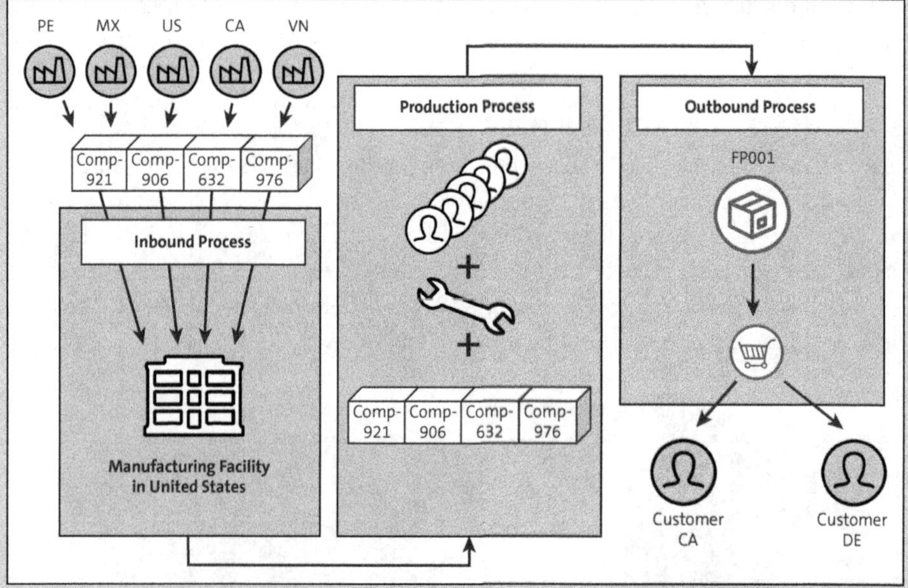

Figure 8.3 Gamma's Inbound, Production, and Outbound Process

To manufacture its high-end men's shoes, Gamma procures the leather, wool, and rubber soles to create unfinished shoes. Next, the second manufacturing process entails adding the procured shoelaces to the unfinished shoe, completing the high-end leather shoe.

Figure 8.4 shows the hierarchy of the final product, its components, and the material prices at which Gamma procures the components sourced from the various vendors.

The selling price is indicated for the finished product.

The example also shows the Harmonized Tariff Schedule (HTS) code for each component; you use these tariff codes to identify which rules of origin apply in a certain FTA. To assess whether the finished product qualifies for preferential treatment, a certifier needs to determine whether the components and the finished product fulfill the criteria detailed in the corresponding FTA's rules of origin.

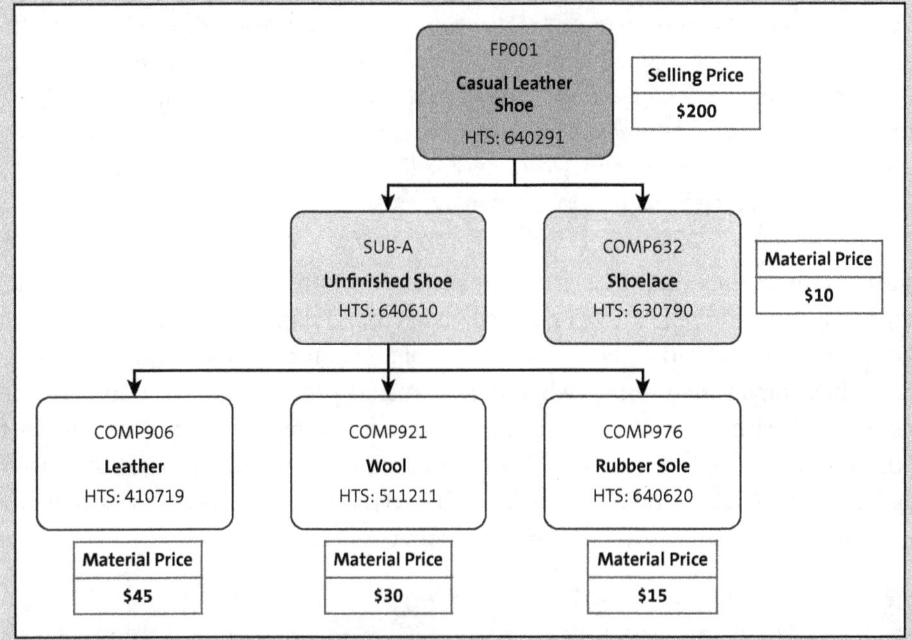

Figure 8.4 Gamma's Bill of Materials (BOM)

8.2 Business Process

The main objective for the company is to determine whether the sold goods are eligible for preferential treatment under a specific trade agreement, store requisite FTA qualification analysis backup (i.e., in case of customs audits or inquiries), and provide customers with the required legal documentation, if applicable.

Note

Companies can also certify goods under preferential agreements for domestic shipments, with the goal of providing domestic customers with proof of origin that those customers can then use for their own preference determination. However, for the objective of this section, we'll focus on the international shipment process.

Prior to determining whether a company can ship goods with preferential status, the company needs to acquire the necessary information about the components used in the finished product and the manufacturing process.

Note

In this chapter, the term *finished product* is used from a business point of view, so it doesn't necessarily imply that the product won't undergo additional manufacturing steps at a later step in the supply chain. Here, the product is considered "finished" for the company selling the product. In the overall supply chain, it might be a semifinished product (e.g., a tire company may consider the tires to be their finished product, but a car company may consider that to be a component).

Figure 8.5 shows the Preference Management business process. In the sourcing phase, a company buys components from third countries and requests supplier declarations to obtain the preferential status and origin of the components. In the production phase, the company determines whether the finished product may be exported under preferential treatment. In the selling phase, the company exports its finished products to third countries, and relevant certificates or statements of origin may be issued when applicable. Depending on existing FTAs and specific rules, third-party customers or intercompany-related entities may benefit from preferential duty rates.

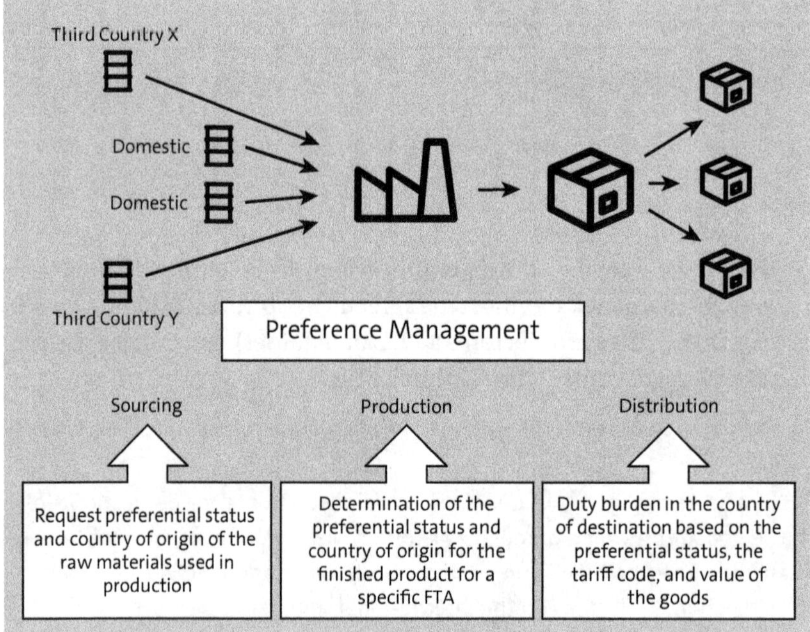

Figure 8.5 Business Processes Relevant for Preference Management

Gamma's Preference Management Process

To manufacture its casual leather shoe, Gamma procures components locally and from third-country suppliers. To calculate the preferential status of its finished product, the leather shoe, Gamma requests supplier declarations from the following partners:

- The supplier in the US for the leather
- The supplier in Canada for the wool
- The suppliers in Peru and Mexico for the shoelaces
- The supplier in Vietnam for the rubber sole

It's up to Gamma's management to decide which suppliers to request supplier declarations from. The company could decide not to request the preferential status of the rubber soles procured in Vietnam because Vietnam isn't a party to the USMCA, thus making it very unlikely that this product would ever have a USMCA preferential status. However, for this business case, Gamma requests supplier declarations from all of its suppliers.

In the production phase, Gamma performs the preference calculation for its finished product. More specifically, the tariff code assigned to the leather shoe informs Gamma about the criteria that need to be fulfilled to obtain preferential status. In this case, the rules of origin of USMCA provide the relevant criteria for each tariff code. Gamma is then able to use the origin-related information stated for the components in the supplier declarations requested to the various suppliers to calculate the preferential origin of its finished product.

If the shoes are of USMCA preferential origin, Gamma can issue proofs of origin to its customers in Canada because they are in a USMCA member country. The Canadian customers can import the shoes without the burden of paying full tariffs and duties. Because there is currently no FTA in place between the US and the EU, no proofs of origin can be issued to Gamma's German customers.

Now, we'll take a closer look at these three main steps in the following sections.

8.2.1 Obtain Preferential Status and Origin from Suppliers

The first step consists of a company requesting proof of origin from their suppliers for externally procured components that they use in the manufacturing process to produce the finished product. Note that companies may not want or need to request the origin-related information from all suppliers. As such, it's a business decision to identify which suppliers to request the proof of origin from. For instance, a company looking to leverage the USMCA FTA may choose to request preferential origin information from only its suppliers located in the US, Canada, and Mexico because suppliers in other locations would likely return negative responses. Likewise, a company looking to leverage the European FTA would only request it from its European-based suppliers.

Suppliers usually provide this preference-related information through specific forms certifying goods for a period of time or for a particular transaction. Forms which certify goods for a period of time are often called long-term supplier declarations (LTSDs) or blanket certificates of origin. Transactional certification is typically printed on existing commercial documentation in the form of a preferential statement of origin on an invoice or packing list. Note that LTSDs and preferential statements of origin on the invoice are more commonly used in European FTAs, whereas blanket certificates of origin are more commonly used in North American FTAs. In some jurisdictions (e.g., ASEAN countries), companies may be required to submit the analysis and backup information to a local authority (either customs or other responsible authorities, such as a chamber of commerce) directly to either have the certification documentation approved prior to certifying goods toward a customer or for the authority to issue the certification themselves. Note that self-certification options (i.e., as exporter or producer) are being introduced gradually in more FTAs globally.

Companies that are regularly working with certain suppliers may find it inefficient to request a supplier declaration on a transactional basis. As such, supplier declarations are often requested/issued for a longer term (e.g., one or two years) depending on the FTA. LTSDs state that the supplier's products will possess identical preferential status and origin throughout the entire validity period and often used for the customer's future preference determination analysis. If the preferential status of a product changes, it's the supplier's responsibility to inform its customers of the changes and provide them with updated and accurate documentation.

Companies may request the preferential status of a product for multiple FTAs if the sourced product or items produced with that product are sent to customers in multiple jurisdictions. If this is the case, the company can request the supplier provide a consolidated LTSD with all the requested FTAs or separate LTSDs for each requested FTA. The ability to consolidate LTSDs across multiple agreements is dependent on whether such an approach is acceptable across the different FTAs requested.

8.2.2 Determine Preferential Status and Origin of Finished Products

The second step consists of determining the preferential status of the company's finished product that can later be provided to its customers. There are two major scenarios:

- **Buy-sell scenario**
  In the case of a buy-sell scenario (e.g., where a product is externally procured and then resold without further manufacturing), the preferential status of the externally procured product doesn't change because the product isn't further processed or changed. In such scenarios, the product may keep its preferential status for future/end customers depending on how the goods are received (e.g., direct shipment,

retained under customs control) and what activities are done to the good after receipt (e.g., repacking, dilution). Assuming the good meets FTA-specific rules for maintaining preferential origin, the preferential status is determined by requesting documentation from the supplier as the basis for preparing one's own certificate of origin (see the prior step for more on obtaining preferential status and origin from suppliers).

- **Manufacturing scenario**
  A more common scenario is that companies often further process the different components they source from suppliers to manufacture new finished products. In such scenarios, the determination of the preferential status for the finished product requires the identification of the applicable rule of origin for a given agreement and the detailed evaluation of the components used within the finished good.

Before explaining the determination process in more detail, we'll define some key terminology and concepts. First, we'll go through some prerequisites to run the preference determination, namely the bill of materials (BOM) and the rules of origin. Second, we'll explain some of the concepts at play during the determination process with the preference determination methods and the BOM calculation methods. Last, we'll go through the definitions of the outcomes of the preference determination with the preference determination status, the preference statements, and the difference between the preferential origin and non-preferential origin.

Bill of Materials

The BOM generally describes how the finished product is manufactured by listing a hierarchical structure of the parent products, intermediate products, raw materials, and associated data (e.g., quantities, values) used in the production process. In Preference Management, the preference determination looks at all materials and subassemblies used in the BOM to determine the finished product's preferential status. BOMs contain components and subassemblies that can be procured externally or produced internally. The BOM is a prerequisite to allow for a preference determination to take place. An example of the BOM was given previously in Figure 8.4 for Gamma. You can also refer to Section 8.6.1 for additional details.

Rules of Origin

When countries establish FTAs, they negotiate a set of eligibility rules for products to receive preferential treatment. In general, there are three types of rules of origin, which may be combined under a given product's rule of origin. While described slightly differently across agreements, the three rule of origin types are generally described as follows:

- **Tariff shift/change of tariff classification**
  This type considers whether a finished product has undergone a tariff classification change in a particular country to claim preferential status for a preference agreement.
- **Regional value content (RVC)/value added**
  Qualifying products must include at least a specified amount of originating content, or a product must not exceed a maximum allowance of non-originating materials.
- **Manufacturing or processing operations/specific rules**
  Particular conditions must be met related to the nature of or manufacturing methods associated with a product's production. These requirements may not be straightforward and may require in-depth knowledge to be validated. When such rules are encountered, consultation with subject matter experts is often necessary.

In SAP GTS, rules of origin can be applicable to the whole agreement or to specific subsections of the tariff schedule:

- *Base rules* apply regardless of tariff classification (i.e., to all or most products under a given agreement) and include tolerance rules, minimal operations, and set of goods rules. The most-used base rules are tolerance rules (also called *de minimis rules*), which state that a certain percentage of a product's value, usually 7% to 15%, can be of non-originating status without disqualifying the product overall.
- *Tariff-specific rules* are product-specific and are applied based on a product's tariff classification (i.e., typically at the chapter, heading, or subheading level) and encompass various requirements that must be met for the goods to be considered qualifying.

Depending on the FTA and the product at stake, one specific rule of origin, a combination of different rules of origin, or even alternative rules of origin may apply. See Section 8.3.5 and Section 8.6.3 for additional details and examples.

Preference Determination Methods

To determine the preference status of a product, rules of origin are translated as different determination methods in SAP GTS. The type of rule and the legal requirements per FTA will dictate what method to use. We'll discuss the important methods next.

Tariff Shift/Change of Tariff Classification

To determine if a tariff shift rule applies, the method requires comparing the tariff code of the finished product against the tariff codes of its components and subassemblies.

A rule of origin may require non-originating components to undergo a *tariff shift*, or a change in the Harmonized System (HS) code (i.e., typically at the levels of chapter, heading, or subheading) of the components as compared with the HS code of the finished product being analyzed. This way, companies prove that the components have been sufficiently processed to qualify for preferential treatment. The extent of the required tariff shift depends on the agreement.

In most agreements, the de minimis base rule is linked to the tariff shift method. If the tariff shift requirement isn't met, the determination typically falls back on the applicable base rules; if the value of non-originating and non-tariff shifting components accounts for less than the allowed de minimis percentage, then the product may qualify for preferential treatment.

Regional Value Content/Value Added

The RVC or value-added rule requires that a finished product must include at least a certain amount of originating content or that a product must not exceed a maximum allowance of non-originating materials. Rules of origin specify the permitted calculation methods for originating content and specify what percentage of originating content a finished product needs to claim preferential status. Each calculation method's description may be worded differently depending on the applicable agreement:

- **Transaction value**
  This method considers the finished product's non-originating content and compares that to the transaction value. For a company to claim preferential treatment on a product, the non-originating content may not exceed a given percentage (i.e., specified by tariff classification within each agreement) of the transaction value. In most EU agreements, the non-originating content is compared against the *ex-works (EXW) sales price*. In contrast, the USMCA also uses this calculation method but compares the non-originating content to the *free-on-board (FOB) price*. Generally, the value compared against the transaction value includes all the incurred costs of production, marketing, profits, and more. Thus, the rules typically require a higher percentage of originating share for this method compared to the net cost method (discussed next).
- **Net cost**
  This method compares the non-originating content against the net cost of the finished product (e.g., production costs, procurement costs). If the non-originating content doesn't exceed the allowed percentage stated in the rules of origin, the finished product contains sufficient originating content (i.e., direct and indirect costs) and is eligible for preferential treatment. The net cost method doesn't depend on a product's sales price because the method compares the originating content against the finished product's standard cost. Additional costs, such as marketing, shipping, and packing costs, aren't included in this comparison method. For those reasons, the required RVC percentage is typically lower than that of the transaction value method.
- **Build-up**
  This method calculates the originating content as the percentage of the finished

product's adjusted value (typically FOB) compared to the value of originating materials, including their production costs and profit.

- **Build-down**
  As opposed to the build-up method, this method calculates the value added as a percentage of the finished product's adjusted value (typically FOB) compared to the value of its non-originating materials.

Bill of Materials Calculation Methods

The BOM plays a key role in determining the preference eligibility of a manufactured product. A BOM can have multiple levels depending on the intermediate processes involved in the creation of a component. These intermediate processes are known as the subassemblies of a BOM, and parent BOMs containing such subassemblies are known as multilevel BOMs. When determining the preferential status of a finished product, the rules of origin for the applicable FTA are applied to the components and/or subassemblies.

A finished product's BOM can be analyzed according to two distinct logics that SAP GTS uses to calculate the preference determination:

- **Top-down method**
  Only the highest (e.g., parent or finished good level) and lowest-level components (e.g., raw materials) of the BOM are considered and compared when performing the preference determination.
- **Bottom-up method**
  The preference determination starts from the lowest-level components of the BOM and works its way up to the finished product. For each subassembly, a preference determination is performed between the subassembly and its subcomponents under the rule of origin applicable to the subassembly. The results of each subassembly's analysis are rolled up in the next subassembly until reaching the ultimate parent product/finished good. In the last stage, the preferential status of the finished product is determined based on the status of the immediate components/subassemblies under the finished product.

Depending on the FTA and the local jurisdiction requirements, companies may either be required to use a single type of BOM calculation method or be allowed the option to use both methods. Additional analysis or proof may be required for certain FTAs depending on the chosen calculation method.

To demonstrate the impact of such BOM calculation methods, see the BOM of Gamma's leather shoe, shown in Figure 8.6. During a top-down preference calculation, the lowest-level components are taken into consideration. This means that for each rule related to the originating content, the preferential status of the lowest-level components (i.e., COMP906, COMP921, COMP976, COMP632) and their values are considered in comparison with the top-level product (i.e., FP001). For each rule related to tariff codes, the tariff codes of these components are compared to the tariff code of the finished product.

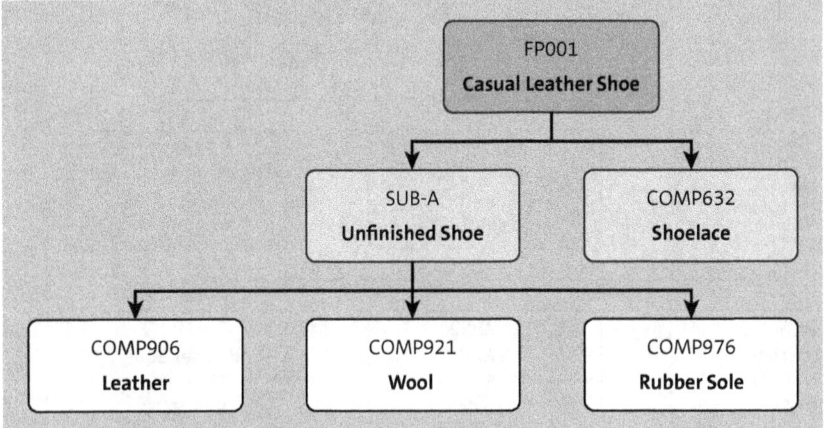

Figure 8.6 BOM for Casual Leather Shoe

For the bottom-up preference determination, the preferential status of each of the subassemblies is determined first. In the example BOM in Figure 8.6, this means a preference determination is performed first for the unfinished shoe (i.e., SUB-A) as compared with its components (i.e., COMP906, COMP921, COMP976) according to the applicable rules of origin linked to the unfinished shoe's tariff code. After the preferential status of the subassembly is determined, the status of the finished product (i.e., FP001) can be determined based on the preferential status and tariff codes of the subassembly (i.e., SUB-A) and shoelace component (i.e., COMP632) without further analysis of the components contained within the subassembly.

Top-down preference determination has an advantage in tariff shift comparison methods because the tariff codes of the lowest-level components, which are more likely to be different from the tariff code of the top-level component, are compared to the tariff code of the finished product. Further, this method disregards all subassemblies, which may have tariff codes closer to the finished product's tariff code.

By contrast, bottom-up explosions are favorable when RVC rules are used. Bottom-up explosion increases the originating value due to subassemblies transforming non-originating components into originating values. The latter is explained in Gamma's comparison of the top-down and bottom-up calculation results.

Gamma's Finished Product Preferential Determination

Consider the scenario for Gamma in which the USMCA rule of origin for the subassembly and the finished product is a net cost rule of 55%. This rule means at least 55% of the finished product's value must be originating. Figure 8.7 shows the BOM, the tariff codes, and the materials' standard costs used as calculation input.

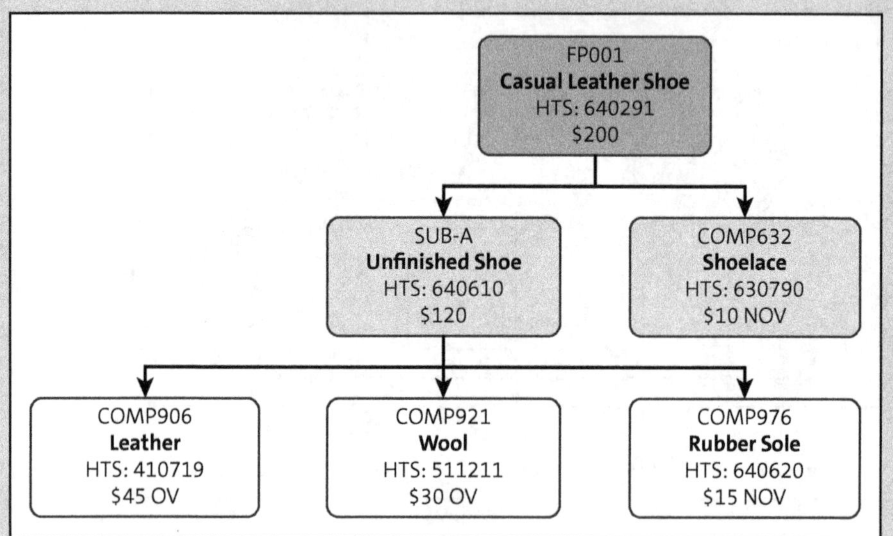

Figure 8.7 BOM for Casual Leather Shoe with Originating Value

Gamma first applies the top-down preference determination. The rule of origin for the finished product is "a change to heading 64.01 through 64.05 from any heading outside that group, except from subheading 6406.10, provided there is a RVC of not less than 55% under the net cost method." All the components met the tariff shift rule as the rubber sole had a 6406.20 classification, the wool had a 5112.11 classification, the leather had a 4107.19 classification, and the shoelace had a 6307.90 classification. The calculation takes the total finished product that is valued at $200 and consists of the following:

- $25 non-originating value, due to the rubber sole and shoelace
- $75 originating value, due to the leather and wool
- $100 originating value, due to manufacturing efforts

The subassembly is disregarded for the analysis, but the finished product is eligible for preferential treatment because the regional value is set at 87.5% (i.e., $175 originating value ÷ $200 total value).

Gamma then performs analysis of the same product using the bottom-up method. The analysis begins with the determination of preferential eligibility of the unfinished shoe (i.e., SUB-A). The rule of origin for the SUB-A is "a change to subheading 6406.10 from any other subheading, except from heading 64.01 through 64.05, provided there is a RVC of not less than 55% under the net cost method."

- $15 non-originating value, due to the rubber sole
- $75 originating value, due to the leather and wool
- $30 originating value, due to manufacturing efforts (i.e., the difference between material price and price of its components, including costs such as tooling and labor)

The net cost rule compares the originating value against the unfinished shoe's material price. This results in an originating share of 87.5% (i.e., $105 originating value ÷ $120 total value), which is well over the required 55% stated in the net cost rule. Additionally, all the components met the tariff shift rule as the rubber sole had a 6406.20 classification, the wool had a 5112.11 classification, and the leather had a 4107.19 classification. As such, the unfinished shoe is qualified for its rules of origin, thus consisting of originating value. In a second step, the preference calculation is performed for the finished product that is valued at $200 and is composed of the following:

- $10 non-originating value, due to the shoelace
- $120 originating value, due to the unfinished shoe (i.e., as the full value is carried forward)
- $70 originating value, due to manufacturing efforts

The net cost rule for the finished product now results in an originating share of 95% ($190 originating value ÷ $200 total value). The finished product thus satisfies the rules of origin and is eligible for export under preferential treatment.

Comparing the two methods, we can see how the bottom-up method yields an additional $15 or 7.5% in originating value. Although the top-down determination results in a lower share of originating value (i.e., 87.5% under top-down versus 95% under bottom-up), it makes no difference to the actual preference result.

BOMs often consist of a multilevel structure with components sourced from various suppliers in different countries and a variety of levels/structures. Due to the complexity of calculating a BOM's preferential status and the variety of BOM structures that may exist, it's nearly impossible to perform the preference determination at scale if preference determination isn't properly automated by a supporting system.

Preference Determination Status

Now that you understand how the rules of origin play a role for preference determination methods and how the BOM can be used in different BOM calculation methods, we'll move on to discuss the possible results of preference determination in SAP GTS:

- **Always preferential**
  In this scenario, a preferential status is always assigned to the finished product. This may be the case if the product is wholly obtained/produced, produced entirely from originating materials, or always satisfies the applicable rules of origin, independent of the price at which the finished product is sold. A product may also be always preferential if it contains enough originating materials to always pass the net cost method.
- **Never preferential**
  This scenario occurs when the finished product can't claim preferential status regardless of sales price. For instance, if a finished product has the same tariff code

as its non-originating components (i.e., tariff shift failed) and there is no applicable RVC rule of origin or general tolerance rule, then the product would be considered never preferential. If a preference determination doesn't return a favorable result, a company's management team can investigate the cause of the determination result. Companies often find that the root cause for such a status may lead to a broader evaluation of their sourcing strategies by selecting alternative suppliers who can provide originating components that positively impact their ability to qualify their products.

- **Threshold value dependent (i.e., to be determined on a transactional basis by comparing the transaction value against a threshold value)**
  This scenario occurs in SAP GTS when a product may be eligible depending on the finished product's sales price or declared customs value. This is a common status when applying the transaction value (but can also happen for other rules of origin, such as a tariff shift that fails and goes through the tolerance rule). For example, a finished product qualifies when sold above a certain value but fails when sold below a certain value (e.g., depending on the profit that was added upon the sale of the good). This status is purely used in SAP GTS and isn't intended to be submitted to customs authorities. The SAP GTS system uses it to store the threshold value to be compared against the transaction value, which is further explained in Section 8.3.5 and Section 8.3.7.

Preference Statements

The *preference statement* (also called preference indicator or preference criterion) presents the per product outcome of the preference determination process and represents the way the goods were determined to qualify under specific FTA terminology. These statements can sometimes be indicated for all products (e.g., on an EU LTSD), but often need to be printed on a per product level of each certificate of origin. While the preferential statements for certain FTAs can be rather straightforward, other FTAs require certain nomenclature and accommodate a wider range of values. Following are some examples:

- Most of the European FTAs will allow for a binary indicator such as "Eligible" and "Not Eligible."
- For USMCA and EU-Japan FTAs, the preference statements are typically presented in letters such as A, B, and C to provide additional insights into how the goods qualified.
- In the ASEAN region, FTAs often require "CTC" or "RVC" to appear on the proof of origin.

The following scenarios represent common preference criteria for establishing a product's preferential treatment eligibility:

- **A product is wholly obtained or produced entirely in the territory of one or more of the parties (wholly obtained or produced goods).**
  This statement is for instances when imported components aren't used. All the products are locally sourced and often supported by vendor certificates of origin or the company's knowledge regarding the source of the goods. As such, the preferential country of origin is determined by the country where the product is obtained or produced. This preferential status is most common for mineral products, animals, and products derived from animals.
- **A product is produced entirely in the territory of one or more of the parties using non-originating materials provided the good satisfies all applicable requirements of (product-specific) rules of origin.**
  This statement is for instances where imported materials that may or may not be originating are used in the production of the goods, but that the rule of origin is still met and the produced item qualifies. We discussed product-specific rules of origin (i.e., rule type linked to classification codes) in Section 8.2.2.
- **A product is produced entirely in the territory of one or more of the parties exclusively from originating materials.**
  This statement applies when the product is fully manufactured using originating materials. These materials can come either from the country where production takes place or using materials that originate from other parties to the agreement and can therefore be considered as originating. For example, under the USMCA, a product produced in the US may also contain components from both Mexico and Canada. The preference statement requires that the necessary proofs of origin be obtained from the respective suppliers.

Note that the preferential status of a specific product may vary across different FTAs due to each agreement's nuances. For instance, a product may be considered as US preferential origin under US-CAFTA-DR rules of origin with the associated FTA-specific statement, but that same product may not fulfill the requirements for the USMCA.

Preferential Origin versus Non-Preferential Origin

As a final note to the outcome of the preference determination process, we'll clarify the difference between preferential and non-preferential origin. This topic is often misunderstood and incorrectly conflated.

Preferential origin and non-preferential origin (also known as economic origin) have different guidelines/rules to determine what should be specified as a product's origin, and each concept is used for different purposes.

As mentioned previously, preferential status and origin concerns the ability to apply reduced or eliminated duties upon importation to a country that is party to a given agreement. As part of the preferential qualification analysis, many FTAs require determining the preferential origin (i.e., country) of the product.

In contrast, non-preferential or economic origin concerns the determination of where the product was either wholly obtained or has undergone significant processing and isn't necessarily the same as the preferential origin. Non-preferential origin is important for antidumping policies, trade embargoes, trade quotas, and other trade-related topics, but it doesn't have an impact on the claim for preferential treatment. This is typically determined based on applying the importing country's non-preferential origin rules and interpretation, which may differ when compared to the exporting country's rules and interpretation. It's important to perform non-preferential origin determination analysis separate from preferential origin analysis and consult local subject matter experts, as needed.

Note

This concept of non-preferential origin isn't relevant under the scope of a preferential agreement. A product can have a non-preferential origin (i.e., "country of origin" or "economic country of origin") that isn't necessarily corresponding to its eligibility to claim a preferential status under a certain preference agreement. These are two separate concepts governed by different sets of rules that don't always correspond and should be considered separately.

8.2.3 Issue Certificates of Origin for Preferential Finished Products

If a finished product has a preferential status, then a company can provide the proof of origin to its customers for shipments of the product. Depending on the FTA, companies can issue the proof of origin transactionally (i.e., along with a specific shipment) or for a given period when allowed under the FTA and per the specific requirements (e.g., LTSD or blanket certification). With the proof of origin, the customer can import the product at a free or reduced-duty amount. Note that in the EMEA region, supplier LTSDs are typically acceptable to use in conducting one's own preferential qualification analysis whereas only transactional (or long term when foreseen by the FTA) statements of origin can be used for the basis of a preferential duty claim upon importation. This is contrasted with the North American preference model where blanket certifications can serve as both the basis for qualification analysis as well as preferential duty claims upon importation. See Section 8.2.1 for more details on the different forms of certification.

The term "customer" as used here refers to any final recipient of the goods. Companies can issue preferential certificates of origin to third-party customers (e.g., end customer) as well as related entities (e.g., affiliated companies). If the certification is issued to a third-party customer, then the duty savings are realized by the third party. The ability to certify items sold to third parties may be part of a broader commercial strategy and may allow the seller to charge the customer more than they would otherwise (i.e., as their costs would be lower). If the certification is issued to a related party, then the company's related entity (and, by extension, the company overall) is the beneficiary. This is particularly important to identify, as the savings are true cost avoidance and can impact a company's cost of goods due to the way duties are typically considered from a cost-accounting perspective.

8.3 System Process

Whereas the previous section covered the Preference Management business process, this section will extensively cover the different steps of Preference Management in SAP GTS from the perspective of system processes. After a general introduction of the various topics, we'll discuss each system step in detail and link it to the business case of Gamma, our fictional shoe manufacturer.

8.3.1 Inbound, Production, and Outbound Processes

Figure 8.8 and Figure 8.9 display the SAP GTS tiles where you can access different processes related to Preference Management. The tiles are categorized into the following different sections under **Preference**:

- **Supplier's Declaration**
  - **Inbound Declarations**
  - **Outbound Declarations**

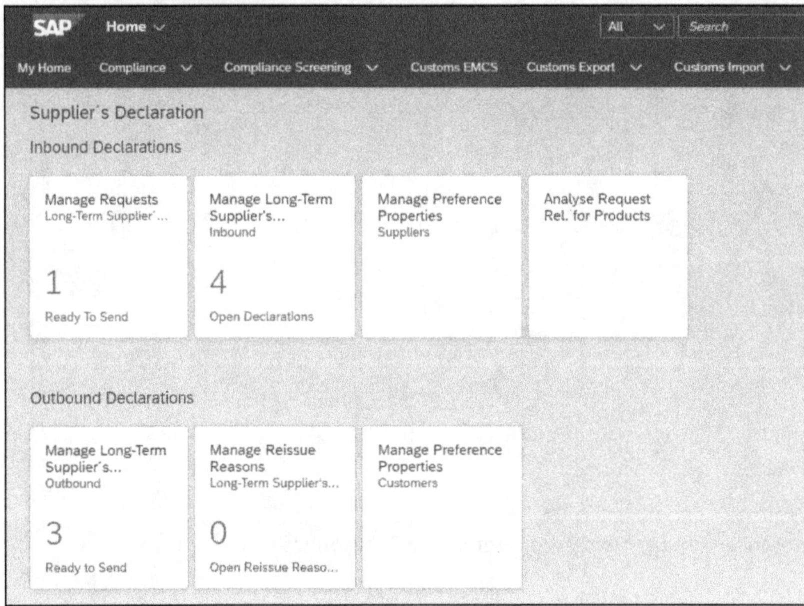

Figure 8.8 Preference Management Tiles (Supplier's Declaration)

- **Preference Determination**
  - **Determination – Products**
  - **Determination – Sales Documents**
  - **Determination – Production Documents**
  - **Aggregation – Products**

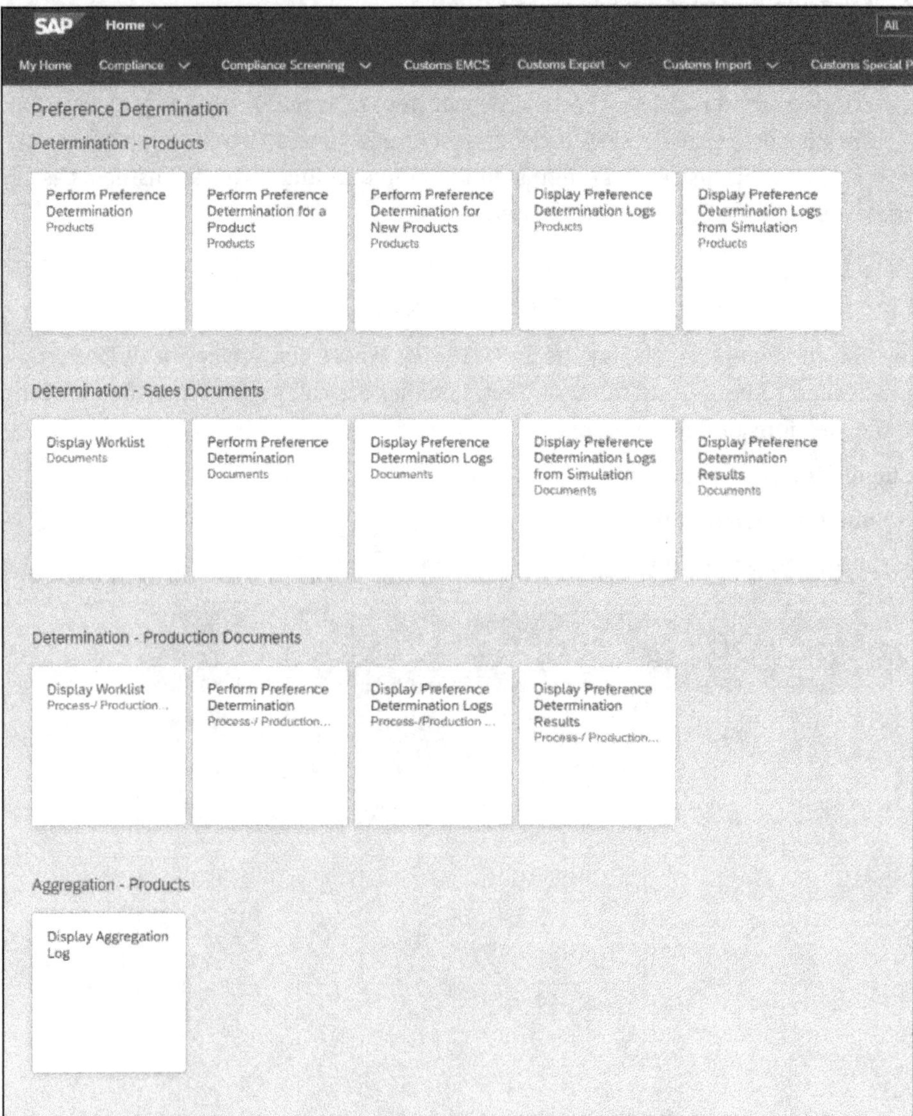

Figure 8.9 Preference Management Tiles (Preference Determination)

Figure 8.10 shows the SAP GTS system process and its different steps that are supported by these applications. You can distinguish among inbound, production, and outbound processes for Gamma.

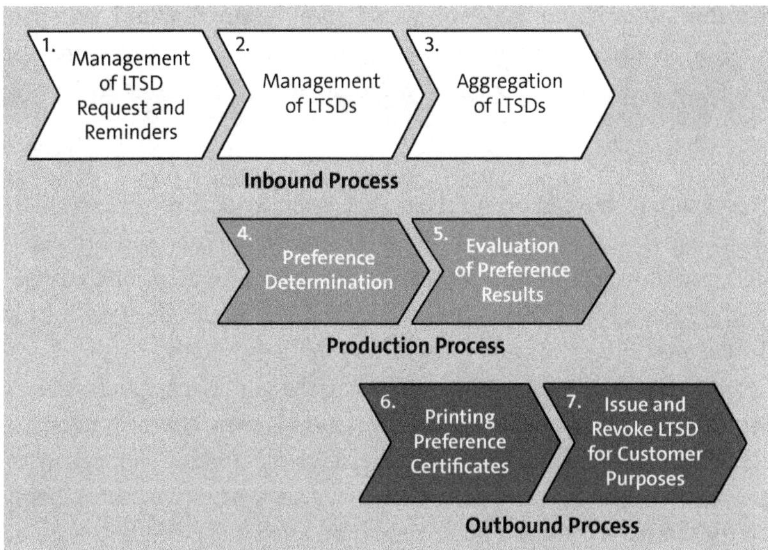

Figure 8.10 Preference Management System Process

The objective of the *inbound process* is to obtain supplier declarations for components that are procured externally, which may have an impact on the preference determination results. The functionalities associated with inbound processes are available in SAP Fiori under the **Inbound Declarations** section, as shown earlier in Figure 8.8. In SAP GTS, this is performed in three major steps:

1. **LTSD request**
   The company requests the origin of each relevant externally procured component from the applicable suppliers. This is the supplier solicitation process for which SAP GTS uses LTSDs. SAP GTS integrates with the relevant purchase orders and/or material documents/goods receipts from SAP S/4HANA in a worklist, which allows the user to request LTSDs from the corresponding suppliers. For each entry in the worklist, SAP GTS will send out an LTSD request document. If suppliers don't respond to the LTSD request, a user can send a reminder to the supplier to respond to the LTSD request. More details are available in Section 8.3.2.

2. **Supplier LTSD maintenance**
   After the suppliers respond to the LTSD requests, the components' preferential information needs to be captured in the system. A component that requires an LTSD doesn't obtain preferential status in the system until it has been expressly marked and entered as such, which is performed at this step. You can store the response from the vendor by maintaining LTSD details in SAP GTS by referencing a purchase order, a goods receipt, or the LTSD request number in the previous step. You would

enter the preferential status based on the response from the supplier and the corresponding declaration for each supplier-component combination. SAP GTS stores the supplier's response as a vendor LTSD entry. More details are available in Section 8.3.3.

3. **LTSD aggregation**
   Components may be dual-sourced or multisourced, meaning that the components are procured from two or more suppliers. The final step of the inbound process is aggregating preferential origin-related information concerning a specific component from the suppliers. Because only one aggregated status can be used during preference calculations, SAP GTS aggregates the preferential status of that component obtained by different suppliers by integrating the goods receipts from SAP S/4HANA and the vendor LTSD maintained in the previous step. The system aggregates the results according to a worst-case principle; if at least one supplier of a product has a negative preferential status (i.e., either due to a negative status being maintained or if the status isn't maintained), then the aggregated preferential status is stored as not preferential in the product's master data. To somewhat alleviate this impact, a goods retention period mechanism has been introduced in SAP GTS, edition for SAP HANA. More details are available in Section 8.3.4.

The *production process* goal is to determine whether the produced good is preferential. Preference determination can be executed for both static BOMs and configurable BOMs. A static BOM for a given product indicates that the BOM doesn't change across sales or production transactions. A configurable BOM is unique for every transaction. Depending on whether you run preference determination for static or configurable BOMs, the tiles that you access would be different. For simplicity, we'll focus on running preference determination on static BOMs going forward. The preference determination process consists of the following two steps, which can be executed under the **Preference Determination – Products**, **Preference Determination – Sales Documents**, and **Preference Determination – Production Documents** tiles, as displayed earlier in Figure 8.9:

1. **Preference determination**
   After you receive and maintain the preferential information for the sourced components and classify the components and finished product with a tariff (i.e., HS) code, you can perform preference determination for the finished product in the fourth system process step (i.e., **Preference Determination**) shown in Figure 8.10. SAP GTS looks at the classification of the finished product to determine the relevant rule of origin and the details of the BOM to understand the relationship between the finished product and the components, their level in the BOM, their values, procurement indicators, preferential status, and HS code. Based on these details, SAP GTS calculates the preferential status based on the specific rule of origin of a given FTA and stores the preference determination result in the product master. More details are available in Section 8.3.5.

2. **Evaluating preference results**
   The second step in the preference determination process is evaluating the preferential results. The detailed steps of the calculation as well as the BOM details used in the calculation can be analyzed in a log. The final preferential results from the preference determination calculation can be displayed in a standard report. Through this standard report, a company's management can look further into finished products with unfavorable preference results and evaluate the causes and possibilities to change the result of the preference determination. For instance, the company can maintain missing components' preferential information or replace non-originating components with originating components from a different supplier. After implementing changes, these finished products may be eligible for preferential treatment. More details are available in Section 8.3.6.

Finally, the *outbound process* covers the preparation and transmission of the proof of preferential origin to customers, related companies, and other third parties (e.g., subcontractors) using the output of the preferential determination performed. It consists of the following steps:

1. **Transactional certification**
   This step of the Preference Management system process is issuing the proof of origin to the company's customers located in member FTA countries. Exporters or certifiers can issue certificate of origin documents on a transactional basis or for a certain blanket period to customers located within a relevant FTA country. If the certificate of origin documents are issued on a transactional basis, then SAP GTS allows you to integrate the Preference Management functionality in SAP GTS with SAP S/4HANA where the results from the preference determination are evaluated per billing document. SAP S/4HANA provides a **Preference Status** checkbox in both the sales order and billing document at the item level to indicate whether the product is preferential. You can set up SAP S/4HANA to retrieve the preferential eligibility of these products from SAP GTS. More details are available in Section 8.3.7.

   Additionally, SAP GTS allows you to integrate the Preference Management functionality in SAP GTS with the SAP GTS Customs Management functionality. This gives you the option to populate the preference eligibility status in export declarations. Thus, you can print the certificate of origin documents via the Customs Management functionality (refer to Chapter 7).

2. **Periodic certification**
   Depending on the applicable FTA, customs authorities may permit importers to use a blanket certificate where the proof of origin is issued for a specified validity period. The names and validity periods allowed for blanket certificates vary across each FTA. For example, in the case of USMCA, the certificate's maximum validity date range (i.e., blanket period), is for one year and may be used to cover multiple shipments of

identical goods imported during the validity period. In the EU, LTSDs can also be provided to EU domestic customers and have a validity period of up to two years, but they can't serve as the basis for a preferential claim at importation. Blanket certificates may be issued via the Preference Management functionality. The latter process can be found in SAP Fiori in the **Outbound Declarations** section, as noted earlier in Figure 8.8 and further explained in Section 8.3.7.

3. **LTSD revocation**
   As displayed previously in Figure 8.10, an additional step may occur. If a finished product's preferential status changes (i.e., preferential products become not preferential or non-preferential products become preferential) during the blanket certificate's validity period, the relevant certificates need to be revoked and reissued. The preferential status may change, for example, due to new or different sourcing strategies, component pricing changes, or updates to the finished product's selling price. In this case, SAP GTS automatically creates entries to reissue the proof of origin with the updated information. More details are available in Section 8.3.8.
4. **Intercompany LTSD issuance**
   Finally, and as introduced before, preferential certificates of origin can be shared across related companies. SAP GTS foresees a mechanism to automatically issue and receive LTSD's across administrative units (AUs). More details are available in Section 8.3.9.

In the Preference Management functionality, all preference-related flows are managed under AU concepts. An *administrative unit* (AU) represents a plant or plant group (i.e., multiple plants) engaged in administration (e.g., point of contact, certifying entity) related to Preference Management and isn't necessarily just a physical representation of a company. An AU is the organizational structure in SAP GTS that enables all the Preference Management processes and reporting. We'll discuss the concept and setup of AUs further in Section 8.4.1.

Before you can perform any Preference Management steps, you need to select an AU. Furthermore, because the preference administration needs to typically be performed at least once a year, it's also important to select the relevant period. Considering the Gamma business case, a US AU is selected with USMCA as an active agreement. Now that we have a glimpse of the system process steps in SAP GTS, let's examine each one closer using the Gamma business case.

8.3.2 Requests and Reminders for Supplier Declarations

Now that you understand the end-to-end system processes for Preference Management, we can deep dive into each individual process step. Let's start with the management of LTSD requests and reminders.

Communication Methods

In the **Manage Requests** app under **Preference Management**, there are two distinct ways to manage the supplier solicitation process to obtain the required origin-related information for externally procured components and subassemblies, as found under **Communication Method** (see Figure 8.11).

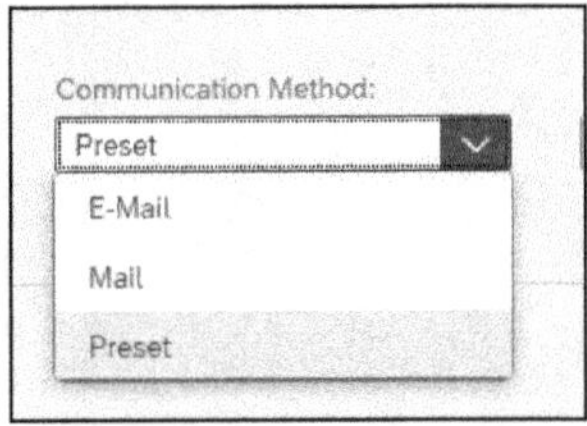

Figure 8.11 Communication Methods for LTSDs

Let's discuss both methods:

1. **Mail**

   For this method, you perform supplier solicitation using forms. You can download these forms from SAP GTS to print them and send them offline to the relevant suppliers. The suppliers can then enter the preferential information of their sold products on these forms and return the information via either postal mail or email.

   Note, SAP has standard certificate of origin forms that can be used. For instance, the EU LTSD form for supplier solicitation is shown in Figure 8.12 and Figure 8.13. Note that the LTSD request form for European suppliers isn't an actual certificate of origin but a template to solicit this information for a supplier.

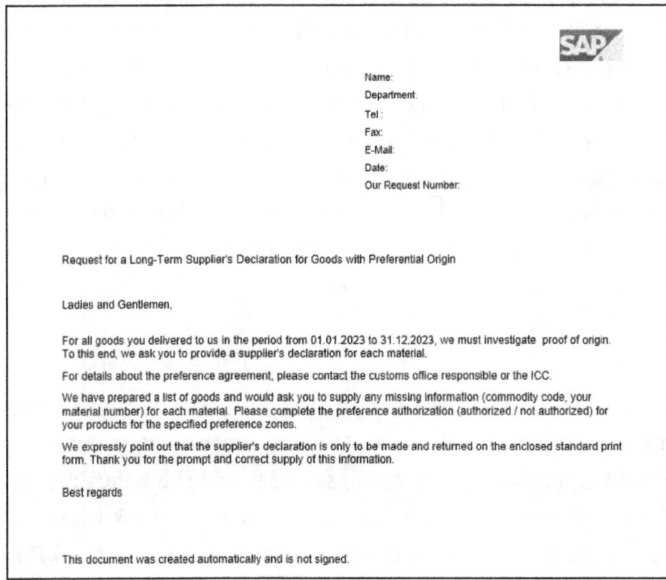

SAP

Name:
Department:
Tel.:
Fax:
E-Mail:
Date:
Our Request Number:

Request for a Long-Term Supplier's Declaration for Goods with Preferential Origin

Ladies and Gentlemen,

For all goods you delivered to us in the period from 01.01.2023 to 31.12.2023, we must investigate proof of origin. To this end, we ask you to provide a supplier's declaration for each material.

For details about the preference agreement, please contact the customs office responsible or the ICC.

We have prepared a list of goods and would ask you to supply any missing information (commodity code, your material number) for each material. Please complete the preference authorization (authorized / not authorized) for your products for the specified preference zones.

We expressly point out that the supplier's declaration is only to be made and returned on the enclosed standard print form. Thank you for the prompt and correct supply of this information.

Best regards

This document was created automatically and is not signed.

Figure 8.12 EU Standard Form Page 1

Issuer: | Consignee:
Our Recipe Number: | Your Request Number:

Product Number:	
Product Description	Product Number of Supplier
Commodity Code:	Cumulation Note: ☐ Cumulation used with ☐ Cumulation Not Applied.
Preference Criterion: ☐ Authorized ☐ Not authorized	
Agreement:	

Issuer: | Consignee:
Our Recipe Number: | Your Request Number:

Preference Agreement Definition

Country Group of Issuer

Agreement	Partner Countries	Agreement	Partner Countries

Figure 8.13 EU Standard Form Pages 2 and 3

2. **E-Mail**
   For this method, SAP GTS sends an email to the supplier with the LTSD request form as an attachment. To identify each supplier's person of contact and email that should receive the LTSD, a user must maintain the email address in the business partner master data and create a relationship between the supplier and the person of contact (see Section 8.6.2 for further details). When the suppliers receive the LTSD request, they can provide the preferential status of the requested products by replying via email with the completed LTSD attached. The supplier provides the validity period of the preferential status of their products, the FTA(s) in scope based on the preferential status between the signatory countries, the commodity code, and other relevant data elements in the LTSD. Note that these forms and texts are editable according to the recipient's needs. Standard forms and text modules exist, but development is often recommended to tailor to specific company needs. Note that while the first email is generated out of SAP GTS, the remainder of the mail communication is done outside of SAP GTS. Therefore, don't expect the email from the supplier to be directly received in SAP GTS.

Integration Methods

To support the LTSD request process and minimize the workload associated with determining relevant vendors and sourced materials, SAP GTS is integrated with purchasing documents from SAP S/4HANA. Specifically, the purchase order and the goods receipt document types can be configured to create entries in the request LTSD worklist of SAP GTS. Purchase orders have the advantage of triggering an entry earlier in the LTSD request worklist, helping with soliciting information from suppliers at an earlier stage in the inbound process. However, such solicitation requests may contain requests for product details that haven't yet been shipped from the vendor and risks over solicitation, which increase the likelihood of partial vendor responses. Note that it's possible to delete entries from the LTSD request worklist. In contrast with purchase orders, goods receipts or material documents have the advantage of triggering the LTSD request worklist for products that are delivered to the company's warehouse. Consequently, LTSDs aren't requested for unnecessary components that may be included in the purchase order but not actually delivered to the relevant plant. The disadvantage of goods receipts or material documents is that LTSDs are requested and maintained at a later stage in the inbound process and can become a reactive solicitation activity. Note that both document types can be set as trigger documents.

For the activated document types, an entry is created in the worklist for requesting supplier declarations if the following prerequisites apply:

- The relevant plant in the transactional document is assigned to the specific AU in SAP GTS.
- The supplier is in a country for which the AU has an FTA.
- The purchased product and supplier are available in SAP GTS and considered relevant for requesting supplier declarations (Section 8.6.2).
- No LTSDs exist for this product-supplier combination for the current period. If an LTSD already exists, there is no need to request a new declaration unless it reaches the end of its validity period.

LTSD Request Process

As mentioned earlier, Figure 8.8 and Figure 8.9 show the SAP Fiori UI for SAP GTS. The **Supplier's Declarations** section contains the relevant tiles for the inbound Preference Management process. The first app, **Manage Requests for Long-Term Supplier Declarations**, contains the worklist of LTSD requests, as shown in Figure 8.14.

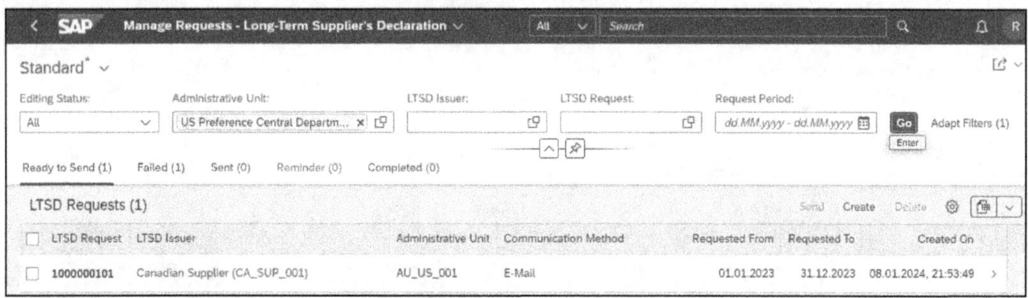

Figure 8.14 Manage Requests: Long-Term Supplier's Declaration

As discussed, the documents set up as triggers (i.e., purchase order, goods receipt) create LTSD request entries in this worklist with a status of **Ready to Send**. These LTSD requests will contain the supplier information and product details from the purchasing document. You can always change, add, or delete the details of these prepopulated entries as long as the status remains as **Ready to Send**. It's even possible to create LTSD requests manually from scratch (i.e., without prepopulation from the underlying documents) in the SAP GTS UI.

Once the LTSD request details have been reviewed, you can send the request. The entry is removed from the **Ready to Send** worklist and transferred to the **Failed**, **Sent**, or **Reminder** worklist, as shown in Figure 8.14. If the data elements of the documents are erroneous or in the event of a system failure (e.g., Adobe Document Server failure), the LTSD request will update to a **Failed** status, and you would need to investigate and resolve the issue before resending the LTSD. If sent successfully, the LTSD request will be issued to the supplier according to one of the two transmission methods explained previously (i.e., **Mail** or **E-Mail**), and the LTSD request document updates to reflect the **Sent** status and moves to the **Sent** worklist.

Reminder periods can be configured (Section 8.4.7) so that a reminder can be sent if no vendor reply is received for the LTSD request (i.e., requires user maintenance to be recognized by SAP GTS). This means that the entry from the **Sent** status will move to the **Reminder** status. You can then easily identify and send out the reminder of the LTSD request. When the reminder is sent, the entry will move again to the **Sent** status. This can be useful for companies making sourcing decisions between different suppliers to see which of them can provide originating materials that would be advantageous for preference determination. Based on that information, companies can make an informed decision about trading with a specific supplier.

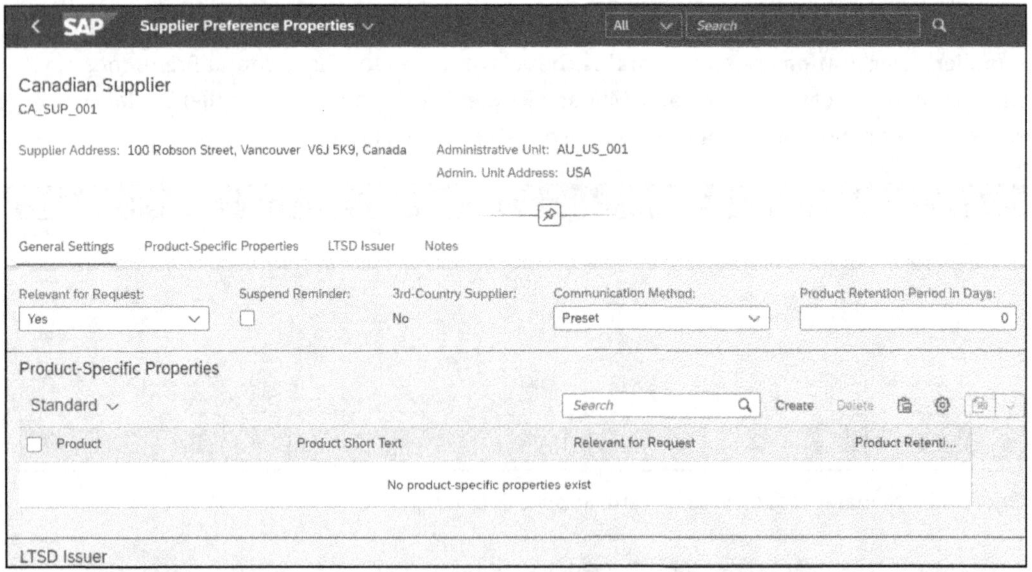

Figure 8.15 Supplier Preference Properties

Reminders can be triggered manually from the **Sent** worklist or automatically via the configuration setting, as explained in Section 8.4.7. Even if automatic reminders are set in configuration, the system allows users to **Suspend Reminders** for a particular supplier under the **Supplier Preference Properties** tile, as shown in Figure 8.15. When reminders are suspended, manual reminders are possible. Any reminder entries are populated in the **Reminder** worklist once issued.

Annual LTSD Request Process

In addition to the reminder functionality, SAP GTS can also create a linkage for LTSD request campaigns across subsequent years via SAP GUI Transaction /SAPSLL/LTSDREQ_2 (Annual LTSD Request), as shown in Figure 8.16. Because LTSDs are typically valid for one year, companies often need to request and maintain the supplier declarations at the start of each year. To facilitate this process and allow for proactive solicitation (i.e., before new purchase orders, goods receipts, and material documents are created in the system), you can configure a transition period in which supplier-component combinations maintained for the current year are automatically created as entries in the worklist for LTSD requests for the following year.

8

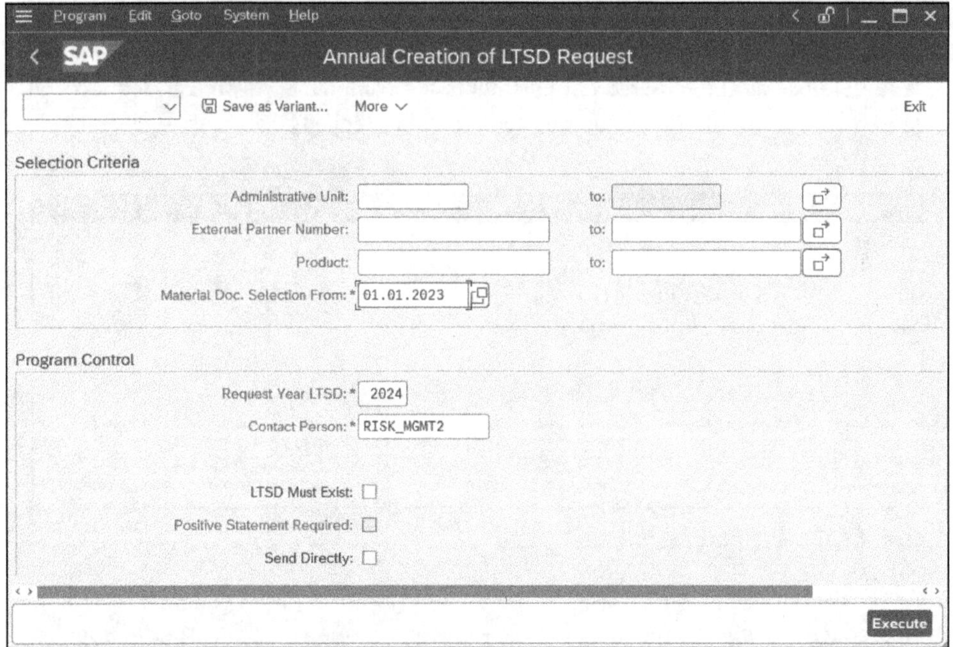

Figure 8.16 Annual Creation of LTSD Request in SAP GUI

Gamma's Supplier Solicitation Process

To manufacture its casual leather shoe, the finished product that will eventually be shipped to its customers, Gamma needs to procure certain components from various suppliers. Gamma doesn't request supplier declarations prior to the purchasing process, but it does request LTSDs that are generated in the worklist after being triggered by purchase and material documents.

Gamma places purchase orders for the following vendors:

- COMP921 (wool): 2,500 pieces from a supplier in Canada
- COMP906 (leather): 1,000 pieces from a supplier in the United States
- COMP976 (rubber sole): 1,750 pieces from a supplier in Vietnam
- COMP632 (shoelaces): 2,490 pieces from a supplier in Mexico
- COMP632 (shoelaces): 120 pieces from a supplier in Peru

Suppose that four purchase orders and five goods receipt documents triggered new entries in the worklist to request LTSDs. In the LTSD request worklist, as shown in Figure 8.17, SAP GTS will create a total of three new entries (i.e., one for each unique supplier related to the relevant purchase order and goods receipt) for the newly created purchase and material documents.

The **Go** button at the top right of the screen allows Gamma's users to trigger LTSD requests once all the required data elements are completed. A user can request the preferential status of multiple unique components supplied by a single supplier in one LTSD request.

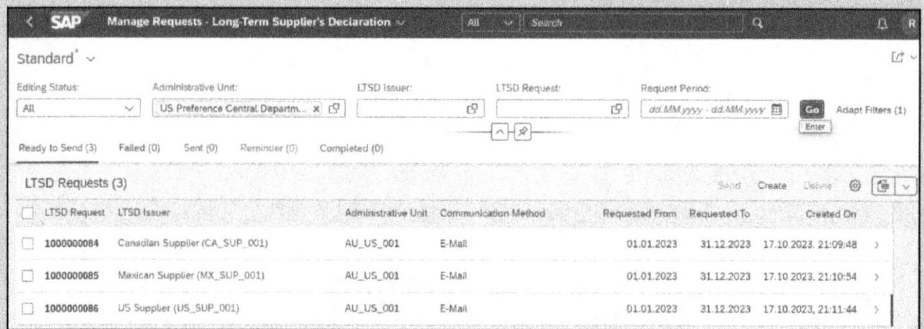

Figure 8.17 Worklist for Requesting LTSDs

Gamma's users execute the LTSD request for each business partner. SAP GTS indicates all products for which a request for origin-related information to one specific supplier is requested, as shown in Figure 8.18.

After clicking **Send**, the supplier declaration is printed as an empty USMCA certificate of origin form with the prepopulated material details based on the underlying materials in the purchase orders and goods receipts.

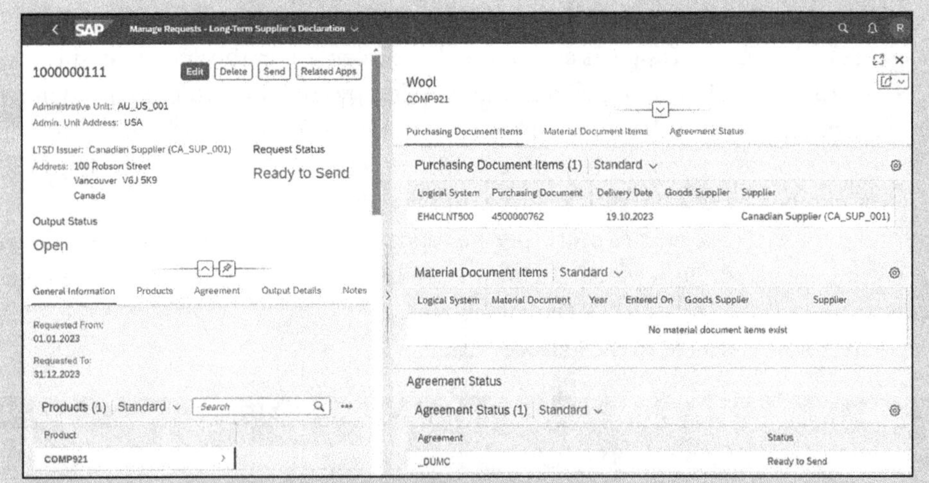

Figure 8.18 LTSD Request to the Canadian Supplier

Furthermore, SAP GTS allows Gamma to attach both a cover letter with a formal request and a list of products that require preferential origin-related information to the supplier.

After all the LTSD requests are sent out, the entries are moved from the **Ready to Send** worklist to the subsequent **Sent** worklist, as shown in Figure 8.19.

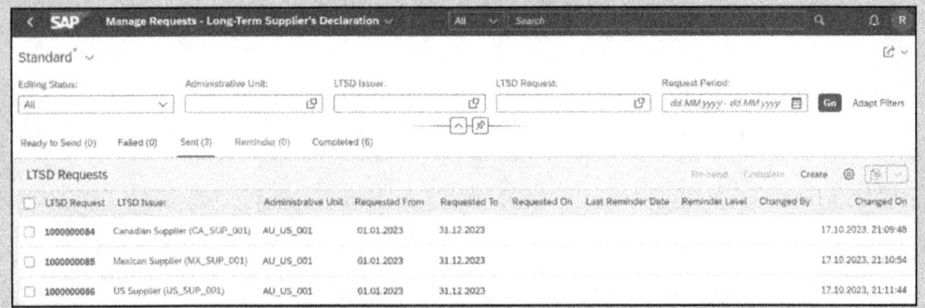

Figure 8.19 Supplier-Based Supplier Declarations Menu after Requesting LTSDs

The next step for Gamma is to maintain the information it receives from the various suppliers.

8.3.3 Maintaining Supplier Declarations

SAP GTS must store the preferential origin-related information for each component to run an accurate preference determination. If no origin is recorded, SAP GTS considers the product to have a *not preferential* status.

When a supplier receives an LTSD request, the supplier should send the completed LTSD back to the requesting company or reply that the products don't qualify for the

FTA. The requesting company needs to manually validate the completeness of the response and, if the data appears accurate, enter the content of the document into SAP GTS. You can maintain responses from the suppliers by creating entries in the **Manage Long-Term Supplier's Declaration – Inbound** app, as shown in Figure 8.20. An entry in the app can be created with reference to the purchase order, goods receipt, LTSD request number, or without reference (i.e., full manual input of the supplier and products references). These entries must then be validated based on the issuing company's received information (e.g., certificate of preferential origin, LTSD, invoice statement) before they are saved in SAP GTS. Once you save the entry by clicking **Release**, the entry moves from the **Open** list to the **Released** list.

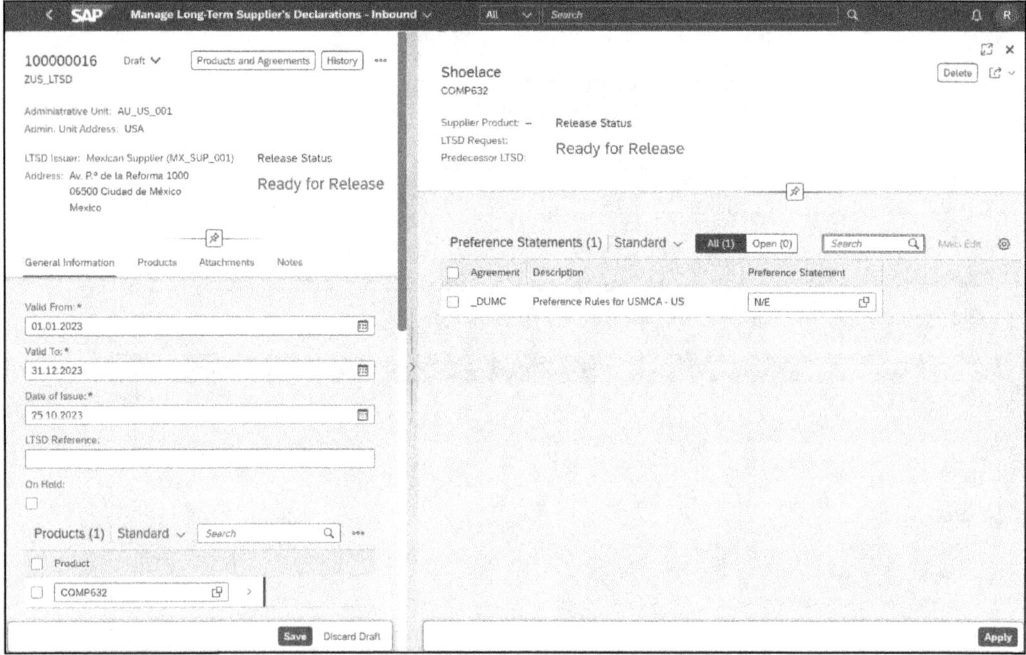

Figure 8.20 Maintain Responses from Suppliers

For each productin the **Open LTSD**, you can enter the *preference indicator* in the **Preference Statement** field. As indicated in Section 8.2.2, the preference statements and criteria vary by FTA. In the case of Gamma, the most common USMCA preference indicators that are set up in SAP GTS are the following (see Figure 8.21):

- **A: Goods wholly obtained or produced entirely**
- **B: Goods satisfy specific rules of origin**
- **C: Goods entirely from originating materials**
- **N/A: No Response Received**
- **N/E: Not Eligible**

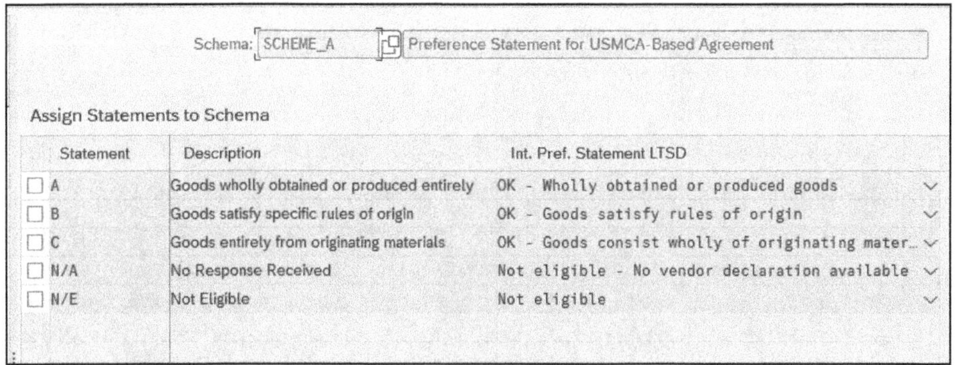

Schema: SCHEME_A — Preference Statement for USMCA-Based Agreement

Assign Statements to Schema

Statement	Description	Int. Pref. Statement LTSD
A	Goods wholly obtained or produced entirely	OK - Wholly obtained or produced goods
B	Goods satisfy specific rules of origin	OK - Goods satisfy rules of origin
C	Goods entirely from originating materials	OK - Goods consist wholly of originating mater...
N/A	No Response Received	Not eligible - No vendor declaration available
N/E	Not Eligible	Not eligible

Figure 8.21 Preference Statement for USMCA-Based Agreement

As another example, EU-based agreements typically have the following preference indicators (see Figure 8.22):

- **-: Not Eligible**
- **N/A: No Response Received**
- **X: Eligible**

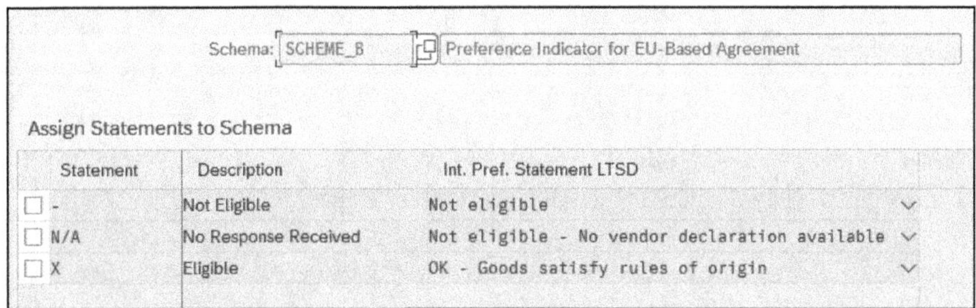

Schema: SCHEME_B — Preference Indicator for EU-Based Agreement

Assign Statements to Schema

Statement	Description	Int. Pref. Statement LTSD
-	Not Eligible	Not eligible
N/A	No Response Received	Not eligible - No vendor declaration available
X	Eligible	OK - Goods satisfy rules of origin

Figure 8.22 Preference Indicator for EU-Based Agreements

Gamma Receives Preferential Information from Its Suppliers

A few weeks after Gamma sent out the LTSD requests, it received responses from all suppliers, complete with the required preferential information, which means that there is no need for Gamma to send out reminders.

The following information is found on the completed USMCA LTSDs, stating whether the component is preferential:

- COMP921 (wool): Goods wholly obtained from Canada
- COMP906 (leather): Goods wholly obtained from United States
- COMP976 (rubber sole): Not preferential

- COMP632 (shoelaces): Goods wholly produced from Mexico
- COMP632 (shoelaces): Not preferential

The Canadian, US, and Mexican suppliers confirmed the preferential status of their components for the USMCA agreement. However, the suppliers from Vietnam and Peru replied that their components don't meet the agreement's requirements.

Gamma users need to enter the information manually in SAP GTS. It can either be created manually from scratch or based on the reference of a purchase order, goods receipt, or LTSD request. Figure 8.23 shows the creation with reference where the Gamma user filtered on AU and supplier number to find the document to be used as reference (in this case, a purchase order was used, but the LTSD request can equally be used). The supplier LTSD is created for the AU, the LTSD issuer (i.e., the supplier) and document of origin type, and is filled with all the product details of the reference document. Note that for a specific product, SAP GTS creates an entry for each active agreement in the selected AU. Consequently, a component's preferential information for the USMCA agreement may differ under other agreements.

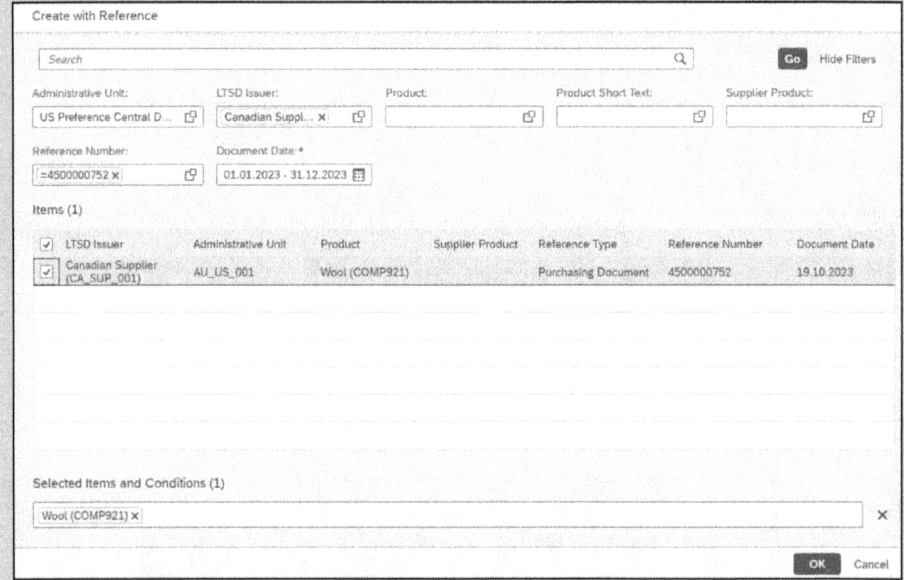

Figure 8.23 Creating an LTSD with a Reference

Gamma's users enter the information received from the various suppliers for the requested LTSDs, as shown in Figure 8.24. Because USMCA is the only active agreement for Gamma's AU in this example, you can maintain each product only with USMCA-specific preferential information. When selected, each save triggers an aggregation logic, which is discussed in the following system step.

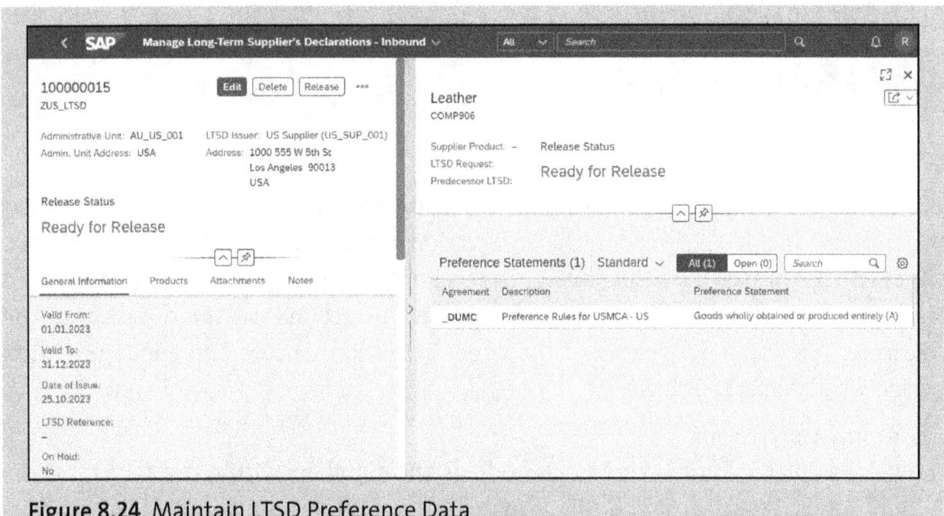

Figure 8.24 Maintain LTSD Preference Data

8.3.4 Aggregating Supplier Declarations

When it comes to dual-sourcing or multisourcing, SAP GTS aggregates LTSDs to consolidate the preferential information received from multiple suppliers into one preferential status for each component for a certain agreement. For example, one supplier may provide a preferential LTSD for a component, while another supplier selling the same component may not. SAP GTS applies the *worst-case principle*, meaning that if one supplier can't provide preferential origin or the preferential status for a component received isn't maintained, then SAP GTS considers the specific component as not having preferential origin across all suppliers.

Per the worst-case principle, SAP GTS will use all the goods receipts and the entries in the Manage Long-Term Supplier's Declaration – Inbound app that are released to aggregate preferential results. By using both categories of document, SAP GTS will check the preferential status (or the lack thereof) of the product for all suppliers who delivered the component to the AU. If two suppliers have LTSDs with conflicting preferential statuses maintained for the same product, the aggregation will apply the worst case of these preferential statuses for the product. Moreover, if a supplier isn't located in a country that is party to an FTA, a company won't typically solicit this supplier, and SAP GTS won't create a supplier LTSD entry. However, this supplier should still be used in the aggregation if the company received the component from this supplier. Therefore, the goods receipt is taken into account in the aggregation logic. (Note that the purchase order would not need to be considered for the aggregation, as it doesn't mean that the goods are received, so it should not yet influence the preference result of the goods.)

In the following sections, we'll take a closer look at the aggregation process for supplier declarations, from the duration and sequence to the resulting log. We'll also explain the results in the product master.

Goods Retention Period

The aggregation logic can be customized (Section 8.4.7) to define the duration of the worst-case principle. The start of the aggregation can be based on the release date of a supplier LTSD or based on the first goods receipt date within the validity of a released supplier LTSD. The duration of the worst case will depend on the duration of the released LTSD (e.g., one year) and the goods retention period. The goods retention period is defined at four levels:

1. **Administrative unit**
   Where a default value is set for all products and suppliers within the AU. This is the only goods retention period that is mandatory in the system as it comes from Customizing (Section 8.4.7).
2. **Supplier or product**
   Where a supplier-specific or product-specific retention period can be maintained. The maintenance is either done via the Manage Products app or the Manage Preference Properties – Suppliers app.
3. **Supplier and product**
   Where a retention period is maintained for a specific supplier-product combination. This maintenance is done via the Manage Preference Properties – Suppliers app.
4. **Material document**
   Where a retention period is maintained for a specific material document via Transaction /SAPSLL/RET_PERI_UPD.

Each of these levels are enabled in sequence and define the duration of the aggregation. For instance, let's assume the following:

1. On the first of January, a supplier LTSD is maintained for a period of one year for a product, stating that the product is preferential for an FTA.
2. On March 1st of that same year, a goods receipt document for the same product is posted in SAP S/4HANA from another supplier. No supplier LTSD was maintained for that supplier product combination in SAP GTS.

According to the aggregation, the worst-case principle should apply for this product, and it will receive a non-preferential status in SAP GTS. The duration of this non-preferential status will depend on the goods retention period maintained in any of the four levels. If the AU retention period is equal to one year, then the product won't be eligible from March 1st to the remainder of the LTSD period. If the company maintained a specific retention period of one month for the product, the supplier, or both, for instance, then the product won't be eligible from March 1st to March 31st, but it will regain its preferential eligibility in April, thanks to the previously maintained LTSD.

Finally, the company can choose to update the product retention period for the goods receipt document item to apply a similar logic and alleviate the aggregation duration according to the worst case. Figure 8.25 provides a visual aid to help you understand the goods retention period impact on the preference status.

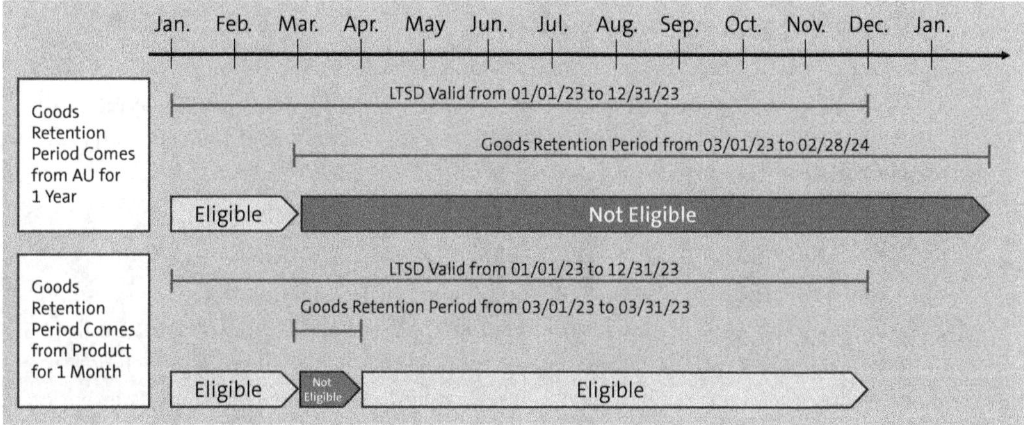

Figure 8.25 Goods Retention Period

Aggregation Log

To understand the aggregation results, the **Display Aggregation Log** app, as shown in Figure 8.26, is available via its tile under **Preference Determination • Aggregation – Products.**

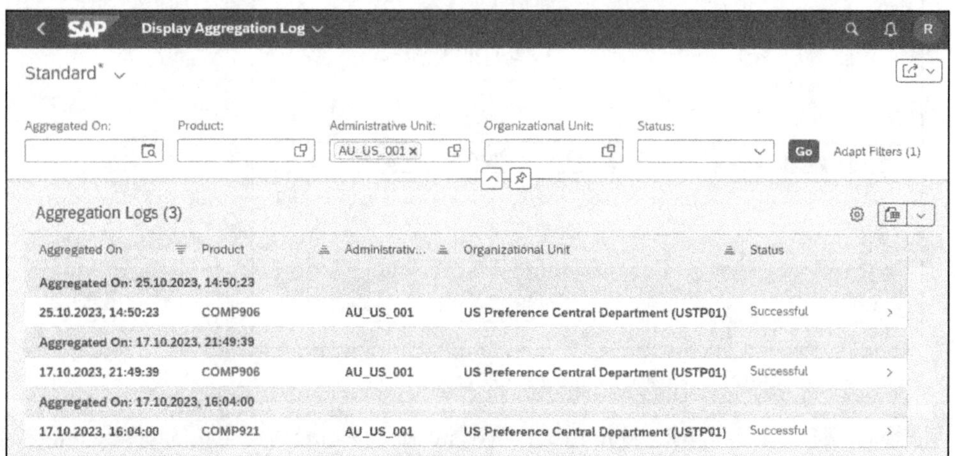

Figure 8.26 Aggregation Log

The aggregation log allows the user to drill down and see the details of the aggregation by AU, product, and date, and it provides information about the preference statements by FTA as well as the LTSDs and goods receipts that are taken into account during the aggregation. Another method to alleviate the worst-case principle is by implementing the Management of Preference Data for Product Identifiers functionality, which is further explained in Section 8.5.

Gamma's Aggregated Preference Data

In the previous step, Gamma received the completed LTSDs from its suppliers. Gamma's users maintained their responses in the SAP GTS system as discussed in the previous step, which performs an aggregation on the components sourced from multiple suppliers.

The shoelaces (COMP632) used in the production process of the finished product are multisourced from suppliers in Mexico and Peru.

The Mexican vendor confirmed their components to be wholly obtained. The vendor in Peru replied that their components aren't preferential. SAP GTS consolidates this information and, according to the worst-case principle shown in Figure 8.27, concludes that the multisourced component isn't preferential for the duration of the goods retention period.

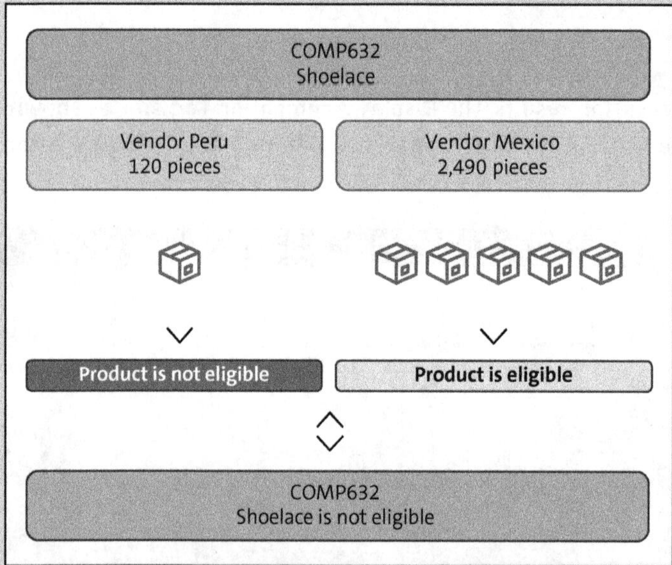

Figure 8.27 Preferential Status Aggregation Logic

Looking back at the finished product's BOM, shown in Figure 8.28, the following information is now obtained to use as input in preference determination:

- COMP632 (shoelaces): Non-originating value ($10)
- COMP906 (leather): Originating value ($45)

- COMP921 (wool): Originating value ($30)
- COMP976 (rubber sole): Non-originating value ($15)

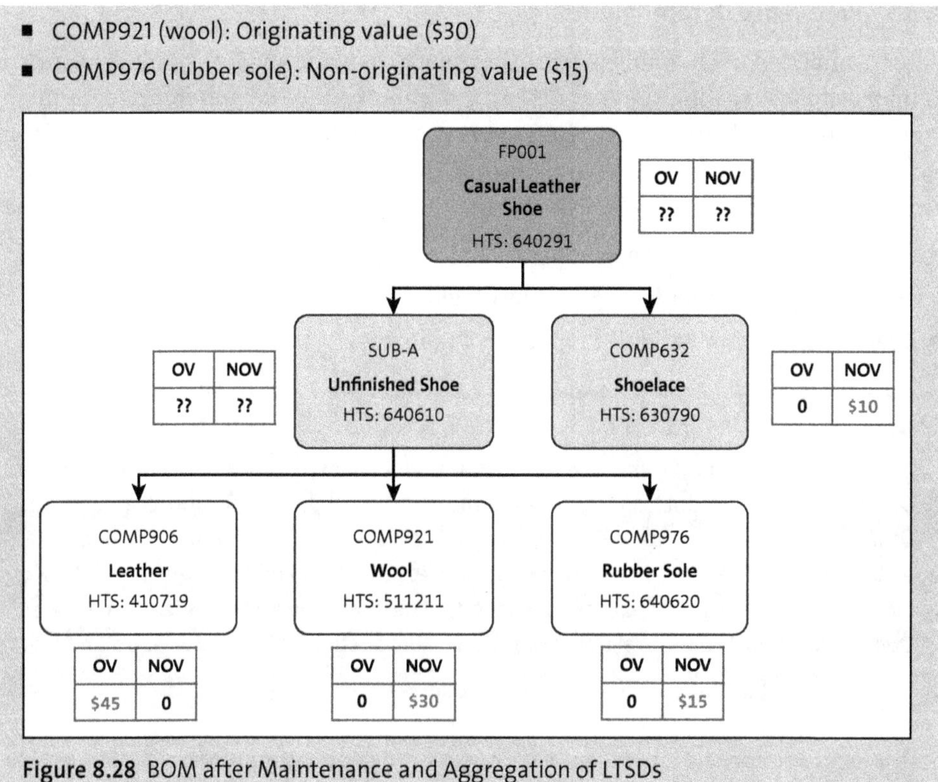

Figure 8.28 BOM after Maintenance and Aggregation of LTSDs

Preferential Master Data

SAP GTS stores the preference result of the LTSD responses after aggregation in the product master in the **Preference Statements** tab. Figure 8.29 shows the preference data for the leather shoe.

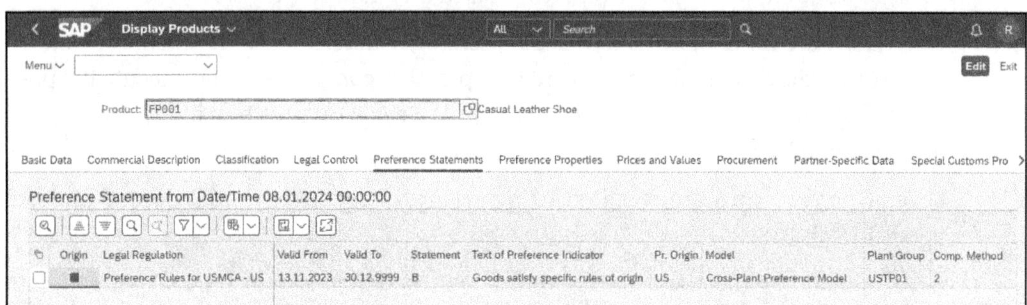

Figure 8.29 Preference Data in the Product Master

SAP GTS uses the preference data stored in the product master in the next step to perform preference determination for the finished product. It may also perform the preference determination for subassemblies depending on the BOM explosion method.

8.3.5 Preference Determination

After completing the inbound process (i.e., requesting [and sending reminders], maintaining, and aggregating supplier LTSDs), you can execute the preference determination for the finished product. *Preference determination* in SAP GTS has its own dedicated tiles, as shown earlier in Figure 8.9. These include the following SAP Fiori sections:

- **Preference Determination – Products**
- **Preference Determination – Sales Documents**
- **Preference Determination – Production Documents**

> **Note**
>
> As mentioned earlier, **Preference Determination – Sales Documents** and **Preference Determination – Production Documents** are SAP Fiori apps specifically designated for configurable BOMs, which will change per transaction. To keep the narrative concise, we'll focus on the **Preference Determination – Products** app intended for static BOMs, which remains constant across transactions. The **Preference Determination – Sales Documents** and **Preference Determination – Production Documents** apps are also addressed in Section 8.5.

In the next sections, we'll explain the process for performing preference determination, including rules of origin, preference status, and BOM explosions.

Preference Determination Process

We'll start in the **Perform Preference Determination – Products** app. As displayed in Figure 8.30, click **Execute** to run preference determination for all in-house-produced products with at least one BOM. The program can be run for a selection of products, FTAs, or different plant groups (i.e., AUs) based on business needs.

Note, there are other options to run the preference determination, which were shown in the other tiles previously in Figure 8.9. In practice, companies often execute the preference determinations on a periodic basis (e.g., weekly or monthly) in the background via automated jobs.

Several prerequisites are required to run the preference determination of finished goods in SAP GTS:

1. AUs, including the plant-based or cross-plant-based preference model, are set up in SAP GTS (Section 8.4.1).
2. Preference agreements relevant for the AU are active in SAP GTS. In addition, the rules of origin for these FTAs are available.
3. Product-related master data is available in SAP GTS, such as the BOMs for the finished product, procurement indicators, costs, and prices, as well as classification.

The classification code of the finished product dictates the applicable rules of origin for the FTA (Section 8.6). Note the importance of the procurement indicator value, as only mixed or in-house-produced goods are relevant for determination, while externally procured goods are relevant for an LTSD request.

4. A BOM calculation model (top-down or bottom-up explosion) is chosen, and the preference determination methods relevant for the FTA are set up appropriately (Section 8.4.1).
5. Optionally, the supplier LTSD has been requested and maintained for the components of the BOMs, so that their preferential eligibility is considered during the preference determination process.

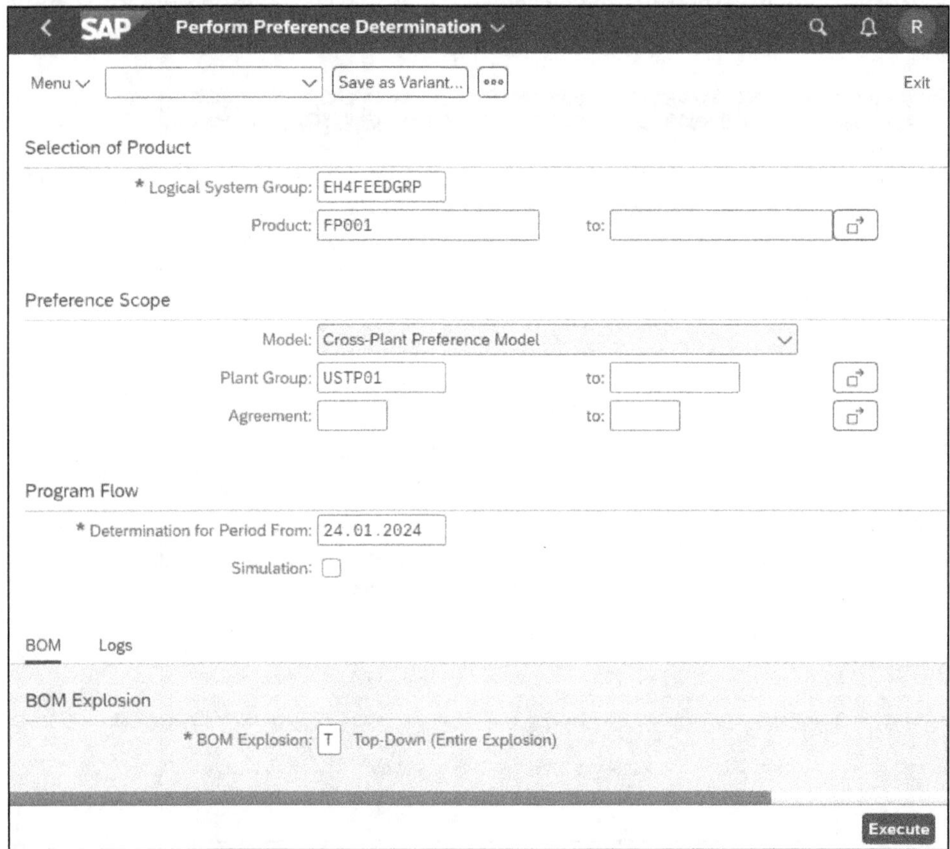

Figure 8.30 Perform Preference Determination

When running the preference determination, the preferential status is determined for each active agreement of all exploded BOMs in the appropriate AU for a specific date. Every preference determination run can be completed for one AU or multiple AUs at once, but all preferential results will be stored/related to a specific AU-FTA-product combination. BOMs transferred from SAP S/4HANA to SAP GTS need to be exploded according to the right calculation method (either top-down or bottom-up), as explained in Section 8.2.2. Finally, the preference determination is done for a given date and for the active FTAs relevant to the AU.

Preferential Rules of Origin

SAP GTS uses the classifications of the finished goods of the BOM to determine the applicable preference rules of origin to run the preference determination. There are different types of rules per FTA as explained in Section 8.2.2, and SAP GTS maintains them as configuration/master data, as explained in Section 8.6.3.

Preference rules of a specific FTA can be displayed at any time in the **Display Preference Rules** tile as long as the rules have been stored in SAP GTS. Considering the Gamma business case, Figure 8.31 displays the applicable rules for Gamma's casual leather shoe.

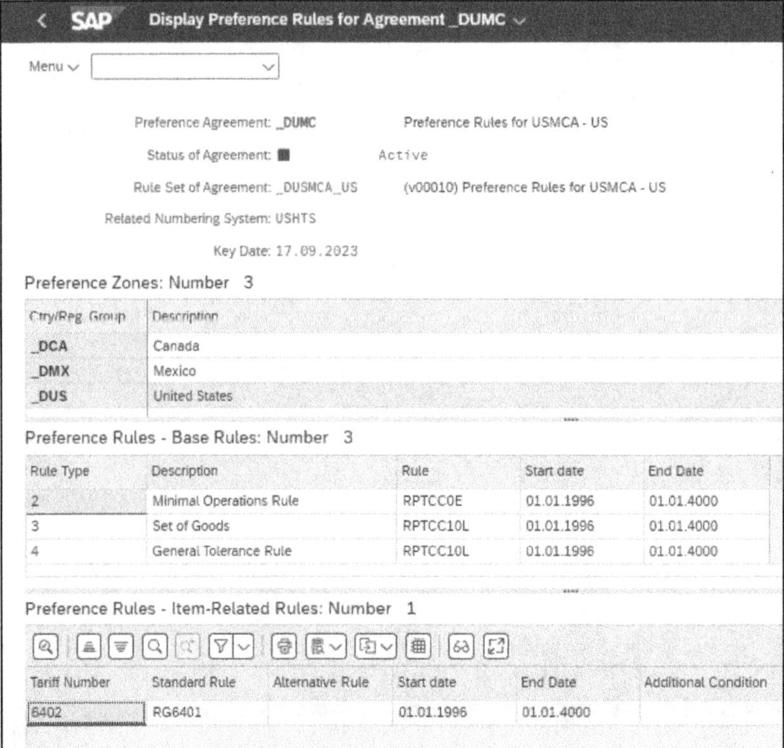

Figure 8.31 Preference Rules for Gamma's Casual Leather Shoe

Recall that there are several ways for a product to qualify as preferential. When a product isn't wholly obtained/produced or produced entirely from originating materials, it can still have preferential status and origin if it has been sufficiently processed. This sufficient processing can be determined by fulfilling the applicable rule of origin for the product and applying a comparison method during the calculation that doesn't necessitate the maintenance of a supplier LTSD in SAP GTS.

The methods and rules of origin that can be automated in SAP GTS are translated into one of seven SAP GTS preference comparison methods as displayed in Figure 8.32 and also further explained in Section 8.2.2:

- **External Preference Document (Vender/Manual) (0)**
  This comparison method is applied when the preferential status and origin are either aggregated by LTSDs or manually maintained in the finished product's master data.
- **Transaction Value Method (1)**
  This comparison method considers the finished product's maximum non-originating content and compares it to the transactional value.
- **Net Cost Method (2)**
  This comparison method considers the finished product's non-originating content and compares it to the finished product's net cost.
- **Tariff Shift (3)**
  This comparison method compares the tariff code of the finished product against the tariff codes of its components and subassemblies.
- **Build-Up Method (4)**
  This comparison method calculates the originating content as a percentage of the finished product's adjusted value (typically FOB) compared to the value of originating materials, including their production costs and profit.
- **Build-Down Method (5)**
  This comparison method calculates the originating content as a percentage of the finished product's adjusted value (typically FOB) compared to the value of its non-originating materials.
- **Transaction Value Method RVC (6)**
  This comparison method considers the finished product's minimum originating content and compares it to the transactional value.

Comparison Method (1)

Search and Select

Items (7)

Comparison Met...	Short Description
0	External Preference Document (Vendor/Manual)
1	Transaction Value Method
2	Net Cost Method
3	Tariff Shift
4	Build-Up Method
5	Build-Down Method
6	Transaction Value Method RVC

Figure 8.32 Comparison Methods

As mentioned in Section 8.2.2, some rules of origin can't be fully automated without the intervention of a user as they require additional details on the nature of the product (e.g., manufacturing or processing operations/specific rules). Conditions, minimal operations, or set specific rules of origin are further explained in Section 8.6.3.

Per Figure 8.31, the rules that apply to the leather shoe under USMCA are (1) any of the base rule for the FTA and applicable to the product, which in this case is only the general tolerance base rule (as the product isn't subject to minimal operations nor is a set of goods), and (2) the rules specific to the tariff classification of the concerned product, in this case, chapter 6402. Figure 8.33 more specifically displays the USMCA general tolerance rule as a de minimis calculation with a threshold of 10% of either (1) the transaction value of the good adjusted to exclude any costs incurred in the international shipment of the good, or (2) the total cost of the good. This means that if a tariff shift rule applies, then the finished product can still qualify with some non-shifting components as long as the value of non-originating and non-shifting components in the finished product's BOM is less than or equal to 10% of the finished product value; conversely, 90% of the good's components by value must be originating or shift in accordance with the rule of origin.

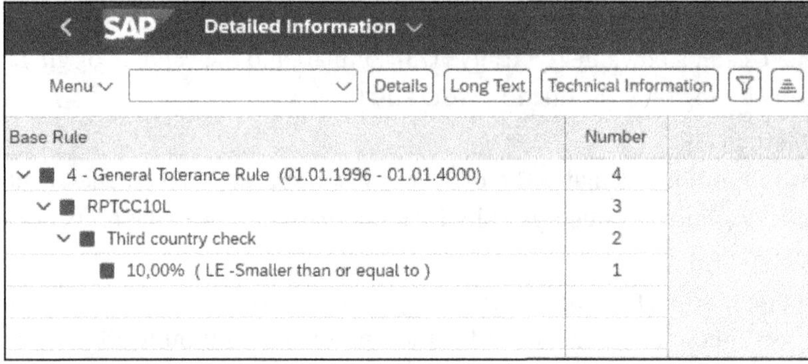

Figure 8.33 USMCA General Tolerance Rule

Figure 8.34 shows the rule for chapter **6402**. In this scenario, the standard rule is an AND operation, signifying that the requirements listed for the AND operation are all required to satisfy the standard rule. See Section 8.6.3 for further details on the rule of origin and the analysis of another example.

For the standard rule in Figure 8.34 to be satisfied, both the following requirements need to be met:

- **HS check**
  The finished product may not include non-originating components with an HS heading ranging from 6401 to 6405. Non-originating components also may not be classified with HS code 6406.10.

- **RVC**
  The net cost method is applied, stating that at least 55% of the finished product's value must be USMCA-originating.

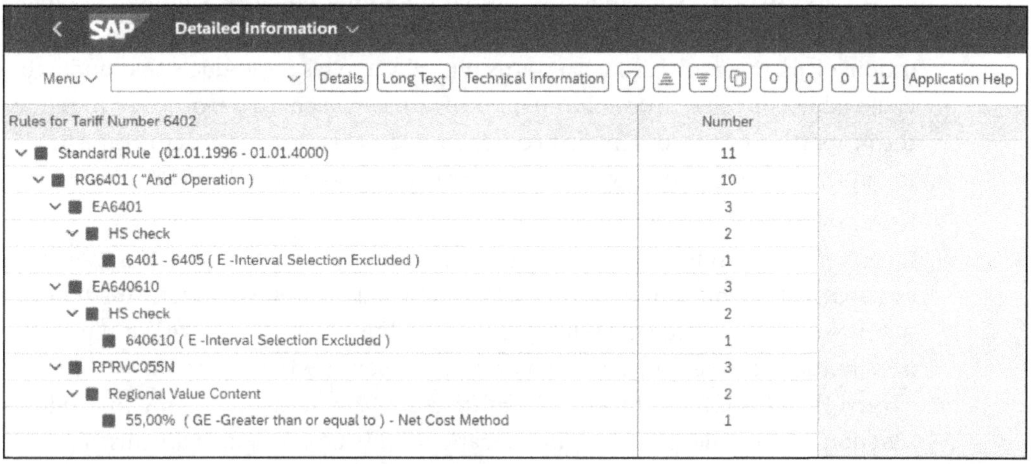

< SAP Detailed Information

Menu | Details | Long Text | Technical Information | 0 | 0 | 0 | 11 | Application Help

Rules for Tariff Number 6402	Number
Standard Rule (01.01.1996 - 01.01.4000)	11
RG6401 ("And" Operation)	10
EA6401	3
HS check	2
6401 - 6405 (E -Interval Selection Excluded)	1
EA640610	3
HS check	2
640610 (E -Interval Selection Excluded)	1
RPRVC055N	3
Regional Value Content	2
55,00% (GE -Greater than or equal to) - Net Cost Method	1

Figure 8.34 USMCA Item-Related Rules for Gamma's Leather Shoe

Thus, the standard rule falls under the net cost method, for which a few additional tariff requirements apply. In both cases, the rule needs to consist of one of the comparison methods discussed in Section 8.2.2.

Preference Status

As a result of the preference determination based on the rules of origin, SAP GTS calculates one of the following three possible statuses, each with a specific SAP GTS symbol and threshold value. Note that these results correspond to the preference determination results discussed in the Preference Management business process in Section 8.2.2.

- **Preferential**
  If the product always satisfies the rule of origin and will always qualify as long as the underlying data used in the preference run remains unchanged, then the result is visualized as a green square symbol and assigned a threshold value of minus infinity, meaning that the product is preferential regardless of the transactional selling price because the price will always be above the minus infinite threshold. For instance, this status can apply to products that undergo a complete tariff shift under a tariff shift rule.
- **Preferential depending on a threshold value**
  A second scenario may occur when a finished product is eligible for preferential treatment depending on the product's transactional sales price. In this scenario, the result of the finished product is visualized as a yellow triangle symbol. The assigned

threshold value is the value at or above which (i.e., greater than or equal to) the non-originating value of a product's components makes up an acceptable share in the finished product, as determined in the preference rules. If the product is sold at a price lower than the threshold value, then the non-originating components make up a larger share in the total value than allowed by the preference rules, and the product isn't preferential. Such a status typically applies to products that undergo a value-added or value threshold/de minimis rule. In such a case, one way to reduce the proportion of non-originating content and decrease the value of the threshold is to request for a supplier LTSD for the sourced components of the BOM.

- **Non-preferential**
  If no preferential status is applicable for the finished product, the result is visualized as a red circle symbol and assigned a threshold value of plus infinite. This signifies that no matter how high the selling price is set, the product can never be eligible for preferential treatment in the given agreement because the selling price can never exceed the plus infinite threshold. For instance, this status can apply to products that don't satisfy the conditions in their applicable rule of origin or are missing a finished goods classification.

When performing the determination, applicable rules are verified in parallel in SAP GTS. This means when a standard rule exists in parallel to an alternative rule, the most advantageous outcome is used as the outcome for the entire preference determination. In contrast to other functionalities in the Preference Management functionality, the worst-case principle isn't applied.

BOM Explosions: Top-Down and Bottom-Up

With the preference determination and preference status calculation complete, it's time to perform the top-down or bottom-up BOM explosion. In this section, we'll explain the resulting preference determination log and take a look at both the top-down and bottom-up calculations for our Gamma use case.

Running a Top-Down Calculation on Gamma's Finished Product

Gamma obtained preferential status for all of its components, so it's now time to perform the preferential determination.

Because USMCA allows both top-down and bottom-up BOM explosions, Gamma performs both calculations. Based on the calculation results, Gamma will determine which BOM explosion method is more favorable to apply.

Figure 8.35 shows the BOM considered in the top-down calculation. SAP GTS neglects the unfinished shoe and considers only the lowest-level components. Recall that this technique is more favorable in tariff shift comparison methods because the lowest-level components are more likely to undergo tariff shifts than the subassemblies.

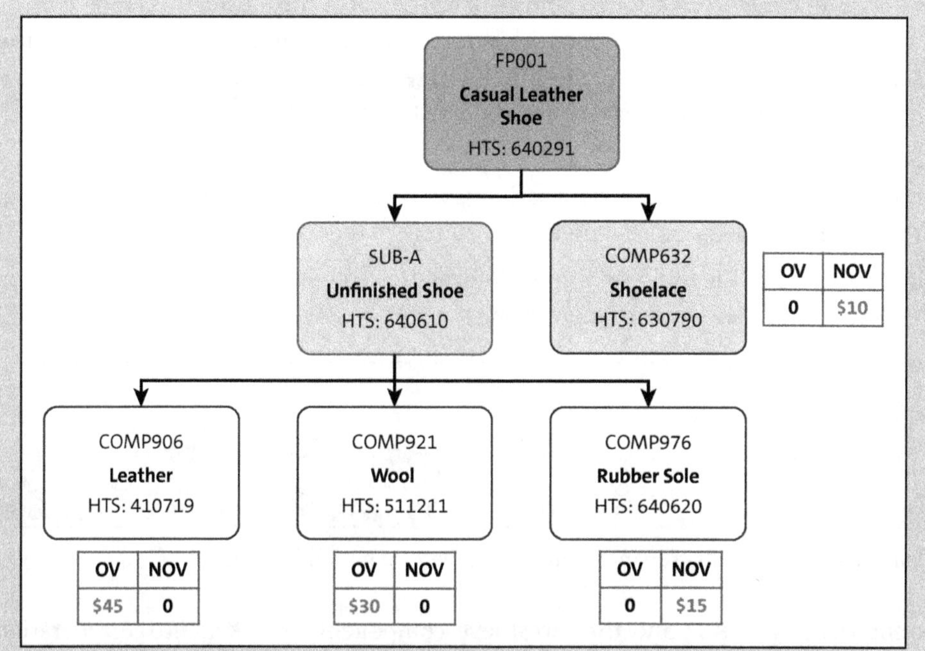

Figure 8.35 BOM Considered in the Top-Down Calculation

Based on the four components, the system applies the preferential rules and determines the preference status for the finished product.

You can see the result of the top-down preference calculation in the calculation log (accessed via the **Display Preference Determination Logs** app), as shown in Figure 8.36, from which Gamma observes that its finished product is eligible for preferential treatment.

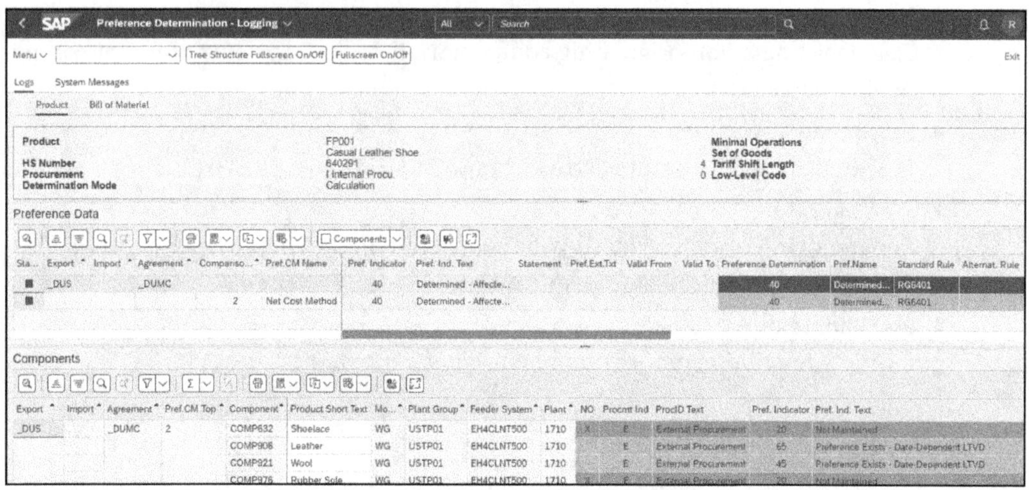

Figure 8.36 Preference Calculation Result (Top-Down)

The preference determination log shown in Figure 8.36 is the most critical screen in the Preference Management determination process from a system point of view. To better grasp the logic and structure attained in the screen, let's break it down into three different sections: the header, components, and finished product.

The header contains basic information concerning the finished product, including product name and number, HTS classification code, and procurement indicator (e.g., procured externally, produced in-house, or a mix of the two).

Figure 8.37 shows the most significant information found in the **Components** section that SAP GTS uses as input for performing the preference determination.

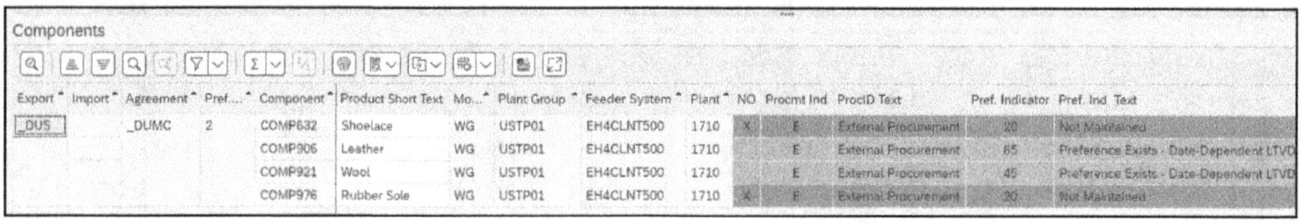

Components

Export	Import	Agreement	Pref....	Component	Product Short Text	Mo...	Plant Group	Feeder System	Plant	NO	Procmt Ind	ProcID Text	Pref. Indicator	Pref. Ind. Text
_DUS		_DUMC	2	COMP632	Shoelace	WG	USTP01	EH4CLNT500	1710	X	E	External Procurement	20	Not Maintained
				COMP906	Leather	WG	USTP01	EH4CLNT500	1710		E	External Procurement	65	Preference Exists - Date-Dependent LTVD
				COMP921	Wool	WG	USTP01	EH4CLNT500	1710		E	External Procurement	45	Preference Exists - Date-Dependent LTVD
				COMP976	Rubber Sole	WG	USTP01	EH4CLNT500	1710	X	E	External Procurement	20	Not Maintained

Figure 8.37 Components Log in the Preference Determination (Top-Down)

As shown in Figure 8.37, only the lowest-level components are taken into consideration under the top-down method, while the calculation ignores the subassembly (i.e., the product SUB-A [unfinished shoe] isn't part of the list of products in the determination log). The system retrieves the following information (accessible via scrolling sideways) to correctly perform the preferential calculation that corresponds to the columns in Figure 8.37:

- **Component**
  Component product number.
- **Product Short Text**
  Component product description.
- **NO**
  Flag depicting a non-originating component (if flagged, the entry for that component is marked red).
- **Procmt Ind**
  Procurement indicator transferred from SAP S/4HANA.
- **Pref. Indicator**
  Preference indicator provided by the supplier of the specific component in an LTSD during the supplier solicitation process.
- **Pref. Ind. Text**
  SAP GTS description of the component's preference indicator.

- **Threshold Value**
  Threshold value at which the component is considered originating. This column applies when you're dealing with subassemblies as a component in the preference calculation. (We'll discuss it in the bottom-up calculation method.)
- **Comparison Price**
  Like the previous value, this column applies to subassemblies, which we'll discuss in the bottom-up calculation method.
- **Orig. Val./NO Value**
  These columns specify the amount of originating or non-originating value, respectively. SAP GTS considers the component's material price and checks the non-originating flag to determine whether the product consists of originating or non-originating value.
- **RVC Benefit**
  This column specifies the RVC benefit that is gained by applying the specific calculation method. This also applies to the bottom-up method, which we'll discuss later.
- **Commodity Code**
  This is the component HTS classification code.

Apart from this information, other data elements are available at the component level, including the quantity and unit of measure (UoM) used in the finished product, preference model, plant, and so on.

The actual result of the finished product preference calculation is shown in detail in Figure 8.38.

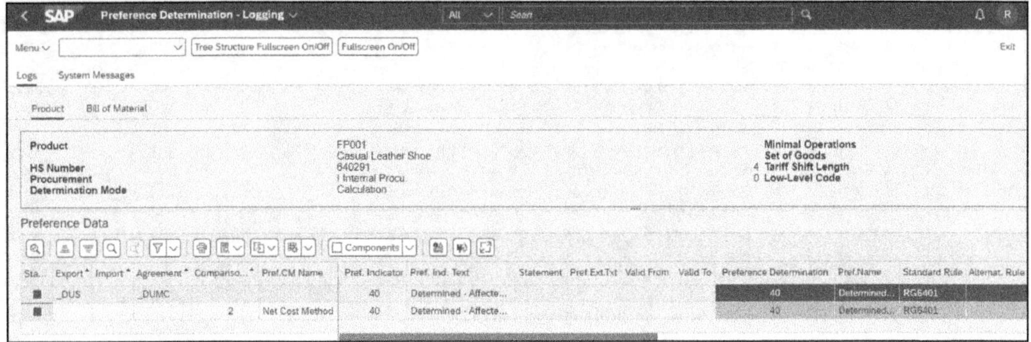

Figure 8.38 Header and Finished Product Log in the Preference Determination (Top-Down)

SAP GTS calculates the applicable rules according to the right preference comparison method for each agreement relevant to the corresponding AU. As shown earlier in Figure 8.31, the net cost method is the sole comparison method included in the preferential rules for Gamma's finished product, but other methods may be allowable under different FTAs (see Section 8.2.2 for further details).

The first column in the calculation log (**Status**) depicts the originating status. This value is aggregated on a best-case principle, meaning that if the rules of origin for a product contain multiple comparison methods, satisfying any rule is sufficient for the finished product to claim preferential status. For instance, if the finished product was subject to both a net cost and transaction value rule, the finished product would only need to meet one of the two rules to be preferential.

The top entry for each agreement is a summary copy of the entry with the most favorable result. For Gamma, the finished product preference determination results in a favorable preference status seen by a green status, signifying that the finished good satisfies the specific rules of origin. As we discussed earlier, the applied net cost method results in either a positive or a negative calculation result, thus leaving the **Threshold Value** column empty.

However, for other comparison methods (e.g., transactional value), the preferential eligibility can depend on the eventual sales price of the finished product. In such situations, SAP GTS calculates the **Threshold Value** at which the finished good is preferential. This threshold can be seen in the **Preference Determination Logging** tile or in the **Display Preference Data for Product** tile, as shown in Figure 8.39 for other illustrative examples.

SAP Display Preference Data for Product

Menu

Preference Statement for Period 20.11.2023 To 20.11.2023

Valid From	Valid To	Statement	Preference Indicator	Pr. Origin	Model	Comp. Method	Threshold Value	Value Basis	C...
08.05.2023	02.01.2024	B	Determined - Affected by Price	US	Cross-Plant Preference Model	3	17.47		USD
08.05.2023	02.01.2024	B	Determined - Affected by Price	US	Cross-Plant Preference Model		8.96		USD
08.05.2023	30.12.9999	B	Determined - Affected by Price	US	Cross-Plant Preference Model		18.93		USD
08.05.2023	30.12.9999	B	Determined - Affected by Price	US	Cross-Plant Preference Model		18.45		USD
08.05.2023	30.12.9999	B	Determined - Affected by Price	US	Cross-Plant Preference Model		18.93		USD
08.05.2023	30.12.9999	B	Determined - Affected by Price	US	Cross-Plant Preference Model		11.21		USD
08.05.2023	30.12.9999	B	Determined - Affected by Price	US	Cross-Plant Preference Model		5.73		USD
08.05.2023	02.01.2024	B	Determined - Affected by Price	US	Cross-Plant Preference Model		9.66		USD
08.05.2023	02.01.2024	B	Determined - Affected by Price	US	Cross-Plant Preference Model		16.59		USD
08.05.2023	02.01.2024	B	Determined - Affected by Price	US	Cross-Plant Preference Model		9.68		USD
08.05.2023	02.01.2024	B	Determined - Affected by Price	US	Cross-Plant Preference Model		16.57		USD
08.05.2023	02.01.2024	B	Determined - Affected by Price	US	Cross-Plant Preference Model		9.67		USD
08.05.2023	02.01.2024	B	Determined - Affected by Price	US	Cross-Plant Preference Model		16.52		USD
08.05.2023	02.01.2024	B	Determined - Affected by Price	US	Cross-Plant Preference Model		5.45		USD

Figure 8.39 Threshold Value of a Finished Good

Furthermore, the amount of originating and non-originating value in the finished product's BOM is specified. These values, together with the manufacturing efforts (e.g., labor and overhead), are used to determine the originating content share. The puzzle icon at the top of the **Preference Data** section shows the detailed calculation, as shown in Figure 8.40.

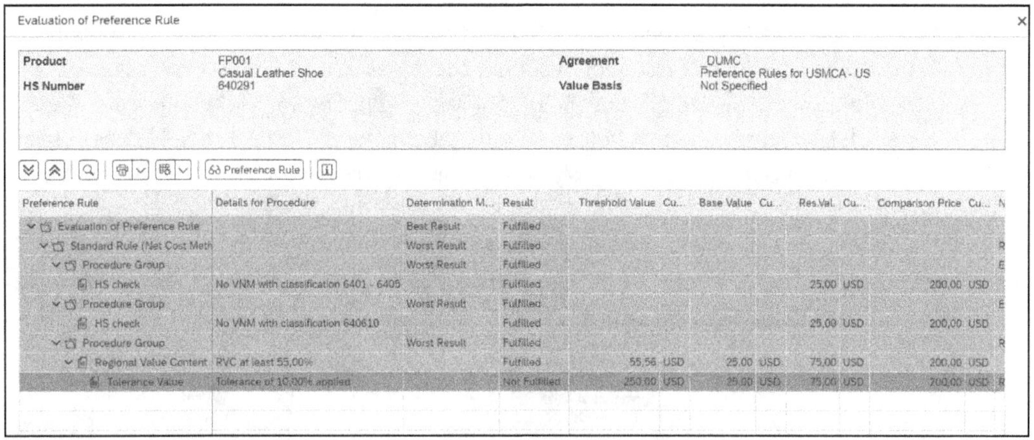

Figure 8.40 Evaluation of Preference Rule (Top-Down)

As previously determined, the various components consist of $75 originating and $25 non-originating value. The system derives an additional $100 in production efforts as the finished product's material price is $200. Combined, the originating share accounts for 87.5%, equal to the value calculated in the analogous business process calculation.

Finally, note that upon completion of the preference determination, the preference result is updated in the finished good's product master in SAP GTS under the **Preference Statements** tab, as explained in Section 8.6.1. However, when the determination is run in simulation mode, the result can be shown in the system but isn't saved in the product master unless a user performs the necessary action to save the results. The simulation run will be explained in Section 8.3.6.

Running a Bottom-Up Calculation on Gamma's Finished Product

After running the top-down determination, Gamma performs a bottom-up determination. The BOM determination now starts at the lowest-level components and works its way up.

The system first determines the preferential status for the subassembly and then combines that information with the shoelace component to determine the finished product's preferential eligibility.

For the subassembly's preference determination, the system considers the BOM, as you can see in Figure 8.41, followed by a preference determination considering the BOM, as displayed in Figure 8.42. Next, the system uses the subassembly's preference result to calculate the preference determination for the finished product based on the BOM, as shown in Figure 8.43. Finally, the result is shown later in Figure 8.44.

Figure 8.42 shows the **Bill of Material** tab for the bottom-up determination. As shown in the **Assemblies** section, two preference determinations were performed: first on the unfinished shoe and second on the finished product. The rule of origin for the unfinished shoe's tariff code also requires a net cost threshold of 55%. Because only $15 of the unfinished shoe's total value ($120) is of non-originating value, the unfinished shoe is eligible for preferential treatment. Thus, the whole unfinished shoe is considered originating, including the previously non-originating $15 component.

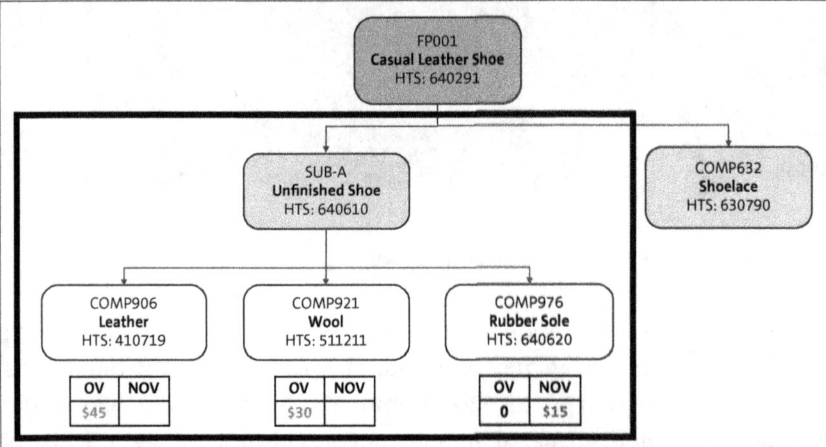

Figure 8.41 BOM Considered in the Bottom-Up Subassembly Determination

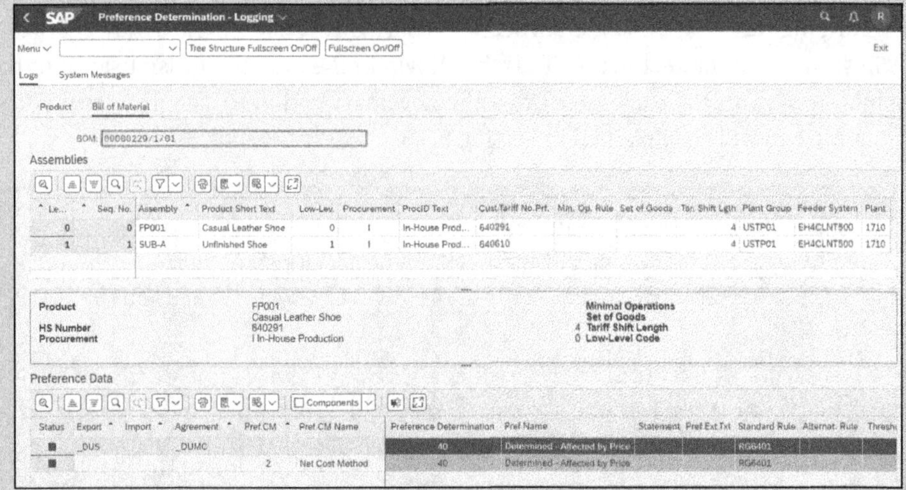

Figure 8.42 Bill of Material Log in the Bottom-Up Preference Calculation

Note that apart from the **Logs** tab with **Product** subtab of the **Preference Determination** screen, which reveals the actual determination relevant to the finished product, the **Bill of Material** subtab gives insight into the determination of each subassembly in the finished product's BOM.

After determining the unfinished shoe's eligibility, the system considers the BOM as a whole, as shown in Figure 8.43.

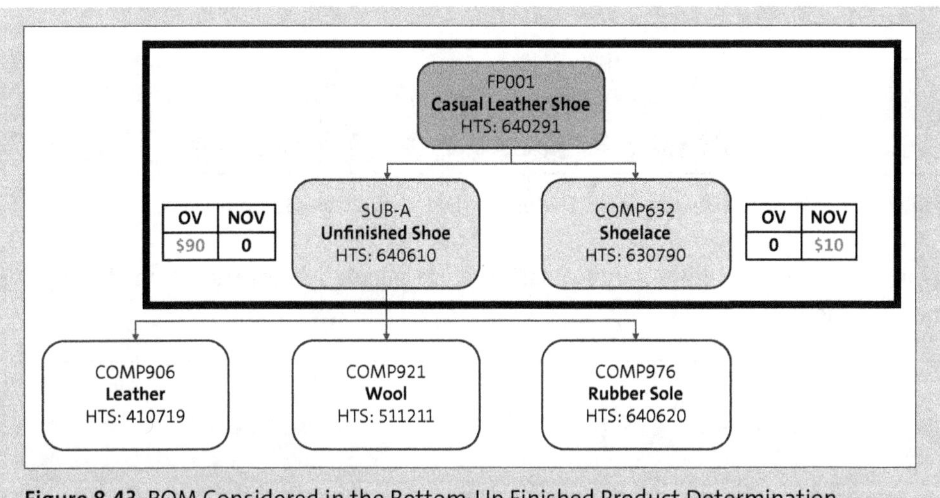

Figure 8.43 BOM Considered in the Bottom-Up Finished Product Determination

The result of the preference determination is displayed in Figure 8.44. As opposed to the result shown earlier in Figure 8.36, which included all four components, only two products are considered in the preference determination: **Unfinished Shoe** and **Shoelace**.

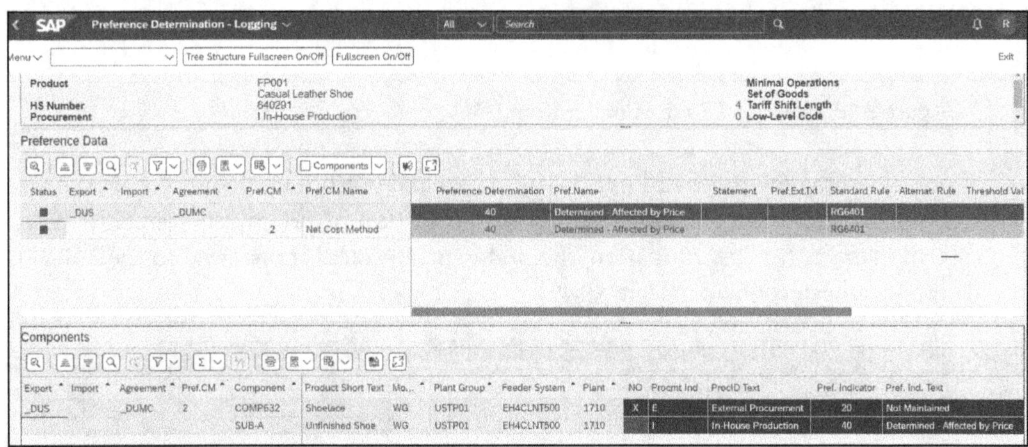

Figure 8.44 Preference Calculation Result (Bottom-Up)

The subassembly is used in the finished product's preference determination as consisting of $90 worth of originating components, as displayed in Figure 8.45, which is an enlarged image of the **Components** section of Figure 8.44. By leveraging the bottom-up calculation method, you include the non-originating rubber sole component as originating value, resulting in an additional $15 RVC benefit as compared with the top-down method.

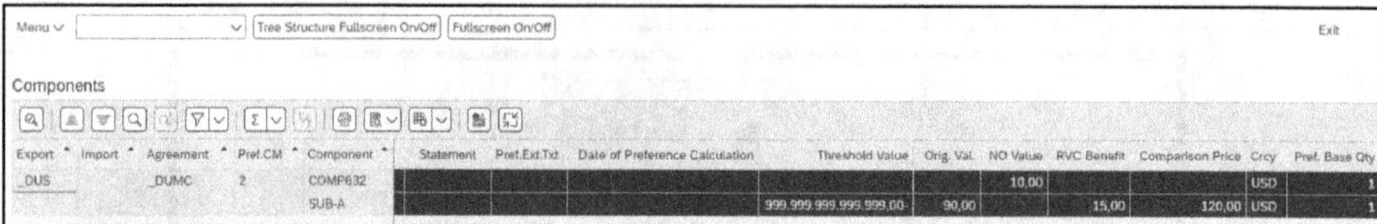

Figure 8.45 Components Log in the Preference Determination (Bottom-Up)

Finally, Figure 8.46 shows the originating share calculation for the finished product's net cost method.

Evaluation of Preference Rule

Product	FP001 Casual Leather Shoe	**Agreement**	_DUMC Preference Rules for USMCA - US
HS Number	640291	**Value Basis**	Not Specified

Preference Rule	Details for Procedure	Determination M...	Result	Threshold Value	Cu...	Base Value	Cu...	Res.Val.	Cu...	Comparison Price	Cu...
Evaluation of Preference Rule		Best Result	Fulfilled								
Standard Rule (Net Cost Meth		Worst Result	Fulfilled								
Procedure Group		Worst Result	Fulfilled								
HS check	No VNM with classification 6401 - 6405		Fulfilled					10.00	USD	200,00	USD
Procedure Group		Worst Result	Fulfilled								
HS check	No VNM with classification 640610		Fulfilled					10.00	USD	200,00	USD
Procedure Group		Worst Result	Fulfilled								
Regional Value Content	RVC at least 55.00%		Fulfilled	22.22	USD	10,00	USD	90,00	USD	200,00	USD
Tolerance Value	Tolerance of 10.00% applied		Fulfilled	100.00	USD	10,00	USD	90,00	USD	200,00	USD

Cancel

Figure 8.46 Preference Evaluation (Bottom-Up)

After Gamma investigated both the top-down and bottom-up determinations, it was obvious that either method resulted in satisfaction of origin requirements. When Gamma runs the determination, the preference result is transferred to the SAP GTS product master shown in Figure 8.47.

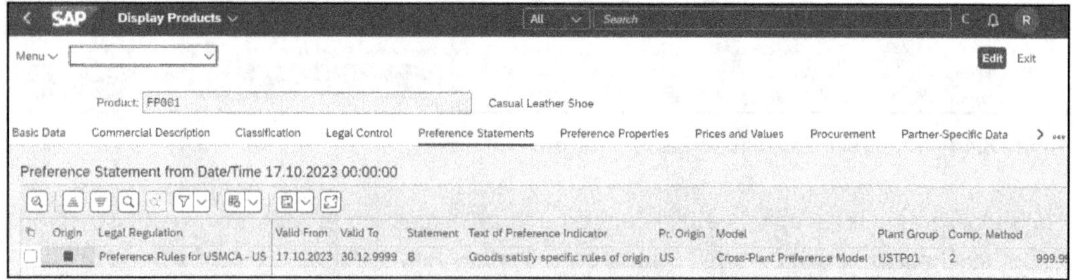

Figure 8.47 Preference Statements Tab on the Finished Product's SAP GTS Product Master

8.3.6 Evaluating Preference Results

In the next step of the Preference Management system process, the results from the preference determination identified in the previous step are stored in a log. The detailed analysis of the determination as well as the BOM details can be reviewed in the preference determination log (we saw this at work in the top-down and bottom-up calculations in the previous section). The final preferential results from the preference determination can be displayed in a standard report. We'll take a closer look at these evaluation options in the next sections.

Preference Determination Logs

SAP GTS provides a standard feature for evaluating the preference data from the company's products. An example of this report is displayed in Figure 8.48, and you can access it through the **Display Preference Determination Logs** tile in the **Preference Determination** section. For more details, refer to the previous section.

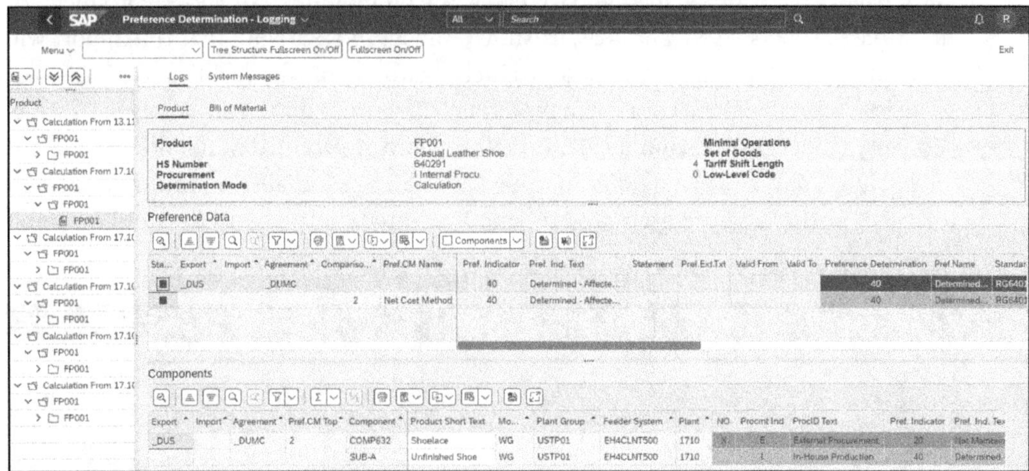

Figure 8.48 Evaluating a Product's Preference Results

Preference Determination Simulation

You can also simulate a preference determination run. These preferential results can help a company understand how the preference determination results would be calculated without saving the data in the product master. This feature can be accessed through the **Display Preference Determination Logs from Simulation** tile in the **Preference Determination** page under the **Determination – Preference** section. After running a preference determination run under simulation, the logs of the simulation will appear as shown in Figure 8.49.

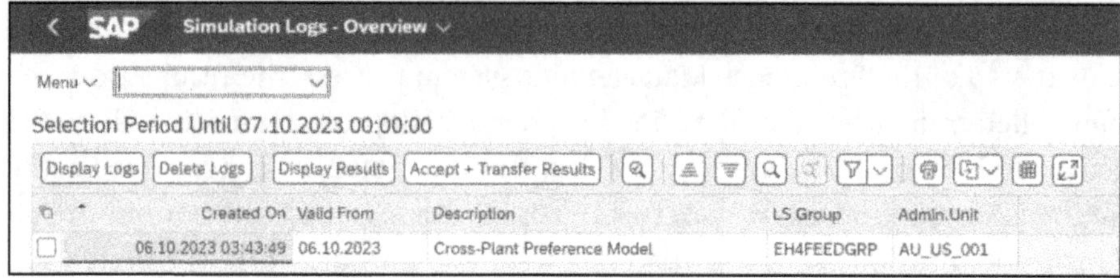

Figure 8.49 Preference Determination Simulation Log

In the simulation preference determination run, you can select **Display Results**. If you're comfortable with the results shown, you can select the **Accept + Transfer Results** button to save the results to the product master.

Preference Data

Also, SAP GTS has a **Display Preference Data – Products** tile in the **Master Data – Preference** page. This report enables you to review the preferential status(es) of products for multiple products, FTAs, and AUs, as shown in Figure 8.50. Note that this report will provide the preferential result of a product regardless of the qualification basis (i.e., based on a supplier LTSD or from a preference determination). You can use this functionality to identify the volume of products that are preferential or not preferential for any given FTA at any given point in time. Based on this initial set of information, you can also further analyze each product and investigate why the product did or didn't qualify.

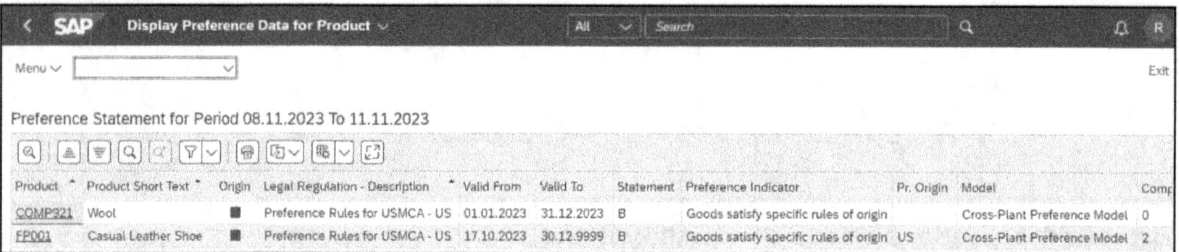

Figure 8.50 Display Preference Data – Products Tile

Gamma Evaluates the Preference Data of Its Products

Gamma's management decides to look into the preference results of its product, as displayed earlier in Figure 8.48.

Gamma's finished product is of preferential origin in the USMCA agreement, visualized with the green symbol and preferential indicator **B**. Note that SAP GTS sets the preferential country of origin for the qualifying finished product as the country of the AU performing the preference determination.

Furthermore, because the preferential origin is independent of the sales price, the threshold value is set at minus infinity as the good is always eligible for preferential treatment when shipping to USMCA customers regardless of sales price.

The next step for Gamma is to issue proof of origin to its customers when applicable, allowing the company to enjoy the benefits of the USMCA FTA.

8.3.7 Issuing Preference Certificates

Once SAP GTS and the company have determined whether a product meets the rule of origin for a given FTA, preferential certificates of origin can be generated and issued to customers. These certificates or proofs of preferential origin can vary depending on the agreement and country of destination. There are two main categories of proofs of origin, which we'll discuss in the next sections.

Transactional Proofs of Origin

Transactional proofs of origin can be issued as a transactional certificate or statement of origin (e.g., certificate of origin or preference statement in an invoice).

From an SAP GTS point of view, two integration options exist to support the transactional issuance of proofs of preferential origin:

- **Integration with Customs Management**
  The first option is the integration of SAP GTS Preference Management with Customs Management, so the certificates of origin can be printed in addition to an export declaration. In this case, the **Preference** tab on the item level in the customs export declaration compares the transaction sales price (e.g., EXW, FOB) from the SAP S/4HANA billing document against the SAP GTS preference threshold value. The result is a preferential status in the export declaration that can be used to print additional documents (e.g., preference certificate of origin) as discussed in Chapter 7. Note the **Preference** tab for customs declarations in SAP GTS needs to be activated for the export declaration document type in Customizing.
- **Integration with sales order or billing document**
  The second option is the integration of SAP GTS Preference Management with the sales order or billing document in SAP S/4HANA so the document receives a transactional preference flag at the item level. For relevant sales order or billing document types, the preference indicator flag at the item level is populated when the product is eligible for the FTA related to the country of departure (i.e., based on the plant of the item) and country of destination (i.e., based on the address of the ship-to party) of the document. An example is shown in the billing document in Figure 8.51, where the **Cus. Preference** indicator is marked. The determination of the preference eligibility of the item is done by comparing the sales price (e.g., EXW, FOB) of the sales order or billing document item with the threshold value in SAP GTS for the FTA and

service rendered date. This preference indicator flag can be used to issue transactional proofs of origin in SAP S/4HANA, such as a preferential statement in the output of an invoice, for instance. The activation of this functionality is explained in Section 8.4.11.

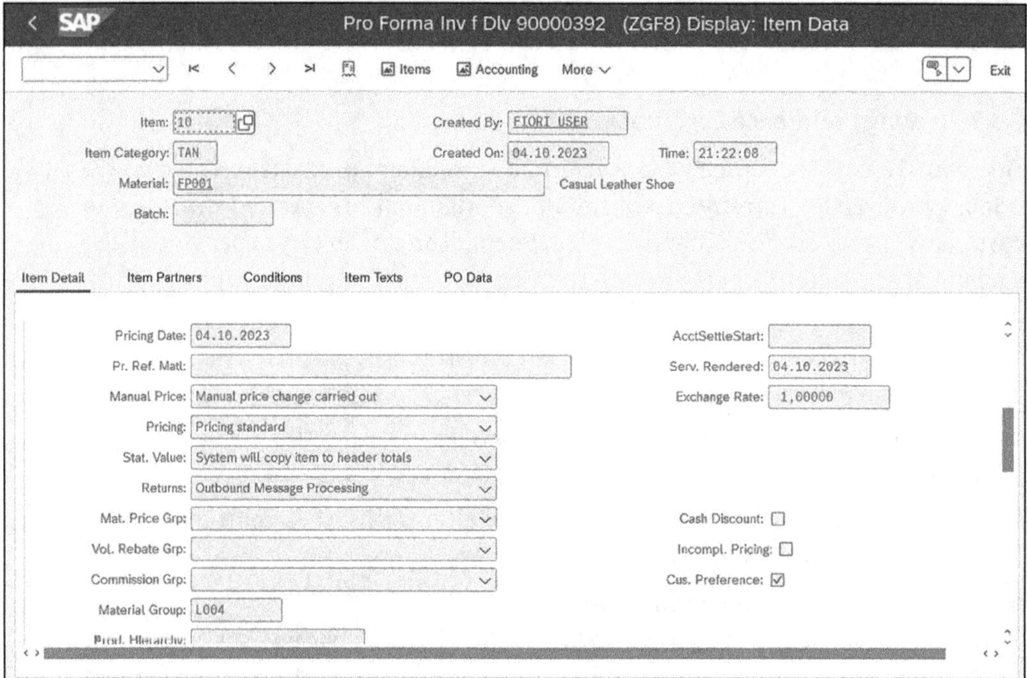

Figure 8.51 Preference Indicator on the Billing Document

Periodic Proofs of Origin and Customer LTSDs

If the proof of origin is issued for a period, such as a blanket certificate (e.g., as done with USMCA), you can issue the proof as an LTSD for the customer's purposes in Preference Management. Customer LTSDs can be issued using one of two methods (i.e., mail or email), as described in Section 8.3.2.

The tiles for issuing, monitoring, and revoking LTSDs for the customer's purposes are provided in the **Outbound Declarations** page of the **Supplier's Declaration** space, as shown in Figure 8.52.

In the **Manage Long-Term Supplier's Declarations – Outbound** app, all relevant billing documents from SAP S/4HANA are transferred to SAP GTS for Preference Management purposes. For the activated billing document types, an entry is created in the worklist for issuing customer LTSDs if the following prerequisites apply:

- The supplying plant in the billing document is assigned to an AU in SAP GTS.
- The customer is located in one of the member countries of at least one of the preferential agreements relevant to the specific AU.

- The sold product and customer are available in SAP GTS and considered relevant for issuing LTSD declarations (Section 8.6.2). Similar to the suppliers for requesting supplier declarations, customers can be manually flagged in SAP Fiori as relevant or not relevant for issuing certificates of origin.
- No customer LTSD already exists or has been sent for this product-customer-FTA-preference status combination for the period.

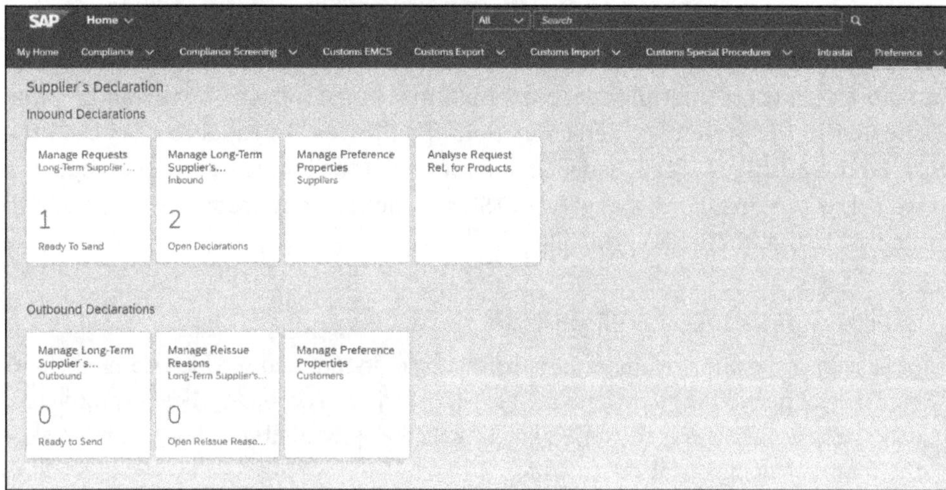

Figure 8.52 Supplier's Declaration: Outbound Declarations

If these conditions are met, then the creation of a billing document in SAP S/4HANA generates a new entry in the worklist for the **Manage Long-Term Supplier's Declaration – Outbound** app. For all entries proposed in the worklist, the company can issue LTSD to its customers, so their customers can claim preferential status for a given product or report non-preferential status for a given product.

Note

The term *customer* is used in a generic fashion and can represent any third-party customer, affiliate companies, or any other recipient of the goods.

Once a company determines that an LTSD is ready to be sent, it can issue the LTSD from the **Ready to Send** worklist. After sending the LTSD, SAP GTS will remove the entry from the **Ready to Send** worklist and transfer the entry to the **Failed** or **Sent** worklist. If the data elements of the documents are erroneous or a system failure occurs (e.g., Adobe Document Server failure), the LTSD issue will update to a **Failed** status, and the issue must be investigated and resolved before resending the LTSD. If sent successfully, the LTSD will be issued to the customer according to one of the two media types previously explained (i.e., mail or email). Additionally, the LTSD entry updates to reflect the **Sent** status and moves to the **Sent** worklist.

Until the customer LTSD is sent, all relevant billing documents for this customer-product-FTA combination are consolidated in the customer LTSD. If a finished product qualifies only when the sales price is equal to or exceeds the SAP GTS calculated threshold value, then the comparison values (i.e., sales price) of these billing documents play a role in determining the preference statement of the product to be issued. The invoice determination method and period of validity for the billing documents can be customized, as explained in Section 8.4.7, so that the last invoice value, the middle invoice value, or the highest invoice value are taken into account to compare with the threshold. Therefore, at the moment of issuing, all available billing documents for that customer-product combination are compared to the threshold value calculated in the preference determination. If all prices are equal to or exceed the threshold, a positive LTSD is issued. If at least one of the billing documents reports a price below the threshold, a negative LTSD for the relevant product(s) is generated according to the worst-case principle.

Gamma Issues Blanket Certificates to Its Customers

After performing the preference determination for its finished product, Gamma created several sales orders from its customers in Canada and Germany. Gamma proceeds with the sales for both its customers and creates the associated sales document in SAP S/4HANA (including billing documents).

Note that only the Canadian importers can benefit from the USMCA agreement as the German customers aren't located within the FTA region.

Figure 8.53 displays the worklist for issuing LTSDs for customer purposes and contains an entry for a Canadian customer.

To send out the LTSDs, Gamma needs to access the **Manage Long-Term Supplier's Declaration – Outbound** app. An employee from Gamma needs to click **Go** at the top-right side to display the LTSD entries pending in the worklist. The **Ready to Send** worklist automatically detects the LTSDs that are relevant to be sent out. This results in the screen displayed in Figure 8.53.

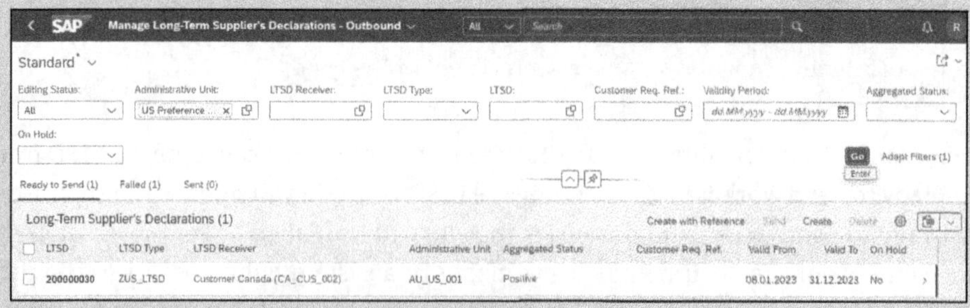

Figure 8.53 LTSDs to be Sent to the Relevant Customers

After selecting the relevant LTSD, the Gamma company can review the data elements in the LTSD and select **Send** when ready to be executed, as shown in Figure 8.54. Once selected, the LTSD PDF and email will be created and sent to the customer's contact person. Because the LTSD specifies that Gamma's finished product is eligible for preferential treatment, Gamma achieves their objective of allowing its customers to benefit from the USMCA, which may significantly increase Gamma's market attractiveness as a supplier of high-end leather shoes. A similar process applies if Gamma looks to issue customer LTSDs to its related entities in the relevant FTA region that aren't created as AUs in SAP GTS; see Section 8.3.9 for more information on the intercompany LTSD process for related entities that are also created as AUs in SAP GTS.

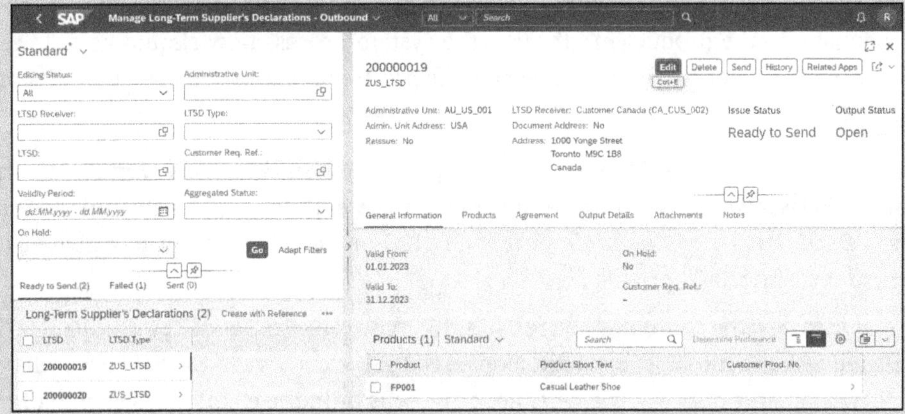

Figure 8.54 Manage Long Term Supplier's Declaration – Outbound Entry

8.3.8 Revoking and Reissuing Supplier Declarations

If a finished product's preferential status changes during an issued LTSD's blanket period, the issued LTSD becomes invalid for the impacted products. Additionally, customs authorities require certifiers to inform customers if a previously certified product is no longer eligible for preferential treatment. SAP GTS can revoke supplier declarations that are no longer valid and reissue them with the updated information.

The preference determination on finished products should be performed on a periodic basis (e.g., monthly, quarterly), as well as whenever the finished product undergoes significant changes (e.g., changes in the BOM's structure, changes to the preferential status of the components, changes to the costing of the components and/or finished product).

It's possible that a finished product for which an LTSD has been sent out with a preferential (or not preferential) status is later assigned a different status due to one of the following events:

- **Change in the preference determination**
  This can occur due to changes in the finished product's BOM, changes in component material prices, or even a different sourcing strategy.

- **New billing documents reporting a price below the threshold value calculated by the finished product's preference determination**
  For the transaction value method, the sales price of a billing document will be compared to the threshold stored in SAP GTS as indicated in the previous section. Therefore, a change in the sales price may change the preference status of a product from eligible to non-eligible (or vice versa).
- **Change in the preference status of the product due to a manual change of the preference status in the product master**
  While it's typically not recommended to change the preference status of a product manually in the product master (i.e., the system process flow should be followed whenever possible), a user can override the preference status for a given product in SAP GTS.

If such a situation occurs, SAP GTS automatically creates an entry in the worklist for the **Manage Reissue Reasons – Long Term Supplier's Declarations** app (located on the **Supplier's Declaration** page) and references the LTSD previously sent out to a given customer for that finished product.

After you review the content in the reissue LTSD, you can click **Send LTSD**. Similar to the LTSD request and the LTSD issuance, a reissued LTSD can move to the **Failed** worklist if the data elements of the documents are erroneous or in the event of a system failure (e.g., Adobe Document Server failure). In this case, you would need to investigate and resolve the issue before resending the LTSD. If the reissued LTSD is sent successfully, a new customer LTSD is created, and the LTSD reissue entry reflects the **Completed** status and moves to the **Completed** worklist.

Note that the newly issued LTSD will make reference to the previously issued LTSD to track and trace the evolution of the preference status in SAP GTS, identify the reason, and aid in communication to the customer.

Instead of clicking **Send LTSD** to reissue the LTSD to the customer, SAP provides another option to mark the LTSD as **Complete**. This functionality allows you to close the reissue reason. If you decide to select this option, the LTSD won't be sent to the customer, and there is no method to reverse the LTSD once it has been completed.

Loss of Preferential Status for a Gamma Vendor

After Gamma performed the preference determination for its finished product and issued preferential LTSDs to its customers, it received new LTSDs from its vendors in Canada and the US, stating that their wool and leather products were no longer of preferential origin.

Gamma updates the preferential status information in the worklist for the **Manage Long-Term Supplier's Declaration – Inbound** app containing the maintained LTSDs and performs the preference determination again. Figure 8.55 displays the updated BOM used in the new preference determination.

Neither the top-down nor bottom-up explosion results in a favorable preference result. Because the finished product contains $100 worth of non-originating components and $100 worth of originating manufacturing costs, the RVC is set at 50%—below the required 55% stated in the rules of origin.

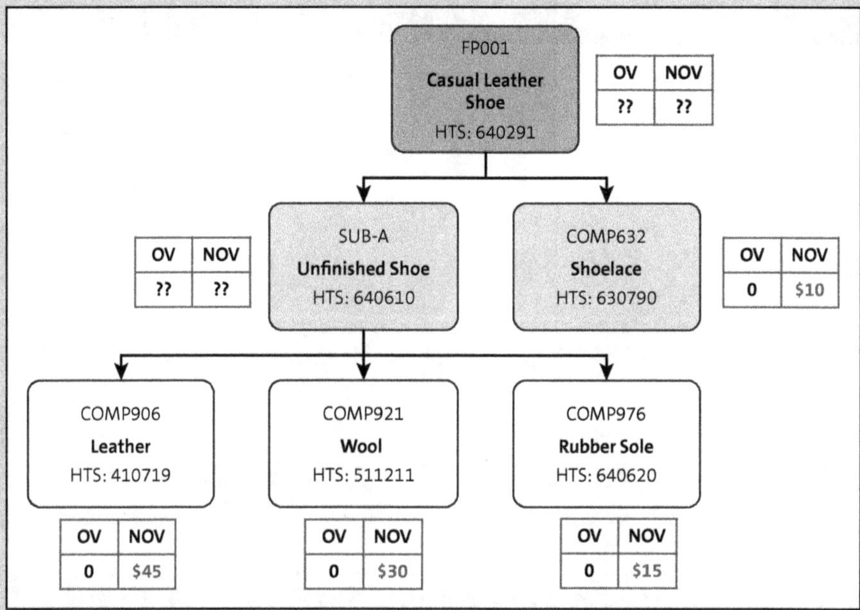

Figure 8.55 Finished Product's BOM after New Vendor LTSDs

As such, Gamma needs to revoke the LTSDs that were sent out to its customers to inform them that the finished product is no longer eligible for preferential treatment. By executing the worklist in the **Manage Reissue Reasons – Long-Term Supplier's Declarations** app displayed in Figure 8.56 and clicking on the **Go** button at the top right, Gamma notices that some of its issued LTSDs must be reissued.

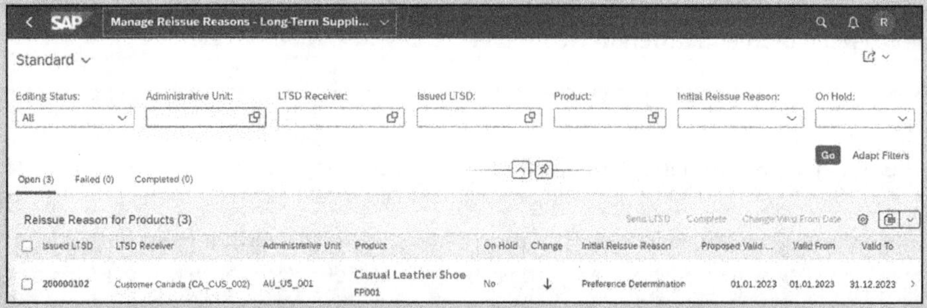

Issued LTSD	LTSD Receiver	Administrative Unit	Product	On Hold	Change	Initial Reissue Reason	Proposed Valid ...	Valid From	Valid To
200000102	Customer Canada (CA_CUS_002)	AU_US_001	Casual Leather Shoe FP001	No	↓	Preference Determination	01.01.2023	01.01.2023	31.12.2023

Figure 8.56 Manage Reissue Reasons: Long-Term Supplier's Declaration Worklist

SAP GTS detects the LTSDs that were previously sent out by Gamma with a positive preferential status. The reference to the previously issued LTSD together with the details of the preference status change is displayed in the **Reissue Reason** document, as shown in Figure 8.57. The **Send LTSD** button triggers the new LTSD to be created and sent to the customers with updated preference information, alongside a cover letter explaining to Gamma's customers the change in its finished product's preferential status.

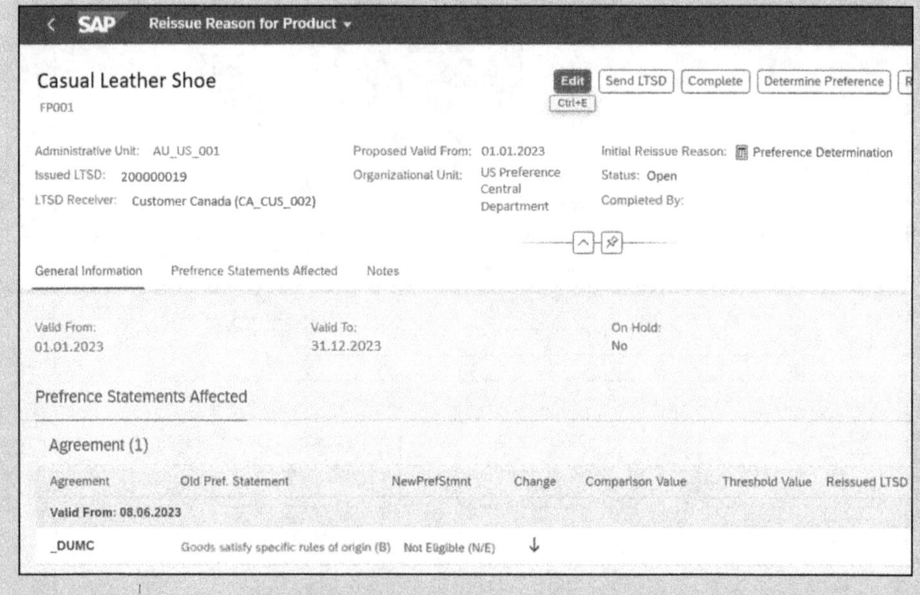

Figure 8.57 Reissuing Vendor Declarations

8.3.9 Intercompany Stock Transfers

Intercompany stock transfer orders relate to movements of goods between plants that are affiliated with the same company. In such scenarios, the movements of LTSDs across plants (issuing and receiving) remain relevant. The goal of this section is to explain how SAP GTS supports the functionality to share preferential status information between plants in different AUs.

Note

The functionality to manage intercompany stock transfer orders will only work if the system has been set up and configured for the intercompany process. The details of this setup are explained in Section 8.4.10. In addition, it's important to state that LTSD handling between two plants is only possible in SAP GTS if both these plants are part of two distinct AUs in SAP GTS. Refer to Section 8.4.1 to see how to create AUs.

If the AUs relevant for the intercompany LTSD process have been identified in Customizing, SAP GTS will initiate the process by creating a customer LTSD for the issuing plant. In this case, the billing document of the stock transfer order flow in SAP

S/4HANA will create a customer LTSD in SAP GTS. The key difference with a normal customer LTSD with unrelated parties is that, for intercompany customer LTSDs, SAP GTS uses the customer LTSD to trigger the creation of a supplier LTSD for the receiving AU and share the preference status for each product in the issuing plant to the receiving plant (i.e., instead of creating a form and/or email to send to the customer).

To provide additional context, a customer LTSD will be created when a relevant intercompany billing document has been transferred to SAP GTS. An entry for the billing document is created in the worklist for **Manage Long-Term Supplier's Declarations – Outbound**, as displayed in Figure 8.58, between the issuing plant of one AU (the **Administrative Unit**) and the receiving plant in another AU (the **LTSD Receiver**). This process is like any other outbound transactional flow for a billing document except for the different action that occurs when the intercompany LTSD is issued, as shown in Figure 8.59. A shipment to the customer requires the system to issue a blanket certificate or another statement of origin (via printing or mail), whereas an intercompany stock transfer will use an electronic exchange to send the LTSD status of finished products automatically between the issuing and receiving AUs.

When you issue the intercompany LTSD, SAP GTS enables the automatic exchange of LTSDs between the two involved AUs, and a customer LTSD generates a supplier LTSD.

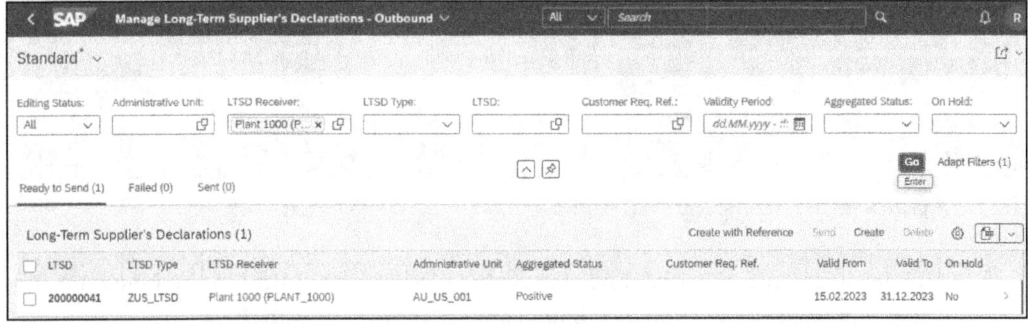

Figure 8.58 Entry Created for the Intercompany Flow

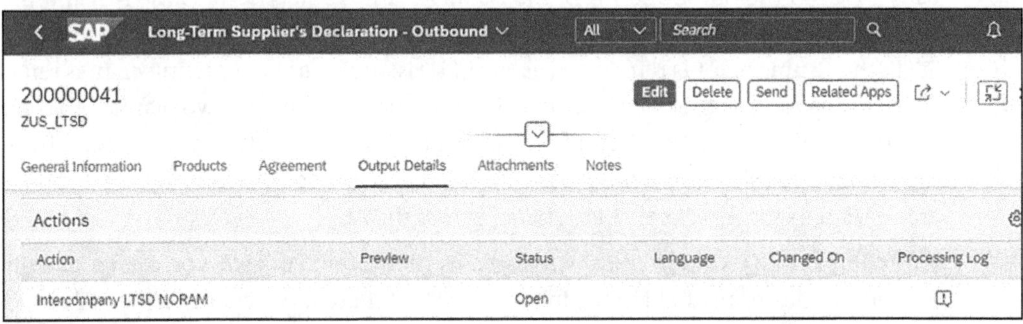

Figure 8.59 Outbound LTSD Actions

Once the LTSD gets issued successfully from the plant on the outbound side, the intercompany LTSD number is generated, and the LTSD's **Issue Status** changes to **Sent with Output Status – Successful**. The intercompany LTSD showcases the inbound LTSD number, which helps you map the outbound LTSD to the inbound LTSD. Note that this field is only provided once the outbound LTSD has been executed. Additionally, as soon as the outbound LTSD is executed, this event automatically leads to follow-on actions of creation, release, and aggregation of the newly created inbound LTSD. On the screen, you'll see a popup message stating that the outbound LTSD was executed successfully and the inbound LTSD was released and aggregated, as shown in Figure 8.60. Furthermore, a PDF of the outbound LTSD will be created as an attachment that can be downloaded and viewed in the **Manage Long-Term Supplier's Declarations – Outbound** app.

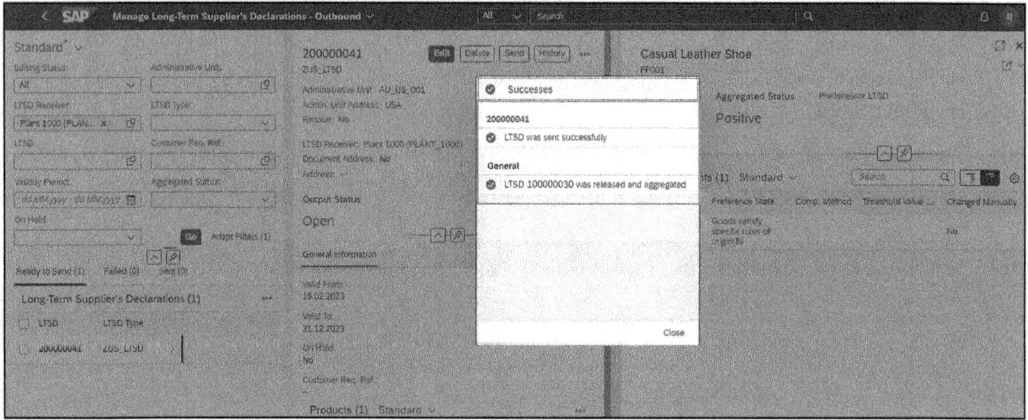

Figure 8.60 Intercompany LTSD Sent and Inbound LTSD Released

The newly created, released, and aggregated LTSD can be found in the **Manage Long-Term Supplier's Declarations – Inbound** app, as shown in Figure 8.61.

The inbound LTSD has the reference number from the associated outbound intercompany LTSD. In the outbound LTSD, the issuing plant is referenced as the AU, and the receiving plant is referenced as the LTSD receiver, while in the corresponding inbound LTSD, the issuing plant is referenced as the LTSD issuer, and the receiving plant is referenced as the AU. In other words, SAP GTS automatically maps the two plants into their appropriate fields in the outbound and inbound LTSDs. In addition, the products from the outbound LTSD are mirrored to the inbound LTSD along with their aggregated preferential status for all FTAs that are shared across the AUs.

The inbound LTSD is automatically released by the system. As such, you can't manually edit, add, or delete any field in the inbound LTSD. If there was a change needed due to a preferential statement change, SAP GTS would create an entry in the **Reissue LTSD** worklist. This reissue process would follow the same structure as the initial LTSD issue.

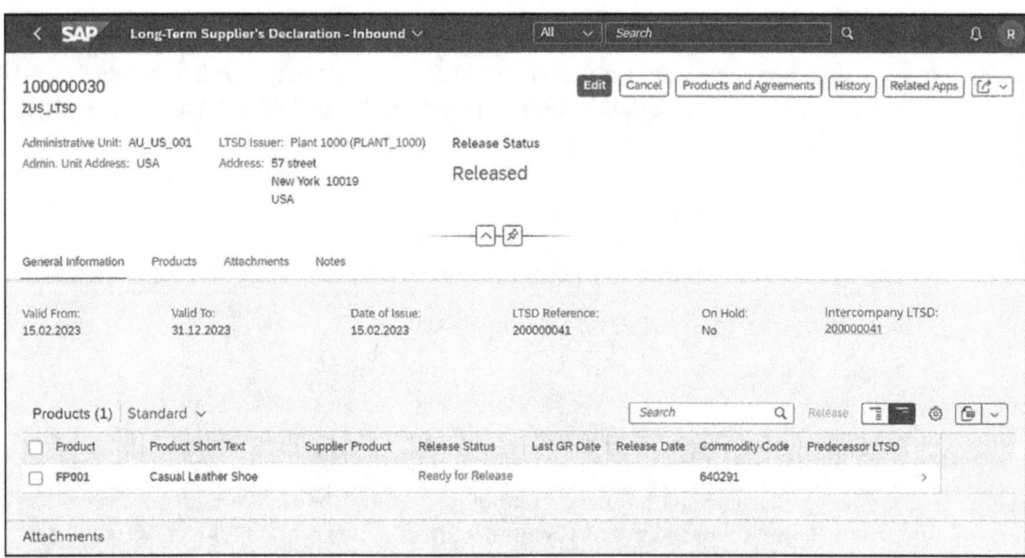

Figure 8.61 LTSD Created That References the Outbound Intercompany LTSD

> **Note**
>
> Users need to maintain preference properties for both suppliers and customers under the **Supplier's Declaration** space via the **Manage Preference Properties – Suppliers** tile or **Manage Preference Properties – Customers** tile for intercompany stock transfer, as shown in Figure 8.62. This process is similar to a normal inbound or outbound transactional flow. Additional details on the configuration setup for intercompany transactions can be found in Section 8.4.10.

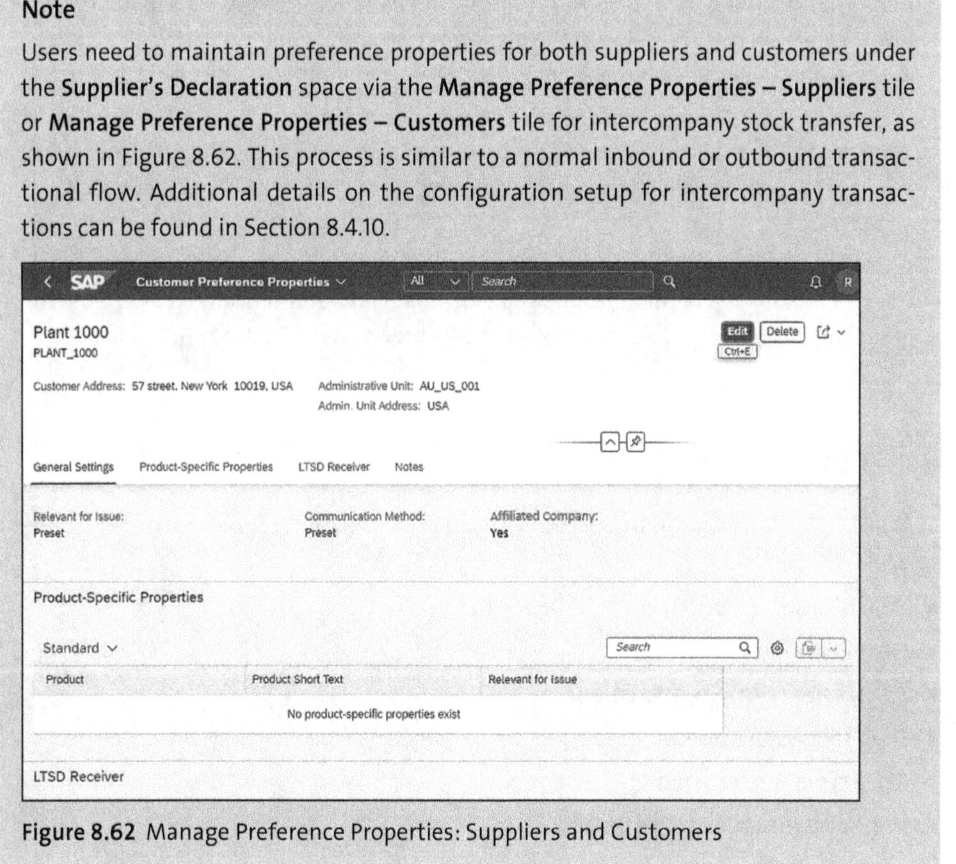

Figure 8.62 Manage Preference Properties: Suppliers and Customers

Additionally, a company can choose to set up the intercompany stock transfer LTSD process to send preferential status information unidirectionally or bidirectionally by means of which **Administrative Unit** is maintained (see Section 8.4.10 for further details). In other words, a company can choose to send preferential status information from AU 1 to AU 2 and not from AU 2 to AU 1, or it can choose to send it from both AU 1 to AU 2 and AU 2 to AU 1.

Issue an Intercompany LTSD to a Gamma AU

Gamma has a stock transfer order between plants in two AUs:

1. AU US (AU_US_001)
2. AU CA (AU_CA_001)

Gamma would like to share the preferential status for its finished product between the two AUs. There is a product that is manufactured in AU US and is sent to AU CA via a stock transfer order. This flow would trigger the intercompany LTSD process.

Due to a billing document created in SAP S/4HANA that is relevant for the intercompany process, SAP GTS creates an entry in the **Manage Long-Term Supplier's Declarations – Outbound** app, as shown previously in Figure 8.58. The AU is the AU US (i.e., AU_US_001). The LTSD receiver is the plant from the AU CA. This outbound LTSD has all the relevant details filled out (e.g., products, FTAs, preferential status, contact person).

Once a user from Gamma clicks **Send** to execute the outbound LTSD, SAP GTS will show a popup stating that the outbound LTSD was sent successfully, as shown previously in Figure 8.60. Additionally, the popup states that the inbound LTSD was released and aggregated.

A Gamma employee can also navigate to the **Manage Long-Term Supplier's Declarations – Inbound** app to find the inbound LTSD that was automatically created from the intercompany stock transfer process, as shown previously in Figure 8.61.

8.4 Customizing

This section covers the main Customizing steps for enabling Preference Management in SAP GTS:

- Organizational structure
- Number ranges
- FTAs, rules of origin, and rule sets
- Preference statements
- Price types agreements
- Price categories for value bases

- Settings for the LTSD process
- Document of origin
- Output management
- SAP S/4HANA plug-in

The Customizing activities for configuring Preference Management functionalities follow the SAP GTS logic, whereby general activities (e.g., defining legal regulations) are grouped in the **SAP Global Trade Services, edition for SAP HANA • General Settings** menu, while others are addressed in the specific section for Preference Management under **SAP Global Trade Services, edition for SAP HANA • Preference Management.**

8.4.1 Organizational Structure

Recall from Chapter 3 and Section 8.3.1 that an additional organizational structure element—an AU—is needed in the Preference Management functionality. An AU isn't necessarily a physical representation of a company; it represents a plant or a group of plants as a single point of contact for all the administration related to Preference Management.

Similar to the settings for FTOs and legal units (LUs) (refer to Chapter 3), one or multiple plant group(s) can be assigned to one AU. This setup is known as a *cross-plant preference model.* This model allows the preference administration of multiple plants to be grouped, enabling only one LTSD to be requested and maintained for specific component(s) from the same supplier for all plants assigned to the group. Preference determination for products produced in multiple plants in the same plant group are only run once. This determination is valid for all plants in the plant group in which the good is manufactured. Finally, if a customer sources the same finished product from multiple plants, a certificate of origin only needs to be issued once if these plants are grouped together. The downside of this preference model is that sourcing non-originating components in one plant affects the other plants in the plant group under the worst-case principle.

As opposed to the cross-plant preference model, the *plant-based preference model* entails linking a single plant to an AU. For this preference model, the supplier solicitation, preference determination, and certificate issuance needs to be performed for each plant individually. The plant-based preference model often requires more administrative efforts from the company's users but can offer a higher degree of flexibility and accuracy in the preference determination because the various steps are performed for the individual plants. Some companies find that further distinction in preference statuses is needed below the plant level, which is explored in more detail in Section 8.5.

Before defining AUs, the BOM evaluation method for the relevant plants needs to be defined. The plants used for the plant-based model in SAP S/4HANA are defined in **SAP**

Global Trade Services, edition for SAP HANA • General Settings • Organizational Structure • Define Plant-Based BOM Evaluation. Note that this setup is defined per logical system.

For a cross-plant-based BOM evaluation method, the plants that belong in a plant group are defined in **SAP Global Trade Services, edition for SAP HANA • General Settings • Organizational Structure • Define Cross-Plant BOM Evaluation**.

The AUs are created via either **SAP Global Trade Services, edition for SAP HANA • Preference Management • Organizational Structure • Define Administrative Units for Vendor Declarations** or the standard business partner transaction (Transaction BP). AUs are created as business partners with the specific business partner role SLLMGR. The country of the AU determines which FTAs are relevant for the assigned plant or plant group.

Once the AU has been created as a business partner, the next step is to define the AU attributes via **SAP Global Trade Services, edition for SAP HANA • Preference Management • Organizational Structures • Define Administrative Unit Attributes**. In this step, depending on the processing model chosen, the assignment of the organizational unit is defined under the **Assign Organizational Unit** folder, as shown in Figure 8.63. If the cross-plant preference model is used, the **Plant Group** must be assigned in this step.

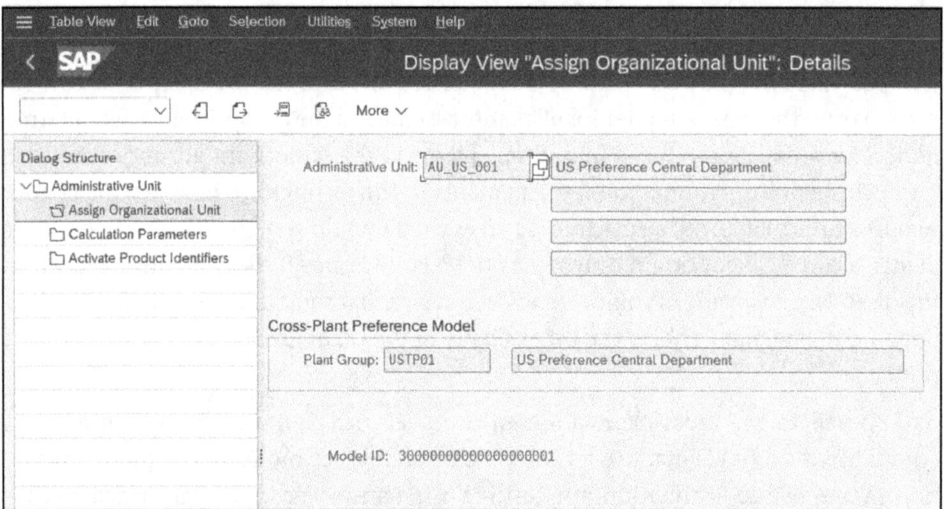

Figure 8.63 Administrative Unit Characteristics

Next, a currency and exchange rate type are assigned by filling in the fields under the **Calculation Parameters** folder.

In the same folder, a numbering system (related to HS) is assigned to the AU that will be used in all preference determinations performed for that AU. It's possible to configure a different numbering scheme for each AU or choose to adopt a global approach where the same numbering scheme is used across all AUs of the company.

In addition, the standard calculation parameter, top-down or bottom-up method, needs to be assigned to each AU. This will be the default calculation method for the preference determination. However, another calculation method can be selected by the user when running preference determination, if required.

Finally, a dedicated folder called **Activate Product Identifiers** exists to enable the Management of Preference Data for Product Identifiers functionality explained in Section 8.5.

8.4.2 Number Ranges

For Preference Management, number ranges must be set up for the following set of data: business partners, product master, LTSD requests, document of origin types, and preference model. Number ranges are set up following menu path **SAP Global Trade Services, edition for SAP HANA • General Settings • Number Ranges.** More details can be found in Chapter 3, Section 3.3.4.

8.4.3 Free Trade Agreements, Rules of Origin, and Rule Sets

When it comes to FTA setup in SAP GTS, the first step is to assign the determination procedure via **SAP Global Trade Services, edition for SAP HANA • Preference Management • Assign Determination Procedure for Active Preference Agreements.** By default, it will be set to **LEPRE** (**Determination of Legal Regulation for Preference Processing**). Other options aren't recommended or aren't used as part of the standard configuration.

As a second step, the preference agreements need to be created. Preference agreements are created as legal regulations under the **General Settings** menu, which includes all regulations for the three SAP GTS functionalities, accessed via **SAP Global Trade Services, edition for SAP HANA • General Settings • Legal Regulations • Define Legal Regulation.** The preference agreements are assigned a specific **Preference Law** legal regulation code (**Legal Code** column), as displayed for the USMCA and EU-Canada agreements in Figure 8.64.

As with the Compliance Management and Customs Management legal regulations, Preference Management regulations need to be activated at the country-group level in both the **General Settings** menu and the Preference Management-specific menu. The Preference Management activation is done via **SAP Global Trade Services, edition for SAP HANA • Preference Management • Activate Preference Agreements.** Preference agreements are activated at the **Country Group** and **Country Group/Country Group** sublevel.

Rules of origin, or preference rules, describe the rules and parameters by which goods can be determined as eligible for preferential treatment. These preference rules are either manually maintained via Customizing or uploaded via XML files.

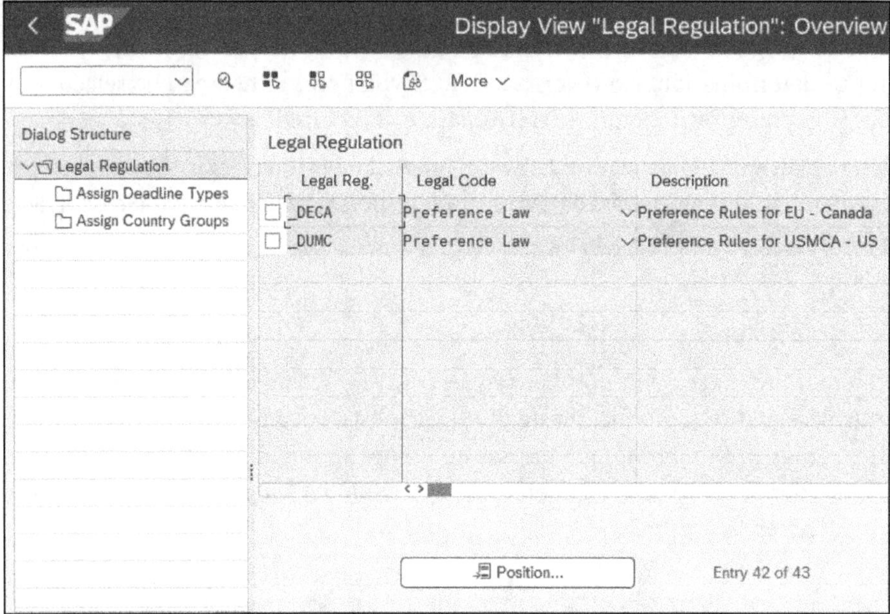

Figure 8.64 Define Preference Agreements as Legal Regulations

In the manual process, the legal regulations and assignment to related country group(s) needs to be done manually. In addition, preference rules are entered via several Customizing steps under **SAP Global Trade Services, edition for SAP HANA • Preference Management • Define Rule Set Attribute for Preference Processing (Manual).** Depending on the number and complexity of rules, this can be a very time-consuming process. This procedure carries the risk that errors in setting up the different rules may impact the Preference Management determination results and potentially result in noncompliance.

When the rules of origin are uploaded via XML files, the files are delivered per FTA and consist of all the rules of origin for that FTA. These files can be purchased from external content providers. To upload such an XML file, execute Transaction /SAPSLL/PREF_RLS_UPL.

Uploading the rules of origin via XML results in two main events: the upload of configuration activities, described next, and the upload of master data, described in Section 8.6.

The configuration for uploading rules of origin requires validating the following activities:

1. The legal regulation together with its related country groups related to the preference agreement are created automatically and then activated in the following configuration step: **SAP Global Trade Services, edition for SAP HANA • General Settings • Legal Regulations • Define/Activate Legal Regulation.**

2. Rule sets are defined under **SAP Global Trade Services, edition for SAP HANA • Preference Management • Define Rule Set**. A rule set is the grouping of all preference rules assigned to one or multiple preferential agreements that can be linked to a common numbering scheme and rule set type (e.g., EU-based agreements).
3. The preference agreements are assigned to a rule set under **SAP Global Trade Services, edition for SAP HANA • Preference Management • Assign Preference Agreement to Rule Set**.

8.4.4 Preference Statements

The preference statements that can be issued to customers in LTSDs or other proofs of origin, or the ones received from suppliers in these declarations and certificates, are based on the requirements of the specific preference agreement. Preference statements are therefore set up per preference agreement. The assignment of preference statements to the system-internal preference indicators will be grouped together in a schema. If several agreements have matching preference statements, one schema can be leveraged for all of those agreements.

The Customizing for defining and assigning preference statements is performed via **SAP Global Trade Services, edition for SAP HANA • Preference Management • Preference Statements • Define Scheme for Preference Statements**.

First, the scheme and statements are defined (see Figure 8.65). For each **Schema** defined, an **Aggregation Method** for preference statements can be chosen, and the specific preferential statements are defined. This is important as the statements will be generated on the LTSD because of the preference determination and can be printed on official communication (e.g., LTSD, blanket certificates, preferential certificate of origin). The statements vary depending on the FTA, as described in Section 8.2.2. Once the scheme and its statements are defined, the next step is to assign them to the rule sets.

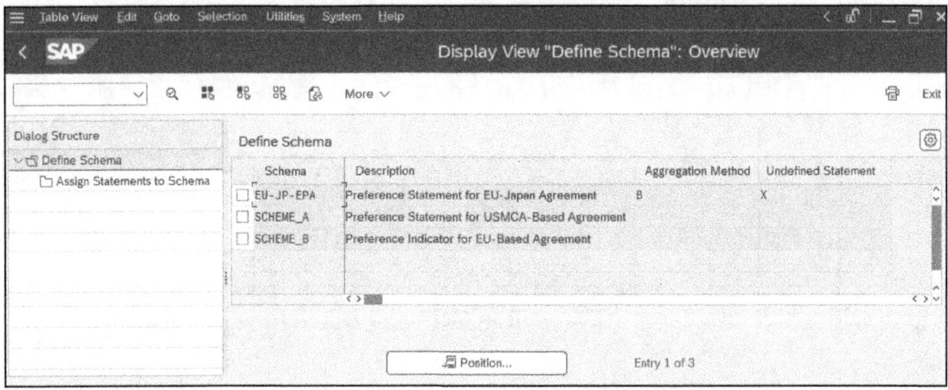

Figure 8.65 Defining the Schema

8.4.5 Price Types Agreements

In this Customizing activity, the price types are assigned per agreement. Price types can be defined under **SAP Global Trade Services, edition for SAP HANA • General Settings • Define Price Types** and can be used to distinguish different values of a given material based on the requirements of the FTA. The price types are used during the preferential determination for a BOM so that components with non-originating and originating content can be evaluated under a value-based rule of origin (see Section 8.2.2 and Section 8.6.3 for additional details).

Price types are assigned to a comparison method via **SAP Global Trade Services, edition for SAP HANA • Preference Management • Assign Price Types Agreement-Specifically for Preference Processing • Assign Price Types to Comparison Methods for Preference Calculation**. A comparison method will be assigned to each FTA, as described in Section 8.2.2, so that the FTA-specific calculation methods can be performed by SAP GTS (e.g., EU-based FTAs can follow the transaction value method or North American FTAs [e.g., USMCA] can follow either the transaction value method or the net cost method). A price type is assigned per FTA to the components with preferential origin, the components without preferential origin, and optionally the finished products whenever the comparison method requires it.

This assignment is necessary for any comparison method. With the net cost method, it's imperative to assign a price type to the finished products of the BOM for the preference determination to take place. In the example shown in Figure 8.66, a net cost (**NET_COST**) is assigned to components while the production price (**PROD_PRICE**) is assigned to the finished product.

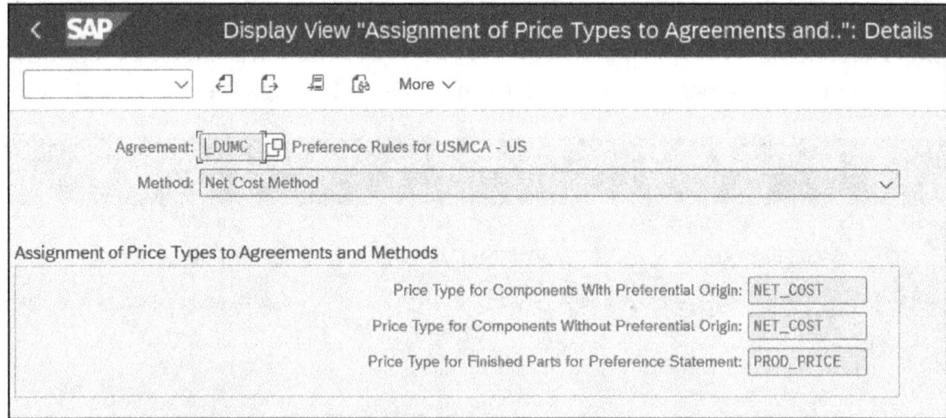

Figure 8.66 Assign Price Types to Comparison Methods for Preference Calculation

This setup can be used to generate a conservative preferential determination result, as non-preferential origin components are given a high value than preferential components, making it more difficult for the finished product to qualify under a value-based rule. In SAP S/4HANA, assigning a value to a price type needs to be defined according to the relevant FTA's regulations and requires an enhancement. The details of this enhancement are specified in the SAP IMG documentation of the previously mentioned Customizing activities.

8.4.6 Price Categories for Value Bases

Another configuration activity that relates to value comparison is the assignment of the price type to a value basis per agreement. This can be done via **SAP Global Trade Services, edition for SAP HANA • Preference Management • Assign Price Types Agreement-Specifically for Preference Processing • Assign Price Categories for Value Bases for Preference Statement.** Depending on the FTA requirement, particularly which comparison method needs to be applied (e.g., transactional value method for EU-based agreements), certain preference determinations lead to a threshold value (Section 8.2.2) that needs to be compared to a transactional value. This Customizing activity allows users to specify what value should be used for comparison per FTA based on a value basis and a value type.

As shown in the example in Figure 8.67, the FTA between EU and Canada is set up so that values from the invoice (or sales order) that have an EXW price value type in SAP S/4HANA (**EXWVAL**) are compared with the threshold value in SAP GTS under the EXW price value basis (**B3 Net Costs**).

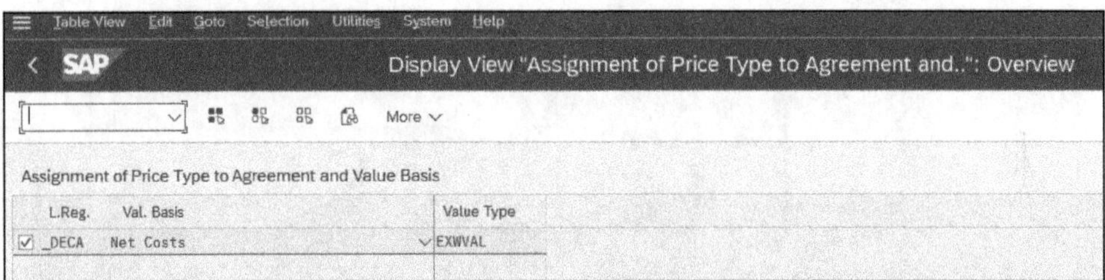

Figure 8.67 Assign Price Categories for the Value Basis of the Preference Statement

8.4.7 Settings for the LTSD Process

This Customizing activity defines which AU is relevant for a request of supplier LTSD and issue of customer LTSD, what aggregation method is used for the summarization of LTSD (Section 8.3.4), and which communication methods are defaulted. The setup of these activities is done via **SAP Global Trade Services, edition for SAP HANA • Preference Management • Supplier's Declaration • Define Settings for LTSD Process.** Figure 8.68 shows the details that can be configured for one AU.

The aggregation details specify both a **Goods Retention Period**—the duration in days of how long a goods receipt or purchase order is taken into account by SAP GTS for the aggregation—and the **Valuation Method**, which details when LTSDs and goods receipts are taken into account for the aggregation and what constitutes the period of validity for the aggregation (e.g., from release date of LTSD to end LTSD validity period, from first goods receipt date to end of LTSD validity period).

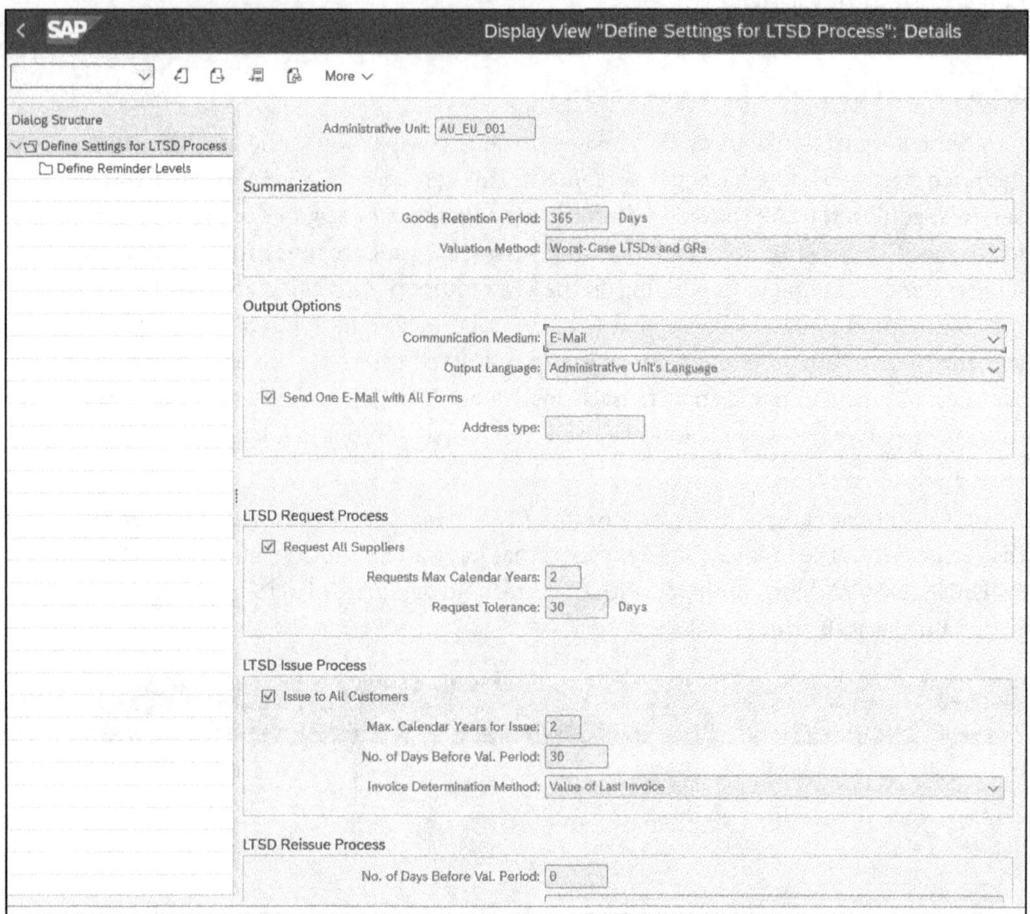

Figure 8.68 Define Settings for the LTSD Process

The **Output Options** area details the default **Communication Medium** (**E-Mail** or **Mail**), in what language, and whether multiple forms should be encompassed in one mail.

The **LTSD Request Process** area details whether all suppliers are defaulted as relevant for the LTSD request or not, the maximum calendar year for the request, and the request tolerance (i.e., how many days a new request can be created before the end of the validity of a LTSD).

The **LTSD Issue Process** area details whether all customers are defaulted as relevant for LTSDs or not, the maximum calendar year for the LTSD issue, the minimum number of days used for billing documents to be accounted for before the LTSD's validity period, and what comparison value should be used out of the billing document (e.g., value of last invoice, middle invoice value, highest invoice value).

Like the **LTSD Issue Process** area, the **LTSD Reissue Process** area details the minimum number of days for billing documents to be accounted for before the validity period of the reissue and what comparison value should be selected from these billing documents.

8.4.8 Documents of Origin

The documents of origin are defined and maintained in this Customizing activity. Each document of origin needs a document of origin type, which is the basis for configuring the document. This document type is automatically determined in the issuing process, and users can select it when entering the document.

The document of origin can be defined via **SAP Global Trade Services, edition for SAP HANA • Preference Management • Supplier's Declaration • Define Documents of Origin.** In this Customizing activity, the document of origin type is created and made relevant for the inbound process, outbound process, or both processes. Once the document type is created, the document types need number ranges and validity periods. The country group(s) and preference agreement(s) are also assigned for each document of origin type. For instance, a document of origin type created for European LTSD processes would be assigned with the EU-based FTAs as well as the country groups corresponding to these FTAs (see Figure 8.69).

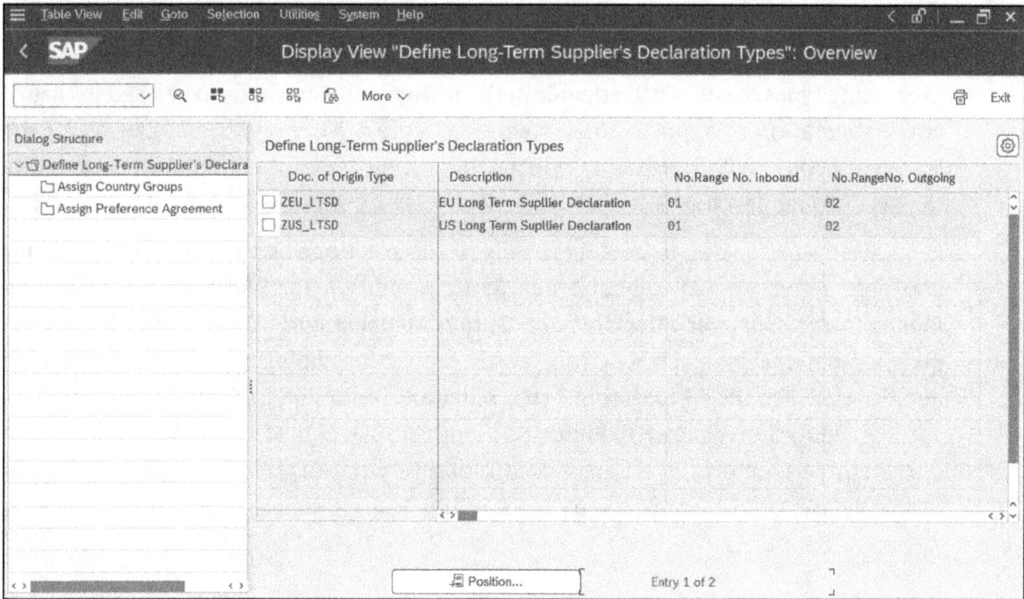

Figure 8.69 Document of Origin Type

The next step is to activate the document via **SAP Global Trade Services, edition for SAP HANA • Preference Management • Supplier's Declaration • Activate Documents of Origin**. In this Customizing activity, each AU and its corresponding document of origin type can be activated for incoming, outbound, and request processes. This activity completes the enablement of the LTSD processes relevant for an AU such that, in our previous example, the document of origin type created for European LTSD can be assigned to EU-based AUs and made relevant for inbound, outbound, and request processes.

8.4.9 Output Management with PPF Actions

The output management configuration defines the settings related to generating email and mail (i.e., offline communication) in the LTSD request, the maintenance of the supplier LTSD, and the issue process. This configuration is done via **SAP Global Trade Services, edition for SAP HANA • Preference Management • Supplier's Declaration • Output Management**.

The first step related to the overall setup consists of defining Post Processing Framework (PPF) actions via **SAP Global Trade Services, edition for SAP HANA • Preference Management • Supplier's Declaration • Output Management • Define Actions (PPF)**. The actions are defined in the action profiles. The standard action profiles differentiate the processes of issuing the LTSD for the EU, issuing declaration of origin between EU and GB, requesting LTSDs for EU, and creating and releasing the LTSD (see Section 8.3.9 for intercompany LTSD details). Note that for countries other than EU or GB, nonstandard entries need to be created. Under action profiles, action definitions primarily generate forms or issue emails containing these forms. Actions are then completed with processing types to specify the details of the forms or the methods to be used in case of email generation or supplier LTSD creation. Figure 8.70 provides an example of the standard **Action Definition** and **Permitted Processing Types of Action** used to generate the EU LTSD request forms.

Next, users specify the processing details of these actions and define the conditions and parameters via **SAP Global Trade Services, edition for SAP HANA • Preference Management • Supplier's Declaration • Output Management • Define Conditions/Output Parameters**. Under each action profile, the action definitions are enriched with details related to their **Processing Type**, **Schedule Condition**, or **Start Condition**. Standard configuration is shown in Figure 8.71 for the LTSD request forms. Note that in case of dynamic use of printer determination for an output (e.g., define a specific output device by user or AU), Transaction /SAPSLL/PRE_LTSD_PRI in SAP GUI should be used.

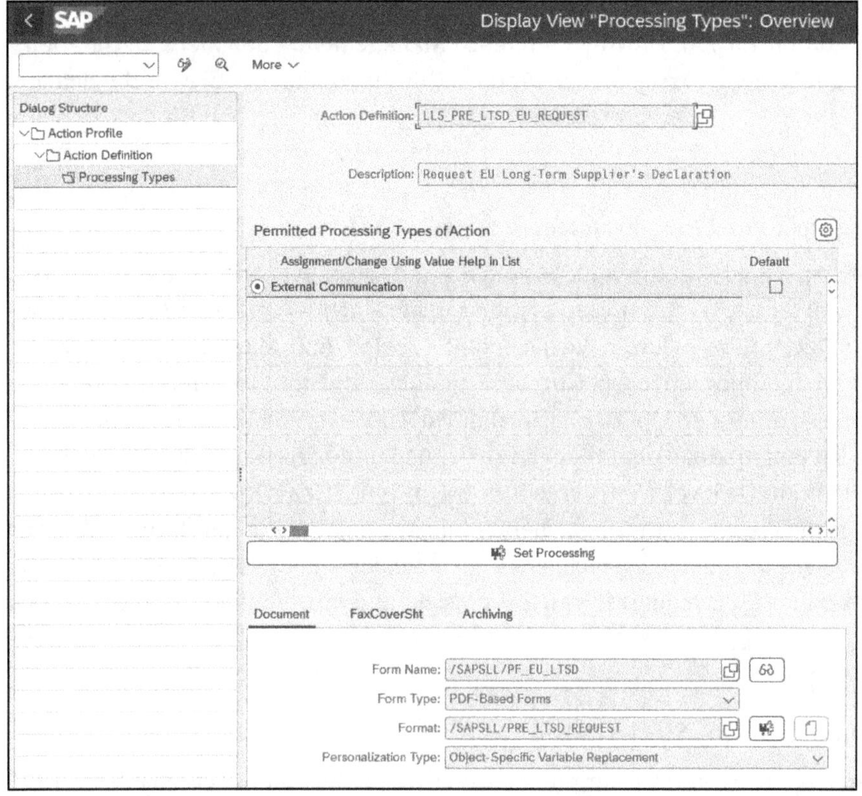

Figure 8.70 Define Actions (PPF)

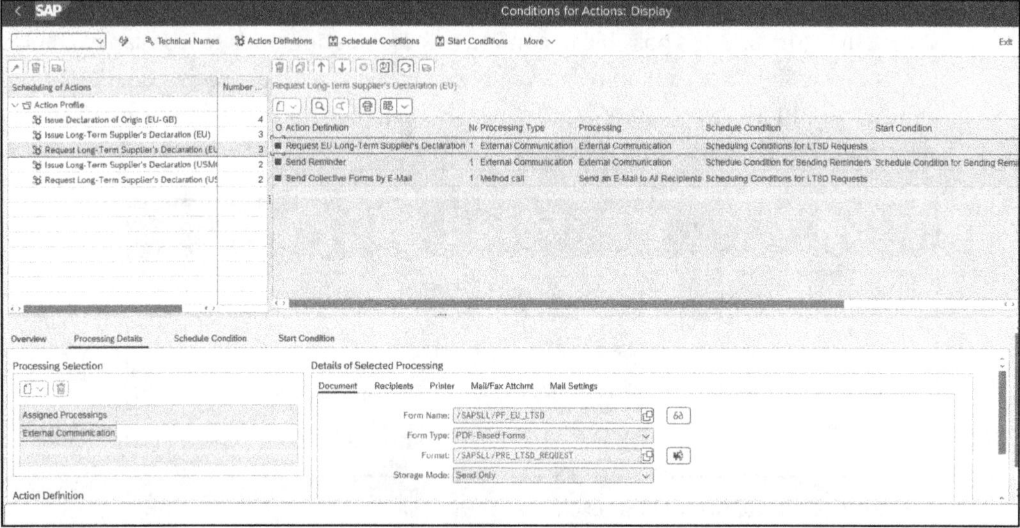

Figure 8.71 Define Conditions/Output Parameters

The third step is to assign action profiles to the LTSD processes via **SAP Global Trade Services, edition for SAP HANA • Preference Management • Supplier's Declaration • Output Management • Assign Action Profiles.** This Customizing activity is divided into two parts: the LTSD request process and the LTSD issue process. For both, an action profile is assigned to an AU and a document of origin type. Keeping with our earlier example, this configuration would allow users to assign the EU LTSD request and to issue forms and emails to the EU document of origin and EU-based AU.

The final step assigns text modules to emails and forms related to the action definitions via **SAP Global Trade Services, edition for SAP HANA • Preference Management • Supplier's Declaration • Output Management • Assign Text Modules**, which is where general text modules, email-specific text modules, and form-specific text modules (LTSD request, reminder, LTSD issue) are defined. The text modules are assigned by AU (e.g., general text modules), but they can be defined more precisely at the action definition level, reminder level, or document of origin type. This allows users to tailor the content of the emails and forms generated during the LTSD request, reminder, and issue processes (e.g., email subject, email text, header and footer details, cover letter and core text for LTSD request forms, text of issue, and text of reissue for the customer LTSD).

8.4.10 Intercompany Stock Transfer

The intercompany stock transfer process described in Section 8.3.9 is enabled via **SAP Global Trade Services, edition for SAP HANA • Preference Management • Supplier's Declaration • Cross-Company-Code Stock Transfer • Activate Administrative Units for LTSD Exchange.** This activity defines which AUs can issue or receive LTSDs from other AUs, as shown in Figure 8.72. It's possible to configure these relationships so AUs can exchange LTSDs bilaterally or so that an AU can only send the LTSD to or receive the LTSD from another AU. In addition, users can define whether the LTSD is sent automatically by selecting the **Automatic Send** checkbox.

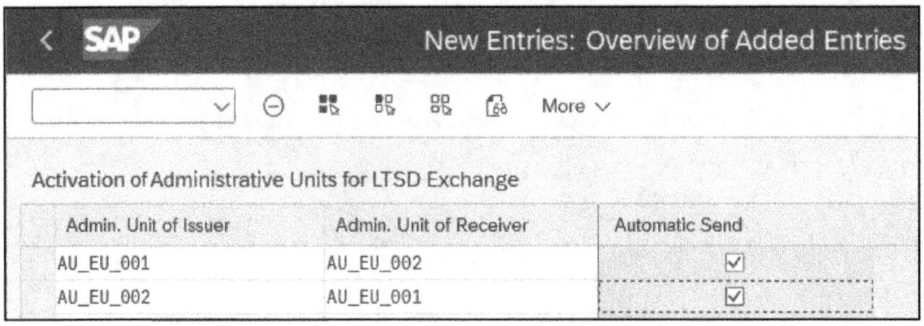

< SAP New Entries: Overview of Added Entries

More ∨

Activation of Administrative Units for LTSD Exchange

Admin. Unit of Issuer	Admin. Unit of Receiver	Automatic Send
AU_EU_001	AU_EU_002	☑
AU_EU_002	AU_EU_001	☑

Figure 8.72 Activate AU for LTSD Exchange

The process to create and release the supplier LTSD automatically is available in an action definition under **SAP Global Trade Services, edition for SAP HANA • Preference Management • Supplier's Declaration • Output Management • Define Actions (PPF)** (see Section 8.4.9 for details).

8.4.11 SAP S/4HANA Plug-In

Preference Management is enabled in SAP GTS through two activities in the SAP S/4HANA plug-in. The first one relates to a BOM transfer and is defined via **SAP Global Trade Services – Plug-In • Control Data for Transfer to SAP Global Trade Services • Control Transfer of Bills of Product for Preference and Re-Export**. In this activity, as shown in Figure 8.73, the category of BOM to be transferred is defined. There are two main categories: the static BOM and the configurable BOM. While the static BOM relates to master data and is the same across multiple transactions, the configurable BOM varies by transaction. Users define what **Category of BOM**, **Type of BOM Expl.** (top-down versus bottom-up), and BOM usage (e.g., **Production BOMs**) are relevant for the transfer. Users also determine the maximum of alternatives transferred for a BOM and whether certain material types should be excluded during the transfer (e.g., packaging materials related to shipping or logistics). This Customizing activity identifies the BOMs that need to be transferred to SAP GTS for use in preference determination. The configuration allows for global-, country-, and plant-level settings. The same configuration details are available at the country and plant levels and allow for the BOMs to be transferred at a more refined level.

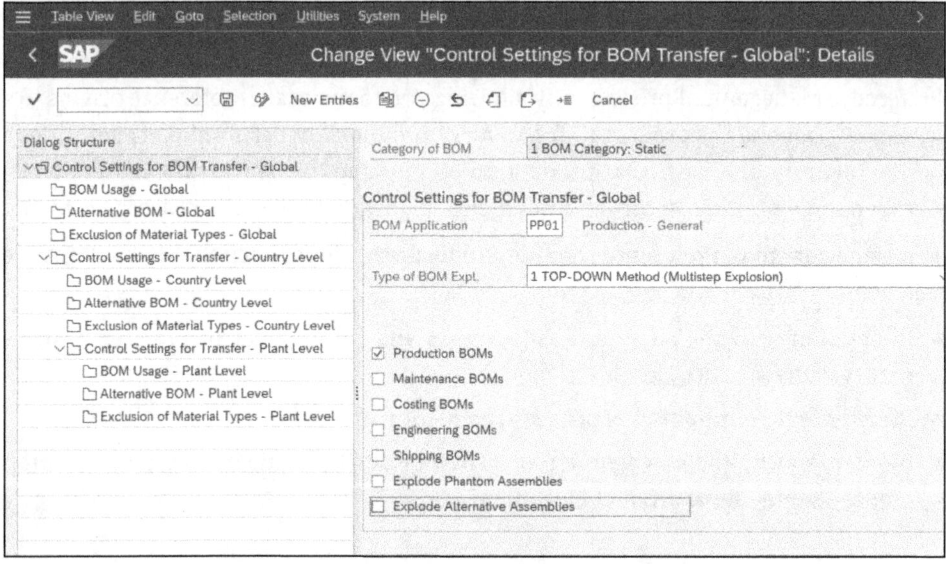

Figure 8.73 Control Transfer of BOM for Preference and Reexport

The second activity relates to the activation of documents to be sent to SAP GTS for Preference Management, which is done via **SAP Global Trade Services – Plug-In • Control Data for Transfer to SAP Global Trade Services • Configure Control Settings for Document Transfer**. This activity requires flagging document types as relevant for Preference Management (the **SD WorkL** checkbox), as shown in Figure 8.74. Purchasing and material document types (application level MMOA and MMOC, respectively) need to be flagged to identify which documents should generate an entry in the Manage Requests app. For the billing document types (application level SDOC), a flag identifies which document should trigger LTSD issuance in SAP GTS or which document should receive the preference indicator from SAP GTS (the **SD WorkL** checkbox and **Pref. Ind.**, respectively).

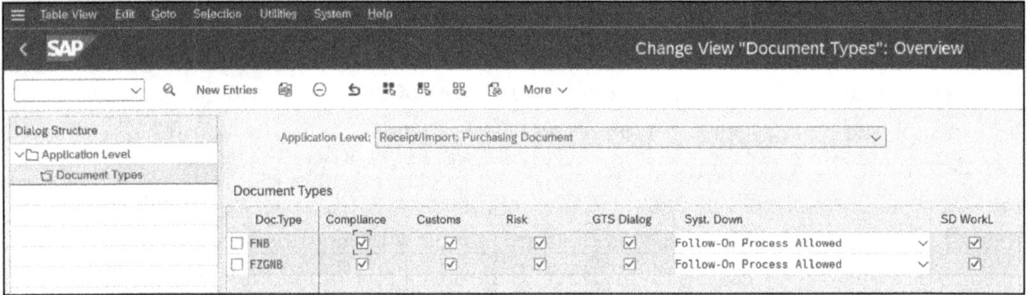

Figure 8.74 Application Level Document Types Activation

8.5 Management of Preference Data for Product Identifiers

The *Management of Preference Data for Product Identifiers* functionality enriches the Preference Management processes by allowing you to maintain preference origins to a level going beyond the product ID for an AU. A product identifier uniquely identifies a partial quantity of a particular production or sourcing element. This can be a batch number, serial number, or any another identifier, depending on the process setup.

The Management of Preference Data for Product Identifiers functionality contains the following:

- Integration with the supplier LTSD process and the transfer of preference statements to the product identifiers
- Preference determination of process/production orders
- Integration with the management of customer LTSDs, taking into account the preference statements of product identifiers

Note

Management of Preference Data for Product Identifiers is integrated as part of the standard functionality as of SAP GTS, edition for SAP HANA version 2023. SAP S/4HANA 2022 is required as a prerequisite.

To fully grasp the benefit of Preference Management for product identifiers in SAP GTS, rigorous integration with the full supply chain is paramount. Compared to the product-based solution, Management of Preference Data for Product Identifiers will increase the integration with the production processes.

The following sections provide details of specific business processes in Management of Preference Data for Product Identifiers and illustrate the enrichments compared to the product-based solution via examples.

8.5.1 Business Process

The standard business processes of Preference Management are still in scope with Managing Preference Data for Product Identifiers. Therefore, we'll still divide the business processes between inbound, determination, and outbound, but we'll also highlight the additional benefits of using this functionality compared to the product-based preference processing. Additionally, a batch will be used as an identifier to illustrate business requirements, but other identifiers such as serial number could be used as well.

Obtain Preferential Status and Origin from Suppliers

The proof of preferential origin from suppliers is still requested and maintained at the product level using the standard process described in Section 8.3.2 and Section 8.3.3. The product identifier details come into play upon the receipt of the goods when more detailed information on the product identity is known. At this stage, the preferential status maintained in a supplier LTSD for the externally procured goods can be carried over to the product identifier (e.g., the batch). The preferential information of the product identifier can also be sent back to SAP S/4HANA, so it's known upon the production or manufacturing process. In turn, this information can help to determine which batch to use in the production process to maximize the returns that an FTA can provide.

With the preferential origins of the products being maintained at a level beyond the product ID, the aggregation process at the product level is no longer obligatory. When multisourcing a product from different suppliers, the worst-case principle can be avoided. This will be possible when suppliers ship the product with different preferential statuses because the preferential origin can be kept at a supplier-product (through the batch) combination level on a transactional basis in this example.

Determine Preferential Status and Origin of Finished Products

The determination of the preferential status of a product for a certain agreement can be performed at the product level, product identifier level, or both. If the determination is performed on the product identifier level, the integration needs to be made with production/process orders. The preference determination principles explained in Section 8.2.2 and Section 8.3.5 remain valid, but the preferential origins of the components used in the production process is more specific as they are captured at the product identifier level. In addition, the preferential result determined for the finished product will only concern the corresponding product identifier (e.g., batch). The preferential information of the product identifier determined from the calculation can be sent back to SAP S/4HANA. In this way, the company can use this information during their sales and distribution processes to prioritize the shipping of specific product identifier combinations (e.g., batch) based on factors such as the country of the customer.

Issue Certificates of Origin for Preferential Finished Products

Similarly, the integration with the sales process remains seamless, and the preference indicator can be received on the item level in the sales document such as the billing document in SAP S/4HANA, as explained in Section 8.3.7.

In addition, an origin-based determination strategy can be used in the outbound deliveries. In this way, the Manage Preference Data for Product Identifiers solution can support the delivery process of a company. A smart delivery process can be enabled to pick the most beneficial batch available for specific customers; however, this isn't SAP standard, and custom code with the specific business rules would need to be created.

With the Manage Preference Data for Product Identifiers functionality, customer LTSDs can be issued in a similar fashion to the product-based solution for Preference Management. If the billing document contains the details of the product identifier (e.g., the batch), the customer LTSD will be issued for this identifier, and the preference status will be for the identifier record. If the billing document doesn't contain the identifier, the preference status in the customer LTSD will be at the product level. Similar to the product-based solution, if the product identifier preferential status changes after a customer LTSD has been sent, a revoke or reissue of the certificate is possible for the product identifier.

Gamma's Preference Management Process Using Product Identifier

Gamma decides to enable the product identifier solution at the batch level and expects that it will improve its capacity to leverage the USMCA benefits. Recall that Gamma sources it's shoelaces from suppliers in Peru and Mexico and has customers in Canada and Germany.

With this Preference Management at batch level, Gamma will identify the batches of the shoelaces that qualify for USMCA upon receipt of the goods from its Mexican supplier.

During the manufacturing process, Gamma can then prioritize the use of USMCA qualifying shoelaces for a first batch of casual leather shoes. This increases the chances for this batch to qualify for USMCA. In parallel, it will take the nonqualifying shoelaces for another batch of leather shoes that are less likely to qualify for USMCA.

Gamma can finally use this information when shipping its finished products to its two main customers. The USMCA qualifying batch is shipped to its customer in Canada, while the other batch is shipped to its customer in Germany.

With a more detailed tracking of the preference status for its casual leather shoe, Gamma can incorporate this information into its sourcing, manufacturing, and distribution processes to benefit more of the FTAs.

8.5.2 System Process

After highlighting the business process of preference processing for product identifiers, the system process will be covered in this section. The functionality undergoes a sequence of individual steps to identify the management for preference data for product identifiers. In SAP GTS, there are three dedicated tiles for management of preference data for product identifiers, which are found under the **Master Data – Preference** section within the **Master Data** space: **Display Bills of Material – Process-/Production Orders, Display Where-Used List – Batches**, and **Manage Preference Data – Product Identifiers**, as shown in Figure 8.75.

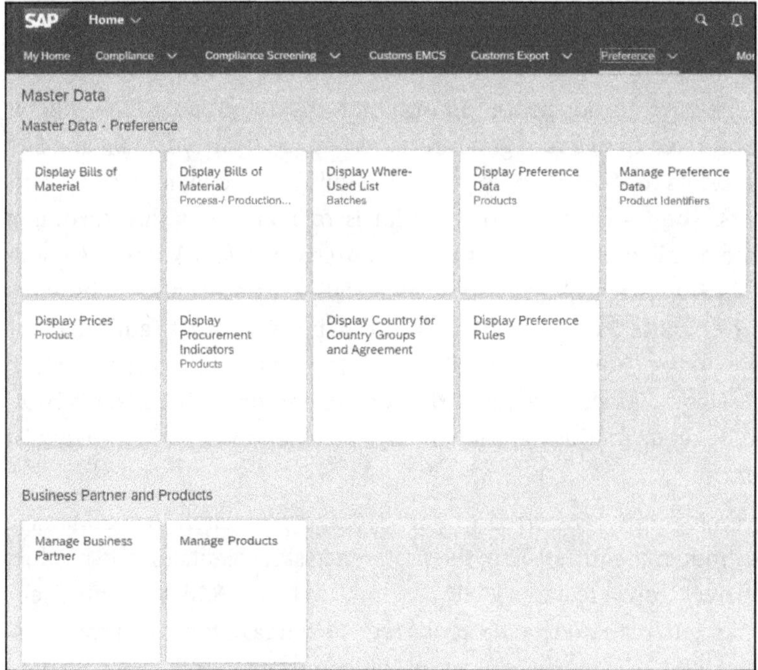

Figure 8.75 Dedicated Tiles for Management of Preference Data for Product Identifier

Additionally, there are also some relevant tiles for the determination of production and process orders. Those are combined in their own set of tiles, **Determination – Production Documents**, under the **Preference Determination** space, as shown previously in Figure 8.9.

In addition, the standard tiles for Preference Management will still be used. The standard system process explained in Section 8.3 and used for preference processing for products (refer to Figure 8.9) is similar to the one for product identifiers. Therefore, the next paragraphs will only highlight the nuances of the product identifier compared to the preference process at the product level.

The most important tile of the functionality is **Manage Preference Data – Product Identifiers**. Within this app, users can visualize preference statements at the agreement level for individual product identifiers. The determination of this preference status at the product identifier level depends on the **Procurement** indicator of the product, being external procurement or in-house production in the app. We'll take a closer look at scenarios for external procurement and in-house production in the following sections, followed by the process of issuing preference certificates/customer LTSDs.

Note

Product identifiers with procurement indicator **Mixed-Origin** (i.e., procurement type X in SAP S/4HANA representing both procurement types, externally procured and in-house-produced) aren't supported by the application.

Externally Procured Goods and Supplier LTSDs

For the externally procured products, the information on the product identifiers is received via the automated transfers of goods receipts between SAP S/4HANA and SAP GTS. These goods receipts contain not only product details but also identification data, including identifiers. The transfer of goods receipt is mandatory as this document often contains more detailed information than the purchase orders. When the goods receipts are posted in SAP S/4HANA, this will automatically create an entry in the worklist of the **Manage Preference Data – Product Identifiers** app. Note that to automate the population of the worklist, Transaction /SAPSLL/PRPID_1 has to be executed, which manages the processing of goods receipts with associated product identifiers. A background job is typically set up for this transaction, so the entries populate in the worklist in a timely manner.

When the goods receipts are transferred to SAP GTS and populated in the **Manage Preference Data – Product Identifiers** app, the preference statements for the product identifier are taken over from a released LTSD, as shown in Figure 8.76. More precisely, the product identifier entry created by the goods receipt will take the preferential status per FTA of a released LTSD for the relevant product and supplier (i.e., LTSD issuer) combination and valid for the entry date of the goods receipts. This is contrasted with the product-level aggregated preference status for all vendors and associated LTSD status within the AU.

If no released LTSD exists for the product supplier combination, the product identifier entry will be marked as not originating, but it will be automatically updated with a new preference status as soon as a new relevant LTSD is released. Additionally, as already mentioned in Section 8.5.1, it should be noted that the aggregation is still done on the product level only. Consequently, a product identifier's preference status would not necessarily be affected by the receipt of the same product from multiple suppliers and the corresponding worst-case principle. An example of this principle is explained for the Gamma business case later in this section.

Lastly, if there are multiple goods receipts for the same product identifier with different preference statements, an alert will be raised to notify the user of the worst case being applied. Users have the option to either close the alert manually or redetermine the preference, even if the alert is automatically removed when only one document remains or the preference statement across product identifiers is identical.

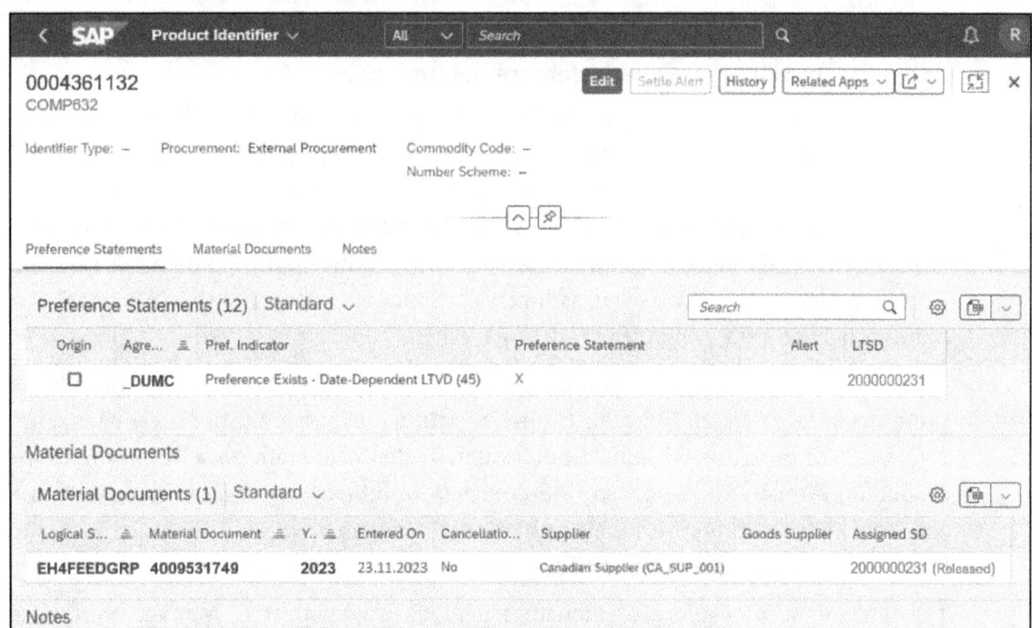

Figure 8.76 Product Identifier for an Externally Procured Product

> **Gamma's Aggregation under Product Identifier**
>
> As mentioned, Gamma decides to enable the product identifier solution at the batch level.

Gamma creates a goods receipt document for its shoelace (COMP632) with batch 001 from the Mexican supplier (MX_SUP_001). Additionally, Gamma has already maintained and released a supplier LTSD for the product COMP632 and LTSD issuer MX_SUP_001 for USMCA. Consequently, the preference status for COMP631 – batch 001 will take over the positive status for USMCA and will be shown accordingly in the **Manage Preference Data – Product Identifiers** app.

Furthermore, Gamma also creates a goods receipt for its shoelace (COMP632) with batch 002 from its Peruvian supplier (PE_SUP_001). Gamma doesn't request or maintain any supplier LTSD from this PE_SUP_001 as it would not be eligible for USMCA. Per the aggregation process explained in Section 8.3.4, the product COMP632 won't be eligible for USMCA according to the worst case.

The goods receipt for COMP632-batch 002 and supplier PE_SUP_001 will show as not eligible for USMCA due to the lack of a positive LTSD for this product supplier combination.

However, the COMP632-batch 001 won't be affected by the worst case and will hold its originating preference status for USMCA.

Goods Produced in House and Preference Determination

To determine the preference status of an in-house-produced BOM at the product identifier level, it's required to integrate the production or process orders from SAP S/4HANA to SAP GTS. The production or process orders allow you to establish a structure and relationship between the finished product and its components, comparably to the BOMs definition given in Section 8.2.2 but with product identifier details for the finished product of the BOM as well as for its components (assuming this information is maintained and kept available in SAP S/4HANA).

Because of the integration of production or process orders, it's possible to establish a preference status for an in-house-produced product at the identifier level when running the determination without simulation in the **Perform Preference Determination – Production/Process** app under the **Preference Determination** space, as shown in Figure 8.9. Note that the preference determination concepts explained in Section 8.3.5 and Section 8.3.6 remain applicable but are enabled to production or process orders.

The determination results at an identifier level are listed within the **Manage Preference Data – Product Identifiers** app, where preference statements per FTA are filled with positive or negative status or with the threshold value in case the determination results in price-dependent statements.

If there are multiple production orders, each production order is displayed with the last determination date within a separate section for production and process orders in the tile, as shown in Figure 8.77.

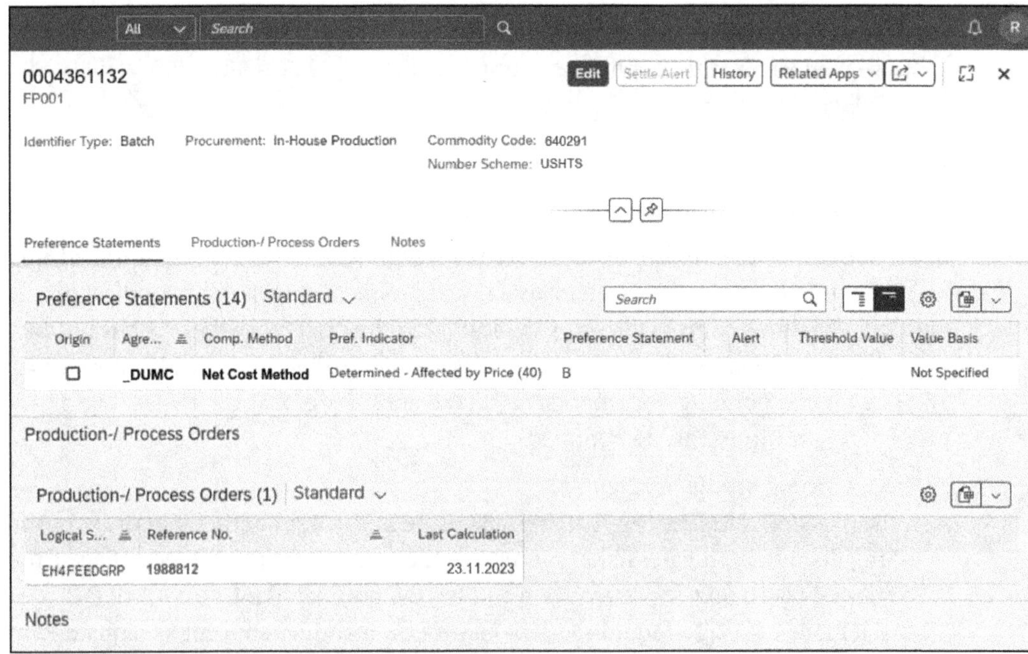

Figure 8.77 Product Identifier for In-House-Produced Product

The preference results for product identifier, whether for externally procured or in-house-produced goods, can be shared back with SAP S/4HANA by means of an enhancement. As mentioned in Section 8.5.1, sharing preferential information at the product identifier level with SAP S/4HANA can be used during production or distribution processes to extend the benefits of FTAs within a supply chain.

Gamma's Production Using a Product Identifier

To optimize its production based on FTA information, Gamma chooses to push back to SAP S/4HANA the preference information at the identifier level for the externally procured goods. Now, Gamma knows that COMP632-Batch 001 is eligible for USMCA, while COMP632-Batch 002 is not, and will consider this during its production of the casual leather shoe.

Gamma runs its production of a first batch of casual leather shoe FP001, which uses COMP632-Batch 001 exclusively. The corresponding production order is sent in SAP GTS and used for determination via the **Perform Preference Determination – Production/ Process** app.

The preference result for the casual leather shoe at the identifier level, say FP001 - Batch 998, is shown in the Manage Preference Data – Product Identifiers app, where the batch indicates preferential for USMCA and refers to the production order.

In another production run, Gamma may choose to use COMP632-Batch 002. The corresponding production order would then be transferred to SAP GTS and used in the preference determination. The result would be reflected in the **Manage Preference Data – Product Identifiers** app, where the new batch, say FP001-Batch 999, would not be preferential for USMCA.

Issue Preference Certificates and Customer LTSD

In Section 8.3.7, we described how preferential certificates of origin can be issued, and we made the distinction between transactional proofs of origin and periodical proofs of origin. The same concepts remain applicable for the preference data for the product identifier functionality, but it implies that the identifier is available in the SAP S/4HANA corresponding documents. Let's take another look at the two categories of proof of origin within this new context:

1. **Transactional proofs of origin**
   If the product identifier (e.g., batch) is available in the item of the sales order or the billing document, it's possible to integrate the transactional preference flag at the item level based on the preference result for the product identifier in SAP GTS. The enablement of such a requirement would follow a similar concept as explained in Section 8.4.11 but would need to be made identifier specific.
2. **Periodic proofs of origin**
   The customer LTSD process remains applicable, and the same tile used for products—**Manage Long-Term Supplier's Declarations – Outbound**—is used with the product identifier. The only difference is that to have a customer LTSD issue a preference status at the product identifier, the billing document needs to contain this identifier (e.g., batch) for the related item. Alternatively, in situations where products lack identifiers, preference statements from the product level serve as the foundational source.

In a similar way, revoking product identifiers follows the same principles as for products. A new entry is created in the Manage Reissue Reasons app if there is a change in the preference status of the identifier (e.g., batch) and this product identifier has already been sent in a previous customer LTSD.

8.6 Master Data

Chapter 4 elaborates on master data received and maintained in SAP GTS. Details on the transfer, recurring delta upload, and overall dynamics of upload and maintenance can be found there. This section focuses on the master data elements required for Preference Management. The Preference Management master data elements fall into three areas, which we'll discuss in the following sections:

- Material-related master data
- Business partner-related master data
- Preference agreements and rules of origin

8.6.1 Material Master Data

To enable Preference Management, several material-related master data elements need to be gathered in SAP GTS, including the following:

- Materials
- Procurement indicators
- Product costs and prices
- BOMs
- HS classification codes

Let's take a closer look at each.

Product Master

Initially, all finished products, components, and semifinished products used in the Preference Management process must be transferred as product master data from SAP S/4HANA to SAP GTS. The product master data in SAP GTS is accessible under the **Manage Products** tile in the **Master Data – Preference** section (refer to Figure 8.75). As shown in Figure 8.78, this app contains basic, but necessary, information about the goods, such as their system identification number, description, weights, and unit of measures.

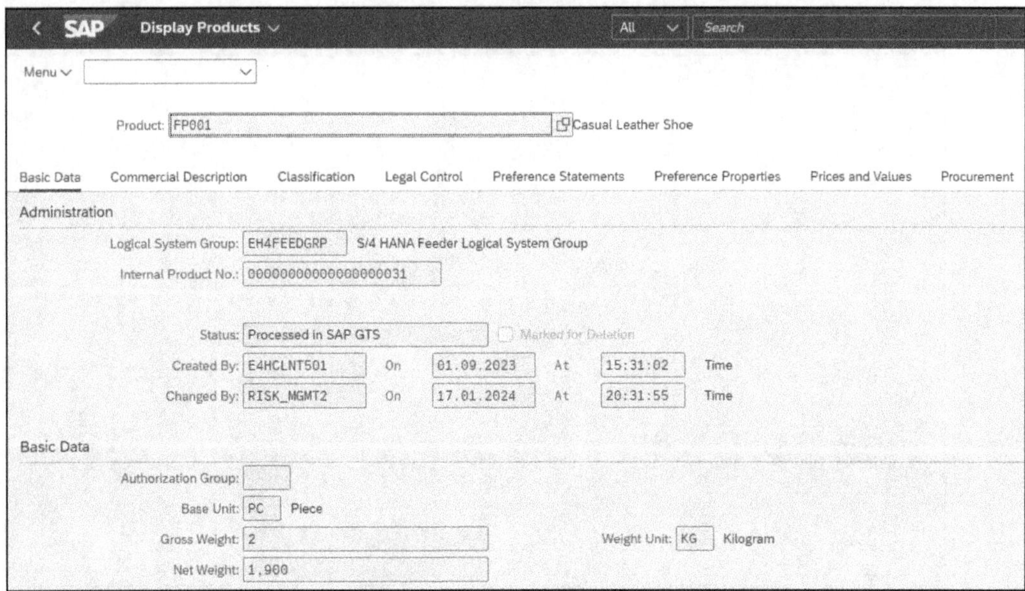

Figure 8.78 Product Master

In addition to the basic data information, the product master contains two additional tabs related to Preference Management:

- **Preference Properties**
  This tab allows users to maintain information specific to inbound LTSD processes or agreement-specific properties for the product.

  Under the **LTSD Process – Inbound** area, details about the product retention period, its relevance for the request, and validity for the request can be maintained by an AU, as shown in Figure 8.79.

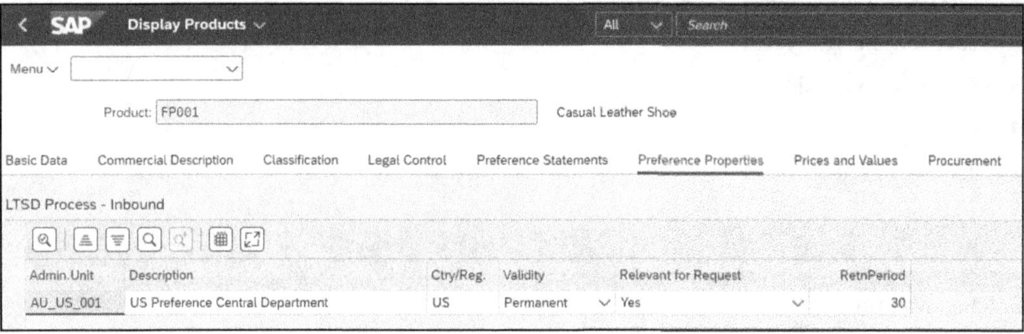

Figure 8.79 Product: Preference Properties

A product's AU- and FTA-specific information, such as the product's relevance to minimal operations rules or whether it relates to a set or not, for instance, is maintained under the **Agreement-Specific Properties**, as shown in Figure 8.80. More information about these rules is provided in Section 8.6.3.

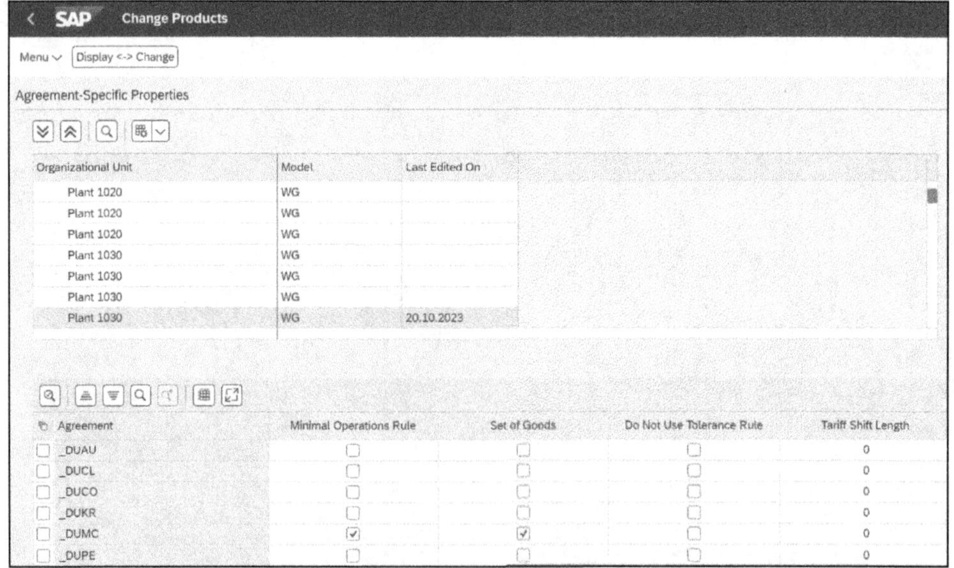

Figure 8.80 Preference Properties

- **Preference Statements**
  This tab provides a product's preferential status per FTA and AU during a specified validity date, as shown in Figure 8.81. This result comes from the preference determination, the supplier LTSD maintenance after aggregation, or a manual input of the preferential statement.

  Details related to the preference status, such as the preference statement or the determination method (Section 8.2.2), as well as a full traceability of changes over time, are displayed under this tab.

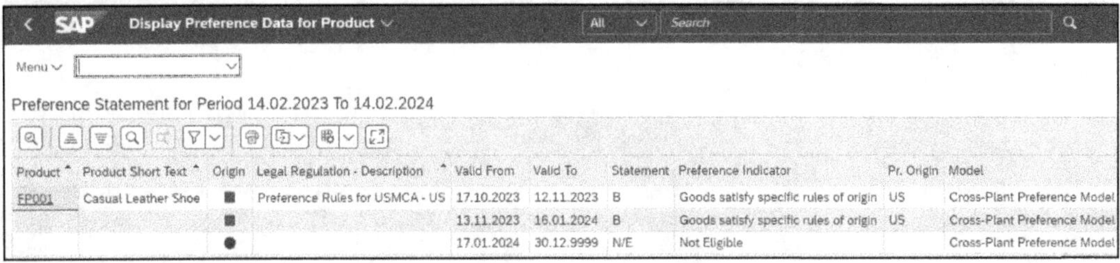

Figure 8.81 Product: Preference Statements

Procurement Indicator

Procurement indicators are maintained in SAP S/4HANA at the plant level and then transferred to SAP GTS at the plant level, as shown in Figure 8.82. A procurement indicator is displayed in the **Display Procurement Indicator – Products** tile under the **Master Data – Preference** section. A procurement indicator can have different values in SAP GTS. For each procurement indicator, a specific system process is relevant to reach a preferential result for the product in SAP GTS. Let's walk through the most common indicators:

- **I – In-House Production**
  Products with this procurement indicator are expected to undergo substantial transformation. Consequently, such products are relevant for the preference determination process in SAP GTS.
- **E – External Procurement**
  Products with this procurement indicator aren't manufactured, grown, or produced by the company but rather purchased or sourced externally. Consequently, such products are relevant for the LTSD request and maintenance process in SAP GTS.
- **M – Mixed Origin**
  Products with this procurement indicator could be procured or produced. For these products to have a positive preferential result, SAP GTS requires them to undergo a preference determination and requires that a supplier LTSD is maintained.

- **G – Internal Group Origin**
  This procurement indicator typically results from the assignment of a special procurement key in SAP S/4HANA, generally related to internal or intercompany movements. If the special procurement key translates to movement of goods between plants that belong to the same plant group, SAP GTS will automatically assign the procurement indicator **G** to the product. The process required to determine a preferential status for such products is dictated by the overall aggregated procurement indicator for the plant group.

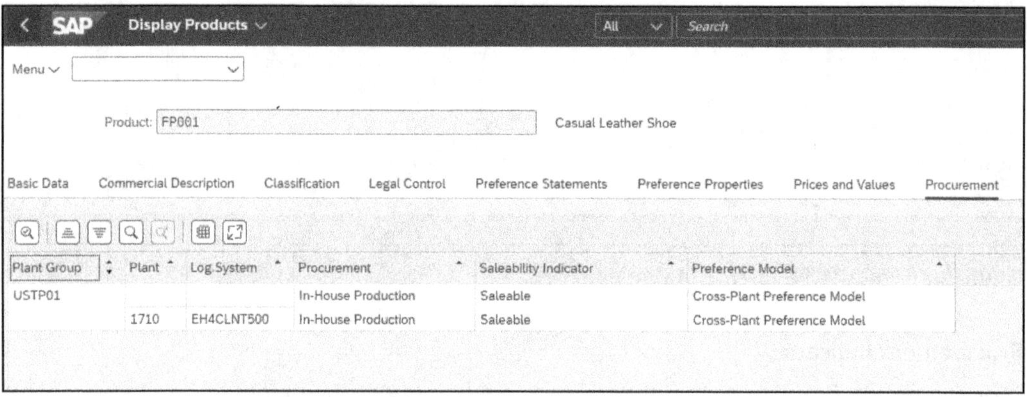

Figure 8.82 Product Procurement Indicator

When the AU is maintained as a plant group in SAP GTS (Section 8.4.1), all the procurement indicators for the plants of the plant group are aggregated under one value. Typically, this aggregation follows an inclusive principle, meaning if one product is externally procured for one plant and produced in-house for another plant within the same plant group, then the aggregated procurement indicator for this AU will be mixed.

Product Costs and Prices

To successfully run the preference determination process, correct values for the BOM products need to be available in SAP GTS. As explained in Section 8.2.2, the preference determination and the FTA dictate the necessary product values. Prices and costs for products are displayed in the **Display Prices – Product** tile (see Figure 8.83) found under the **Master Data – Preference** section.

In standard integration, the product standard price is maintained in SAP S/4HANA at the plant level and transferred to SAP GTS. Transferring the prices to SAP GTS can be enhanced so that different values are made available when calculating for different FTAs (Section 8.4.5). Regardless, the costs or prices will be used during the determination process and split as originating or non-originating values.

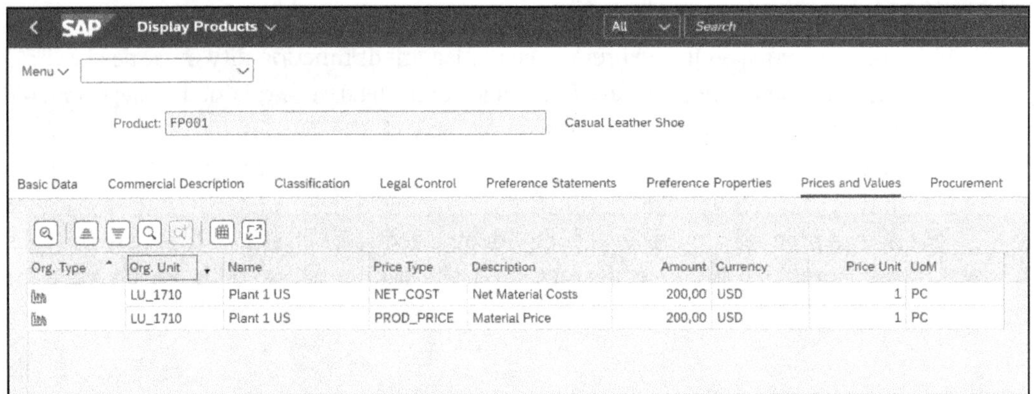

Product: FP001 Casual Leather Shoe

Org. Type	Org. Unit	Name	Price Type	Description	Amount	Currency	Price Unit	UoM
	LU_1710	Plant 1 US	NET_COST	Net Material Costs	200,00	USD	1	PC
	LU_1710	Plant 1 US	PROD_PRICE	Material Price	200,00	USD	1	PC

Figure 8.83 Product Costs and Prices

Bill of Materials

The BOM is also transferred from SAP S/4HANA to SAP GTS. The BOM explains the relationship between the finished product, subassemblies, and raw materials, and it identifies the level of each of the BOM's products for the preference determination process. In general, BOMs are maintained at the plant level in SAP S/4HANA, so they are represented per plant in SAP GTS. Therefore, in a cross-plant-based setup, the calculation for one finished product will be done for all BOMs from all plants within the same AU, and the worst-case principle will be taken as a result for the overall AU. BOMs are displayed in the **Display Bill of Material for Product** tile (see Figure 8.84) under the **Master Data – Preference** section.

Display Multilevel Bill of Material Structure

Product/Logical Feeder System: FP001 / EH4CLNT500
Description: Casual Leather Shoe
Bill of Material Number: 00000229
Usage/Alternative: 1 / 01
Plant / Plant Group: 1710 / USTP01
Key Date: 14.02.2024 17:28:20

BOM Level	Product	Product Short Text	Plant Group	Plant	Quantity	UoM	Amount	Crcy	Procurement Indicator	Tariff Number	Tariff Number
0	FP001	Casual Leather Shoe	USTP01	1710	1	PC	0,00	USD	In-House Production	640291	Other footwear:: Covering the ankle
.1	SUB-A	Unfinished Shoe	USTP01	1710	1	PC	120,00	USD	In-House Production	640610	Uppers and parts thereof, other than stiffene
..2	COMP906	Leather	USTP01	1710	1	PC	45,00	USD	External Procurement	410719	Whole hides and skins:; Other
..2	COMP921	Wool	USTP01	1710	1	PC	30,00	USD	External Procurement	511211	Containing 85% or more by weight of wool or
..2	COMP976	Rubber Sole	USTP01	1710	1	PC	15,00	USD	External Procurement	640620	Outer soles and heels, of rubber or plastics
.1	COMP632	Shoelace	USTP01	1710	1	PC	10,00	USD	External Procurement	630790	Other

Figure 8.84 Display Bill of Material for Product

Any change to these master data elements in SAP S/4HANA can impact the preferential results in SAP GTS. For instance, if a finished product undergoes changes to its BOM structure or component material prices, the preference determination needs to be redone for the finished product, which could lead to a new preferential result. Consequently, it's important to keep the master data in SAP GTS up to date. Chapter 4 discussed the mechanism to transfer data to SAP GTS and automatically enable change pointers to keep SAP S/4HANA and SAP GTS synchronized.

HS Classification Code

All products managed in the Preference Management functionality must be classified according to the applicable classification schemes (HS). As explained in Section 8.4.1, it's possible to select a numbering scheme per AU. The HS classification code of the product undergoing preferential determination dictates the relevant rule of origin to use during the preference determination. Moreover, if there are tariff shift-based rules of origin, a product's lower-level components need to be classified to satisfy the rule. Chapter 5 provides mode details on the classification process and is equally valid for Preference Management.

8.6.2 Business Partner Master Data

In addition to product-related master data, the handling of LTSDs in SAP GTS requires a number of business partner master data elements. The following sections will explain what business partner data is needed and for what purpose.

Suppliers and Customers

Supplier information must be in SAP GTS to request or maintain supplier LTSDs and customer details must be in SAP GTS to issue customer LTSDs. For these processes to function, both suppliers and customers must be transferred from SAP S/4HANA to SAP GTS.

The **Manage Preference Properties – Suppliers** and **Manage Preference Properties – Customers** tiles (see Figure 8.85) allow users to define the relevance of a business partner for the LTSD request or issue process (supplier and customers, respectively) as well as define specific communication methods (i.e., mail or email).

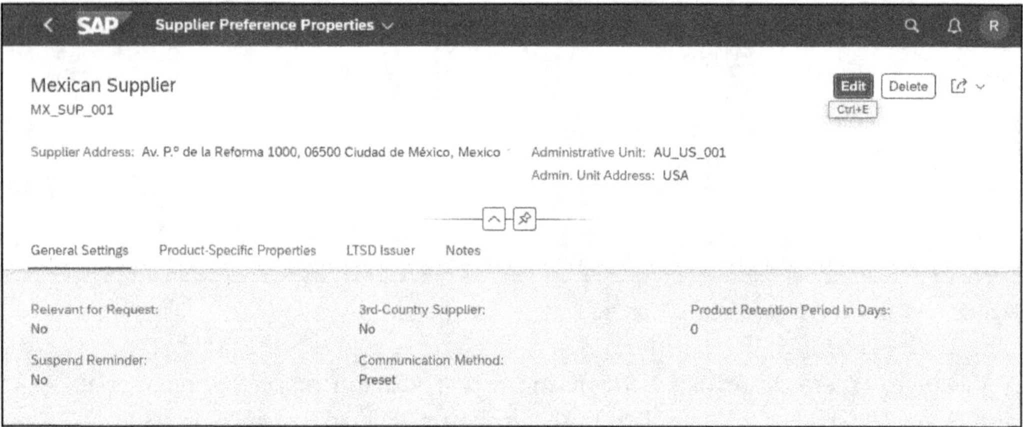

Figure 8.85 Supplier Preference Properties

Users can also specify the relevance of the issue or request for specific business partner–product combinations or create relationships among business partners to bundle the issue or request among multiple LTSDs at once. For instance, a specific supplier could be marked as irrelevant for the LTSD request process, a specific customer could always be issued an LTSD by mail for select products, and the LTSD created for customer X could be issued to customer Y.

Additional Business Partners

In addition to the suppliers and customers, another set of business partners needs to be created in SAP GTS to enable the LTSD request and issue processes:

- **Contact person**
  Supplier LTSDs and customer LTSDs always require a contact person to be created. The contact person acts as the person of contact from the AU. This contact person can be defaulted in the LTSD document when it's created and can be used in the corresponding forms or emails generated. For a contact person to be populated in the LTSD document, a contact person needs to be created as a user in SAP GTS.
- **Recipient**
  Supplier LTSD requests and customer LTSDs also require a recipient so these documents can be sent. This recipient contains the contact details (i.e., name, address, email, etc.) linked with the LTSD's business partner (supplier or customer). Practically, this recipient is seen as the actual contact person within the company who receives/issues the LTSD. For instance, the LTSD recipient in a LTSD request could be the person in charge of handling preferential documentation within the company of the supplier. This recipient needs to be created in SAP GTS as a business partner with the **Responsible for Preference** role in the first step. In the second step, this recipient needs to be created as a relationship to the supplier or the customer of the LTSD under the **Has Employee Responsible for Preference (Customer/Supplier)** option in the **Relationship Cat.** field, as shown in Figure 8.86.

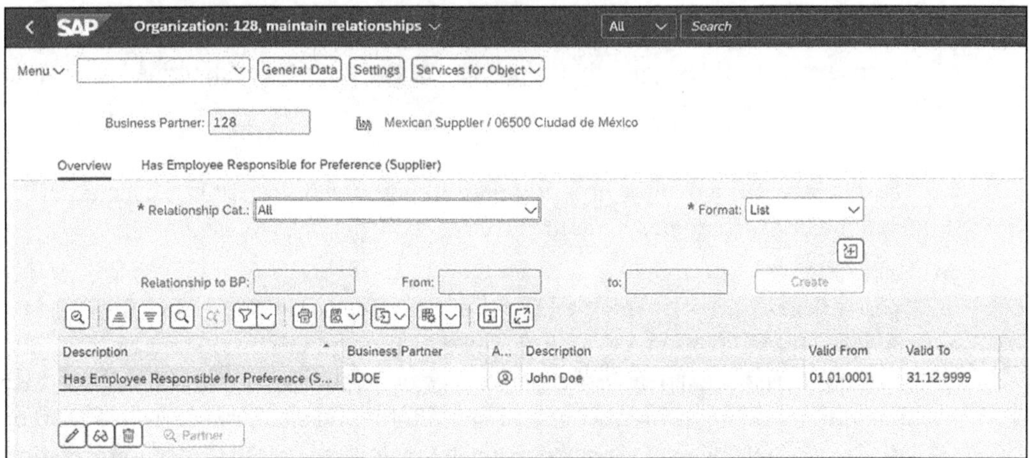

Figure 8.86 Business Partner Relationship

8.6.3 Preference Agreements and Rules of Origin

To perform the preference determination as explained in Section 8.3.5, rules of origin need to be maintained in SAP GTS per the relevant FTAs. Even though rules of origin could be maintained manually in SAP GTS, it's typically recommended to load these rules based on the XML file received from a content provider. These XML files will update the configuration, as described in Section 8.4.3, and contain the master data. The master data content of the rules of origin distinguishes between base rules and tariff code-related rules.

Base rules apply to the whole agreement and all included products. There are three types of base rules typically encountered:

- **Tolerance rules**
  The tolerance rule applies to all products that fail to fulfill the item-related rules and are typically similar to a value-added rule (discussed later in this section).
- **Minimal operations rules**
  Minimal operations rules are typically assigned to products that don't go through sufficient processing operations or undergo minimal operations (e.g., simple packaging). When a product isn't substantially transformed, it's rarely granted preferential status, and specific rules should apply: either very restrictive or completely preventing the preferential eligibility of the good. When a product doesn't undergo sufficient processing, it's important to flag the corresponding products in SAP GTS. This can be done via the product master, as shown previously in Figure 8.80.
- **Set-specific rules**
  Similar types of rules can be granted to products bundled as sets. The FTA-specific definitions of set can vary, so checking the relevant FTA can show whether the company's definition of set matches the FTA interpretation. When a product meets the FTA's definition of a set, it's important to flag the corresponding products in SAP GTS. This can be done via the product master, as shown in Figure 8.80.

Tariff code-specific rules typically fit into one of the three types described in Section 8.2.2:

- **Tariff shift**
  These rules require that the non-originating component's classification codes differ from the finished product's classification code.
- **Value-added**
  These rules require that the total preferential value contained within the finished product (either component, tooling, overhead, or other costs based on the FTA) must be at least equal to a certain percentage of the finished product's value or that the non-originating content contained within the finished product is less than or equal to a certain percentage of the value of the finished product. The interpretation

of these rules and what values may be used vary by FTA. This rule type includes RVC rules, transaction value method rules, value-added rules, and tolerance rules.

- **Manufacturing or processing operations/specific rules or conditions**
These rules relate to specific processing operations, manufacturing mechanisms or rulings, product details and characteristics, or other conditions needed to satisfy the rule of origin. They vary greatly across FTAs and HS codes, but, for all types, SAP GTS needs user intervention to interpret them. SAP GTS represents these requirements as additional conditions, which are text descriptions of the imposed requirement. If a condition type rule of origin applies to a product, the company must review the details of the condition and assess whether the product meets that condition. After this assessment, users must flag whether the product fulfills the condition in SAP GTS through the **Classification** tab of the **Manage Products** app (selecting the **Preference Condition Applies** checkbox), as shown in Figure 8.87. This enables SAP GTS to account for this result during preference determination. A product that isn't assigned a condition in SAP GTS will automatically fail that rule of origin.

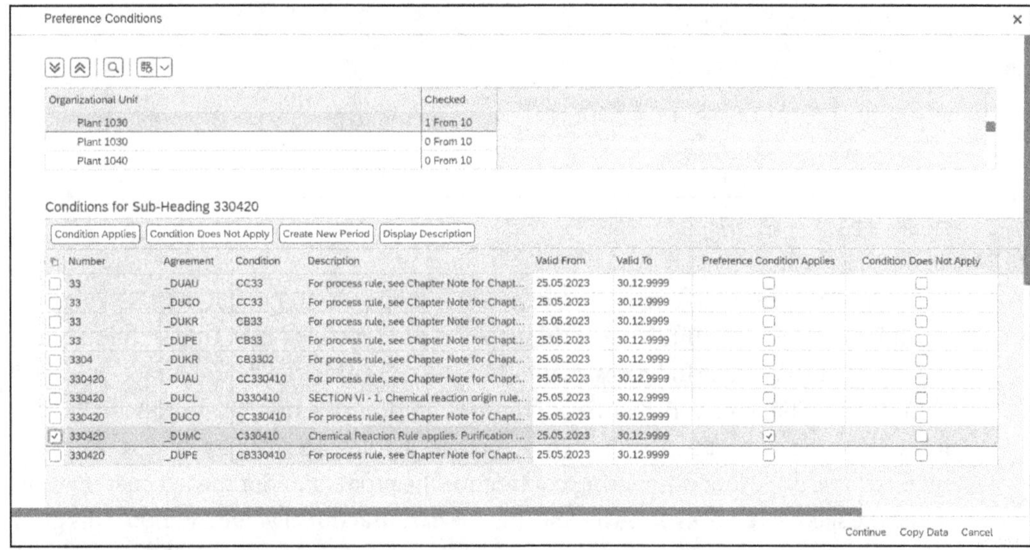

Figure 8.87 Special Condition

Tariff code-related rules of origin can contain one or more of these rule types. These rules can become complex structures that combine simple rules with alternative and standard rules as well as AND and OR statements. Due to the complex and diverse nature of rules of origin, it's typically recommended to purchase them from a content provider. In this case, the rules are received in the form of XML files that can be loaded in SAP GTS, as explained in Section 8.4.3. Figure 8.88 shows a rule of origin from USMCA, as an example.

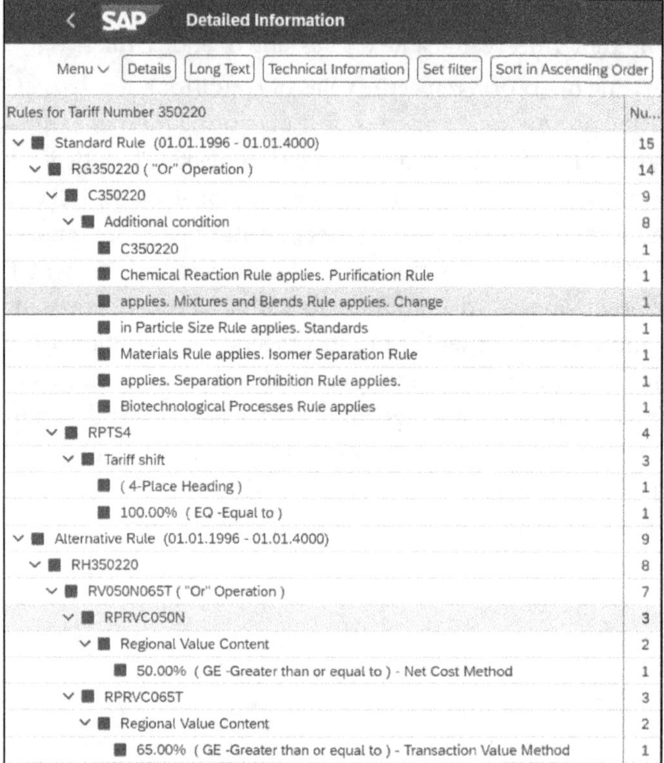

Figure 8.88 Rule of Origin

The rule allows users to choose between a standard rule and an alternative rule, and products only need to fulfill one of these for the product to qualify. Under the standard rule, the OR statement indicates that either the condition needs to be met or a tariff shift needs to be observed. If either of these conditions are met, the product will qualify. Under the alternative rule, two value-based rules are displayed. The first one requires at least 50% of originating content for the product under the net cost method, while the second requires at least 65% under the transactional value method. This gives the company the option to select their calculation method (Section 8.2.2), and the product would qualify under either of these methods according to this rule of origin. Another example of rule of origin analysis is shown in Section 8.3.5 where an AND rule is analyzed for the leather shoe.

8.7 Summary

This chapter addressed the Preference Management functionality of SAP GTS in a comprehensive manner by covering the key processes (i.e., from both the business and system point of view), master data considerations, Customizing steps, and extended features that come with SAP GTS, edition for SAP HANA.

The functionality supports requirements related to the request and determination of the preferential origin of goods to be able to issue preferential proofs of origin so that products may be exported under preferential treatment.

SAP GTS sequences its processes in three main steps. The first focuses on the request and maintenance of preferential origin of sourced goods via the LTSD request and supplier LTSD maintenance processes. The second determines the eligibility of products per FTA during the preference determination process where rules of origin and BOMs are combined to reach a preferential result. The third allows preferential proofs of origin to be issued by not only leveraging the customer LTSD and revoke processes of the Preference Management functionality but also by integrating SAP GTS with sales documents of SAP S/4HANA or the export declaration of the Customs Management functionality.

To support these processes, key integration and maintenance of master data is paramount. The master data integrated from SAP S/4HANA (e.g., business partners, BOMs, procurement indicators, prices) needs to be comprehensive and complete, and the master data created or uploaded in SAP GTS (e.g., rules of origin, preference properties for suppliers, customers or products, conditions) needs to be looked at thoroughly. The Preference Management processes are only as good as the master data quality that can be provided by SAP S/4HANA. Moreover, transactional document integration with SAP S/4 HANA (i.e., purchase orders, goods receipts, billing documents) is of equal importance, as it will support the automatic identification of LTSDs to be requested, aggregated, issued, and revoked, making the overall process more robust.

Finally, this chapter elaborated on some of the key innovations of the Preference Management functionality in SAP GTS such as the new SAP Fiori apps and processes, the latest Customizing steps, and the new features, including the preference data for product identifiers, the goods retention periods, and the intercompany LTSD processes.

In the next chapter, we'll discuss the standard SAP GTS reporting capabilities.

Chapter 9
Global Trade Reporting

SAP GTS allows users to use large amounts of global trade data for strategic decision-making by leveraging its comprehensive reporting capabilities, which we'll discuss in this chapter.

Companies need to be able to monitor, measure, and analyze the significant amount of data that is created for each global trade transaction. With SAP GTS, users can create the necessary reports to manage trade compliance, and these standard reports can be further enhanced to address specific business or legal requirements. The reporting capabilities also allow companies to monitor the effectiveness of global trade operations based on key performance indicators (KPIs).

This chapter has been structured into three sections: Section 9.1 covers the SAP GTS reporting capabilities that are standard and readily available; Section 9.2 addresses embedded analytics capabilities; and Section 9.3 focuses on the analytical reporting functionalities.

9.1 SAP GTS Reporting

This section covers key SAP GTS reporting functionalities: SAP Fiori-based reporting and the SAP GTS reports per functional area (note that these reports use embedded analytics, which is further clarified in Section 9.2).

SAP GTS can generate various reports related to different modules. Users can access two main types of reports:

- **Standard or descriptive reports**
  These reports collect data from the system retrospectively; for example, users may leverage these reports to display an audit trail for business partners screened against sanctioned party lists (SPLs).
- **Transactional or operational reports**
  These reports collect data on transactional activities, for example, displaying a list of all inbound movements.

Both types of reports will be further referenced throughout this chapter.

9.1.1 SAP Fiori-Based Reporting

SAP Fiori provides a consistent and holistic user experience (UX) for SAP GTS. SAP Fiori is part of the SAPUI5 frontend library, which includes numerous user interface (UI) controls and various layout options. SAP Fiori apps have the same layout, control behavior, and visual appearance.

SAP Fiori is primarily designed based on these five key principles:

- **Role-based**
  Designed for users' business needs and how users work.
- **Adaptive**
  Enables users to adapt based on the type of device used and ways of working.
- **Simple**
  Provides easy-to-use essential functions and user personalization options.
- **Coherent**
  Provides the same consistent experience for users to enjoy across the enterprise.
- **Delightful**
  Enriches the user's work experience, increasing user engagement.

In SAP GTS, edition for SAP HANA, SAP Fiori is enriched with the power of SAP HANA and provides a faster application response and query executions for reports. One of the important improvements provided by SAP GTS, edition for SAP HANA is the embedded analytics in the SAP Fiori apps.

Embedded analytics is a key new feature in SAP GTS, edition for SAP HANA that is exclusively based on SAP Fiori and SAP Analytics UIs. The solution comes with many predefined analytical apps based on SAP Fiori for business users. Users can create specific KPIs, reports, and dashboards.

Embedded analytics can be defined as built-in SAP Fiori apps in SAP GTS and can be accessed by using filters in the SAP Fiori reports. The analysis can be visualized in terms of bar charts and pie graphs using visual or compact filters in SAP Fiori reporting. For example, embedded analytics used in SAP Fiori report Manage Blocked Documents can be used to evaluate several blocks per foreign trade organization (FTO), number of blocks by legal regulation, and/or number of documents by business document category. Figure 9.1 illustrates a comprehensive insight into the business usage. On the top half of the screen, an overview of the statistics is provided from where the user can opt to drill down to the relevant individual transaction or SAP GTS document by selecting one of the statistics.

While in previous SAP GTS versions, most reports were accessible through the standard SAP GTS cockpit or existing menu paths, in SAP GTS, edition for SAP HANA, users can access reports through the SAP Fiori app.

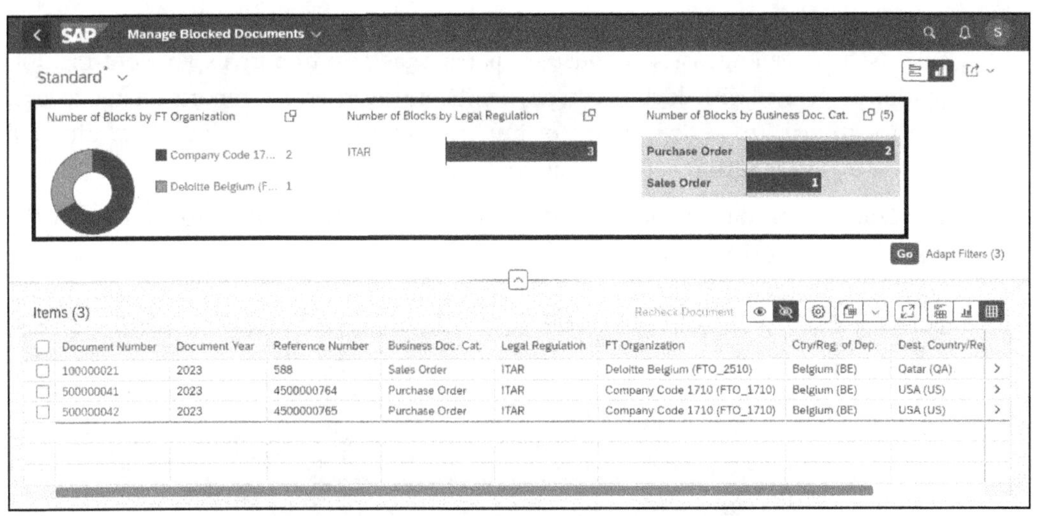

Figure 9.1 Example of Embedded Analytics in SAP GTS

Reporting in SAP GTS has now shifted from SAP List Viewer (ALV)-based applications to SAP Fiori apps, which provide a better UI. This new SAP GTS version has released an extensive number of SAP Fiori apps for the following main functional areas of SAP GTS: Products and Classification, Compliance Management, Customs Management, Preference Management, and Intrastat.

SAP GTS has introduced and deployed a wide range of extensive SAP Fiori apps; combined with their embedded analytics capabilities, these apps are set to revolutionize operational reporting in SAP GTS. To use the new SAP Fiori apps, the following product versions are required alongside SAP GTS, edition for SAP HANA: SAP S/4HANA 2022 and SAP Fiori front-end server (FES) 2022 for SAP S/4HANA.

SAP GTS reporting through SAP Fiori is structured around the three main SAP GTS functional areas: Compliance Management, Customs Management, and Preference Management. In each functional area, the reports available are divided into different subfunctionalities such as master data, general reports, and other subareas specific to the functional area, further categorizing the relevant reports (e.g., starting with general compliance reports and then moving to SPL screening category followed by embargo and legal control reports). In each category, the most common reports will be explained in further detail next.

9.1.2 Compliance Reporting

Both operational and descriptive reports are available in the Compliance Management functionality. General compliance reports are used in all subareas (e.g., the Display Documents report can be used to review the documents under all compliance checks for SPL, embargo, and legal control).

Descriptive reports present views of historical data and serve as audit trails for topics such as the screening history of business partners and the import/export control information associated with documents, payments, and licenses. The reports related to legal control are used to review export or import control licenses, license determination strategies, and assignment of licenses to the documents.

The Compliance Management reports are listed in Table 9.1 and grouped under the respective categories.

Category	SAP Fiori Reports
SPL screening	Manage Positive List – BPs Manage Negative List – BPs Manage Business Partner Manage Blocked Partners Display Audit Trail for Documents Display Audit Trail for Partners Display Audit Trail for External Addresses Display Audit Trail for General Addresses Analyze Reasons for Release Manage Sanctioned Party List Master Display Sanctioned Party Lists Display Expiring Sanctioned Party List Entries Display SPL Change History
Embargo	Display Released Embargo Documents Analyze Reasons for Release Display Partner in Embargo Countries
Legal control	Display Existing Licenses Display Assigned Documents Display Determination Strategies Display Classified Products (Legal Control) Display Products – Re-Export Display BOMs for Re-Export Display Assigned Documents/Licenses ITAR Tracking
General	Manage Blocked Documents Display Documents – Compliance Management

Table 9.1 Compliance Reports

Next, the most common compliance reports will be explained under each category.

Sanctioned Party List

To prevent trading with sanctioned entities and sanctioned parties, SAP GTS provides the ability to perform SPL screening on business partners. Besides the actual screening process explained in Chapter 6, SAP GTS provides a number of reports to support business decisions and facilitate compliance with regulatory requirements (e.g., audits). SPL reporting is generally divided into monitoring, audit trail, and master data analysis. The monitoring reports focus on business partner screening against SPLs, whereas the audit trail reporting is more focused on the chronology of screening results and any user action that has been performed.

In this section, we'll take a closer look at the following important SPL reports:

- Manage Blocked Partners
- Display Audit Trail for Partners
- Analyze Reasons for Release

One of the key SPL reports used relates to managing blocked partners. The Manage Blocked Partners report provides a highly detailed view of blocked customers, suppliers, and other business partners in SAP GTS. From a business perspective, users can analyze their business partners in this report and decide whether to release (e.g., due to a false positive) or confirm the block. This report is available in the **Partner Screening** section of the **Sanctioned Party List** page in the **Compliance** space of the SAP Fiori launchpad.

To enable an analysis of the companies' business partners, users can specify the FTO (**FT Organization**) and business partner category (**Bus.Part.Categ.**; e.g., customer, supplier, bank, etc.) fields as well as other selection criteria, as shown in the selection screen in Figure 9.2.

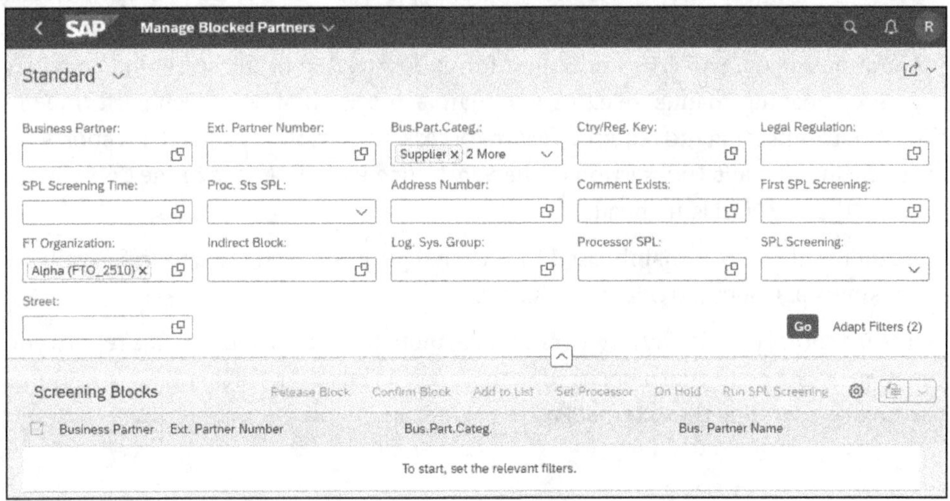

Figure 9.2 Manage Blocked Partners Selection Screen

The report (see Figure 9.3) lists all blocked business partners (i.e., customers or supplier) by external and internal business partner number. This report also shows information specific to the SPL screening, such as the legal regulation against which the block occurred, and data used for the screening process, such as name and address details of the business partner.

The total number of **Screening Blocks** (blocked partners) is displayed in the header of the report, and this gives an indication on the workload of the review. Users can take further action on the blocked partners using the functions available on the result screen (e.g., **Release Block**, **Confirm Block**, **Add to List**, etc.) (see Chapter 6, Section 6.2).

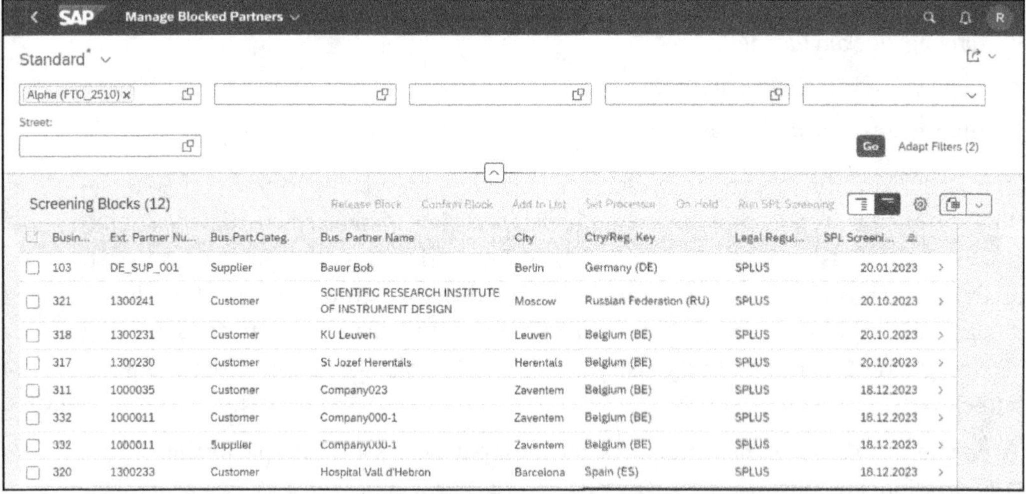

Figure 9.3 Manage Blocked Partners Report

When a business user has taken action, it's important for companies to keep historical SPL screening results regarding business partner, documents, and payments available for audit purposes. SAP GTS stores all information related to the screening activities (e.g., SPL screening, manual releases, and manual blocks) that have been performed in the system, rendering the required records accessible in case of audits. The reports are available in the **Audit Trail** section of the **Sanctioned Party List** page in the **Compliance** space of the SAP Fiori launchpad.

The audit trail report is mainly used for internal purposes, and one report is available for business partners and one for documents.

Figure 9.4 shows the **Display Audit Trail for Partners** selection screen. In this report, you can request the audit trail associated with a specific business partner (i.e., or multiple partners) by entering the **External Partner Number**.

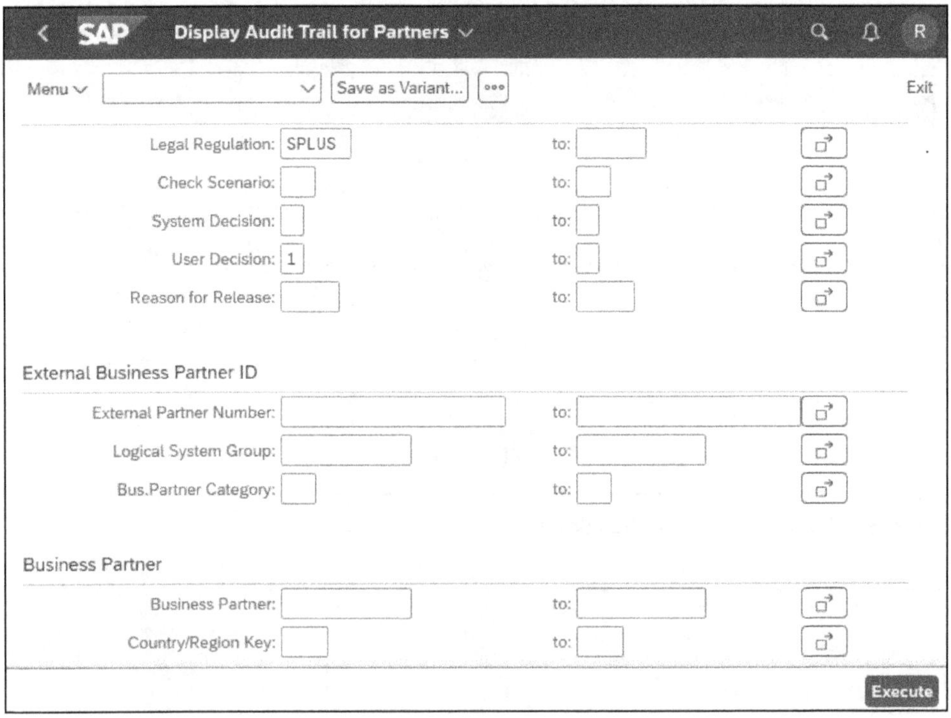

Display Audit Trail for Partners

Menu | Save as Variant... | Exit

Legal Regulation: SPLUS to:
Check Scenario: to:
System Decision: to:
User Decision: 1 to:
Reason for Release: to:

External Business Partner ID

External Partner Number: to:
Logical System Group: to:
Bus.Partner Category: to:

Business Partner

Business Partner: to:
Country/Region Key: to:

Execute

Figure 9.4 Display Audit Trail for Partners: Selection Screen

The report output is displayed in three parts. The top part of the screen shows a list of the selected partners. A user can select a blocked entity in the top section. In doing so, the system will present the SPL entries that caused a particular block in the **Matched Sanctioned Party List Entities** area. When selecting an entry in the list of matched SPL entities, the system opens a list of address components that matches the SPL entries resulted in the block. This is displayed in the **Matched Terms in Address** area, which is further illustrated in Figure 9.5. In this list, the release reason and any additional comments entered by the user can also be consulted, as displayed in Figure 9.6.

The audit report contains two SPL screening results: **SCR** (system check result) and **UDR** (user check result). For the consecutive system checks, the business partner failed on the system check (signified by the red symbols). However, for the highlighted line, the audit trail reveals that the business partner has been manually released, with a specified reason **False positive release**.

Business partners released against an SPL block require careful consideration from a legal and compliance perspective. Therefore, the ability for companies to review employees' past actions and justifications for those actions (e.g., reasons for release) is important. Users can access historical information on those actions in the Analyze Reasons for Release report.

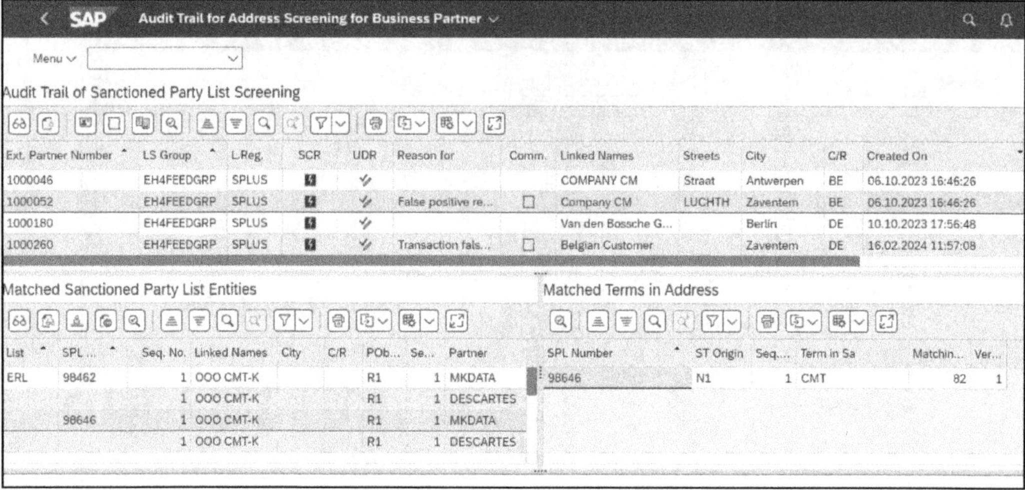

Figure 9.5 Matched Terms in the Address

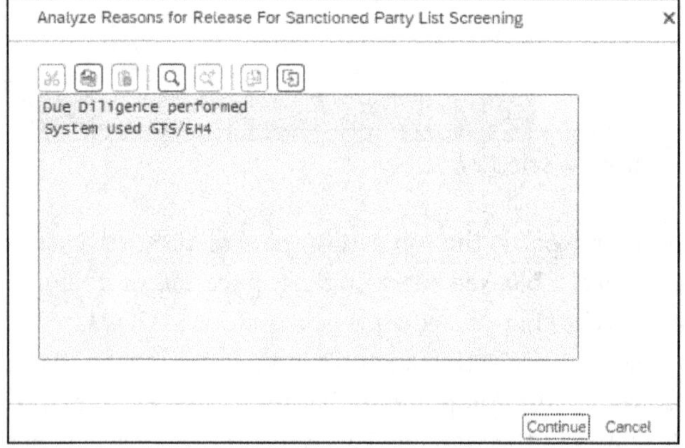

Figure 9.6 Analyze Reasons for Release: Comment Box

Figure 9.7 illustrates the **Analyze Reasons for Release** selection screen. The most relevant inputs to run the report are the specific **Legal Regulation** and the **Reason Object**. The latter will determine whether the report displays a list of released documents or a list of released business partners.

In Figure 9.8, the report was executed with **Business Partner** as the **Reason Object**, which will result in displaying all previously blocked business partners that have been released along with the associated reasons for release.

In addition, the report illustrates where business partners have been manually released for either a false-positive match after escalation review (i.e., **R2**) or when sanctions were not applicable to the transaction (i.e., **TR1**). The report is also capable of displaying business partners that were manually released for a false positive match on address (i.e.,

ADDR) or name (i.e., NAME). The report outlines release description (e.g., due to incorrect match of an address) and business partner information, as well as the time of the release.

Analyze Reasons for Release

Menu | Save as Variant...

Compliance Block

* Legal Regulation:

Reason for Release: to:

Date of SPL Screening: to:

Document

Reference Number: to:

Object Type: to:

Logical System: to:

Created By (Feeder Sys.): to:

Changed By (Feeder Sys.): to:

Created By: to:

Created On: to:

Layout

Layout:

* Reason Object: Customs Document

Figure 9.7 Analyze Reasons for Release Report Selection Screen

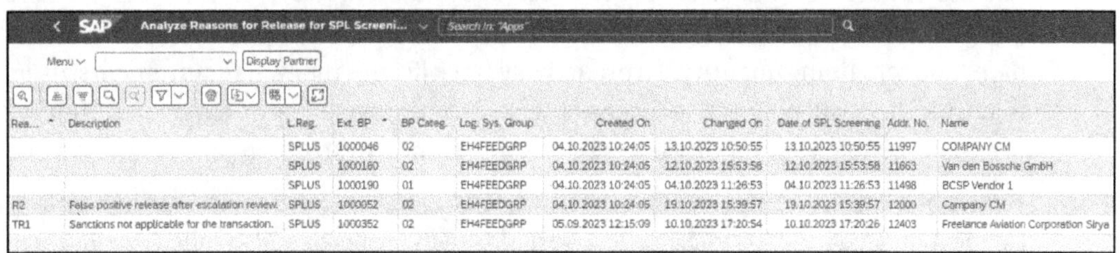

Analyze Reasons for Release for SPL Screeni...

Menu | Display Partner

Rea...	Description	L.Reg.	Ext. BP	BP Categ.	Log. Sys. Group	Created On	Changed On	Date of SPL Screening	Addr. No.	Name
		SPLUS	1000046	02	EH4FEEDGRP	04.10.2023 10:24:05	13.10.2023 10:50:55	13.10.2023 10:50:55	11997	COMPANY CM
		SPLUS	1000160	02	EH4FEEDGRP	04.10.2023 10:24:05	12.10.2023 15:53:58	12.10.2023 15:53:58	11663	Van den Bossche GmbH
		SPLUS	1000190	01	EH4FEEDGRP	04.10.2023 10:24:05	04.10.2023 11:26:53	04.10.2023 11:26:53	11498	BCSP Vendor 1
R2	False positive release after escalation review.	SPLUS	1000052	02	EH4FEEDGRP	04.10.2023 10:24:05	19.10.2023 15:39:57	19.10.2023 15:39:57	12000	Company CM
TR1	Sanctions not applicable for the transaction.	SPLUS	1000352	02	EH4FEEDGRP	05.09.2023 12:15:09	10.10.2023 17:20:54	10.10.2023 17:20:26	12403	Freelance Aviation Corporation Sirya

Figure 9.8 Analyze Reasons for Release for SPL Screening Report

Embargo

Governments also can impose trade restrictions with parties in embargoed countries. For a more detailed overview of the SAP GTS embargo functionality, refer to Chapter 6, Section 6.3.

The SAP GTS embargo functionality applies at a country level and isn't specific to business partners. All embargo-related reporting is linked to the country of the business partner as maintained within the transaction. This is applicable for both outbound and inbound transactions.

One example of an embargo report is the Display Partner in Embargo Countries report, which allows a company to analyze its business partners in embargo situations. Figure 9.9 shows the result of which business partners are residing in an embargo country. In this example, the company has decided not to do business with Russia (RU) and therefore has flagged **RU** as an embargoed country.

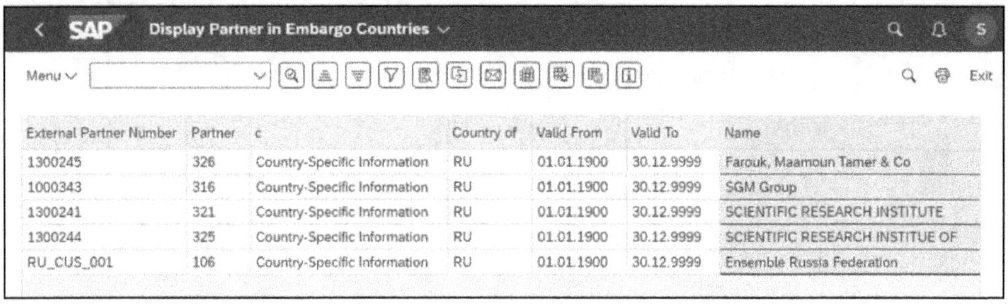

Display Partner in Embargo Countries

External Partner Number	Partner	c	Country of	Valid From	Valid To	Name
1300245	326	Country-Specific Information	RU	01.01.1900	30.12.9999	Farouk, Maamoun Tamer & Co
1000343	316	Country-Specific Information	RU	01.01.1900	30.12.9999	SGM Group
1300241	321	Country-Specific Information	RU	01.01.1900	30.12.9999	SCIENTIFIC RESEARCH INSTITUTE
1300244	325	Country-Specific Information	RU	01.01.1900	30.12.9999	SCIENTIFIC RESEARCH INSTITUE OF
RU_CUS_001	106	Country-Specific Information	RU	01.01.1900	30.12.9999	Ensemble Russia Federation

Figure 9.9 Display Partner in Embargo Countries Report

Legal Control

For some specific controlled products, governments can impose import or export restrictions that may trigger license requirements. Similar to SPL reporting, legal control reports provide information which will enable the user to monitor blocked transactions due to missing licenses. When a license is missing, users can leverage a report to retrieve the transactions where a license needs to be assigned. Similar to embargo reporting, there are reports for inbound and outbound transactions.

The Display Existing Licenses report is useful to maintain an overview of all licenses present in the system. The overview presents relevant information such as validity dates and remaining quantity. To proactively track license renewal requirements, users may run the Display Existing Licenses report to retrieve a list of export and import licenses with expiring validity dates in the near future, near-zero quantity, and others.

Depending on the company's business, various types of licenses and authorizations can be managed in SAP GTS. Each license has its own properties, also called *attributes* in SAP GTS. The Display Existing Licenses report can be used to obtain an overview of all licenses maintained in the system with its properties (attributes). The selection criteria for this report consists of the **Legal Regulation** for which the report displays the applicable licenses, as shown in Figure 9.10. In addition, users can select other criteria such as **License Type**, **Status** (e.g., **Expired**), **Product Number**, and **Valid From/Valid To** dates to enable a more detailed search.

Figure 9.11 shows the result of running the report for ITAR legal regulation from the US. The existing licenses are grouped by license type and display information, such as the corresponding external license number (i.e., typically the actual number assigned by the authority that granted the license), quantity and value data, and validity date.

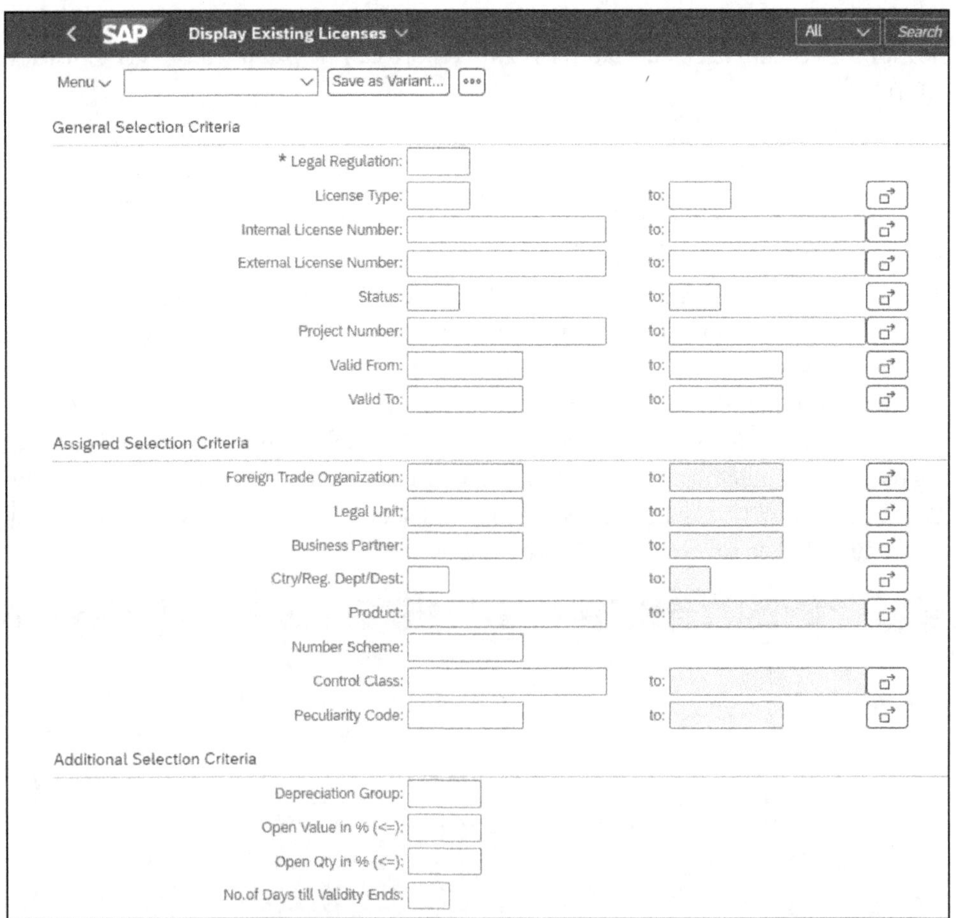

Figure 9.10 Display Existing Licenses Report: Selection Screen

L...	Licens...	Internal License Number	LicenseNo.	Status	Valid From	Valid To	I...	Product	Product Short Text	Quantity	Unit	Maximum Value	Currency
ITAR	DSP-5	15000000000000000021	ITAR A 01	004	01.10.2023	31.12.2024	3			10.000	PC	100.000,00	EUR
	DSP-5	15000000000000000057	ITAR-B-003	004	01.01.2023	31.12.2024	3			20	EA	50.000,00	USD
	DSP61	15000000000000000016	610281990	004	01.01.2023	31.12.2023	3	GB1005	Aircraft Engine Parts			196.000,00	EUR

Figure 9.11 Display Existing Licenses Report

In addition, the Display Assigned Documents to Licenses report provides detailed information about which documents are assigned to a certain license in SAP GTS, with information such as item, weight, and decrementing quantity.

Another key master data report for Compliance Management is Display Classified Products (Legal Control). For companies that handle large numbers of materials, manually classifying one product at a time can be a complex task. SAP GTS enables users to display a selection of materials that haven't yet been classified and enables mass classification of certain selected materials. This report can help reduce the classification effort.

Users can leverage the Display Classified Products (Legal Control) report in a number of ways. Note that this report covers the analysis of compliance classification codes and not customs classification codes (i.e., commodity and tariff codes). The classification of customs codes is covered in Section 9.1.3.

By entering a certain product (or range of products) in the selection screen, users can analyze the classification over the various legal regulations. Alternatively, users could look at all classifications for products created in the past month. It's also possible to search for products that are similarly classified. Figure 9.12 provides a sample output of such a report. Products are listed with their classification code, alongside corresponding data (e.g., validity date, grouping, peculiarity code, **No Control** checkbox, and **IndivMaint** checkbox). This is an effective report for users or companies wanting to display all products linked to a certain existing control class.

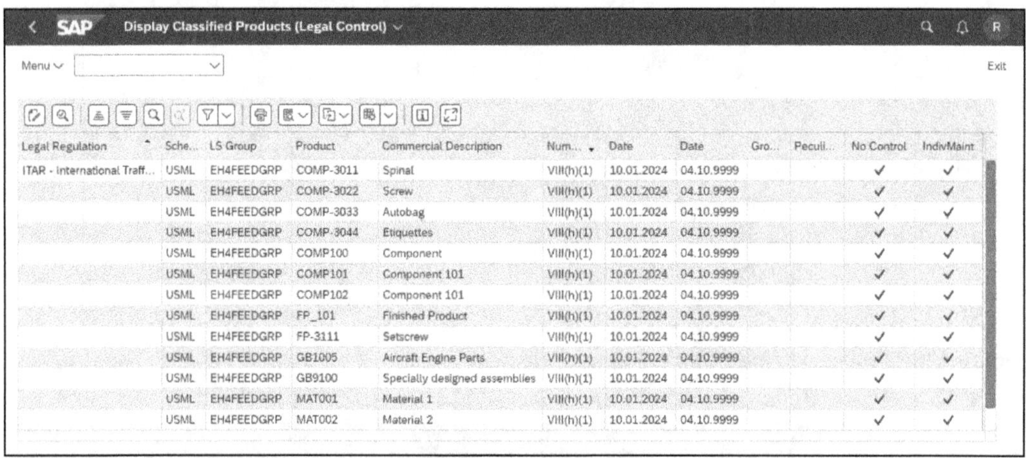

Legal Regulation	Sche...	LS Group	Product	Commercial Description	Num...	Date	Date	Gro...	Peculi...	No Control	IndivMaint
ITAR - International Traff...	USML	EH4FEEDGRP	COMP-3011	Spinal	VIII(h)(1)	10.01.2024	04.10.9999			✓	✓
	USML	EH4FEEDGRP	COMP-3022	Screw	VIII(h)(1)	10.01.2024	04.10.9999			✓	✓
	USML	EH4FEEDGRP	COMP-3033	Autobag	VIII(h)(1)	10.01.2024	04.10.9999			✓	✓
	USML	EH4FEEDGRP	COMP-3044	Etiquettes	VIII(h)(1)	10.01.2024	04.10.9999			✓	✓
	USML	EH4FEEDGRP	COMP100	Component	VIII(h)(1)	10.01.2024	04.10.9999			✓	✓
	USML	EH4FEEDGRP	COMP101	Component 101	VIII(h)(1)	10.01.2024	04.10.9999			✓	✓
	USML	EH4FEEDGRP	COMP102	Component 101	VIII(h)(1)	10.01.2024	04.10.9999			✓	✓
	USML	EH4FEEDGRP	FP_101	Finished Product	VIII(h)(1)	10.01.2024	04.10.9999			✓	✓
	USML	EH4FEEDGRP	FP-3111	Setscrew	VIII(h)(1)	10.01.2024	04.10.9999			✓	✓
	USML	EH4FEEDGRP	GB1005	Aircraft Engine Parts	VIII(h)(1)	10.01.2024	04.10.9999			✓	✓
	USML	EH4FEEDGRP	GB9100	Specially designed assemblies	VIII(h)(1)	10.01.2024	04.10.9999			✓	✓
	USML	EH4FEEDGRP	MAT001	Material 1	VIII(h)(1)	10.01.2024	04.10.9999			✓	✓
	USML	EH4FEEDGRP	MAT002	Material 2	VIII(h)(1)	10.01.2024	04.10.9999			✓	✓

Figure 9.12 Display Classified Product (Legal Control) Report

General

SAP GTS also offers some additional general compliance reports. Companies can use the Managed Blocked Documents report to monitor transactions and subsequent documents that are blocked due to compliance reasons. Trade compliance users can review the Manage Blocked Documents report and decide whether to release the document or keep it blocked. This reporting and subsequent processing is applicable for multiple compliance functionalities so users can apply filters (e.g., filtering on SPLUS legal regulation).

Figure 9.13 shows the **Manage Blocked Documents** output screen. From this report, a user can take corresponding actions allowing for further processing of the blocked document. Through this report, a user can navigate to a blocked document.

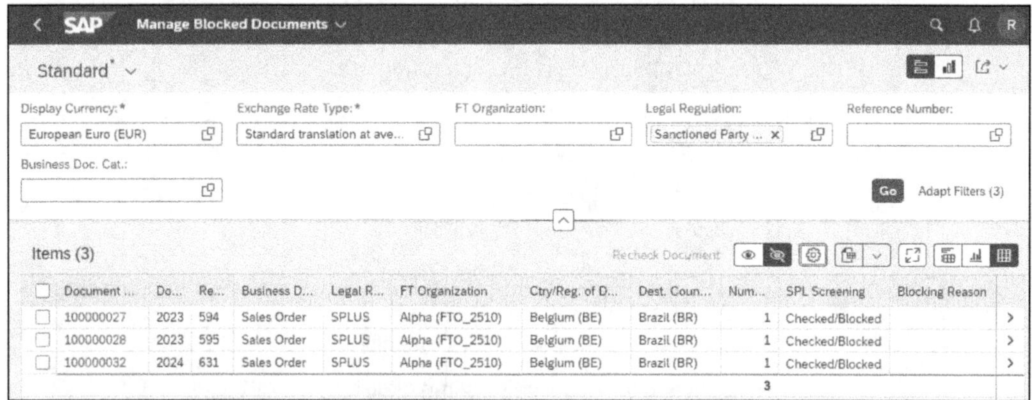

Figure 9.13 Manage Blocked Documents Report

9.1.3 Customs Reporting

Similar to Compliance Management reporting, SAP GTS provides reports to support the customs processes discussed in detail in Chapter 7. A full list of available reports is given in Table 9.2.

Customs Management reporting is grouped into the following three subcategories:

- Customs documents providing data on the general import and export functionalities
- Reports related to special customs procedures, where the focus of the reports is mostly on inventory levels
- Reports concerning customs master data

Category	SAP Fiori Reports
Customs general	Display Messages in Communication with Customs Brokers
	Display Performance Reporting for Customs Brokers
	Display Completed Transactions
Customs exports	Manage Export Declarations
	Display Export Declarations
	Overdue Exit Confirmations
	Display Documents for Customs Export Declarations
	Display Product Catalog Based on Customs Declaration

Table 9.2 Customs Reports

Category	SAP Fiori Reports
Customs imports	Display Inbound Activities Display Existing Customs Import Declarations Display Existing Customs Shipments Display Documents for Customs Import Declarations Monitor Tax Statements Display Daily Statement Display Monthly Statement Display Existing Customs Waybills Display Unassigned Customs Waybills Display Existing Customs Invoices Completed Transactions from Worklist Search Logistics Documents in Worklists Compare Customs Declaration/Goods Receipt Quantity Differences Compare Delivery and Declaration Quantities Display Differences for Delivery and Declaration Quantities Display Completed Customs Declar./Goods Receipt Quantity Comparisons Display Completed Differences for Delivery/Declaration Quantities Display Customs Declaration Quantities and Goods Receipt Quantities Compare Transferred/Cleared Worklist Quantity Display Documents Assigned to Securities
Special customs procedures	Display CWH Individual Stock Display CWH Stock overview Display CWH Receipts & Issues Display CWH Start and End Stocks Display CWH Opening/Closing Stock – Extended Display Display Domestic Goods Issues from CWH Display Completed CWH DGIs Display CWH DGI to Be Declared Display Declared CWH DGI Display Internal CWH Movements Display Customs WH Stock Information Display Collective Completion of CWH Display IP Individual Stock Display IP Stock Overview

Table 9.2 Customs Reports (Cont.)

Category	SAP Fiori Reports
Special customs procedures (Cont.)	Display IP Receipts/Issue List Display IP Settlement Run Monitor IP Processing Period Display PUCC Individual Stock Display PUCC Stock Overview Display PUCC Receipts/Issue List Display PUCC Settlement Run Monitor PUCC Processing Period Display OP Individual Stock Item Display OP Stock overview Display OP Receipts and Issues Manage OP Processing Period Processing Trade: Display Individual Stock Processing Trade: Display Stock Overview Processing Trade: Display Receipts and Issues Display Individual Stock of Foreign-Trade Zone Display Stock Overview of Foreign-Trade Zone Display Receipts and Issues in Foreign-Trade Zone Display Forecast of Foreign-Trade Zone Process FTZ Annual Board Report Manage Annual Reconciliation Report Manage Harbor Maintenance Fee Quarterly Summary Report Display Customs Declaration Worklist After GR
Master data	Display Existing Authorizations Transit/Presentation Display Products Report Display Existing Import/Export Authorizations

Table 9.2 Customs Reports (Cont.)

Customs Documents

The reports for customs documents provide information on the general inbound and outbound declarations in SAP GTS.

As discussed in Chapter 7, SAP GTS has an operative cockpit from which you can access all documents that require an action grouped by activities. Users can access these reports from the SAP Fiori launchpad under their respective activity groups. Note that these reports exist for both inbound and outbound declarations. An example of each is provided here:

- Display Inbound Activities
- Display Outbound Activities
- Display Documents for Customs Import Declarations Report
- Display Documents for Customs Export Declarations Report

Figure 9.14 illustrates the selection screen for the Display Inbound Activities report, applicable for all inbound activities requiring attention. If a user runs the report but leaves all data fields blank, the user will get a list of all inbound declarations across the different FTOs that haven't yet been completed. To narrow the search output, a user can enter different selection criteria such as a specific **Foreign Trade Organization**, resulting in the report shown in Figure 9.15. From the report, a user can access and further process customs declarations.

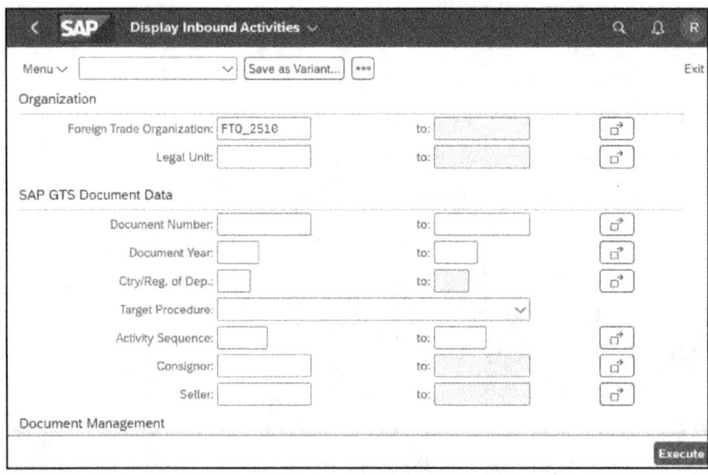

Figure 9.14 Display Inbound Activities Selection Screen

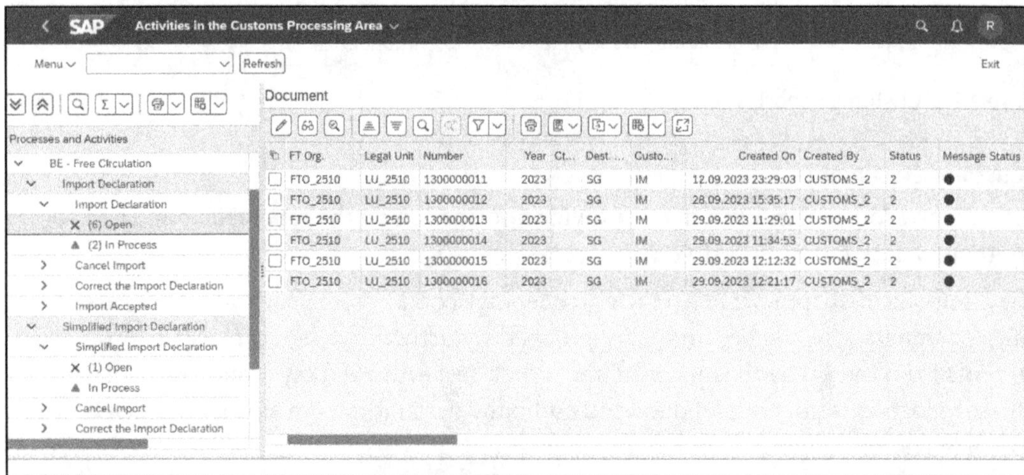

Figure 9.15 Activities in the Customs Processing Area Result

In addition to these general import reports, similar reports exist for the inbound activities related to transit, presentation to customs authorities, and pre-declaration. These reports follow a similar approach as explained previously.

The customs monitoring functionality in SAP GTS also provides an alternative way of displaying an overview of the customs declarations. It lists all customs declarations, independent of the activity with which they are associated. The Display Export/Import Declarations app in SAP Fiori also enables users to track fully processed, completed, and canceled documents.

Figure 9.16 displays the result of executing the Display Export Declarations report, which lists all existing export declarations for the specified FTO. This report displays all export transactions for the selected legal entity as well as all relevant information to declare goods to customs authorities such as the local reference declaration number (**Document Number**), the date of declaration, the country of departure, the country of destination, and all item information to be declared. 9

In addition, the report leverages embedded analytics, enabling the user to obtain additional information related to the requested entries such as a list of used classification codes (**Commodity Code**) throughout the selected documents.

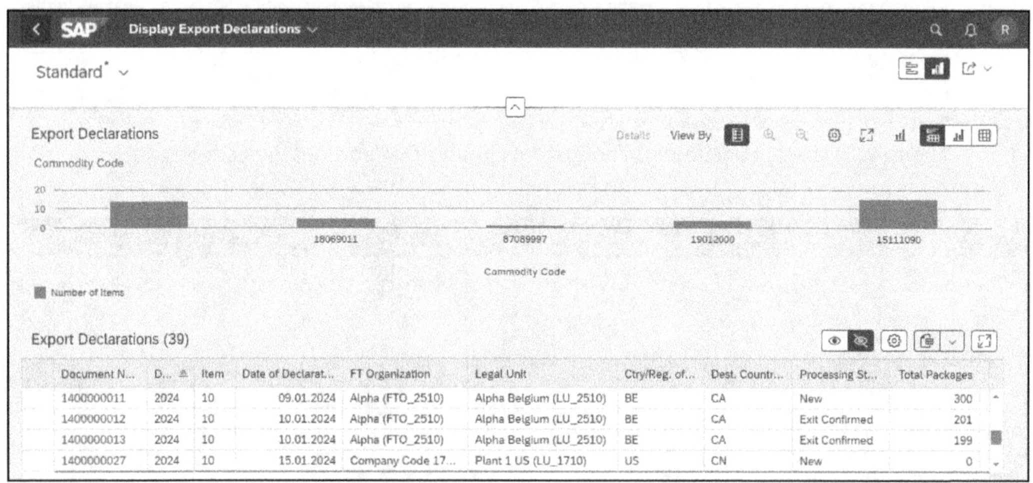

Figure 9.16 Display Export Declarations Report

Another example is the Documents for Customs Declaration report. This report enables users to retrieve, for a given FTO, all customs declarations in which documents have been maintained at the item level (e.g., import/export documents/exemptions, submitted documents).

The search can be further narrowed down by filling in different fields in the selections criteria. As an example, in Figure 9.17, we've selected a specific **FT Organizational Unit**, the **Year 2023**, the document **Category SUMA** (previous document), and the **Document Number** to find all import declarations created in 2023 for FTO_2510 that have items with document category SUMA and the document number 1874039. The result of the search is displayed in the report shown in Figure 9.18. Here, you can see the documents assigned to different customs declarations.

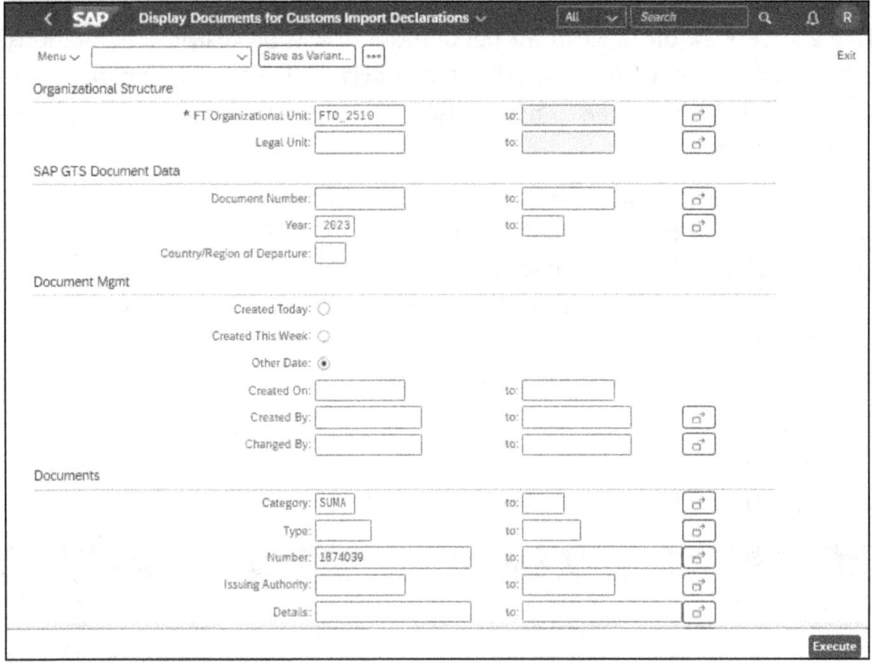

Figure 9.17 Display Documents for Customs Import Declarations: Selection Screen

Display Documents for Customs Import Declarations

Year	Number	FT Org.	Ctry/Reg.	Item	Product	Prod.Short Text	Category	Type	Type	Seq. No.	Number	N...	Created On
2023	1300000011	FTO_2510	SG	1	COMP632	Shoelace	SUMA	Z270	Loading List	1	1874039	X	12.09.2023 23:29:03
	1300000012	FTO_2510	SG	1	COMP632	Shoelace	SUMA	Z270	Loading List	1	1874039	X	28.09.2023 15:35:17
	1300000013	FTO_2510	SG	1	COMP632	Shoelace	SUMA	Z270	Loading List	1	1874039	X	29.09.2023 11:29:01
	1300000014	FTO_2510	SG	1	COMP632	Shoelace	SUMA	Z270	Loading List	1	1874039	X	29.09.2023 11:34:53
	1300000015	FTO_2510	SG	1	COMP632	Shoelace	SUMA	Z270	Loading List	1	1874039	X	29.09.2023 12:12:32
	1300000016	FTO_2510	SG	1	COMP632	Shoelace	SUMA	Z270	Loading List	1	1874039	X	29.09.2023 12:21:17
	1300000017	FTO_2510	SG	1	COMP632	Shoelace	SUMA	Z270	Loading List	1	1874039	X	29.09.2023 12:26:10
	1300000018	FTO_2510	SG	1	COMP632	Shoelace	SUMA	Z270	Loading List	1	1874039	X	04.10.2023 13:53:37
	1300000019	FTO_2510	SG	1	GB1001	Flywheel	SUMA	Z270	Loading List	1	1874039	X	04.12.2023 12:08:19

Figure 9.18 Display Documents for Customs Import Declarations Report

Special Customs Procedures

Special procedure reports provide information on specific customs procedures. Such procedures include but aren't limited to customs warehouse (CWH), inward processing relief (IPR), outward processing relief (OPR), and processing under customs control (PUCC). For more information on other special customs procedures and on the reporting related to the foreign trade zone (FTZ), refer to Chapter 7, Section 7.3.

When using special customs procedures, companies need to provide specific reports to the customs authorities on a regular basis (i.e., often depending on the program, country, or local customs office requirements). If the company uses SAP GTS to track these special customs procedures, the standard SAP GTS reports can be used or customized (i.e., enhanced). Such reporting requirements typically include overviews of the stock movements. Stock list reports can be divided into several groups: Overview of Individual Stock Items, Stock Overview, or Receipts and Issues.

The Overview of Individual Stock Items reports show the receipt and issue postings that have an economic impact on the special procedure products. Because SAP GTS processes the products on a first-in, first-out (FIFO) basis, SAP GTS links each receipt posting with the corresponding issue postings. This relation between receipt and issue postings is also included in those reports. In addition to the FIFO procedure, the use of the stock separation ID allows different tracking options.

Stock Overview reports show the total stock of customs products at a given time, independent of the stock postings period. From this report, users can directly access the goods receipt/issue or FIFO list for a product over a period of up to 12 months. Further, this report allows users to move duty-unpaid stock directly into free circulation or place duty-paid or duty-unpaid stock in the respective reserved stock for scrapping.

Finally, the Receipt and Issues reports show all receipt and issue postings per product for a specific period, which allows users to analyze the goods movements and related stock changes for up to five products relevant to certain special customs procedures. Unlike the individual stock item reports, these reports list all flows, regardless of the stock type, in chronological order.

> **Note**
>
> The method for declaring duty-paid and duty-unpaid stock levels depends on the country and applicable customs regulations.

Users can also display a full list of the products that are managed in a CWH.

Master Data

The last section of Customs Management reporting covers reports related to master data. As many master data topics have already been discussed, this section is limited to reports for authorization/security management and product classification to illustrate the report capabilities for master data. For more details on master data, refer to Chapter 4.

Customs Authorization and Security

To apply certain customs procedures, a company needs to have the customs authorization granted by the relevant customs authority. When the company uses SAP GTS to record the customs declarations/instructions and/or relevant stock overviews, it requires the master data maintenance of these customs authorizations. In other words, users can enter authorizations and the specific attributes of the authorizations (e.g., status, FTO, customs offices, texts) granted by the authorities as master data in SAP GTS.

In addition, certain countries require companies to deposit securities or bonds when importing goods. This procedure helps customs authorities ensure that companies pay the amount of duties owed. Users can enter securities as master data in SAP GTS in the same way as authorizations or licenses. The specific type and attributes of the securities depend on Customizing settings and regulations received from or communicated by the customs authorities. Figure 9.19 shows the selection screen for displaying the existing authorizations. In this example, a user ran a report for the Belgian Customs Legal Regulation (**CUSBE**), searching for authorizations that possess an active **Status** (**0002**).

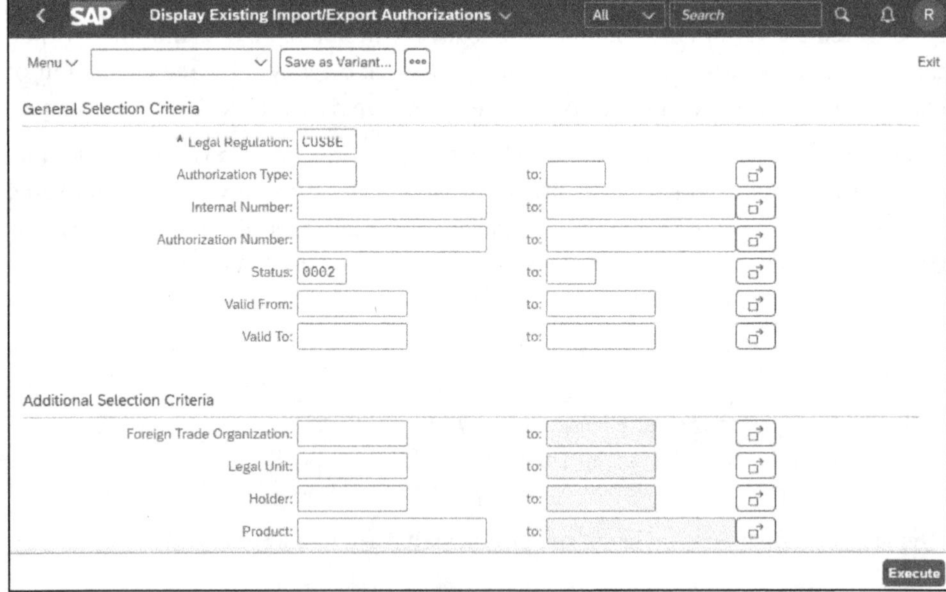

Figure 9.19 Display Existing Import/Export Authorizations Selection Screen

Figure 9.20 illustrates a list of import/export authorizations maintained with an active status in SAP GTS. For further details about the authorization, you can double-click on the line in the report to display the detailed attributes maintained in SAP GTS.

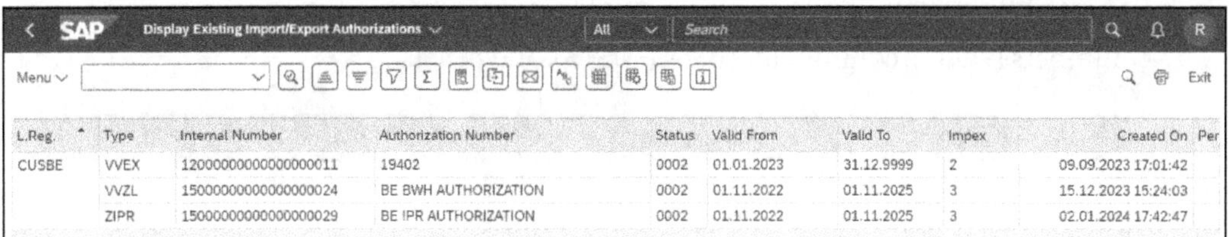

L.Reg.	Type	Internal Number	Authorization Number	Status	Valid From	Valid To	Impex	Created On	Per
CUSBE	VVEX	12000000000000000011	19402	0002	01.01.2023	31.12.9999	2	09.09.2023 17:01:42	
	VVZL	15000000000000000024	BE BWH AUTHORIZATION	0002	01.11.2022	01.11.2025	3	15.12.2023 15:24:03	
	ZIPR	15000000000000000029	BE IPR AUTHORIZATION	0002	01.11.2022	01.11.2025	3	02.01.2024 17:42:47	

Figure 9.20 Display Existing Import/Export Authorizations Report

Product Classification

The second set of Customs Management master data reports is related to the customs product classification. Note that these reports consider customs classification (i.e., tariff and commodity codes) only and are distinct from the legal control classification discussed in Section 9.1.2.

Figure 9.21 shows the selection screen of the Display Products report. In the **Product** field selection, users can choose the range of products to check the assigned classification. Furthermore, users can select the **Output Format** to be displayed **By Numbering Scheme** or **By Legal Regulation**, enabling both customs- and compliance-relevant classification codes. This is different from the report discussed earlier, Display Classified Product (Legal Control), which only creates a report for compliance classification codes.

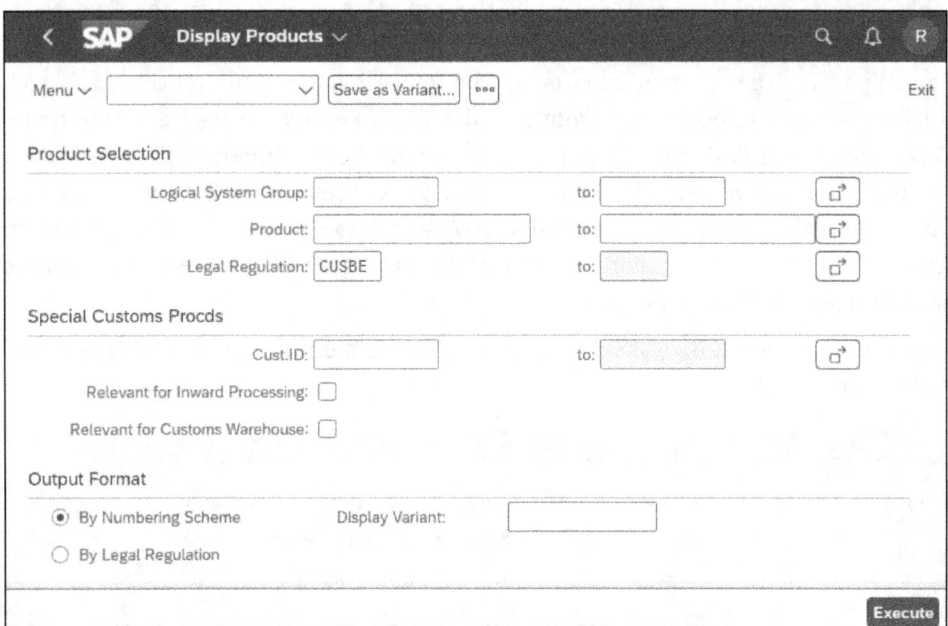

Figure 9.21 Display Products Report: Selection Screen

In this example, the report retrieved all products classified for a Belgian legal regulation. The output of this selection screen results in the report shown in Figure 9.22, which is a list of all products with corresponding data such as **Product Short Text**, tariff numbers (**TARIC**), commodity codes (**EUSTA**), and others.

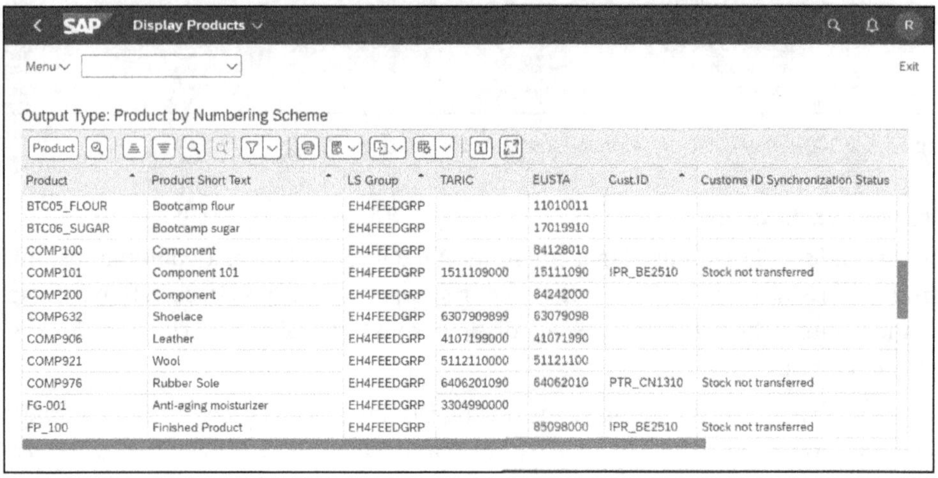

Product	Product Short Text	LS Group	TARIC	EUSTA	Cust.ID	Customs ID Synchronization Status
BTC05_FLOUR	Bootcamp flour	EH4FEEDGRP		11010011		
BTC06_SUGAR	Bootcamp sugar	EH4FEEDGRP		17019910		
COMP100	Component	EH4FEEDGRP		84128010		
COMP101	Component 101	EH4FEEDGRP	1511109000	15111090	IPR_BE2510	Stock not transferred
COMP200	Component	EH4FEEDGRP		84242000		
COMP632	Shoelace	EH4FEEDGRP	6307909899	63079098		
COMP906	Leather	EH4FEEDGRP	4107199000	41071990		
COMP921	Wool	EH4FEEDGRP	5112110000	51121100		
COMP976	Rubber Sole	EH4FEEDGRP	6406201090	64062010	PTR_CN1310	Stock not transferred
FG-001	Anti-aging moisturizer	EH4FEEDGRP	3304990000			
FP_100	Finished Product	EH4FEEDGRP		85098000	IPR_BE2510	Stock not transferred

Figure 9.22 Display Products Report

9.1.4 Trade Preference Reporting

This section provides an overview of the reporting capability in the Preference Management functionality. There are logical groupings provided in the SAP Fiori launchpad for the relevant reports in Preference Management processing, which include **Supplier's Declarations from Supplier/Customer**, **Preference Determination – Sales Documents**, **Preference Determination – Products**, and **Preference Determination – Production Documents** (for identity-based preference processing). In addition, a number of relevant reports are available in **Master Data – Preference Management**. In this section, we cover the reports for each of these categories. For more details on these standard reports, refer to Chapter 8.

A comprehensive list of available Preference Management reports in SAP GTS is provided in Table 9.3.

Category	SAP Fiori Reports
Supplier declaration	Manage Requests – Long-Term Supplier's Declaration
	Manage Long-Term Supplier's Declarations – Inbound
	Manage Preference Properties – Suppliers
	Analyze Request Rel. for Products
	Manage Long-Term Supplier's Declarations – Outbound
	Manage Reissue Reasons – Long-Term Supplier's Declarations
	Manage Preference Properties – Customers

Table 9.3 Preference Management Reports

Category	SAP Fiori Reports
Product identifier	Display Worklist of Process-/Production Orders Display Where-Used List of Batches Display Bills of Material for Orders
Master data	Manage Preference Data – Product Identifiers Display Country Assignment to Country Group/Agreement Display Preference Rules

Table 9.3 Preference Management Reports (Cont.)

All of these reports depend on the specified administrative unit (AU), which a user must enter before running any of the following reports. In this section, we use an AU in Germany for illustration purposes only.

The following reports are described in more detail in the following sections:

- Manage Requests – Long-Term Supplier's Declarations
- Manage Long-Term Supplier's Declarations
- Display Worklist of Process/Production Orders
- Display Country Assignment to Country Group/Agreement

Long-Term Supplier Declarations from Supplier

Long-term supplier declarations (LTSD) proposals are created based on the feeder system documents transferred (purchase orders and/or goods receipts). Users can manage requests for LTSDs to their suppliers in addition to reminders for open requests. With the Manage Requests – Long-Term Supplier's Declaration report shown in Figure 9.23, users can manage LTSD requests either triggered from the feeder system or created manually.

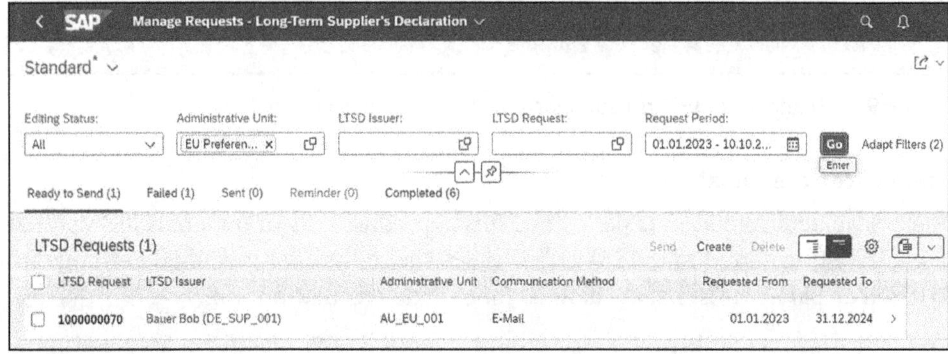

Figure 9.23 Manage Requests: Manage Long-Term Supplier's Declarations Report

After selecting the corresponding **EU Preference Central Department** (**Administrative Unit**) and period (**Validity Period**), a list of relevant LTSD requests is shown. From here, the user can perform the necessary follow-on actions.

The other report, Manage Long-Term Supplier's Declarations – Inbound, is used to perform activities (change, display, release, delete, or cancel) on inbound LSTDs that users receive from suppliers.

Long-Term Supplier Declarations for Customers

Outbound LTSDs are created based on billing documents from the feeder system. In addition, LTSDs can be checked and, if required, processed before submission.

Upon selection of relevant criteria **EU Preference Central Department** (**Administrative Unit**) and period (**Validity Period**), the Manage Long-Term Supplier's Declarations – Outbound report provides an overview of relevant documents, as shown in Figure 9.24. From this report, users can create, display, change, delete, and send outbound LTSDs to their customers. In this example, the company can issue one LTSD for a customer in Belgium.

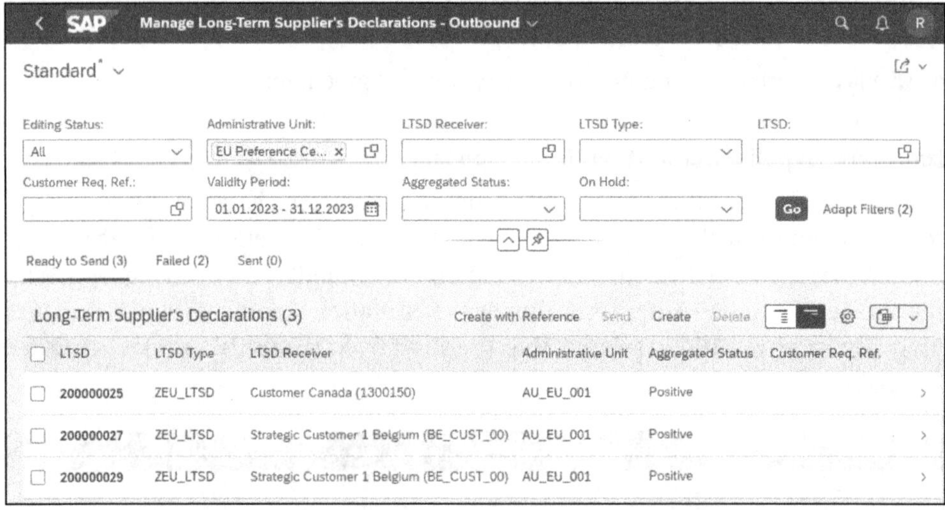

Figure 9.24 Manage Long-Term Supplier's Declarations: Outbound Report

Preference Determination

Preference determination is performed for all product bills of materials (BOMs) that users have transferred to the SAP GTS system. The result of preference determination is based on parameters that are defined for each determination run.

To further facilitate this process for product identifier-specific preference determination, SAP GTS offers the Display Worklist of Process/Production Orders report. A user can display this report to get an overview of all process or production orders that have been transferred to SAP GTS and are ready for determination. Typically, users accessing the worklist do this by filtering on the **Worklist Status** field (see Figure 9.25) on the selection screen, which allows users to only display process orders with a specific status.

Display Worklist of Process-/Production Orders
All Search
Menu Save as Variant...
Selection Conditions
* Logical Feeder System:
Document: to:
Item: to:
Identifier: to:
Status
Worklist Status: 10 to:

Figure 9.25 Display Worklist of Process-/Production Orders: Selection Screen

By filtering on the **Ready for Determination** status, the system will provide an overview of process orders that have been transferred from the feeder system and are ready for calculation, as shown in Figure 9.26. Users can also schedule the determination as an automated background job.

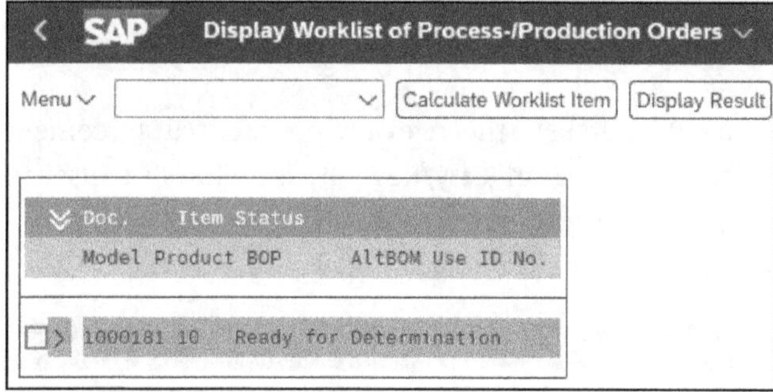

Figure 9.26 Display Worklist of Process-/Production Orders: Overview of Process Orders

Master Data

Preference Management master data is critical to the compliant execution of preference processes. Key data elements include the product master with specific data fields for Preference Management such as procurement indicator, prices, and preferential data. In addition, the BOMs in SAP GTS are used as the basis for preferential calculation.

The origin and the use in production of a batch across several production levels are displayed in a list, which you can find in the SAP Fiori report called Display Where-Used

List of Batches. The composition of semifinished and finished products must be confirmed to determine the preferential status of a batch.

Transfer logs help users ensure that the SAP GTS system is in sync with the applicable feeder systems. In addition, rules of the agreement, uploaded or manually created, can be consulted in the Display Preference Rules report. Users can create preference rules manually or upload them to your system from a data provider. Users can display all the preference rules for a given agreement. For more details, refer to Chapter 8, Section 8.5.

There are also transactions linked to the country setup. In SAP GTS, users can assign countries to country groups via Customizing (**SAP Global Trade Services, edition for SAP HANA • General Settings • Legal Regulations • Assign Countries/Regions to Country/Region Group**) to manage, for example, the activation of legal regulations or the setup of the determination strategy at a country group level. Entering a country of departure/destination in the selection screen (see Figure 9.27) produces a list of country groups to which the specified country has been assigned.

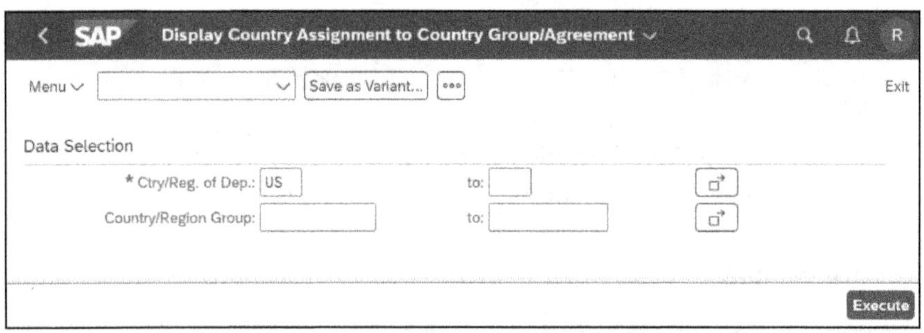

Figure 9.27 Display Country Assignment to Country Group/Agreement: Selection Screen

The result is illustrated in Figure 9.28. It lists the relevant preferential agreement (**Agreement**) for each relevant country or region (**Ctry/Reg.**). It also allows the user to efficiently monitor which agreements are configured in the system and which corresponding geographic location they adhere to.

Country/Region for Ctry/Reg. Groups and Agreement

Ctr...	Ctry/...	C/R ...	Partner...	Description	Agreem...	Description	Import Preference	Export Preference
US	USA	_DUS		United States	_DUMC	Preference Rules for USMCA - US	Check: Receipt/Import (including domestic)	Check: Dispatch/Export (Including Domestic)
			_DCA	Canada	_DUMC	Preference Rules for USMCA - US	Check: Receipt/Import (including domestic)	Check: Dispatch/Export (Including Domestic)
			_DMX	Mexico	_DUMC	Preference Rules for USMCA - US	Check: Receipt/Import (including domestic)	Check: Dispatch/Export (Including Domestic)

Figure 9.28 Results Showing Relevant Preferential Agreement per Country/Region

Next, we'll focus on more technical details, specifically related to embedded analytics and reports.

9.2 Embedded Analytics

It's important to understand the fundamental difference between operational and analytical reporting. *Operational reporting* aims to support the day-to-day organizational functions, for example, obtaining information on the invoice value of a certain set of documents. *Analytical reporting*, by contrast, aims to support strategic trade analysis. The primary goal of analytical reporting is understanding why things are happening and how to use that information to optimize or change day-to-day operations.

When it comes to operational reporting, SAP GTS, edition for SAP HANA has provided a new set of technologies that can be leveraged to gain insight into your day-to-day processes: embedded analytics. Embedded analytics leverages the concept of SAP HANA Live and extends it to the concept of the virtual data model (VDM), which becomes the single linking layer between the database tables and the various frontend clients and applications. From a technology point of view, embedded analytics serves as the successor of SAP HANA Live.

The power of the solution primarily lies in addressing a typical problem with traditional SAP ERP installations and landscapes: disjointed and disparate data. Traditionally, transactional data needs to undergo extraction, transformation, and loading (ETL) to a data warehouse before it can be consumed by users. This process isn't instantaneous and has an inherent shortcoming: ad hoc analysis often requires switching data sources and, in some cases, waiting for necessary data sets to become available. Embedded analytics mitigates these issues by bridging the transactional and analytical realms to increase efficiency while providing new ways of enhancing and improving existing business processes.

Embedded analytics leverages the new SAP HTML5-based UI technologies and principles (e.g., SAP Fiori for UI). This means that the UX is intuitive and consistent across all devices and that there is a single point of entry for accessing both transactional and analytical data sources.

Embedded analytics benefits from a simplified architecture that eliminates redundant overhead and throughput, allowing for real-time data consumption. In this case, simplified architecture means minimal layers between the data persistency layer and the exposure of the data to particular users. Using embedded analytics is the right choice when it comes to ad hoc, real-time, and operational reporting.

As indicated in Figure 9.29, on the frontend side, SAP S/4HANA offers tools that can be grouped into two main categories:

- **Tools to display (i.e., suitable for business users)**
  These tools include analytical list page, SAP Smart Business KPIs and analysis path, multidimensional reports, Query Browser, and analytical SAP Fiori apps.

- **Tools to create and manipulate (i.e., suitable for power users)**
  These tools include View/Query Designer, View/Query Browser, SAP Fiori KPI Modeler, and ABAP for Eclipse.

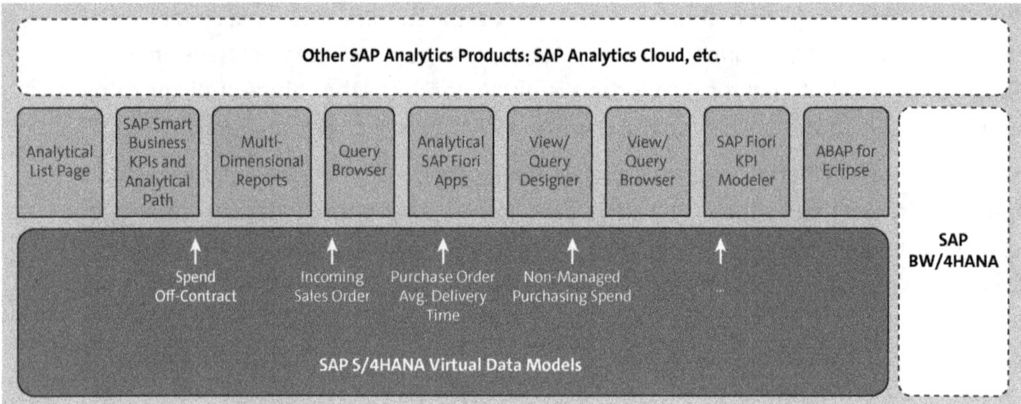

Figure 9.29 Logical Overview of Embedded Analytics

Embedded analytics comes with predefined and ready-to-use business content in many core business areas, such as finance, sales and distribution, and production, which is provided with the help of the VDM. With every new release of SAP S/4HANA, SAP is further enhancing and enriching this content.

For complex data warehousing scenarios that require extensive data manipulation and separate storage, SAP recommends the newest enterprise data warehousing solution: SAP BW/4HANA. SAP BW/4HANA can be used in combination with embedded analytics. SAP BW/4HANA is discussed later in this chapter.

As shown in Figure 9.30, in the backend side, the technology behind the VDM is core data services (CDS). CDS is an innovative technology from SAP that creates semantically rich database views by using a data definition language (DDL). The DDL facilitates the "code pushdown" paradigm introduced with the SAP HANA in-memory database and allows for designing complex real-time reporting scenarios.

The two types of CDS views—SAP HANA CDS and ABAP CDS—are used by developers to model analytical scenarios by applying various business intelligence (BI) concepts such as cubes, dimensions, and KPIs. Another distinguishing feature of the CDS views is that they enable users to leverage already-existing SAP authorizations and control data access.

The VDM (CDS views) can be directly linked to an SAP Fiori frontend by exposing the view as a Representational State Transfer (REST) web service (OData) or linking to an SAP analytical product (SAP Lumira or SAP Analytics Cloud) through a meta-object (transient analytical query).

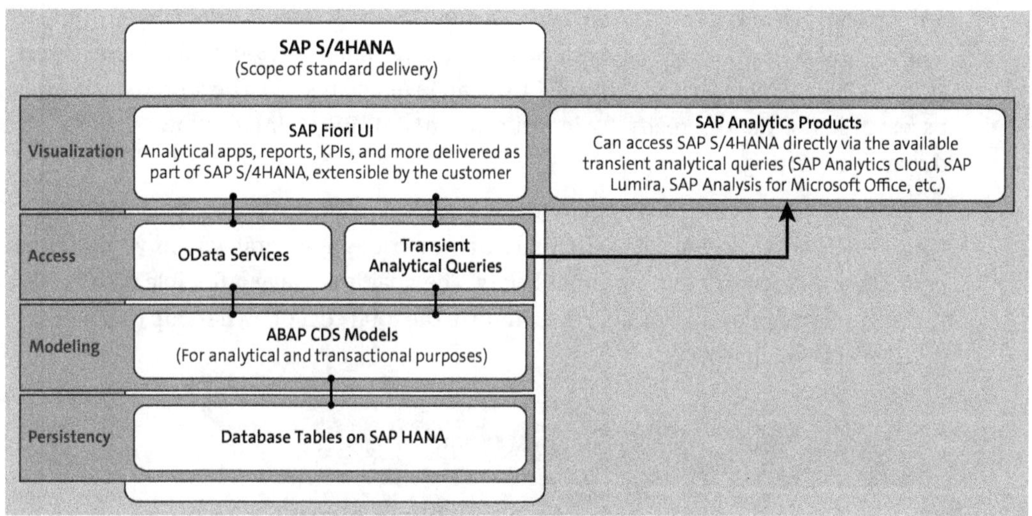

Figure 9.30 Layers in the Embedded Analytics Architecture

Even though embedded analytics is the technological successor of SAP HANA Live, no standard content is available in this technology for the SAP GTS platform yet. However, for users who want to create operational reports in SAP GTS, using the embedded analytics toolset is still the preferred method. If a current SAP GTS instance doesn't have SAP HANA as a database, it's already possible to create ABAP CDS views to support operational reporting.

9.3 Analytical Reporting

Unlike operational reporting, which focuses on the day-to-day processes, analytical reporting supports strategic trade initiatives. There are three main solutions that are used to support analytical reporting: (1) SAP Datasphere, (2) SAP BW/4HANA, and (3) SAP HANA for SQL data warehousing. The primary difference between these three solutions is how they are deployed.

While a data warehouse is at the heart of any analytical reporting solution, you typically also need other capabilities such as to achieve adequate data visualization. While this chapter won't cover visualization tools, SAP recommends the use of SAP Analytics Cloud.

9.3.1 SAP BW/4HANA

SAP BW/4HANA serves as a data warehousing system that brings significant capabilities and benefits to enhance standard SAP GTS reporting. SAP BW/4HANA is the successor of SAP BW on SAP HANA; however, the developments come from a different codebase.

SAP BW/4HANA can be leveraged to harmonize and structure data from multiple sources and systems into data models. It can handle data from SAP and non-SAP feeder systems. Not all data that SAP BW/4HANA can leverage needs to be stored in the database. After the data is ready for consumption, SAP BW/4HANA allows for fast and dynamic reporting.

The general architecture of a data model is based on the Layered and Scalable Architecture ++ (LSA++) from SAP. This architecture prescribes a general way of working that consists of mandatory and optional layers. The goal is to have a flexible architecture that is both scalable and reusable to minimize the total cost of ownership (TCO). Figure 9.31 depicts this approach.

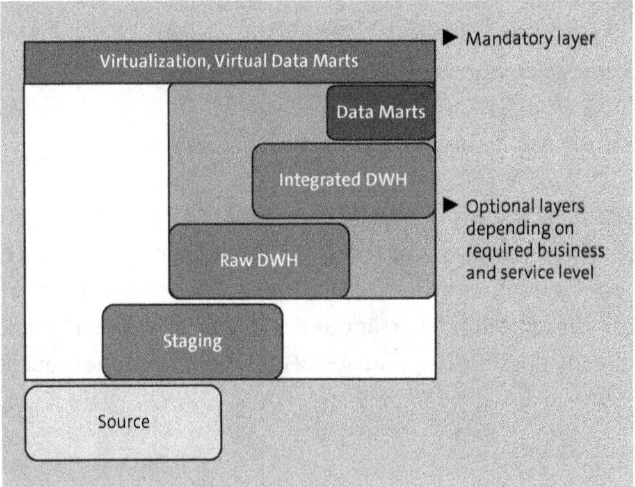

Figure 9.31 Simplified Illustration of LSA++ Architecture

The virtualization layer is the only mandatory layer and also allows access to all the layers beneath it to maintain a flexible approach. All the data needed for analytical reporting is consumed through the virtualization layer. If anything were to happen to the underlying data model or if any of the other layers were to change or disappear, this would not impact reports built on top of the virtualization layer.

All other layers in the LSA++ architecture are optional. It's important to note that optional doesn't mean unnecessary; it means that while each of these layers has its own uses and advantages in a data warehouse approach, you can select the layers you need based on the data model you're developing.

To use SAP GTS data for analytical reporting, you need to integrate SAP GTS with SAP BW/4HANA. The integration between SAP GTS and SAP BW/4HANA can be established as part of their standard functionality. First, set up a remote function call (RFC) connection between SAP BW/4HANA and SAP GTS. Then, enable the data transfer from SAP GTS to SAP BW/4HANA via Customizing by activating the data transfer to SAP BW/4HANA on the FTO, document type, and item category levels.

Figure 9.32 shows the integration between SAP GTS and SAP BW/4HANA. As previously mentioned, SAP BW/4HANA can leverage data from both SAP and non-SAP systems.

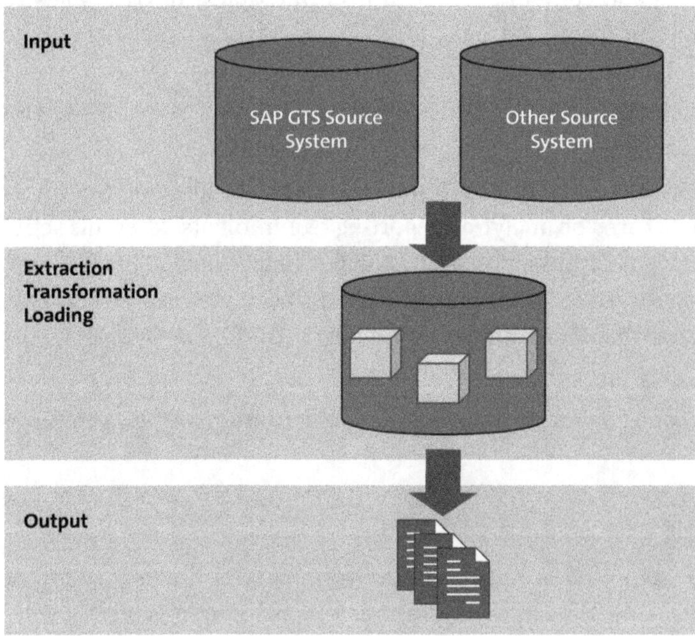

Figure 9.32 Integration of SAP BW/4HANA and SAP GTS

The objects that make data available for use in SAP BW/4HANA are called data sources. Data sources can use different technologies to extract or virtualize data to SAP BW/4HANA. The following are the most common methods:

- Data extraction through views or function modules that have been created in the source SAP system and where the data will be persisted in SAP BW/4HANA
- SAP HANA views (i.e., all views from SAP HANA Live for SAP GTS as discussed in the previous sections)
- CDS views as discussed in Section 9.2

In this section, we focus on the first method: extracting from SAP GTS to SAP BW/4HANA. Currently, this is the most efficient way to leverage data maintained in SAP GTS.

Note

The other methods (i.e., SAP HANA views and CDS views) can play an important role, but as the provided content for SAP GTS doesn't suit all scenarios, data extraction to the data warehouse is highly recommended. One advantage is that any SAP HANA views or CDS views (i.e., that you would use or create for operational reporting) can also be used to create a data source for your data warehouse.

At a high level, data extraction from SAP GTS to SAP BW/4HANA works as follows. First, the SAP BW/4HANA system fetches the changed (i.e., delta) information from the data sources on a regular basis and combines the data from multiple SAP GTS tables to transform them into a specific format. As many SAP GTS tables contain globally unique identifiers (GUIDs) as unique keys, it's important to review that the extractors don't forward this information to your SAP BW/4HANA system to minimize complexity and storage costs.

After SAP BW/4HANA finalizes extracting data and making it available, you can transform and enhance the data based on analytical reporting requirements. After this activity is complete, master data and transactional data will be available in SAP BW/4HANA, allowing multidimensional analyses from various business perspectives. For example, you can access consolidated data from different modules (SAP GTS, SAP S/4HANA, etc.) via a single report.

The following sections cover the process for extracting data from SAP GTS in more detail.

SAP GTS Queries

When you're setting up SAP BW/4HANA to integrate with SAP GTS, there are 22 standard queries with corresponding dataflows and extractors, as listed in Table 9.4. Queries contain *key figures*, which are numeric values or quantities (e.g., net price or quantity) used in the output reports as objects for analysis. Queries relate to actual reports and specify what data is leveraged.

Queries	Use Case	Key Figures
/IMO/V_SLL10_Q0001	This query provides a report on export declarations per product and the invoice net amount being exported.	▪ Invoice net amount in invoice currency ▪ Invoice net amount in customs currency
/IMO/V_SLL10_Q0002	This query provides an overview per country—based on export declarations—about the invoice net amount and weights being exported.	▪ Invoice net amount in invoice currency ▪ Invoice net amount in customs currency ▪ Gross weight ▪ Net weight
/IMO/V_SLL10_Q0003	This query provides an overview of the number of line items—based on export declarations—being exported in conjunction with the invoice net amount.	▪ Invoice net amount in invoice currency ▪ Invoice net amount in customs currency ▪ Number of items

Table 9.4 Standard Queries from SAP BW/4HANA for Integration with SAP GTS

Queries	Use Case	Key Figures
/IMO/V_SLL10_Q0004	This query provides an overview per customs office of the number of line items—based on export declarations—being exported in conjunction with the invoice net amount.	■ Invoice net amount in invoice currency ■ Invoice net amount in customs currency ■ Number of items
/IMO/V_SLL10_Q0005	This query provides an overview per product on preference statements—based on export declarations—being leveraged.	■ Difference (%) between preference value and threshold value preference ■ Threshold value for preference ■ Value preference
/IMO/V_SLL20_Q0001	This query provides an overview per product about the invoice net amount and customs value—based on import declarations—being imported as well as duties paid.	■ Invoice net amount in invoice currency ■ Invoice net amount in customs currency ■ Customs value ■ Total duty
/IMO/V_SLL20_Q0002	This query provides an overview per country about the invoice net amount, weights, and customs value being imported—based on import declarations—as well as duties paid.	■ Invoice net amount in invoice currency ■ Invoice net amount in customs currency ■ Customs value ■ Total duty ■ Gross weight ■ Net weight
/IMO/V_SLL20_Q0003	This query provides an overview about the number of line items being imported in conjunction with the invoice net amount and duties paid based on import declarations.	■ Invoice net amount in invoice currency ■ Invoice net amount in customs currency ■ Total duty ■ Number of items
/IMO/V_SLL20_Q0004	This query provides an overview per customs office about the number of line items being imported in conjunction with the invoice net amount based on import declarations.	■ Invoice net amount in invoice currency ■ Invoice net amount in customs currency ■ Total duty ■ Number of items

Table 9.4 Standard Queries from SAP BW/4HANA for Integration with SAP GTS (Cont.)

Queries	Use Case	Key Figures
/IMO/V_SLL30_Q0001	This query provides an overview about the number of line items being declared to the transit procedure based on transit declarations.	▪ Number of items
/IMO/V_SLL40_Q0001	This query provides an overview about the number of line items being discharged from the transit procedure based on transit discharge declarations.	▪ Number of items
/IMO/V_SLL110_Q0001	This query provides the overview of SPL screening results of business partner addresses.	▪ Number of records ▪ Number of releases ▪ Number of blocks ▪ Number of checks
/IMO/V_SLL110_Q0002	This query provides the number of releases of manually released business partner addresses by release reason.	▪ Number of records ▪ Number of releases
/IMO/V_SLL110_Q0003	This query provides the number of blocks of open blocked business partner addresses. These blocks need to be resolved.	▪ Number of records ▪ Number of blocks
/IMO/V_SLL120_Q0001	This query provides an overview of SPL screening results of document addresses.	▪ Number of records ▪ Number of releases
/IMO/V_SLL120_Q0002	This query provides the number of releases of changed addresses in manually released documents by release reason.	▪ Number of records ▪ Number of releases
/IMO/V_SLL120_Q0003	This query provides the number of blocks of open blocked document addresses. These blocks need to be resolved.	▪ Number of records ▪ Number of releases
/IMO/V_SLL130_Q0001	This query provides the overview of compliance check results of customs documents.	▪ Number of records ▪ Number of releases ▪ Number of blocks ▪ Number of checks

Table 9.4 Standard Queries from SAP BW/4HANA for Integration with SAP GTS (Cont.)

Queries	Use Case	Key Figures
/IMO/V_SLL130_Q0002	This query provides the number of SPL releases of manually released customs documents by release reason.	▪ Number of SPL releases ▪ Number of records
/IMO/V_SLL130_Q0003	This query provides the number of embargo releases of manually released customs documents by release reason.	▪ Number of embargo releases ▪ Number of records
/IMO/V_SLL140_Q0001	This query provides the overview of compliance check results of customs document items.	▪ Number of records ▪ Number of releases ▪ Number of blocks ▪ Number of checks
/IMO/V_SLL150_Q0001	This query provides the license usage overview in customs document items.	▪ Gross weight ▪ Net weight ▪ Number of records ▪ Number of document items ▪ Number of docu-ments ▪ Number of licenses ▪ Net amount ▪ quantity

Table 9.4 Standard Queries from SAP BW/4HANA for Integration with SAP GTS (Cont.)

Reporting Options

The strength of the SAP BW/4HANA reports lies in their customizable nature. Analysts can continuously make modifications such as adding or removing variables and inserting metrics to analyze large sets of data. This allows users to make informed business decisions.

You can apply SAP BW/4HANA content in three ways:

- **Without modifications**
  This option covers the basic reporting requirements—you activate the content and can use it as is.
- **With modifications**
  This option entails changing delivered BI content by modifying the installed objects.
- **Using BI content as a template to build further customized reporting**
  This is the most frequently used approach, in which you install the standard content and leverage the pieces that you need to comply with naming conventions, coding standards, and reporting requirements.

Advantages and Usage

SAP BW/4HANA offers several advantages for reporting:

- Users can make reports fit their business needs better by via customizations such as adding and removing variables and inserting filters. As such, valuable data may be filtered and extracted from large sets of transactional data.
- Data visualization through various frontend reporting tools (e.g., SAP Analysis for Microsoft Office, SAP Lumira, SAP Analytics Cloud) can help companies create insightful dashboards that showcase metrics tailored to different users.
- SAP BW/4HANA content contains a range of objects that simplify the implementation of an SAP BW/4HANA system and facilitate the development of an effective reporting system.
- When you use SAP BW/4HANA, you can combine data from various systems (i.e., SAP or non-SAP) into a single report. This removes the need to manually combine data from multiple systems into a spreadsheet because data from both SAP and non-SAP systems may be imported in SAP BW/4HANA.

9.3.2 SAP Datasphere

In 2023, SAP announced SAP Datasphere as a next-generation analytics platform and the natural successor for SAP Data Warehouse Cloud. This comprehensive data service is built on SAP Business Technology Platform (SAP BTP), which aims to deliver seamless and scalable access to key business data. By creating a unified experience for data integration, data cataloging, semantic modeling, data warehousing, data federation, and data virtualization, SAP Datasphere allows organizations to distribute mission-critical business data easily across the data landscape.

Additionally, SAP Datasphere is the long-term evolution of SAP BW/4HANA, while currently it can be leveraged as an extension to SAP BW/4HANA. This cloud solution can harmonize and structure data from multiple sources and systems into unified data models. It accepts data from SAP and non-SAP feeder systems. It also can integrate harmoniously with non-SAP data lakes.

Figure 9.33 highlights the primary concepts of SAP Datasphere. From the bottom to the top, it demonstrates the possibility to connect with a variety of data sources to *integrate*, *operate*, and *manage*. SAP Datasphere offers a range of different services dedicated to fulfilling these distinct functions.

The following sections explore SAP Datasphere in more detail, including its integration with SAP GTS.

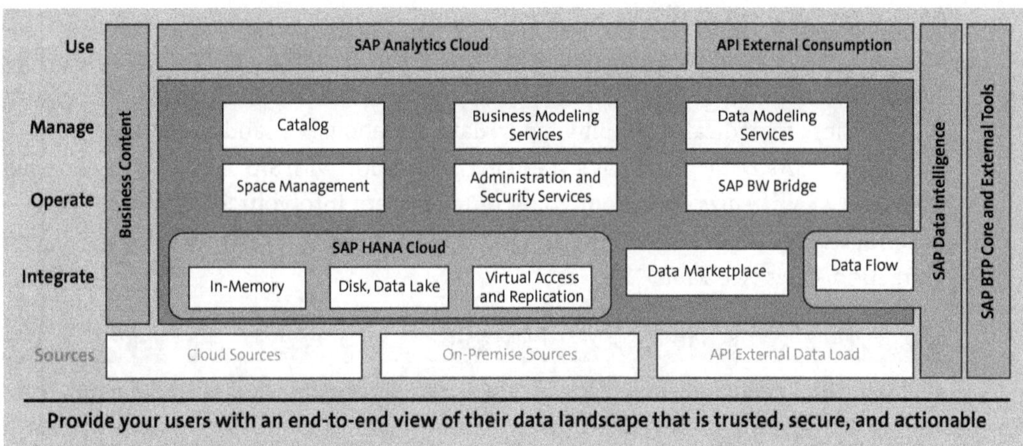

Figure 9.33 SAP Datasphere Architecture

Modeling Layer

Datasphere offers two modeling layers designed to support different user groups:

- **Data layer**
  In this layer, data engineers can create models based on a technical approach. This layer allows for a focus on the data and data modeling itself.
- **Business layer**
  For business users, SAP has designed a separate layer to build models using a more intuitive approach. This allows business users to work independently from data engineers while still being able to collaborate and share data.

The modeling layer approach fundamentally transforms the collaboration between business users and data engineers. It allows data engineers to concentrate on data consolidation and provisioning, while granting business users the freedom to refine and optimize the business models.

Spaces

SAP Datasphere uses the concept of spaces to organize and manage resources within the platform. Spaces act as virtual work environments with their own databases. They can be organized hierarchically to create a specific structure that reflects an organizational or project hierarchy.

This concept provides a logical separation structure and an access control mechanism. System administrators are responsible for assigning members to spaces.

Advantages and Integration

SAP Datasphere is able to blend data coming from SAP and third-party on-premise and cloud environments into a single, fully managed cloud environment to allow your organization to radically simplify your data warehousing landscape. As with SAP BW/4HANA, SAP Datasphere has a standard integration with SAP GTS. CDS can be used to extract or virtualize data from your source system into your SAP Datasphere data warehouse. SAP has provided a list of standard CDS views that can be leveraged for data extraction views (see Table 9.5).

CDS Views for Data Extraction	Use Case
C_ExportDeclarationItemDEX	Data extraction for export declaration items
C_ImportDeclarationItemDEX	Data extraction for import declaration items
C_CstmsCmplncDocItmLglCLicDEX	Licenses/customs document items data extraction
C_CstmsCmplncDocumentItemDEX	Customs document items data extraction
C_CustomsComplianceDocumentDEX	Customs documents data extraction
C_LegalControlLicenseDEX	Licenses in legal control data extraction
C_SPLScrngDocAddrChkResultDEX	Document addresses SPL screening result data extraction
C_TransitDeclarationItemDEX	Data extraction for transit declaration items
C_SPLScrngPartAddrChkResultDEX	Business partner addresses SPL screening result data extraction
C_TrnstDschrgDeclnItemDEX	Data extraction for transit discharge declaration items
I_CstmsCmplncBusTransDocTypeT	Business document category text
I_CustomsLegalRegulationText	Customs legal regulation text
I_CustomsDeclarationTypeText	Customs declaration type text

Table 9.5 CDS Views for Data Extraction

While there is no requirement to integrate SAP GTS with SAP Datasphere, there are four main advantages that should be considered:

- **Data integration**
  SAP Datasphere can be used to ETL data from various sources, including global trade-related sources such as supply chain data, customs information, and trade partner data.

- **Data enrichment**
  SAP Datasphere can enrich trade-related data with additional information or context, making it more valuable for compliance checks or trade analytics.
- **Data governance**
  Ensuring data quality and governance through SAP Datasphere is crucial for maintaining accurate trade-related data in SAP GTS.
- **Reporting and analytics**
  Data prepared and enriched using SAP Datasphere can be used for advanced analytics and reporting within SAP GTS to gain insights into trade operations.

9.3.3 SAP HANA for SQL Data Warehousing

The effectiveness of any analytical solution depends on the capabilities of the database and data warehouse. If SAP BW/4HANA doesn't fit in your company's IT strategy, SAP HANA for SQL data warehousing is another option.

SAP HANA is a platform containing multiple components, as illustrated in Figure 9.34 and described here:

- **SAP HANA database**
  This in-memory database can process both transactional and analytical data in the same tables (table HTAP); provides high-availability, advanced compression of data; and includes various other benefits. This is further illustrated at the bottom of Figure 9.34 under database management.
- **SAP HANA analytics component**
  SAP HANA has integrated capabilities to support different analytics options. These options can be both standard (e.g., strategic reporting) and advanced (e.g., predictive analytics) that can be executed via an SAP enterprise data warehouse (EDW) application or via SAP HANA for SQL data warehousing (i.e., advanced analytical processing), as shown in the advanced analytical processing part of Figure 9.34.
- **SAP HANA application server**
  SAP HANA can create native code-based data processing models for both enhancing the analytics component and providing additional web-based applications (referred to as application development in Figure 9.34).
- **Enterprise information management (EIM) component**
  SAP HANA also has a powerful data integration component that can stream, digest, and integrate data both virtually and persistently. This provides data quality and validation checks in real-time scenarios (shown as data integration and quality in Figure 9.34).

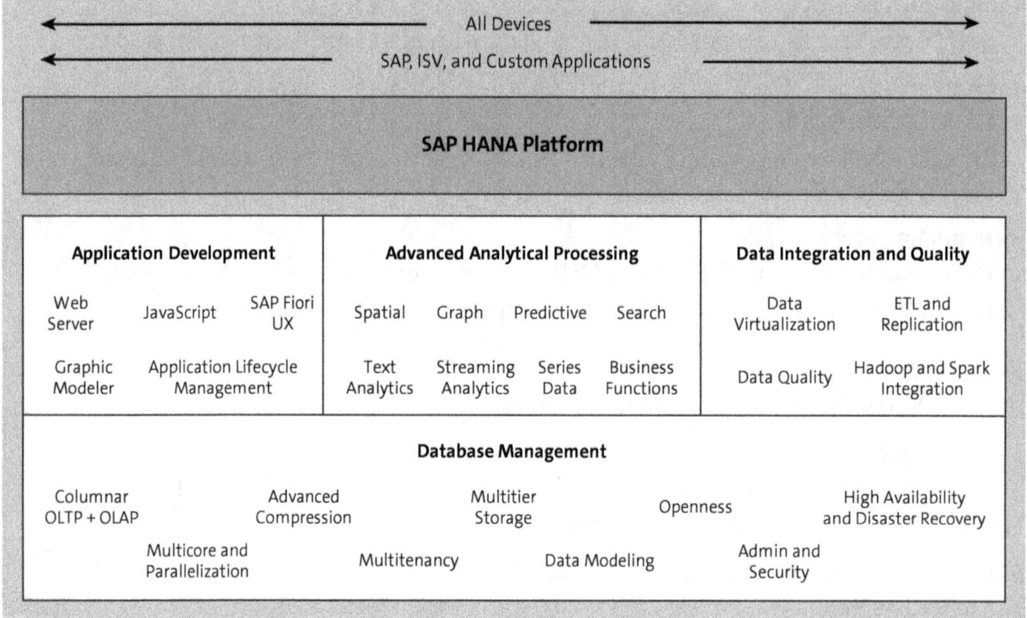

Figure 9.34 SAP HANA Platform

The following sections focus on the second component, SAP HANA analytics, and strategic reporting.

SAP HANA Analytics

SAP HANA analytics offers two options:

- **A prepackaged data warehouse (i.e., SAP Datasphere, SAP BW/4HANA)**
  This is recommended for users who prefer a guided data warehouse approach and want to leverage the out-of-the-box content available in SAP's EDW solution (refer to the previous sections).
- **Custom data warehouse (i.e., based on native SAP HANA for SQL data warehousing)**
  This is recommended for complex analytical use cases or self-built EDWs. This option is native to and included in the SAP HANA platform.

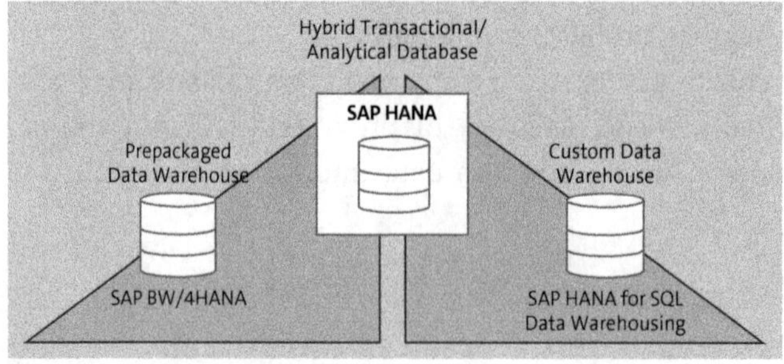

Figure 9.35 Analytics Component Options in SAP HANA

Combined, both options provide a powerful solution for addressing all possible use cases (see Figure 9.35).

Strategic Reporting

EDWs are employed for complex/advanced analytical scenarios, large amounts of varying data sets, or strategic reporting. See Figure 9.36 for some examples of how they could be used.

One use case example is a global trade dashboard that combines structured customer data from multiple transactional systems, social media unstructured data, an internet scraping algorithm, a customer relationship analysis, and a prediction algorithm. In one strategic report, you can include all the SAP HANA capabilities shown in Figure 9.36.

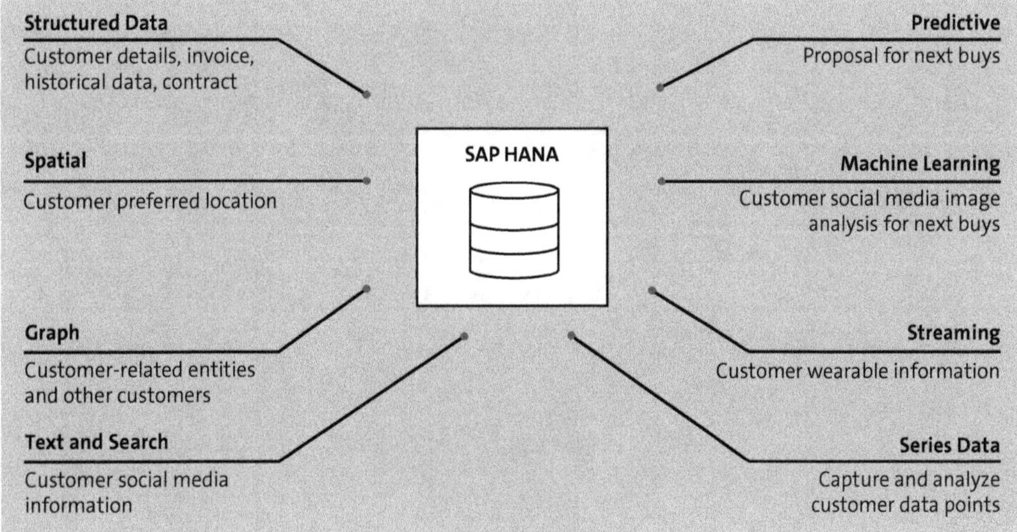

Figure 9.36 Capabilities of Strategic Reporting in SAP HANA

In conclusion, SAP HANA is a platform that provides high-grade support for multiple use cases, and its analytics capabilities offer both extensive standard functionality and options for customization. The strategic reporting supported by SAP HANA's analytical component, including global trade dashboards, leverages the power of advanced analytics to unlock strategic value tailored to the user's organization.

9.4 Summary

In the beginning of this chapter, we highlighted the importance of global trade reporting. The first section dove deep into the various standard reports available in SAP GTS to support the different business processes. These reports can be either descriptive (i.e., for analysis and audit trail reasons) or operational (i.e., allowing practitioners to take further actions on the shown reports). SAP GTS reporting capabilities are available for each of the three functional areas: Customs Management, Compliance Management, and Preference Management.

The second and third sections on embedded analytics and analytical reporting, respectively, illustrated that additional reports in embedded analytics, SAP BW/4HANA, SAP Datasphere, and more can be highly customizable, allowing very specific and detailed reports for analysis purposes. This ensures companies can extract the most valuable details from large amounts of data to optimize their business intelligence strategy.

In the next chapter, we'll discuss the migration pathway to SAP GTS, edition for SAP HANA.

Chapter 10
Migrating from SAP GTS 11.0 to SAP GTS, Edition for SAP HANA

In this chapter, we provide an overview of the changes between SAP GTS 11.0 and SAP GTS, edition for SAP HANA 2023. We also explore the key migration activities to upgrade your system and the simplification list.

The chapter offers a comprehensive overview of the key changes and activities involved in upgrading your SAP GTS system to the latest edition built on the SAP HANA technology. The migration is driven by several compelling factors, such as harnessing the power of SAP HANA, improving user experience (UX) with SAP Fiori, and ensuring a sustainable long-term solution as the existing SAP GTS 11.0 reaches the end of its lifecycle by December 2025.

The chapter delves into the drivers for upgrading (Section 10.1) and the upgrade process (Section 10.2), including crucial decision-making aspects such as choosing between brownfield and greenfield upgrade strategies. It also discusses the technical requirements, prerequisites, and considerations for SAP Fiori deployment and the importance of readiness checks. The chapter provides insights into the execution of the migration process, including the role of the maintenance planner and the Software Update Manager (SUM), as well as the process of custom code conversion.

To conclude, in Section 10.3, the chapter provides a detailed simplification list outlining obsolete functions, reorganized features, and changes in various areas of SAP GTS. This information serves as a valuable guide for readers embarking on the journey of migrating to SAP GTS, edition for SAP HANA 2023, emphasizing the need for careful planning and a commitment to embracing the enhanced capabilities of the latest edition.

10.1 Drivers for Upgrading to SAP GTS, Edition for SAP HANA

There are several key drivers for upgrading to the latest version of SAP GTS. Let's walk through them:

- **Harnessing the power of SAP HANA technology and platform**
  SAP GTS, edition for SAP HANA is built on SAP HANA, a cutting-edge, multimodel database technology known for its in-memory data retention capabilities. This

innovation not only boosts operational efficiency but also paves the way for a superior UX and opens new doors for innovation.

Furthermore, it seamlessly integrates not only with SAP S/4HANA but also with previous versions of SAP ERP and various non-SAP applications.

- **Elevated UX with SAP Fiori**
  When it comes to UX, SAP Fiori plays a pivotal role in making SAP GTS more user-friendly and efficient. Users will find themselves navigating the SAP Fiori interface exclusively via the SAP Fiori launchpad. While adopting SAP Fiori may necessitate some change management, the SAP Fiori UX is a big improvement.
- **Sustainable long-term solution**
  The most compelling reason to migrate to SAP GTS, edition for SAP HANA 2023 is that by the end of December 2025, the existing SAP GTS 11.0 will reach the end of its lifecycle, rendering it ineligible for mainstream maintenance and support.

  The shift to the SAP HANA platform means that SAP GTS will now receive updates every two years, aligned with the release schedule of the SAP HANA platform itself.

  All future enhancements, as well as any additional legal requirements, will be exclusively incorporated into SAP GTS, edition for SAP HANA.
- **Keep up to date with trade regulatory changes**
  For countries where SAP GTS is localized, SAP will release SAP Notes to adapt the product to follow regulatory changes.

10.2 Upgrade Process

The upgrade should consider technical requirements (see Figure 10.1), driven by the change in the underlying product version, and the specific database version, SAP HANA DATABASE 2.0. Additionally, it's imperative to meet application-specific prerequisites within SAP GTS. As shown in Figure 10.1, the upgrade process can be broken into a few phases:

- **Planning and preparation phase**
  The first phase is planning and preparing for your migration. The following tools are involved:
  - SAP Readiness Check: This tool assists customers in planning and preparing for the conversion of an SAP GTS system to SAP GTS, edition for SAP HANA. It executes checks to verify various aspects of an existing SAP GTS system. For more details, see Section 10.2.1.
  - Simplification list: This tool compiles a list of all incompatible or disruptive changes that occur in SAP GTS during the transition from a previous version of SAP GTS. It identifies potential incompatibility issues within the system. For more details, see Section 10.3.

- Maintenance planner: As the successor to the maintenance optimizer in SAP Solution Manager, the maintenance planner serves as the central tool for planning updates, upgrades, or new installations in a system landscape. For more details, see Section 10.2.1.

- **Realization (execution) phase**
  Finally, you're ready to execute your migration using SUM. SUM is the comprehensive tool responsible for the entire conversion process. It covers system validation, shadow instance creation, data migration, software updates, and data conversion. For more details, see Section 10.2.2.

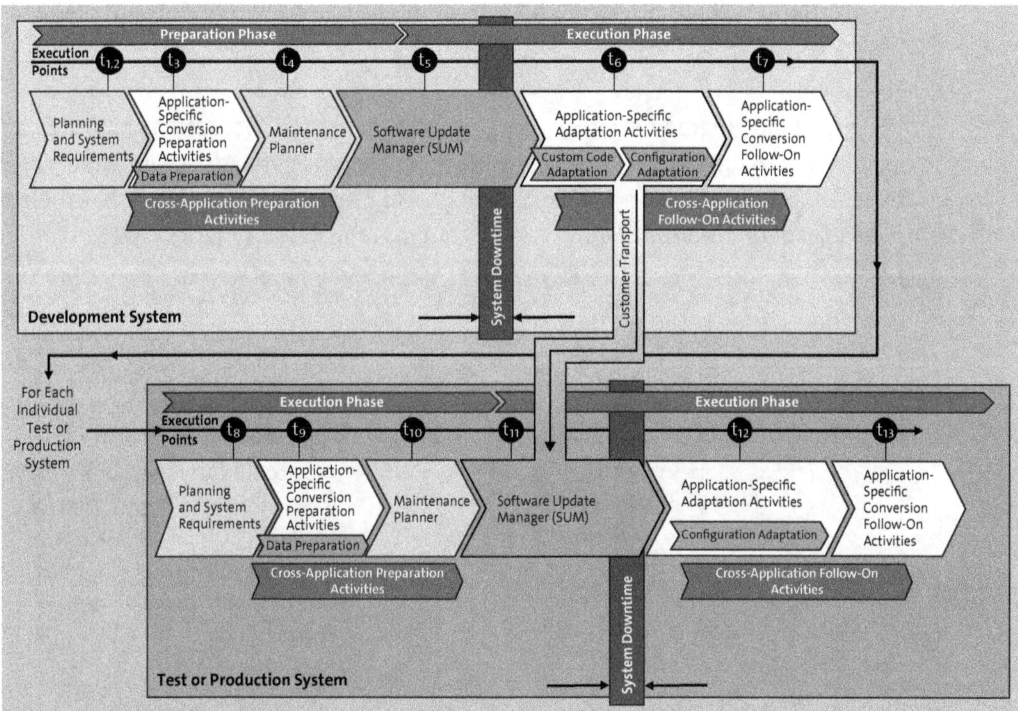

Figure 10.1 Overview of the Conversion Process

We'll take a closer look at the important steps in the upgrade process in the following sections.

10.2.1 Prepare for Your Migration

First, there are some important steps to consider in preparation for an upgrade. We'll examine them next.

Identify the Migration Strategy

The different strategies to execute an upgrade are described here (see Table 10.1):

- **Brownfield**
  In a brownfield upgrade (also called a *system conversion*), the previous system, along with all data, customizations, and custom developments, will be converted to the latest version.

  Custom code and configuration must be adjusted during the conversion due to differences in application processing between SAP GTS 11.0 and SAP GTS, edition for SAP HANA. When switching to SAP GTS, edition for SAP HANA, most application data is left untouched. However, there are some situations where application data is impacted.
- **Greenfield**
  The greenfield upgrade strategy (also called a *new implementation*) provides an opportunity to fully transform your trade processes and systems. New custom code must also be created, if necessary. You may also use a data migration tool to selectively load data from the old SAP GTS system into the new SAP GTS system.

Migration Strategy	Pros	Cons
Brownfield	■ Depending on how satisfied you are with the current solution, brownfield is frequently the quickest and least expensive alternative. ■ With this method, there is less need for change management.	■ Given that the brownfield method typically requires a "big bang" go-live in the new system, it can carry a greater risk. ■ You could carry over some old codes that aren't optimized for SAP GTS, edition for SAP HANA. ■ This method provides less innovation leveraging new capabilities available in SAP GTS, edition for SAP HANA.
Greenfield	■ If the prior implementation contained a lot of complex custom code, some of which was poorly documented, a greenfield approach might be the simplest option to go back to standard. ■ There's an opportunity to standardize processes and systems globally. ■ More flexibility is provided to define deployment and go-live strategies.	■ This method tends to be more expensive. ■ More business stakeholder involvement and change management are required.

Table 10.1 Migration Strategy Pros and Cons

Identify the Prerequisites and System Requirements

Prior to commencing the realization phase for converting an individual system, certain preparatory tasks are essential. It's important to create a checklist of all the preparatory tasks, including hardware and software requirements, backup procedures, and system checks, as shown in Table 10.2 and Figure 10.2.

System Requirements	Description
Unicode	The system needs to be a Unicode system.
Application server (ABAP only)	The system has to be an SAP NetWeaver Application Server (AS) for ABAP-only system. Dual-stack systems (SAP NetWeaver AS for ABAP and SAP NetWeaver AS for Java combined in one system) aren't supported for the conversion.
SAP Readiness Check	The tool checks whether your system meets the prerequisites for conversion to SAP GTS, edition for SAP HANA. It will also point you to SAP Notes to solve the incompatibilities.
Simplification list	To enable customers to better plan and estimate their path to SAP GTS, edition for SAP HANA, there is a simplification list. It describes in detail, on a functional level, what happens to individual transactions and solution capabilities.
System requirements	To upgrade to SAP GTS, edition for SAP HANA, you need to meet some system requirements such as minimum kernel version, SPAM version, table declustering, table depooling, obtain migration key, update source *R3trans* and *tp* executables, and so on.
Add-ons decision	You must check compatibility of SAP and third-party add-ons with target software and eventually uninstall them or find a compatible version.
Remove client 066	Client 066 is the SAP EarlyWatch client, which was set up during the installation of your system. This client isn't used in SAP GTS, edition for SAP HANA. To prevent issues, for example, with job scheduling, you have to remove it before starting the conversion.
Business functions	You need to validate the status for the SAP GTS, edition for SAP HANA target release
SAP Fiori embedded deployment	You must decide whether to go on an embedded deployment (recommended) or hub deployment.
Maintenance planner	The maintenance planner generates the download files (add-ons, packages, database drivers (DBDs), and the stack configuration file) that SUM uses to do the conversion.

Table 10.2 Prerequisite Checklist for SAP GTS, Edition for SAP HANA

System Requirements	Description
Run SUM prerequisite check	This check will be executed the first time you start the tool to perform a conversion and make necessary checks to validate the readiness of the operating system, database, and SPAM.
Create sandbox fresh copy from production	To test the upgrade, a sandbox with similar specifications as production should be created.

Table 10.2 Prerequisite Checklist for SAP GTS, Edition for SAP HANA (Cont.)

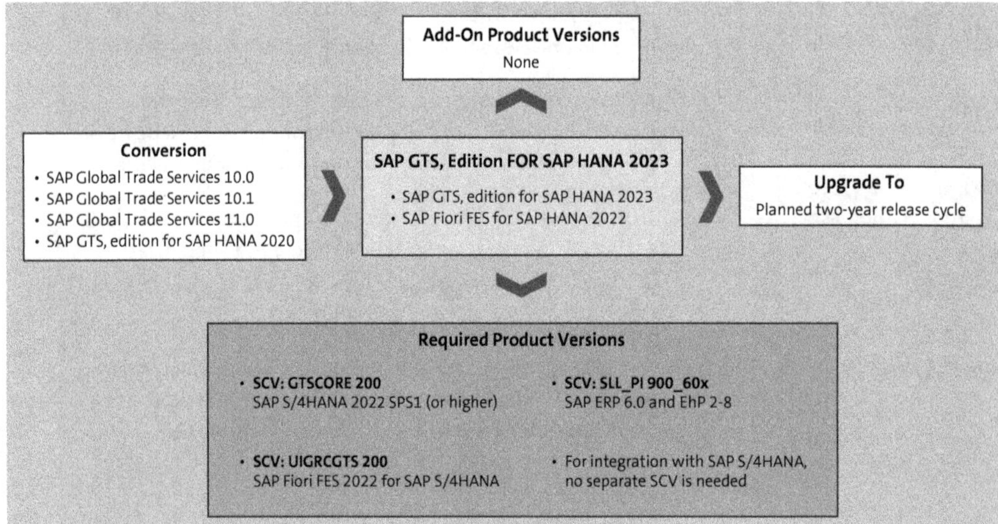

Figure 10.2 Prerequisite Checklist for SAP GTS, Edition for SAP HANA

Create an N + 1 Transit Landscape

Migrating SAP GTS to a new version can entail a comprehensive re-architecture effort. To execute an upgrade, you have to create a dual maintenance landscape, which has two distinct development and test systems. This will be needed until go-live, so you can still support your production system after you upgraded your development and test system. The following systems are needed in addition to the existing systems:

- **DEV**
  Original system with the current version of SAP GTS.
- **DEV + 1**
  Copy of DEV system where you can run the development system upgrade. Post upgrade, the version of this system will be the new version of SAP GTS.
- **TEST**
  Original system with the current version of SAP GTS.

- **TEST + 1**
  Copy of TEST/PRD system where you can run the test system upgrade. Post upgrade, the version of this system will be the new version of SAP GTS.
- **PROD**
  During production, the upgrade can be run directly on the original system, whose version will turn from the current version of SAP GTS to the new version of SAP GTS.

These environments are critical to the success of the upgrade efforts.

Choose Your SAP Fiori Deployment

SAP Fiori is an integral part of your SAP GTS, edition for SAP HANA upgrade and consists of the following add-ons:

- SAP Fiori for SAP GTS, edition for SAP HANA
- SAP Fiori frontend server (FES) for SAP S/4HANA

You must choose a deployment model for the SAP Fiori for SAP GTS, edition for SAP HANA user interface (UI) product. There are two deployment models to choose from:

- **Embedded deployment**
  Embedded deployment (see Figure 10.3) means you install the SAP Fiori required ABAP add-ons directly in your SAP GTS system. If your current system doesn't have any SAP Fiori apps installed previously, you can opt for the installation of the required ABAP add-ons during the conversion process.

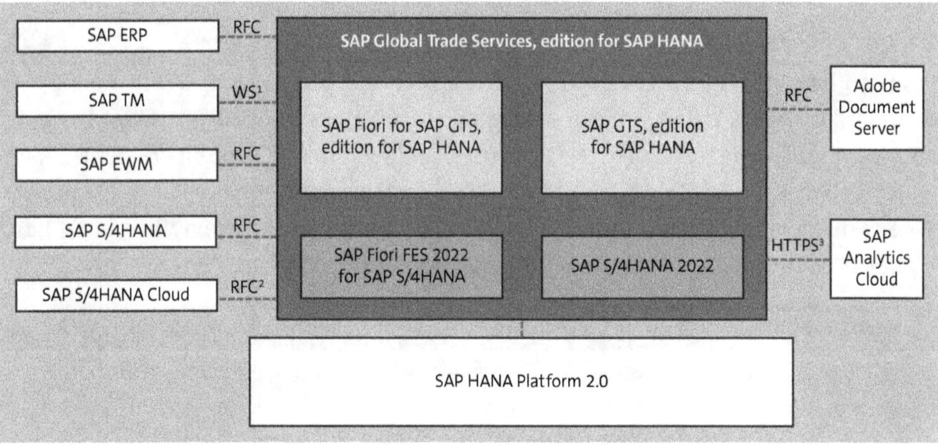

Figure 10.3 Embedded Deployment with SAP Fiori

- **Hub deployment**
  In this deployment option, the SAP Fiori apps run on a centrally installed frontend SAP Fiori launchpad. This means that you separate the backend and frontend components. Hub deployment (see Figure 10.4) can be used for SAP Fiori apps related to

SAP GTS, edition for SAP HANA. The UI for SAP GTS, edition for SAP HANA component of SAP Fiori for SAP GTS, edition for SAP HANA 2023 will be installed on the SAP Fiori FES. There are certain limitations and dependencies to consider when choosing the hub deployment option, including upgrade dependencies between the frontend and backend versions of SAP Fiori apps. For more detailed information, refer to SAP Note 2590653 and SAP Note 3237135.

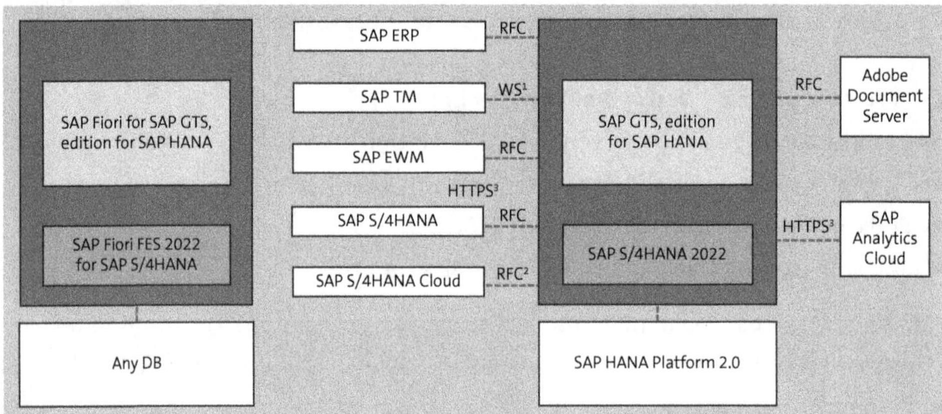

Figure 10.4 Hub Deployment with SAP Fiori

SAP recommends the embedded deployment option for SAP Fiori related to SAP GTS, edition for SAP HANA 2023.

Choose a Co-Deployment or Dedicated Instance

From an infrastructure standpoint, there are two options for SAP GTS: co-deploy or dedicated instance. When you can co-deploy SAP GTS with the SAP S/4HANA instance, you're typically able to reduce the total cost of ownership (TCO), but it creates a dependency between SAP S/4HANA and SAP GTS regarding maintenance and downtimes. When you have a dedicated instance for SAP GTS, you gain more flexibility, but it has a higher TCO. Table 10.3 summarizes the pros and cons to consider.

Instance Type	Pros	Cons
Dedicated instance	■ This is available for SAP GTS, edition for SAP HANA 2023, or future versions. ■ SAP GTS maintenance and changes can be made with less dependency on other systems.	■ There is a higher TCO.

Table 10.3 Pros and Cons of Dedicated Instance versus Co-Deployment

Instance Type	Pros	Cons
Co-deployment	■ There is a lower TCO.	■ There is identical system availability of SAP S/4HANA and SAP GTS, which can add complexity to maintenance and changes. ■ If the SAP GTS part becomes large, it will significantly increase the database size.

Table 10.3 Pros and Cons of Dedicated Instance versus Co-Deployment (Cont.)

Perform Readiness Check

SAP Readiness Check determines if your system satisfies the requirements to upgrade to SAP GTS, edition for SAP HANA. This is done with report /SAPSLL/READINESS_CHK_V8000, as shown in Figure 10.5. Prior to the technical migration, the report should be run in your present SAP GTS environment.

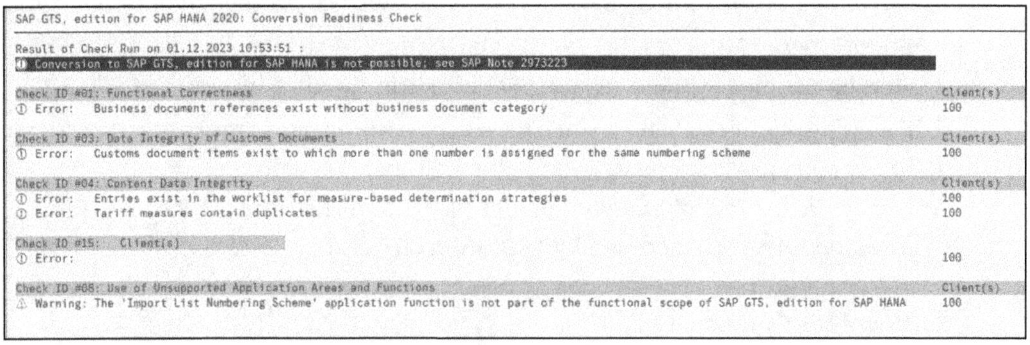

```
SAP GTS, edition for SAP HANA 2020: Conversion Readiness Check

Result of Check Run on 01.12.2023 10:53:51 :
Conversion to SAP GTS, edition for SAP HANA is not possible; see SAP Note 2973223

Check ID #01: Functional Correctness                                                    Client(s)
Error:   Business document references exist without business document category          100

Check ID #03: Data Integrity of Customs Documents                                       Client(s)
Error:   Customs document items exist to which more than one number is assigned for the same numbering scheme   100

Check ID #04: Content Data Integrity                                                    Client(s)
Error:   Entries exist in the worklist for measure-based determination strategies       100
Error:   Tariff measures contain duplicates                                             100

Check ID #15:   Client(s)
Error:                                                                                  100

Check ID #08: Use of Unsupported Application Areas and Functions                        Client(s)
Warning: The 'Import List Numbering Scheme' application function is not part of the functional scope of SAP GTS, edition for SAP HANA   100
```

Figure 10.5 Sample of Report /SAPSLL/READINESS_CHK_V8000

For more information on the readiness check, see SAP Note 3033398.

Maintenance Planner

The maintenance planner tool (see Figure 10.6) evaluates the current SAP GTS components, including add-ons, and offers potential conversion paths to SAP GTS, edition for SAP HANA. Furthermore, it generates the stack file required for the actual conversion process, which is executed using the SUM tool.

This conversion process combines the uninstallation of the old components and the installation of the new components. Specifically, the primary software component will transition from SLL-LEG to GTSCORE, and SAP_AP will be uninstalled as part of the conversion process within the context of converting to SAP S/4HANA.

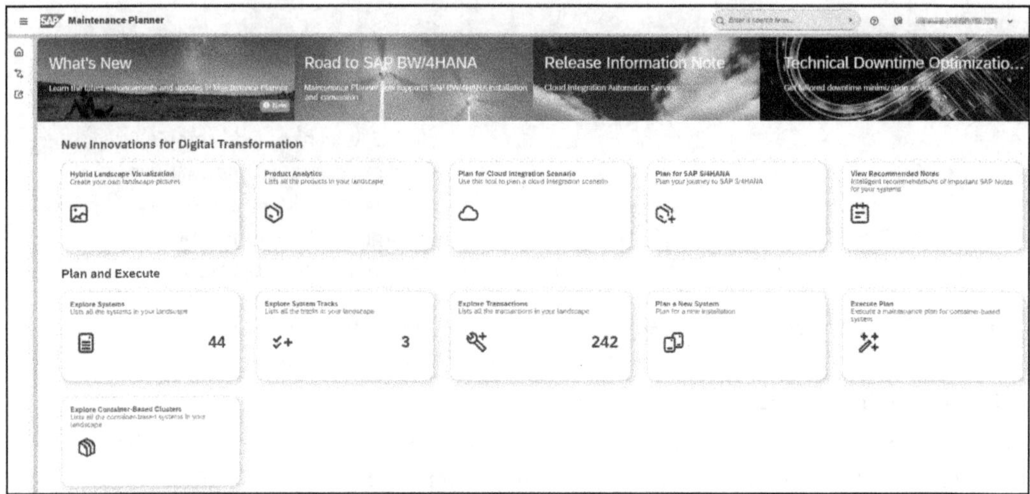

Figure 10.6 Maintenance Planner Overview

The planning process for migration in the maintenance planner should follow this sequence:

1. Start by planning the deinstallation of the existing product instances of SAP GTS.
2. Following the deinstallation, you'll need to plan the update of the base server release to SAP S/4HANA 2022.
3. Finally, in the last phase, you should plan the installation of the new product instances for SAP GTS, edition for SAP HANA 2023.

Custom Code Conversion

SAP GTS, edition for SAP HANA involves similar phases as any other SAP S/4HANA conversion. You'll need to adapt custom code/configuration (see Figure 10.7) for SAP functionality to perform business processes after the conversion process.

You can also delete custom code/configuration where it's not required as much of this custom code may have become obsolete or redundant, streamlining your custom extensions to align with the overarching goal of simplifying your system (see Figure 10.8). Those changes are saved in transport requests, which can be incorporated in the SUM process of the conversions of the other systems such as test and production.

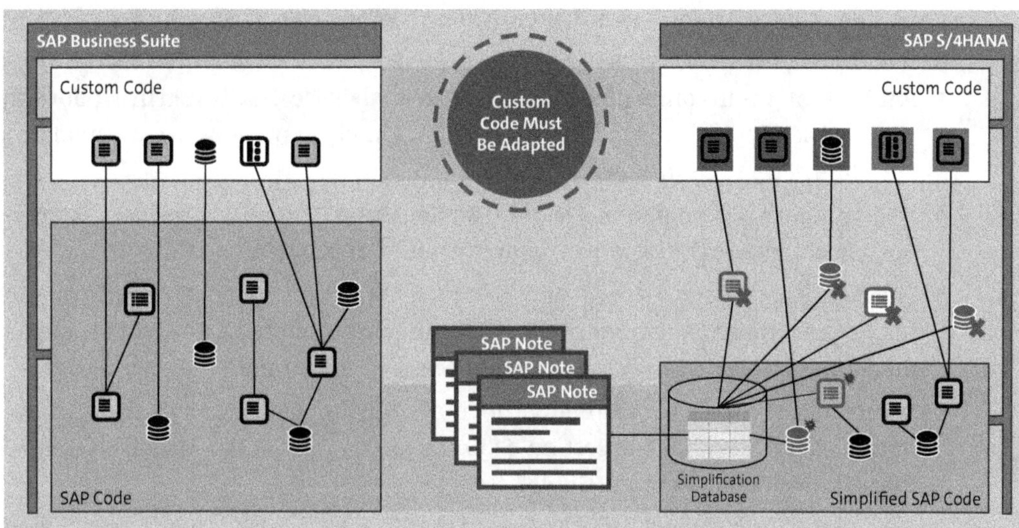

Figure 10.7 Custom Code Migration Elements

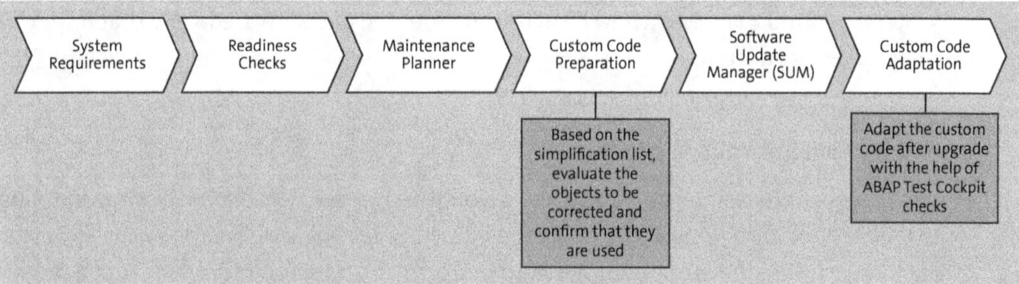

Figure 10.8 Overview of the Custom Code Conversion Process

Business Add-Ins

Modifications to the data model in SAP GTS, edition for SAP HANA can impact existing business add-ins (BAdIs). To ensure seamless integration with minimal adjustments after a technical upgrade, a BAdI compatibility mode has been introduced. This mode ensures that BAdI implementations remain functional, even if certain underlying Data Dictionary (DDIC) tables or table fields become invalid. This primarily affects BAdI methods referencing the following DDIC structures: /SAPSLL/CUHD_OBJ_S, /SAPSLL/LCLIC_OBJ_S, and /SAPSLL/PR_OBJ_S.

It's important to note the following considerations:

- The primary key, such as a globally unique identifier (GUID), of a DDIC table that becomes invalid due to changes in the data model can't be reconstructed.
- The GUID reference to a classification has been replaced with a reference number. Updating the numbering schema content, for example, through an upload function or schema content exchange, may lead to issues when accessing data using the old

GUID reference. To address this, modify the program logic of your BAdI implementation to exclusively use the classification reference number for access.

- BAdI definitions in compatibility mode will eventually be deactivated in a follow-on release. Therefore, it's advisable to migrate BAdI implementations to corresponding new BAdI definitions in a timely manner.
- To accomplish this migration, create an implementation for the corresponding new BAdI, and replicate the logic from your previous BAdI implementations.
- The names of the BAdI methods and their parameters remain the same, and you can implement empty methods if needed. Keep in mind that once a new BAdI is implemented and activated, only that BAdI will be called at runtime.
- To manage BAdIs in compatibility mode or their implementations, you can use Transaction SE18 and Transaction SE19. Successor BAdIs can also be accessed through the Implementation Guide (IMG).
- For classic BAdIs, kernel-based successor BAdIs are available. These kernel-based BAdIs are separate from BAdI migrations, so you'll need to copy the implementation as well.
- Keep in mind that once a new BAdI is implemented and activated, only that BAdI will be called at runtime.

10.2.2 Execute Your Migration

With these preparation steps completed, you'll be ready to proceed with your upgrade to SAP GTS, edition for SAP HANA. In the following sections, we'll explain the key tools and activities that must be completed to perform a successful migration from a preexisting version of SAP GTS, such as SAP GTS 11.0.

Software Update Manager

To upgrade to SAP GTS, edition for SAP HANA, you also need to run SUM. SUM (see Figure 10.9) enables essential tasks such as database migration (if necessary), the actual software update, and data conversion.

The software component SLL-LEG will also undergo a change to GTSCORE, and SAP_AP will be retrofitted into S4CORE. It's important to note that for this process, you must have SUM 2.0 SP16 at a minimum, and it's recommended to always use the latest available patch version.

The SUM process encompasses the following steps:

1. **Database migration (optional)**
   If your source system isn't already running on the SAP HANA database, you can use SUM's Database Migration Option (DMO) to migrate your database to SAP HANA as part of the conversion process. For more information on the DMO, see SAP Note 3338951.

2. **Installation**
   Install the software stack for SAP GTS, edition for SAP HANA 2023.
3. **Data conversion**
   Convert your data into the new data structures employed by SAP GTS, edition for SAP HANA. This step is automated as part of the process.

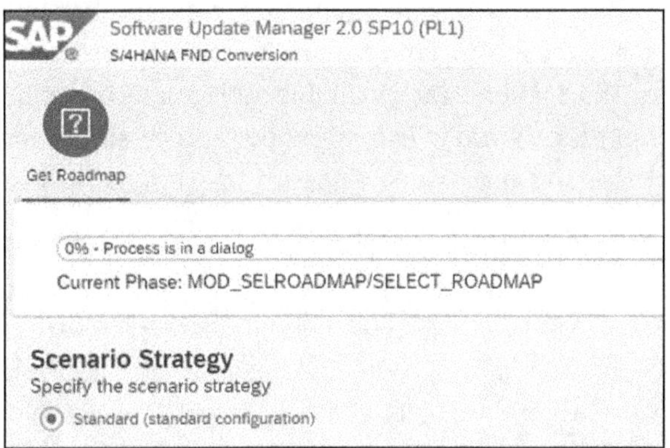

Figure 10.9 Software Update Manager Roadmap Selection

During this phase of the SUM process, the system is intentionally shut down to implement the new software stack and perform all the required technical procedures. It's important to note that no business operations can be conducted while the system is in this downtime state.

Note

It's important to note that system conversion is only supported for systems with SAP GTS versions 10.0, 10.1, or 11.0 installed. The conversion process is specifically supported for systems where SAP GTS, edition for SAP HANA 2023 (with technical release 1.0), is installed on SAP S/4HANA 2022 as the base product version. These systems are referred to as standalone systems.

Given the differences between SAP GTS 11.0 and the SAP GTS, edition for SAP HANA, you must adjust some of your custom code and configurations to enable efficient business processes. It's also important to include the removal of custom code or configurations in cases where the functionality provided by SAP's standard features has been removed in the new product or where custom code/configurations are no longer necessary.

All these software modifications, whether involving changes, additions, or deletions in custom code and client configuration, should be documented within transport requests. This documentation will let you incorporate these changes as customer transports during the SUM process when converting other systems transitioning to the same product.

Execution of Programs after Inputs

The execution of specific programs, known as the execution of programs after inputs (XPRAs), plays a crucial role during the conversion to SAP GTS, edition for SAP HANA. These XPRAs are initiated to perform essential conversion tasks concurrently, expediting the data migration process.

The main XPRA is report /SAPSLL/XPRA_V8000, which generates a comprehensive processing log that lists individual tasks, facilitating an assessment of their successful execution or any encountered errors. This log serves as a valuable tool to determine the overall success of the data conversion process and whether any post-processing steps are necessary.

> **Note**
>
> There may be additional XPRAs included with the SAP GTS, edition for SAP HANA 2023 release.
>
> The XPRA includes built-in checks to assess the necessity of data conversion based on the current system data.

10.3 Simplification List

SAP has developed a simplification list specifically tailored to SAP GTS, edition for SAP HANA. This comprehensive list provides, at a functional level, the changes that individual transactions and solution capabilities undergo when transitioning to SAP GTS, edition for SAP HANA. We'll provide an overview of the key simplification list items in the following sections.

10.3.1 Obsolete Functions

With the migration to SAP GTS, edition for SAP HANA, the following functions are no longer relevant:

- **SAP Easy Access menu**
  The traditional SAP Easy Access menu has been replaced with the modern SAP Fiori launchpad. SAP Fiori launchpad serves as a versatile platform for hosting SAP Fiori apps and enhancing their functionality with essential services such as navigation, personalization, embedded support, and application configuration.

  The launchpad serves as the central hub for accessing SAP Fiori apps. Within this page, users can personalize their experience by adding, removing, or grouping app tiles according to their preferences. This personalization ensures that users interact only with the apps aligned with their role profiles. Given the change in UX, it's important to conduct adequate change management and educate end users about this new UI.

- **Web Dynpro applications for SAP GTS**
  The Web Dynpro applications such as the Creating and Managing SPL Master Data by XI Input Processing app and the Electronic Data Exchange XI of Long-Term Vendor Declarations app are no longer part of SAP GTS, edition for SAP HANA.
- **Data aging**
  The data aging functionality, which is a data management concept for reducing the SAP HANA memory footprint, is no longer supported in SAP GTS, edition for SAP HANA.
- **Case Management**
  Case Management was a functionality that aided in the decision-making process when business partners or documents are blocked as a result of a sanctioned party list (SPL) screening and embargo screening in SAP GTS. A case enabled a secure decision-making process by storing relevant documents and background information. Case Management also traditionally aided in the product classification functions in SAP GTS. However, this functionality is no longer a supported feature in SAP GTS, edition for SAP HANA.

10.3.2 Risk Management Reorganized

The Risk Management application area has been discontinued, leading to a reorganization of its associated functions. Preference Management has now emerged as a standalone application area. In Preference Management, the use of the long-term vendor declaration (LTVD portal isn't supported anymore. Letter of credit and restitution are removed. Letter of credit is now part of SAP S/4HANA Finance for treasury and risk management.

10.3.3 Product and Classification

There are a number of simplifications in the area of product and classification. Here's a breakdown:

- **Changes in upload**
  The interfaces for numerical data, textual information, and customs duties have undergone significant modifications. It's essential to verify that your data provider is compatible with these updated interfaces and to test the interfaces.

  Additionally, the previous upload reports such as /SAPSLL/SLLNS_XML_LOCAL_UPL101 have been replaced with new programs such as /SAPSLL/NSC_NUM_UPLOAD_101.
- **Introduction of the numbering scheme content**
  In versions up to SAP GTS 11.0, numbering schemes determined the origin, structure, and usage of numbering systems in documents. Any changes to a numbering

scheme required the replacement of the existing scheme in documents. To streamline this process and avoid alterations in numbering schemes within documents, SAP GTS, edition for SAP HANA, introduced a new numbering scheme content.

Numbering scheme content encompasses all the necessary definitions for a numbering system and is activated for a designated period within a numbering scheme. Documents created during this period adopt the numbering scheme in place. When a need arises to create new numbering scheme content, it can be assigned to the same numbering scheme for the subsequent period. This ensures that documents may exhibit different numbering structures, while the numbering scheme itself remains unchanged.

The following transactions are no longer required:

- Transaction /SAPSLL/CHK_CLA
- Transaction /SAPSLL/CHK_CONT
- Transaction /SAPSLL/CHK_MIG
- Transaction /SAPSLL/CODENUM_INV
- Transaction /SAPSLL/PR_CL_CPY
- Transaction /SAPSLL/PR_CL_DEL
- Transaction /SAPSLL/PR_CL_MON
- Transaction /SAPSLL/PR_CMLI_CPY

- **Legal control classification**
  Import control classification numbers (ICCNs) are no longer supported in SAP GTS, edition for SAP HANA 2023, and consequently, the following transactions are now obsolete:
  - Transaction /SAPSLL/PR_PRCONI_02
  - Transaction /SAPSLL/PR_PRCONI_03
  - Transaction /SAPSLL/PR_RET_A_001
  - Transaction /SAPSLL/PR_RET_M_001
  - Transaction /SAPSLL/PRCL_001_DEL
  - Transaction /SAPSLL/PR_CLWB_001
  - Transaction /SAPSLL/LLNS_001
  - Transaction /SAPSLL/LLNS_UPL001
  - Transaction /SAPSLL/LLNS_UPLI001

  You can continue to manage import control using the available numbering schemes.

10.3.4 Customs Management

Next, let's take a look at the simplifications for Customs Management, as follows:

- **Dashboards for Customs Management**
  The dashboards for Customs Management that served as an entry page in SAP Business Client are no longer delivered. No substitution for the dashboards is provided.

  The functionality of pinning export declarations to the pinboards also no longer exists.
- **Tax transaction engine**
  The tax transaction engine previously used for importing turnover tax calculations is no longer accessible, and, for that reason, the duty category labeled as **TO: Import Turnover Tax for EU (TTE) - Don't Use** has been removed.

  To calculate import turnover tax, you can use the duty category **NT: Import Turnover Tax for EU** within the duty type definition.

  It's important to review and update your defined duty types as needed, and ensure the necessary Customizing settings are checked for a successful import turnover tax calculation in the customs import declaration duty calculation process:
  - **Define Tax Types and Product Tax Groups**
  - **Enter Tax Rates for Import Turnover Tax Calculation**
  - **Assign Material Tax Classes to Product Tax Groups from the Feeder System**

10.3.5 Intrastat

Finally, there are a few Intrastat simplifications to mention:

- **/ECRS/ namespace**
  In SAP GTS, edition for SAP HANA, the namespace of all objects, for example, tables, authorization objects, and so on, was changed from /ECRS to /ECRG/. The /ECRS/ namespace is obsolete.
- **Data aging for Intrastat declarations**
  Data aging for Intrastat declarations is no longer available in SAP GTS, edition for SAP HANA.

10.4 Summary

As described in this chapter, upgrading to SAP GTS, edition for SAP HANA is more than just a software migration. This conversion isn't merely about upgrading technology—it provides new, broader capabilities that enhance and modernize trade compliance programs.

As you embark on this journey, remember that a successful SAP GTS upgrade requires careful planning, dedicated resources, and a commitment to change.

In the next chapter, we'll close the book with a discussion of what's planned for the future of SAP GTS, edition for SAP HANA.

Chapter 11

The Future of SAP GTS

In conclusion, let's take a look at the future enhancements planned for SAP GTS.

The SAP GTS future roadmap focuses on four major buckets:

- Maintaining and expanding global trade localizations supporting legal changes
- Delivering continued innovations in the latest versions of SAP GTS, edition for SAP HANA
- Integrating global trade into the intelligent enterprise and business networks
- Extending global trade cloud capabilities by developing new web services on top of SAP Business Technology Platform (SAP BTP)

Although these four areas sometimes conflict, they are also tremendously helpful, which is why SAP GTS has remained a top-ranked solution for almost two decades.

Legal Disclaimer

The information contained herein may be changed without prior notice. Some software products marketed by SAP SE and its distributors contain proprietary software components of other software vendors. National product specifications may vary.

These materials are provided by SAP SE or an SAP affiliate company for informational purposes only, without representation or warranty of any kind, and SAP or its affiliated companies shall not be liable for errors or omissions with respect to the materials. The only warranties for SAP or SAP affiliate company products and services are those that are set forth in the express warranty statements accompanying such products and services, if any. Nothing herein should be construed as constituting an additional warranty.

In particular, SAP SE or its affiliated companies have no obligation to pursue any course of business outlined in this document or any related presentation, or to develop or release any functionality mentioned therein. This document, or any related presentation, and SAP SE's or its affiliated companies' strategy and possible future developments, products, and/or platforms, directions, and functionality are all subject to change and may be changed by SAP SE or its affiliated companies at any time for any reason without notice. The information in this document is not a commitment, promise, or legal obligation to deliver any material, code, or functionality. All forward-looking statements are

There are numerous influencing factors that balance these areas, including the following:

- Instituting new legal regulations and regimes while maintaining those already in existence, whether at global, regional, or local levels
- Responding to demands and requirements from the market as identified by users, user groups, and other influence channels
- Identifying new strategies and technologies, including those within SAP, such as artificial intelligence (AI), which are driving major changes in IT

Although changes in legal regulations are continuous, market developments challenge the SAP GTS Product and Solution Management team to openly listen to the market, to dive deeply into the business needs of various organizations in different industries, and to jointly elaborate together with SAP GTS customers on potential solution approaches. Influencing opportunities such as participation in SAP user groups (i.e., ASUG, DSAG, VSNG) or ideation programs such as SAP Customer Innovation offers a variety of options to bring in topics.

SAP strategies are predetermined by global invention and evolution in IT, wherein SAP plays a key, leading role. Such inventions and evolutions have a strong influence on existing solutions, such as SAP GTS, and might require not only rethinking but even redesigning the existing and future functionality. SAP will leverage SAP BTP to add additional value to existing processes.

The inventors of SAP GTS have pursued the strategy of thinking ahead to design, develop, and deliver a modular, exceptionally flexible solution based on a platform idea—deeply integrated into SAP ERP, SAP S/4HANA, and SAP S/4HANA Cloud Public Edition—and capable of being integrated into non-SAP environments. Today, this forms the basis for the success of future releases, the invention of new functions and modules, the development of enriched functionalities, and the path of transformation.

11.1 Planned Innovations

At SAP, all Product and Solution Management teams follow clear and well-defined rules to ensure prime quality and reliability for any solution delivered to the market. This applies to the entire lifecycle of any solution in what is called mainstream maintenance. Following these rules includes properly scheduling release cycles, which typically last two years, depending on influencing factors.

In the following sections, we'll provide insight into the future SAP GTS release.

11.1.1 The Next Release of SAP GTS

Following the official release of the current version of SAP GTS, edition for SAP HANA 2023, SAP plans to deliver the coming releases in a two-year release cycle. The underlying foundation will be aligned with the SAP S/4HANA product to assure co-deployment capability of SAP GTS on SAP S/4HANA.

11.1.2 Planned Innovations for the Next Release

The SAP GTS Product and Solution Management team aims to always provide innovation across all product capabilities (i.e., screening, trade compliance, trade preference, export/import management, special customs procedures). However, the influencing factors—as outlined previously—could impact the development to focus more on specific areas while keeping pace with innovation.

In the following sections, we'll discuss the planned updates and functionalities on the horizon for SAP GTS.

Legal Changes and Continuous Improvement and Modernization

All ongoing activities are aimed at improving existing capabilities, enhancing compliance, and supporting new and changing business processes as well as legal requirements related to global trade operations.

The solution enhancements are planned to include the following:

- Making legal changes for all supported, localized countries
- Adopting latest changes of the European Union Custom Codex Reform (including new reform proposal COM(2023) 258) pushing for self-assessment and simplified central clearance
- Adopting the latest changes of the Swiss customs IT modernization initiative (Passar, coming 2023–2025)
- Modernizing communication technology by replacing Application Link Enabling (ALE)/IDocs
- Delivering cloud-ready applications and services using the ABAP RESTful application programming model, whether for new or existing capabilities

The key benefits of these planned enhancements are as follows:

- Ongoing compliance with global trade regulations
- Single platform supporting trade requirements globally
- Standard support in SAP software for processes to implement legal changes to global trade processes
- Assurance of continued benefits from customer investments in SAP GTS
- Rapid time to value, less implementation time, and lower cost of operations

Origin Management

SAP plans to extend the origin management functionality to include the management and determination of the economic/commercial or non-preferential country of origin for products. Whereas SAP GTS already supports the determination of preferential origin, the economic origin arose as a requirement that was seen to be best supported in SAP GTS. During a co-innovation engagement with SAP GTS user groups, requirement gathering and evaluation workshops first took place between SAP GTS users and SAP GTS product management.

Non-preferential country of origin can be a complex topic, especially in the manufacturing and chemical industry. This topic is highly integrated into logistical processes and often controlled by the logistic process owner. Nevertheless, the handling and determination of the non-preferential country of origin is seen as a task for the trade department, making this a topic for SAP GTS.

Non-preferential country of origin can be covered by an entirely batch-managed process.

With the upcoming release of SAP GTS, SAP is planning a set of functionalities that will cover the following:

- Requests for the non-preferential country of origin from different suppliers
- Management of the non-preferential country of origin
- Calculation of the non-preferential country of origin

The key benefits of these planned enhancements are as follows:

- Single system supporting origin management requirements globally.
- Provision of non-preferential country of origin to invoice printing, export controls, and antidumping determinations from a single source of truth.

Enhanced User Experience

SAP Fiori apps improve the user experience (UX) and enable a smooth, nondisruptive upgrade path. The solution enhancements are planned to include the following:

- The most important metrics for global trade services presented in an easy-to-understand form
- Customization of information from key performance indicators (KPIs) to match business needs and user role
- The ability to drill down and take action where needed.

The key benefits of these planned enhancements are as follows:

- Improved user productivity through dramatic increases in speed and usability
- Customizable KPI tiles to keep the most important information up front
- Integration with transactional SAP Fiori apps to make insights actionable

Integration with SAP S/4HANA

To guarantee all SAP GTS customers a smooth transition from SAP ERP to SAP S/4HANA, SAP GTS is integrated with SAP S/4HANA. This integration is available from SAP S/4HANA release 1511. The integration with SAP S/4HANA Cloud Public Edition, is also available and will continue as new processes become available.

With this level of integration, the recent investments of all SAP GTS customers are safe while moving from SAP ERP to SAP S/4HANA regardless of which deployment option the customer chooses.

One new process is advanced shipping and receiving (ASR), which is the new simplified integration between SAP Transportation Management (SAP TM), SAP Extended Warehouse Management (SAP EWM), and SAP GTS. Additional process variants (i.e., decentralized deployments of SAP TM/SAP EWM) are planned.

Sustainability and the Role of International Trade

Foreign trade data and processes are key parts of various sustainability requirements. The need to manage the company's carbon footprint and its environmental and social responsibilities doesn't stop at borders. Tariff classification plays a core role in defining the scope of reporting requirements. SAP GTS can and will deliver the necessary data to fulfill the requirements imposed by regulations over time.

11.2 The Road Ahead for SAP GTS

Although—generally speaking—plans are subject to change, and nobody is in the position to predict the future, this chapter aims to provide a preview of the future direction for the SAP GTS solution beyond the coming release.

The following areas are being considered for the future direction of SAP GTS development:

- **SAP BTP services**:
  - Integrating into logistics business networks (i.e., blockchain technology)
  - Providing classification support based on AI technology
  - Enabling deeper drilling into industry-specific scenarios
- **Embedded analytics**:
  - Reducing replication of data for global trade
  - Leveraging SAP HANA performance for improved global trade analytics
- **Intuitive UX**:
  - Highlighting the most important information up front with customizable KPI tiles
  - Customizing information from KPIs to match business needs and user roles
  - Having the ability to drill down and take action where needed

Again, all business capabilities are in scope for the future of SAP GTS, and the SAP GTS Product and Solution Management team will put all efforts into combining the future direction with innovation by leveraging new technologies.

The Authors

Nick Moris is a partner at Deloitte's consulting practice and part of the global leadership of Deloitte's SAP GTS center of excellence. He focuses on support and advice regarding global trade solutions. He has built up a lot of expertise within SAP (including SAP GTS and SAP S/4HANA for international trade) and, with his team, supports several multinationals in the setup of international trade processes in enterprise resource planning (ERP).

During his career, Nick has combined technical tax skills with ERP implementation experience. He leads mixed teams of tax/business resources and ERP consultants during the implementation of the global trade processes.

Nick holds a master's degree in applied economics and a master's degree in tax law from the Catholic University of Leuven.

Pablo LeCour is a principal at Deloitte's Global Trade Advisory (GTA) practice in the USA with more than 23 years of experience in regulatory compliance. His focus is on export controls, sanctions, customs, and anti-bribery/anti-corruption (ABAC). Pablo leads the Global Trade Technology practice in the Americas, where he oversees global trade compliance transformation initiatives. Prior to joining Deloitte's US GTA practice, Pablo led the Regulatory Risk practice for Deloitte Advisory in the UK.

Pablo has worked with clients in the aerospace and defense, technology, media and telecommunications, oil and gas, and retail sectors. He has experience working with companies in North and South America, Asia, the Middle East, and Europe. Pablo leads global trade digital transformation programs for large enterprises focused on trade automation, organizational redesign, and the implementation of effective trade processes.

Pablo holds a bachelor's degree in arts in economics and international relations from Claremont McKenna College and a Kellogg-WHU Executive MBA from Northwestern Kellogg School of Management (USA), and WHU – Otto Beisheim School of Management (Germany).

Li Yu is a specialist leader within Deloitte's SAP GTS center of excellence. She has more than 25 years of experience in SAP ERP implementation with a particular focus on SAP GTS implementation. Li has worked with clients globally, including the Asia-Pacific (APAC), Europe, and North America regions. Her expertise includes solutioning and delivering large-scale global projects that are customized to fulfill business needs and regulatory requirements while applying leading practice methodologies.

Li holds a master's degree in applied economic sciences from the University of Ghent.

Deloitte GTS Leadership

Maren Mintjens is the NSE SAP GTS capability leader for NSE One SAP within Deloitte. Since she joined Deloitte in 2010, Maren has been focusing on the deployment of the SAP GTS solution for various international clients, together with her colleagues from Global Trade Advisory. She has been a key member of several SAP GTS (compliance, customs, and risk) implementation projects in the consumer industry and in the pharmaceuticals sector.

Maren has guided large program teams for crown jewel clients around the globe. She holds a master's degree in commercial engineering and has lived in four different countries. She's passionate about life sciences, her team, and the quality delivered to clients.

Dries Bertrand is a partner in Deloitte Belgium's Global Trade Advisory practice, based in Brussels. He has about 20 years of experience in the area of global trade and worked with clients in the Americas, Asia, and Europe, providing global trade strategy, import/export compliance, and global trade automation advice.

He is a subject matter specialist when it comes to organizing/optimizing supply chain activities and operations from trade compliance perspective, looking at people, organization, processes, procedures, and trade automation. He has more than 15 years of experience in implementing SAP GTS (as well as other SAP solutions, such as SAP S/4HANA for international trade, SAP Excise Tax Management, SAP industry solutions, SAP Master Data Governance [SAP MDG], etc.) and guiding companies through their transformation trajectory.

Dries holds a master's degree in business economics and a postgraduate in customs sciences.

Luc Scheiff is a director in Deloitte's SAP practice in Brussels, Belgium. He joined the firm in 2010, and his focus is technical management consulting and the implementation of IT strategies concerning foreign trade topics. Although certified in SAP Sales and Distribution (SD), his main knowledge and project experience lies in the SAP GTS solution.

Luc is part of Deloitte's global SAP GTS leadership team, focusing on go-to-market activities and project delivery, and leading SAP GTS implementations covering all modules with a focus on consumer industry clients.

Luc holds a master's degree in customs administration, law and policy from Münster University, and a master's degree in economics from ICHEC Brussels Management School. His experience in multinational environments, his exposure to multicultural environments, and his extensive international working experience, make him a good choice for large-scale global trade transformation projects.

Gertjan De Boever is a director in Deloitte's global SAP GTS center of excellence and has more than 14 years of experience in global trade and implementing SAP GTS. Gertjan leads the SAP GTS efforts for the automotive industry. He has served as trusted advisor to clients across Europe, North America, and Latin American, where he has focused on delivering tailored technology solutions that help manage complex international trade requirements.

Gertjan also has extensive experience in large-scale digital transformations for clients across sectors. His program management expertise allows efficient transformations focused on enabling leading SAP GTS functionalities. In addition, Gertjan has optimized processes by integrating DevOps methodologies for SAP GTS programs.

Before joining Deloitte, Gertjan was the general manager for his family-owned recycling business, focused on overseeing all aspects of business operations.

Gertjan holds a master's degree in business engineering and operational excellence from the University of Gent in Belgium.

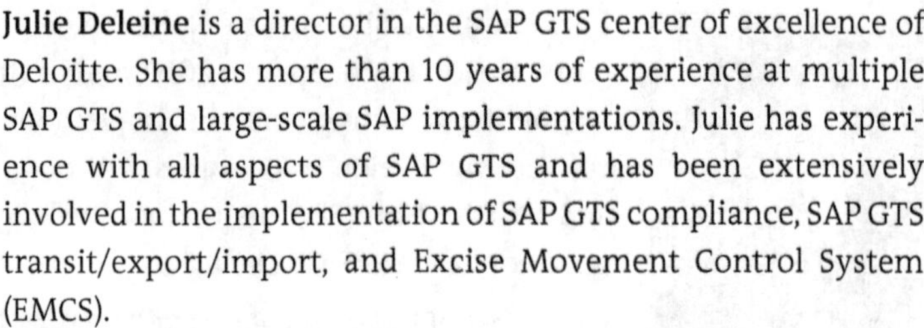

Julie Deleine is a director in the SAP GTS center of excellence of Deloitte. She has more than 10 years of experience at multiple SAP GTS and large-scale SAP implementations. Julie has experience with all aspects of SAP GTS and has been extensively involved in the implementation of SAP GTS compliance, SAP GTS transit/export/import, and Excise Movement Control System (EMCS).

Next to her functional knowledge, Julie has also experience in integration with other SAP systems. Julie has extensive experience in the chemical, natural resources, and consumer industries.

Julie holds bachelor's and master's degrees in business administration, as well as a master's degree in financial economics.

Yannick Schroeyen is a director in the Deloitte Belgium SAP GTS practice and has 12 years of SAP GTS implementation experience with a focus on customs management and special customs procedures. As part of these 12 years, Yannick has also worked 2 years as a customs and global trade manager in the Deloitte US firm.

On top of his SAP GTS experience, Yannick is also the product owner of Deloitte's trade automation solution, TradeClassifier. This solution has translated Deloitte's multidisciplinary experience into an AI-driven classification workflow solution supplementing and automating classification processes for global trade organizations.

Brian Carmitcheal is a managing director in Deloitte's Global Trade Advisory Practice where he specializes in business transformation and automation of global trade processes through the implementation of SAP GTS solutions. He has more than 15 years of experience in delivering SAP technology-based transformations across a variety of clients and industries, including aerospace and defense, semiconductor, and life sciences. He has also worked with clients on various types of engagements, including large-scale SAP S/4HANA business transformations, SAP GTS system implementations/upgrades, and other trade technology improvement projects.

Prior to joining Deloitte, Brian graduated with distinction from the University of Kansas with a BS in management and leadership and a concentration in international business and Spanish.

Filip Cooreman is a specialist leader in the SAP GTS center of excellence of Deloitte.

Filip worked as an SAP logistics consultant on different SAP implementations for a variety of clients before joining the SAP GTS center of excellence in 2010, and he not only brings in-depth SAP GTS knowledge but also broad functional SAP knowledge and a technical background.

He participated in several international projects as an integration and project/program manager or as an SAP GTS subject-matter expert, which allowed him to gain experience in all phases related to a project.

Filip holds a master's degree in business and information systems engineering

Caroline Boeckx is a senior manager in Deloitte's Global Trade Advisory practice in the USA. She has more than 10 years of experience with implementing SAP GTS for numerous clients in various industries. She has substantial experience in all three modules of SAP GTS. Her areas of expertise are customs management, including special customs regimes and self-filing, and trade preference. Consumer business and life science are her core industries in which she has supported multiple clients.

Caroline has extensive experience in SAP GTS end-to-end implementations (including integrations with SAP ERP, SAP S/4HANA, and non-SAP feeder systems), global template rollouts to multiple geographies, transformation programs, and SAP GTS upgrades. She has worked with clients around the world, including the Americas, EMEA, and APAC.

Caroline holds a master's degree in commercial engineering – management and IT from the University of Antwerp, Belgium.

Gabe Kitamura is a senior manager in Deloitte's Global Trade Advisory practice in the USA. He has more than 10 years of experience in customs, export controls, and sanctions regulatory compliance with a focus on global trade technology and analytics related to cross-border trade compliance. From an SAP GTS perspective, Gabe is responsible for the incorporation of global trade regulatory requirements and business process nuances across various standalone and SAP S/4HANA-aligned implementations. He also leads various large global trade operations engagements with teams in various countries focused on classification, free trade agreement qualification, and other operational services. Gabe's specialty is leveraging client and regulatory data to achieve business insights, strategic decisioning, and operational efficiencies.

Gabe works with clients across the chemical, consumer goods, industrial goods, and life sciences sectors. He has experience working with companies in the Americas, Europe, and Asia. Gabe holds a bachelor of arts in political science with a concentration in international relations from the University of Chicago and is a licensed US customs broker and a Certified US Export Control Officer (CUSECO).

Hitesh Mayekar is a senior manager and solution architect based in Seattle, Washington, with a special focus on SAP GTS. With certifications in SAP Activate project management and SAP S/4HANA Cloud, he brings 16 years of robust experience in consulting to the table. His expertise lies in designing and implementing SAP GTS solutions that integrate seamlessly with supply chain and logistic processes, improving business operations.

Hitesh has a wide range of experience in all SAP GTS modules across various industries, including life sciences, chemicals, technology, and aerospace. His work has been pivotal in streamlining trade compliance solutions across these industries and in different parts of the world, including the US, Europe, and APAC. Proficient in using the distributed agile scrum methodology, Hitesh has successfully delivered major releases and has hands-on experience in project management and lifecycle activities. He has led multiple complex SAP GTS transformation programs and has assisted companies in establishing their trade compliance roadmap.

Hitesh consistently uncovers new saving opportunities and builds product teams to support clients' SAP GTS solutions beyond the transformation program. His strategic approach, commitment to efficiency, and comprehensive understanding of the industry make him an invaluable asset in the field of trade compliance.

Contributors

Christian Stadler is chief product manager for SAP GTS, international trade management, and SAP NFE, where he coordinates development projects with SAP customers from around the world. With more than 25 years of experience at SAP, Christian has worked in different roles as a developer, development project manager, and product manager since 1994, and more recently focusing on international trade in SAP S/4HANA Cloud Public Edition.

Christian did dual study at DHBW Mannheim in Germany and spent time at Texas Instruments Inc., writing a diploma thesis focusing on US customs processes and certifying SAP software with the US Automated Export System.

Vishal Verma is the head of Risk, Compliance and Tax Solution Management at SAP. In his role, Vishal and his team are responsible to formulate solution strategy and driving go-to-market for SAP governance risk and compliance solutions, SAP GTS, and SAP solutions for global tax management.

Vishal has 22 years of experience across multiple areas of software industry. Over these years, he worked on managing products from cradle to grave (identifying growth opportunities, justifying investments, developing products, formulating pricing, proposition and positioning, ecosystem relationship, and business development).

He holds an executive MBA from the University of Mannheim.

Index

A

Account group 76, 77
Action profile 232, 235, 418
  assign 420
Activity 194, 239
  counter 241
  sequence 241, 242
Added value method 280
Address comparison 135, 187
Administrative Reference Code (ARC) 315
Administrative transactions 49
Administrative unit (AU) 57, 58, 364
  define 410
  retention period 376
Admission document 291
Advanced shipping and receiving (ASR) 25, 36, 41, 507
Aggregate reconciliation 320
Aggregation log 377
Agreement 168, 169, 172
  define 187
Agreement-license assignment 169
Alias terms 143
Analytical reporting 469, 471
Annual customs form 285
Annual report 285
App Finder 44
Application level 60
Application Link Enabling (ALE) 35
Architecture 32
Archiving 49
Arms embargo 144
Asynchronous screening 128, 131, 132, 134
Audit trail 127, 180, 443, 448
Authorization 50, 89, 152, 160, 255, 261, 302, 310, 313
  PTR 300, 302
  status 256, 263
  type 242, 256

B

Base rule 350
Baseline settings 55
Batch job 74, 124, 132, 134
Bilateral agreement 340
Bill of discharge 278
Bill of materials (BOM) 75, 83, 86, 188, 276, 313, 349, 435
  actual 308
  calculation methods 352
  configurable 362, 421
  control settings for transfer 421
  evaluation method 409
  explosion 158, 386
  external 276
  manually maintained 86
  static 276, 362, 421
  transfer 87
Billing document 124, 405
Blanket reconciliation flagging 319
Blocked document 146, 455
  clear 169
  import 155
  worklist 147, 170
Bonded warehouse (BWH) 267
  closing portion 272
  entry document 269
  exceptional procedures 274
  inbound movements 268
  master data 313
  outbound movement 270
  stock overview 272
Bottom-up calculation 391
Broker enablement 326
  bidirectional communication 331
  enhancement 334
  example 329
  setup 331
Broker filing 326
Brownfield migration 488
Buffering 205
Build-down method 352
Build-up method 351
Business add-in (BAdI) 60, 260, 334, 495
Business partner 71, 72, 76, 91, 120, 123, 124, 129, 149, 166, 167, 261, 313, 436, 443
  address information 187
  AU 57
  audit trail 448
  blocked 126
  FTO 56, 57
  function level 134
  release reasons 449
  role mapping 69

Business partner (Cont.)
roles ... 76
screening ... 118, 124, 130
Business role ... 44, 51
Business scope account (BSA) ... 302
Buy-sell scenario ... 348

C

Case Management ... 26, 499
Centralized declaration ... 308
Certificate of origin ... 348
Change pointer ... 72, 74, 128
Classification ... 27, 81, 82, 93, 118, 151, 160, 173, 188
analysis ... 110
business/system process ... 94
Compliance Management ... 102
Customizing ... 111
Customs Management ... 106
individual product ... 85, 103, 107
reporting ... 463
simplifications ... 499
worklist ... 85, 105, 107
Classification code ... 93
uploading ... 95
Client ... 34
Closing portion ... 272, 307
determine ... 304
FTZ ... 296
Co-deployment ... 26, 32, 492
Commodity ... 71, 81
Commodity code ... 81, 93, 98
Communication message ... 265
Communication method ... 365
Communication process ... 231, 309
control settings ... 238
define messages ... 237
Company code ... 56, 57, 59, 166
Comparison method ... 383, 414
Comparison procedure ... 135
Compliance document ... 120
Compliance Management ... 20, 39, 44, 62, 117, 120, 187
classification ... 102
reporting ... 445
spaces ... 44
Compliance screening ... 39
Compliance web services ... 38
Composite role ... 52
Comprehensive trade embargo ... 145
Concurrence message ... 286
Contact person ... 437
Contract manufacturing ... 299
Control class ... 81, 96, 188
attribute ... 166
maintain ... 97
upload from file ... 96
Control settings ... 185
Conversion provider ... 36
Core data services (CDS) ... 470
views ... 473
Cost distribution ... 222
Cost insurance and freight (CIF) ... 209
Country group ... 468
Country of origin ... 84
Country-group level ... 149, 184
Cross-check indicator ... 137
Cross-plant preference model ... 409
Custom code ... 494
Custom data warehouse ... 482
Customer data ... 436
Customer maintenance ... 77
Customizing ... 33, 55, 62, 141, 227
general settings ... 62
master data ... 71
number range ... 70
Customs ... 22, 40
Customs and Border Protection (CBP) ... 99, 114, 282
Customs authority ... 189, 199, 226, 343
authorizations and securities ... 255
connect ... 36
reporting ... 462
Customs broker ... 326
Customs clearance ... 207
Customs code list ... 250
assign to legal regulations ... 252
procedures ... 252
Customs declaration ... 197
header level ... 197
item level ... 198
Customs document ... 120, 161, 171, 193
monitoring ... 196
reporting ... 455, 457
Customs duty ... 257
Customs export ... 191, 193, 222, 224
Customs ID ... 57
define ... 309
Customs import ... 191, 205, 211
Customs invoice ... 215, 217
item data ... 216
Customs Management ... 21, 189
classification ... 106

Customs Management (Cont.)
communciation 199
Customizing 227
electronic communication 321
excise duty monitoring 315
export processes 222
general settings 228
import processes 205
master data 261
monitoring 196
preference 397
reconciliation 319
reporting 455
simplifications 500
space 46
special procedures 267
supported countries 335
Customs office 71, 76, 261
maintenance 79
Customs office of exit 245
Customs procedure 210
Customs processing 193
Customs product 311
Customs value 208, 220
Customs warehouse (CWH) 267, 309
Customs warehousing procedure 41
Customs waybill 219, 291
selection screen 220
Customs web services 39

D

Data aging 499
Data Dictionary (DDIC) 495
Data record 321
Data source 473
Database Migration Option (DMO) 497
De minimis rule 173, 177, 188, 350
Dedicated instance 492
Default 249
Default procedure 246
Default rule 248, 265
Defaulting data 225, 244, 263
procedure 264
Delimiter 143
Descriptive report 443
Determination procedure 63, 134, 149, 182, 184, 411
define 183
Determination strategy 62, 91, 162, 188
define 163
levels 183
Direct delivery 287
Distribution type 258
Document number 167
Document type 39, 55, 65, 67, 69
Documents of origin 417
Domestic license 180
Domestic sales 270, 307
Duty 189
Duty category 258
Duty framework 257, 260
Duty suspension 212
Duty type 258
external 260
Duty-paid goods 268
Duty-paid stock 271, 282, 289
transfer 312
Duty-suspended 207
Duty-unpaid goods 267
Duty-unpaid stock 282

E

Economic origin 357
E-filing 201, 228
Electronic communication 201, 321
transaction 323
Electronic Data Interchange (EDI) 41, 195
partner profile 323
port definition 323
E-manual 304
Embargo 20, 28, 45, 117, 118, 144, 155
block 146
business process 145
control settings 149
Customizing 148
master data 150
reporting 451
screening 39, 146
system process 145
Embedded analytics 25, 196, 444, 459, 469, 508
Embedded deployment 491
E-message 238
Enhancement framework 334
Enterprise data warehouse (EDW) 481, 483
Enterprise Information Management (EIM) 481
Enterprise Qualification Application (EQA) 302
Enterprise search 25
Enterprise Services Repository (ESR) 331
Entry for consumption 288, 296

Entry for domestic consumption 301
Entry summary declaration 298
Entry-by-entry flagging 319
Entry-by-entry principle 320
Equivalent goods 279
Exchange rate 265
manual maintenance 266
Excise duty monitoring 315
business process 315
Customizing 318
inbound activities 317
master data 318
outbound activities 318
system process 317
Excise duty suspension 315
Excise Movement Control System (EMCS) 191, 315
Execution of programs after input (XPRA) 498
Export 222
business process 223
navigation 224
system process 224
Export accompanying document (EAD) 195, 199
Export before import (EX/IM) 279
Export control 20
Export control classification number (ECCN) 82, 96, 151, 154
Export declaration 37, 40, 158, 193, 195, 225
BWH 270
reporting 459
submit to customs authority 226
Export licensing 157
business process 157
system process 158
Exporter-centric scenario 328
Extended warehouse management (EWM) 37
customs warehousing 41
External communication 233
Externally procured product 426
Extractor 474

F

Facilities Information and Resources Management System (FIRMS) code 309
Fallback procedure 204, 241
Feeder system 31, 39, 55, 67, 69, 72, 77, 81, 82, 124, 161, 166, 171
transfer master data 72
Field category 231
Field combination 246, 248, 250
Field type 231
Finished product 346
preferential 358
First-in, first-out (FIFO) 273
Foreign trade document 243
Foreign trade organization (FTO) 55, 56, 59, 166, 174, 458
define 57
Foreign trade zone (FTZ) 282
benefits 283
business process 285
closing portion 296
FIFO rule 294
forecast 297
inbound movement 286, 289
master data 314
outbound movement 287, 295
procedure 283
production process 294
reporting 298
system process 289
Free on board (FOB) 209
Free trade agreement (FTA) 22, 339, 342
setup 411
Free trade zone 22, 282
Free-circulation declaration 271
Freight order 42
Function module 228
Future direction 503

G

General settings 55, 57, 62, 70, 411
Generalized System of Preferences (GSP) program 22
Global trade dashboard 483
Global trade management 19, 23
Global trade reporting 443
Goods retention period 376, 415
Greenfield migration 488
Grouped shipments 205
Grouping strategy 40
GTSCORE 32, 493

H

Harbor maintenance fee (HMF) 299
Hard block 231
Harmonized System (HS) code 93, 206, 436
Hub deployment 32, 491

I

IDoc ... 36, 195, 226, 236, 321
data records ... 322
list ... 325
monitoring ... 325
output options ... 325
standard ... 321
statistics ... 325
status ... 326
I-message ... 195, 201, 203, 238, 324
Import ... 205
business process ... 206
duty calculations ... 208
duty framework ... 220
navigation ... 211
system process ... 210
Import control classification number (ICCN) ... 96
Import declaration ... 154, 206, 213
admission ... 291
after goods receipt ... 207, 212
before goods receipt ... 207, 213
broker enablement ... 329
BWH ... 269
PTR ... 305
reporting ... 459
Import licensing ... 153
business process ... 153
system process ... 154
Importer-centric scenario ... 327, 332
Inbound declaration ... 361
Inbound delivery ... 212
Inbound process ... 361
Incompleteness check ... 139, 213, 225
setup ... 228
Incompleteness procedure ... 228, 237
field assignment ... 229
Incoterm ... 77, 206
Individual classification ... 85
Integration ... 38
method ... 366
Intercompany sales screening ... 178
advanced check ... 180
audit trail ... 180
business process ... 179
system process ... 179
Intercompany stock transfer ... 404
enable ... 420
Internal number ... 78
Internal Transaction Number [ITN] ... 40
Intrastat ... 27, 49, 501
Inverted tariff duty relief mechanism ... 283
Inward processing ... 275
inbound movements ... 277
outbound movement ... 278
Inward processing relief (IPR) ... 83, 86, 189, 275
procedure ... 276, 277
settlement ... 279
Item category ... 67
Item group ... 40

K

Key figure ... 474

L

Layered and Scalable Architecture ++ (LSA++) ... 472
Legal control ... 21, 28, 40, 45, 83, 117, 118, 151, 152, 158, 161, 171, 180
applications ... 170
block ... 155
classification ... 103
control settings ... 185
Customizing ... 182
export licensing ... 157
import licensing ... 153
intercompany sales screening ... 178
master data ... 187
monitoring ... 170
reporting ... 452
simplifications ... 500
US reexport ... 172
Legal regulation ... 55, 56, 59, 62, 69, 104, 105, 136, 160, 161, 173, 448, 452
activate ... 134, 149, 183
assign customs code list procedure ... 252
assign process schema ... 243
control settings ... 136
country group for reexport ... 186
define ... 411
determination procedure ... 134, 149, 182
embargo ... 150
numbering scheme ... 113
Legal unit (LU) ... 56, 58, 59, 166
Letter of Credit Processing ... 26, 74
License ... 75, 89, 91, 164, 188, 462
depreciate ... 187
manage ... 164
License determination ... 146, 151, 152, 167
inbound ... 155

License determination (Cont.)
outbound 158, 160
License master 156, 161, 164
License type 163, 164, 183
define 187
License-document assignment 169
List type 140, 142
Local clearance document 291
Logical system 33, 34
Logistics costs 220
Logistics service provider (LSP) 37
Long-term supplier declaration (LTSD) 348
aggregation 362
annual request process 369
customer 398
externally procured products 426
intercompany 405
maintenance 361
management 364, 367
reissue 402
reporting 465, 466
request 361, 367
response 379
revocation 364
settings 415

M

Maintenance level 259
Maintenance planner 487, 493
Management of Preference Data for Product Identifiers 422
business process 423
system process 425
Mapping 65
Mass classification 105
Master data 31, 34, 46, 50, 71, 90, 111, 125, 187
authorization 89
BOM 86
business partner 76, 436
country of origin 84
customer 77
Customizing 91
embargo 150
excise duty monitoring 318
import and export 261
initial transfer 72
legal control 187
license 89
material 431
Preference Management 430

Master data (Cont.)
preferential 379
product 80, 82
reporting 461, 467
screening 128
source system 72
special customs procedures 311
SPL screening 141
transfer 39
Merchandise processing fees (MPFs) 284
Merge rule 310
Message ID 237
Message scheme 236
Method call 233
Migration 485
execution 496
preparation 487
system requirements 489
Minimal operations rule 438
M-message 201, 238, 324
Mode of transport 253
Movement reference number (MRN) 208
Multilateral agreement 339
Multisourcing 375, 423

N

Navigation 42
Net cost method 351, 414
New Computerized Transit System (NCTS) 41, 212
New implementation 488
Nondirect delivery 286, 291
Non-preferential origin 357
Non-preferential status 386
Nonprivileged foreign (NPF) goods 283
Number range 70, 78, 411
business partner 77
Number set 114
Numbering scheme 96, 97, 107, 188, 499
assign 113
content 112
setup 111
structure 113

O

OData 470
Offline screening 132
Open transit procedure 227
Operational reporting 443, 469

Organizational structure 55
best practices 59
Customizing 57
Origin management 506
Outbound declaration 398
Outbound delivery 124, 146, 161, 225
Outbound process 363
Outbound transit 207
Output management 418
Outward processing 280
inbound movement 281
outbound movement 281
Outward processing relief (OPR) 83, 86, 280

P

Paper-based declaration 199
Partner function 126, 166, 167, 186
Partner Government Agency (PGA) 99
codes 100
setup 114
Partner groups 69
Partner profile 323
Partner role 55, 69
Penalty 20
Periodic basis 205
Periodic certification 363
Permit to transfer (PTT) 286, 292
Planned innovations 505
legal changes 505
SAP S/4HANA integration 507
user experience 506
Plant 56–58
Plant-based preference model 409
Plug-in 31, 33, 39, 55, 73, 311, 421
settings 59
P-message 238, 244
Post Processing Framework (PPF)
action 232, 418
Preference agreement 438
create 411
Preference calculation 387
Preference certificate 397
Preference comparison method 383
Preference data 396
Preference determination 349, 362, 380
applications 359
components log 388, 393, 395
evaluating results 395
goods produced in house 428
methods 350
perform 380

Preference determination (Cont.)
prerequisites 380
reporting 466
simulation 395
status 355, 385
top-down versus bottom-up 352, 386
Preference Management 22, 25, 58, 62, 74, 83, 339
business process 345
Customizing 408
master data 430
organizational structure 409
reporting 464
simplifications 499
space 47
system process 359
Preference rule 382, 390
define 412
Preference statement 83, 356, 372, 433
define 413
product identifiers 426
Preference status 363, 385
change 402
Preferential origin 357
Preferential status 347, 385
Preferential trade agreement 339
Presettlement 308
Price type 414
assign 415
Printer specification 235
Prior export 279
Privileged foreign (PF) goods 283
Pro forma invoice 225, 295
Process schema 238, 243
Processing proposal 194
Processing status 197
Processing trade regime (PTR) 299
authorizations 300, 302, 310
business process 300
component stock 306
contract 302
inbound process 301, 305
master data 315
outbound process 301, 306
settlement 308
system process 302
Processing type 233
Processing under customs
control (PUCC) 189, 275, 279
Procurement indicator 83, 433
Product 82, 91

Product classification 84, 95, 188
reporting 463
space 48
Product cost 434
Product identifier 425
Product master 431
Production process 362
Profile generator 51, 52
Proof of origin 347, 358
periodic 398, 430
transactional 397, 430
Proposal statement 202
Purchase order 212

Q

Query 474
Quotation 124, 146

R

Reason object 450
Receipt/issue list 272
Receipts and issues 293, 298, 461
Recheck 171
Recipient 437
Reclassification 108
Reconciliation 192, 308, 319
annual 299
reason 319
Reexport 306
Reference type 141
Regional value content (RVC) 350, 351
Registration number 208
Release procedure 128
Relevance code 100
Remote function call (RFC) 32, 39
destinations 34, 35
Reporting 443
analytical 471
classification 463
compliance 445
customs 455
customs document 457
embedded analytics 469
FTZ 298
LTSDs 465
master data 461
Preference Management 464
real-time 469
SAP Fiori 444
special procedure 460
Reporting (Cont.)
stock overview 461
strategic 483
trade pereference 464
Restitution 74
Restitution Management 26
Risk Management 74, 499
Role 50, 51
Rule of origin 349, 438
preferential 382
upload 412
Rule set 413

S

Sales order 122, 124, 146, 179, 224
Sales tax 209
Sanction 20, 144
Sanctioned party list (SPL) screening 20, 39, 45, 77, 90, 117, 118, 122, 443
audit settings 138
audit trail 127, 449
block 125
business process 123
control settings 136, 139, 142
Customizing 133
identity card types 136
master data 141
predefined scenarios 130
reporting 447
scope 137
SPL entries 142
system process 123
SAP Analysis for Microsoft Office 478
SAP Analytics Cloud 478
SAP Business Technology Platform (SAP BTP) 478, 508
SAP BW/4HANA 470, 471, 477
advantages 478
SAP Datasphere 478, 479
advantages 480
SAP Easy Access menu 498
SAP Extended Warehouse Management (SAP EWM) 37
SAP Fiori 24, 469, 470, 486
activate apps 52
authorization 50
deployment 491
page 43
reporting 444
section 43
space 43

SAP Fiori (Cont.)
tile ... 43
SAP Fiori launchpad ... 42, 53, 498
activate spaces ... 53
SAP Global Trade Services (SAP GTS) ... 23
11.0 ... 24, 26, 485
edition for SAP HANA ... 24
SAP GTS, edition for SAP HANA ... 24
migration ... 485
obsolete functions ... 498
SAP GUI ... 49
SAP HANA ... 481, 483, 485
analytics ... 482
search ... 135
SAP HANA for SQL data warehousing ... 481, 483
SAP HANA Live ... 473
SAP HANA search ... 143
SAP HANA view ... 473
SAP Lumira ... 478
SAP Master Data Governance (SAP MDG) ... 32
SAP Readiness Check ... 486, 493
SAP S/4HANA ... 31, 493, 507
BOM transfer ... 87
plug-in ... 59, 421
reporting ... 469
SAP S/4HANA for international trade ... 26
classification ... 27
embargo ... 28
export legal control ... 28
integration ... 29
Intrastat ... 27
SAP Transportation Management (SAP TM) ... 37
SAP Watch List Screening ... 29
SAP_FIORI_CONTENT_ACTIVATION ... 52
SAP_FIORI_FOUNDATION_S4 ... 52
/SAPSLL/CUHD_OBJ_S ... 495
/SAPSLL/LCLIC_OBJ_S ... 495
/SAPSLL/NSC_NUM_UPLOAD_101 ... 499
/SAPSLL/PR_OBJ_S ... 495
/SAPSLL/SLLNS_XML_LOCAL_UPL101 ... 499
Scrapping ... 41, 274, 308
document ... 274
Screening ... 123
Search algorithm ... 135, 142
Search strategy ... 138
Securities ... 255, 261, 310, 313, 462
SEEBURGER ... 36
Sequence number ... 184
Service-oriented architecture (SOA) ... 38
Set-specific rule ... 438
Settlement ... 308
Simplification list ... 486, 498
Simplified procedure ... 239
Single role ... 51
SLL_PI ... 32
SLL-LEG ... 493, 497
Smart picking ... 270, 294
Soft block ... 231
Software as a service (SaaS) ... 29
Software Update Manager (SUM) ... 487, 496
Source system ... 72
Special customs procedures ... 189, 191, 267
BWH ... 267
Customizing ... 308
FTZ ... 282
inward processing ... 275
master data ... 311
PTR ... 299
reporting ... 455, 460
Start conditions ... 235
Status records ... 321
Stock ... 267
maintenance ... 304
monitoring ... 293, 296
update ... 290, 295
Stock correction – inventory count ... 275
Stock count ... 275
Stock items ... 461
Stock overview ... 272, 461
FTZ ... 294
inward processing ... 277
outward processing ... 281
PTR ... 306
Stock-correcting procedure ... 274
Storage location ... 56–58, 269
Supplementary customs declaration (SCD) ... 205, 231
Supplier data ... 436
Supplier declaration ... 359, 364
aggregate ... 375
maintain ... 371
outbound ... 398
reissue ... 401
reporting ... 466
revoke ... 401
Supported countries ... 335
Synchronous screening ... 128, 131, 133, 134

System connection 33, 36
System conversion 488

T

Target field 248
accesses 248
Target procedure 239, 242
TARIC 107
Tariff code 81
Tariff number 98
attributes 109
display data sheet 110
numbering scheme 112
Tariff shift 350, 438
Tariff-specific rule 350
Tax transaction engine 501
Technical messages 236
Text module 420
Threshold value 385, 389, 390
Tolerance rule 384, 438
Tolling manufacturing 299
Top-down calculation 386
Trade classification specialist 94
Trade regulation 20
Transaction
/SAPSLL/BOMMAT_DIRR3 88
/SAPSLL/CDS_CUS 110
/SAPSLL/CUHD_IC_EXP 171
/SAPSLL/CUSTOFF_UPL 80
/SAPSLL/LTSDREQ_2 369
/SAPSLL/MENU_LEGALR3 .... 34, 73, 75, 88
/SAPSLL/PRE_LTSD_PRI 418
/SAPSLL/PREF_RLS_UPL 412
/SAPSLL/PRPID_1 426
/SAPSLL/RET_PERI_UPD 376
/SAPSLL/SPL_ULO1 90
/SAPSLL/TOOO_SPI 33
/SAPSLL/TBDLS_SPI 33
/UI2/FLP 53
/UI2/FLP_CUS_CONF 53
BD87 326
PFCG 51
SM59 34
SPRO 33, 66, 67, 69, 228
STC01 52
WE05 325
WE07 325
WE09 325
WE21 323
Transaction value method 351
Transactional certification 363
Transactional data 31, 37, 73, 91
Transactional document 65
Transactional report 443
Transactional screening 128, 129, 133
Transactional value 208
Transient analytical query 470
Transit declaration 212, 226, 291
inbound 268
manage 293
Transit procedure 292
Transportation cost 222
Transportation management (TM) 37
export declaration 40
web services 38
Transportation unit (TU) 42

U

Unilateral agreement 340
Union Customs Code (UCC) 189
Unit of measure (UoM) 112, 464
United Nations Conference on Trade and (UNCTAD) 21
Upgrade 486
US reconciliation process 192, 319
Customizing 320
US reexport 118, 172
business process 173
calculate 176
logs 176
system process 173
User authorization 31

V

Valuation method 415
Value basis 415
Value of the goods 257
Value-added tax (VAT) 209
Vendor maintenance 78
Virtual data model (VDM) 469

W

Waybill number 214
Web services 38, 334
Work in process (WIP) stock 294
Worklist 101, 102, 128, 213
blocked documents 147
classification 105
World Customs Organization (WCO) 93

World Trade Organization (WTO) 22
Worst-case principle 375, 400

Z

Zone-to-zone transfer 287

- Implement basic and advanced TM in SAP S/4HANA
- Master transportation planning, freight order management, carrier selection, and more
- Calculate and settle charges with LSPs and customers

Lauterbach, Gottlieb, Helwig, Sürie, Benz

Transportation Management with SAP S/4HANA

The definitive book on TM is back! See how transportation management fits into your SAP S/4HANA landscape and find out what's new for basic and advanced shipping. Walk through master data to get your transportation network up and running. Then follow step-by-step instructions to set up and use embedded TM functionality in SAP S/4HANA for transportation planning, freight order management, transportation execution, charge management, and more. Round out your implementation with detailed integration scenarios, migration tips, and best practices.

1088 pages, 4th edition, pub. 04/2023
E-Book: $94.99 | **Print:** $99.95 | **Bundle:** $109.99

www.sap-press.com/5575